International Coalbed and Shale Gas Symposium 2005

Tuscaloosa, Alabama, USA
17 - 19 May 2005

ISBN: 978-1-5108-6369-9

Printed from e-media with permission by:

Curran Associates, Inc.
57 Morehouse Lane
Red Hook, NY 12571

Some format issues inherent in the e-media version may also appear in this print version.

Printed by Curran Associates, Inc. (xxxx)

University of Alabama
Box 870388
Tuscaloosa, Alabama 35487-0388
USA

Phone: (205) 348-6222
Fax: (205) 348-3029

dkeene@ccs.ua.edu

Additional copies of this publication are available from:

Curran Associates, Inc.
57 Morehouse Lane
Red Hook, NY 12571 USA
Phone: 845-758-0400
Fax: 845-758-2633
Email: curran@proceedings.com
Web: www.proceedings.com

2005 INTERNATIONAL COALBED METHANE SYMPOSIUM

May 17-19, 2005
Bryant Conference Center
The University of Alabama
Tuscaloosa, Alabama, USA

TABLE OF CONTENTS

- **RESERVOIRS**

- **GEOLOGY/RESOURCE ASSESSMENT**

- **DIVERSITY**

2005 INTERNATIONAL COALBED METHANE SYMPOSIUM

MAY 16-20, 2005
EXECUTIVE COMMITTEE

Charles M. Boyer, II
Principal Consultant, Unconventional Reservoirs
Schlumberger Data and Consulting Services
Pittsburgh, PA

Charles W. Byrer
Project Manager, National Energy Technology Laboratory, U.S. Department of Energy
Morgantown, WV

Peter E. Clark, Ph.D.
Associate Professor, Chemical Engineering
College of Engineering, The University of Alabama
Tuscaloosa, AL

Richard A. Esposito
Senior Engineering Geologist
Southern Company Services
Birmingham, AL

Richard A. Gates
District Manager, District 11
Mine Safety and Health Administration
U.S. Dept. of Labor, Birmingham, AL

Charles D. Haynes, Ph.D.
SYMPOSIUM CHAIR 2003
Professor, Civil Engineering, College of Engineering, The University of Alabama
Tuscaloosa, AL

J. Read Holland, Ph.D.
President
J.R. Holland & Associates, Inc.
Tuscaloosa, AL

George G. Karkashadze
Professor
Moscow State Mining University
RUSSIA

Donna Keene
Associate Director, Professional Development & Conference Services, College of Continuing Studies, University of Alabama, Tuscaloosa, AL

Dennis Lathem
Executive Director
Coalbed Methane Association of Alabama
Birmingham, AL

Eddie Martin
Mining Engineer
Chevron, Texaco, Pittsburg & Midway Coal Mining Company – North River Mine, Berry, AL

Charles R. Nelson, Ph.D.
Energy & Environmental Research Center
University of North Dakota, Grand Forks, ND
Golden, CO

Jack C. Pashin, Ph.D.
SYMPOSIUM CHAIR 2004 and 2005
Head, Energy and Minerals Section
Geological Survey of Alabama
Tuscaloosa, AL

Scott Reeves
Executive Vice President
Advanced Resources International, Inc.
Houston, TX

John S. Richardson
Vice President, Acquisitions & Engineering
Energen Resources Corporation, an Energen Company, Birmingham, AL

Berry H. "Nick" Tew, Jr., Ph.D.
State Geologist and Director; Geological Survey of Alabama, Supervisor, State Oil & Gas Board
Tuscaloosa, AL

Rhonda Tinsley
Senior Geologist
Southern Company Services
Birmingham, AL

J. Neil Walden, Jr.
Vice President
GeoMet, Inc.
Birmingham, AL

Charles Willis
President & General Manager
Black Warrior Methane Corp.
Brookwood, AL

Shinji Yamaguchi, Dr. Eng.
Associate Professor, Petroleum and Geothermal Engineering, Dept. of Earth Science and Technology, Akita University, JAPAN

May 16-20, 2005
Exhibits and Sponsors

ABB Inc., Totalflow Products
Bartlesville, OK

Advanced Resources International, Inc.
Arlington, VA

Alabama Power, A Southern Company
Birmingham, AL

Athena Industries, LLC
Birmingham, AL

Baker Oil Tools
Houston, TX

Black Warrior Methane Corp.
Brookwood, AL

CDX Gas, LLC
Dallas, TX

Coalbed Methane Association of Alabama, Inc.
Birmingham, AL

Control Microsystems, Inc.
Ontario, Canada

DOWDCO Coring & PDC Company
Oklahoma City, OK

Eastern Gulf Region of the Petroleum Technology Transfer Council, The University of Alabama
Tuscaloosa, AL

Geological Survey of Alabama
Tuscaloosa, AL

GeoMet, Inc.
Birmingham, AL

Hughes Christensen, A Division of Baker Hughes
Houston, TX

International Paper Mineral Resources
Houston, TX

Lufkin Automation
Houston, TX

National Energy Technology Laboratory, U.S. Department of Energy
Morgantown, WV

Patco Electric Contractors
Northport, AL

PCM Technical, Inc.
Tulsa, OK

Pense Bros. Drilling Company, Inc.
Fredericktown, MO

Southern Illinois University
Carbondale, IL

SPOC Automation, Inc.
Trussville, AL

TAM International, Inc.
Columbia, IL

The University of Alabama
College of Continuing Studies
College of Engineering
Tuscaloosa, AL

Thermo Electron Corporation
Houston, TX

TICORA Geosciences, Inc.
Arvada, CO

WellDog, Inc.
Laramie, WY

Woodward Governor Co.
Fort Collins, CO

Improvements in Coalbed Methane and Enhanced Coalbed Methane Reservoir Simulators

W.D. (Bill) Gunter & David H.-S. Law, Alberta Research Council (ARC) Inc., Edmonton, Alberta, Canada; and Matt Mavor, Tesseract Corporation, Park City, Utah, U.S.A.

Abstract

Deep, unmineable coalbeds are a very attractive geological reservoir for the storage of carbon dioxide (CO_2), a greenhouse gas. CO_2 can be stored and at the same time that the recovery of coalbed methane (CBM) is enhanced. The revenue from methane production can offset the cost of CO_2 storage.

Coalbeds form complex gas reservoirs characterized as dual porosity systems with both primary and secondary porosity systems. The primary porosity system contains the vast majority of the gas-in-place volume that is stored by adsorption. Mass transfer through the coal matrix for each gas molecular species is dominated by diffusion driven by concentration gradients. The secondary porosity system, generally natural fractures, provides the conduit for mass transport to production wells. Flow through the secondary porosity system is dominated by Darcy flow driven by pressure gradients. Simulation of these reservoirs requires comprehensive simulation software that incorporates the various storage and flow mechanisms and the changes in reservoir properties caused by changes in gas composition and gas content.

Early CBM numerical simulators, which were developed for primary CBM recovery processes included many important features such as: (1) Darcy flow through the natural fracture system; (2) pure gas diffusion through and sorption in the primary porosity system; and (3) coal matrix shrinkage due to pure gas desorption. However, these simulation models lacked the sophistication required to model more complex processes such as CO_2 injection and storage and flue gas (CO_2 and N_2) injection and storage. Additional features were required such as (1) sorption in and diffusion of multiple gas species through the coal matrix and (2) coal matrix swelling and shrinkage with associated permeability changes caused by changes in gas composition and gas content, and (3) matrix shrinkage with associated permeability changes caused by high injection pressures.

This paper describes recent improvements in numerical simulation of the enhanced CBM recovery/storage process and a comparison study of various reservoir simulation software packages. The comparison study was performed during the GEO-SEQ Project, funded by the National Energy Technology Laboratory of the U.S. Department of Energy. This study evaluated newly improved simulation software including (1) GEM, developed by the Computer Modelling Group Ltd., Canada; (2) ECLIPSE, developed by Schlumberger Information Solutions, U.K.; (3) COMET, developed by Advanced Resources International, U.S.A.; (4) SIMED II, modified by Netherlands Institute of Applied Geoscience TNO and maintained by the Commonwealth Scientific and Industrial Research Organization (CSIRO), Australia; (5) GCOMP, developed by BP-Amoco, U.S.A.; (6) METSIM 2, developed by Imperial College, U.K.; (7) COALCOMP, developed by Pennsylvania State University (PSU), U.S.A. and (8) MoReS, developed by Shell International, The Netherlands. The paper also addresses additional improvements to future simulation models that have not been incorporated into any commercially available software.

Improvements in Coalbed Methane and Enhanced Coalbed Methane Reservoir Simulators

W.D. (Bill) Gunter[1], David H.-S. Law[1] and Matthew J. Mavor[2]

[1] Alberta Research Council (ARC) Inc., Edmonton, Alberta, Canada
[2] Tesseract Corporation, Park City, Utah, U.S.A.

ABSTRACT

Deep, unmineable coalbeds are a very attractive geological formation for the storage of carbon dioxide (CO_2), a greenhouse gas (GHG); the CO_2 is stored and at the same time the recovery of coalbed methane (CBM) is enhanced. The revenue of methane (CH_4) production can offset the expenditures of the storage operation.

Coalbeds form complex gas reservoirs characterized by dual porosity; they contain both primary and secondary porosity systems. The primary porosity system contains the vast majority of the gas-in-place volume while the secondary porosity system provides the conduit for mass transfer to production wells. Primary porosity gas storage is dominated by adsorption. Mass transfer for each gas molecular species is dominated by diffusion that is driven by the concentration gradient. Flow through the secondary porosity system is dominated by Darcy flow that relates flow rate to permeability and pressure gradient. A full understanding of all of the process mechanisms is essential to performing numerical simulation of the CO_2 enhanced CBM (ECBM) recovery/storage process in which CO_2 or flue gas (i.e., a mixture of CO_2 and nitrogen (N_2) gases) is injected into the coalbed to replace the adsorbed CH_4.

Existing CBM numerical simulators which were developed for the primary CBM recovery process, have many important features such as: (1) a dual porosity system; (2) Darcy flow in the natural fracture system; (3) gas diffusion and sorption in the primary porosity system; and (4) coal matrix shrinkage due to gas desorption. However, process mechanisms become more complex with CO_2 or flue gas injection. Improvement of the numerical simulators were needed to consider additional phenomena such as: (1) coal swelling due to CO_2 sorption into coal; (2) mixed gas sorption; (3) mixed gas diffusion; and (3) non-isothermal effects during gas injection.

This paper describes the improvements in the numerical simulation of the ECBM recovery/storage process that have occurred over the past few years. A summary of a simulator comparison study performed during the GEO-SEQ Project, funded by the National Energy Technology Laboratory (NETL) of the U.S. Department of Energy (DOE) is also included. This comparison evaluated newly improved simulators including: (1) GEM, Computer Modelling Group (CMG) Ltd., Canada; (2) ECLIPSE, Schlumberger Information Solutions (SIS), U.K.; (3) COMET, Advanced Resources International (ARI), U.S.A.; (4) SIMED II, Netherlands Institute of Applied Geoscience TNO, The Netherlands / Commonwealth Scientific and Industrial Research Organization (CSIRO), Australia; (5) GCOMP, BP-Amoco, U.S.A.; (6) METSIM 2, Imperial College, U.K.; (7) COALCOMP, Pennsylvania State University (PSU), U.S.A. and (8) MoReS, Shell International, The Netherlands.

INTRODUCTION

The injection of carbon dioxide (CO_2), a greenhouse gas (GHG), in coalbeds is probably the most attractive option of all underground CO_2 storage possibilities: the CO_2 is stored and at the same time the recovery of coalbed methane (CBM) is enhanced [1]. The revenue of methane (CH_4) production can offset the expenditures of the storage operation [2,3].

Coalbeds are characterized by a dual porosity system: they contain both primary (i.e., micropore (< 2 nm) and mesopore (2 - 50 nm)) and secondary (i.e., macropore (> 50 nm) and natural fracture (1 mm or less)) porosity systems. The primary porosity system (or coal matrix) contains the vast majority of the gas-in-place volume while the secondary porosity system provides the conduit for mass transfer to the wellbore. Primary porosity gas storage is dominated by adsorption. The primary porosity system is relatively impermeable due to the small pore size (on the order of microdarcy, 10^{-6} D, range). Matrix mass transfer of each gas molecular species is dominated by diffusion that is driven by concentration gradients. Flow through the secondary porosity system is dominated by Darcy flow that relates flow rate to permeability and pressure gradient.

Figure 1 illustrates the overall process of gas storage and the movement through coalbeds. The conventional primary CBM recovery process begins with a production well that is often stimulated by hydraulic fracturing to connect the wellbore to the coal natural fracture system via an induced fracture. When the pressure in the well is reduced by opening the well on the surface or by pumping water from the well, the pressure in the induced fracture is reduced which in turn reduces the pressure in the coal natural fracture system. Gas and water begin moving through the natural and induced fractures toward the well. When the natural fracture system pressure drops, gas molecules desorb from the primary-secondary porosity interface and are released into the secondary porosity system, reducing the adsorbed gas concentration in the primary porosity system near the natural fractures. This reduction creates a concentration gradient that results in mass transfer by diffusion through the micro and mesoporosity. Adsorbed gas continues to be released as pressure is further reduced.

When CO_2 (which is more strongly adsorbed than CH_4) is injected into the coal natural fracture system during the CO_2 storage process, it is preferentially adsorbed into the primary porosity system. Upon adsorption, the CO_2 drives CH_4 from the primary porosity into the secondary porosity system. The secondary porosity pressure is increased due to CO_2 injection and CH_4 flows to production wells. The CO_2 is stored in-situ and is not produced unless the injected gas front reaches the production wells. The process is, in general, terminated at CO_2 breakthrough.

An alternative to pure CO_2 injection is to inject flue gas, a mixture of CO_2 and nitrogen (N_2) gases, to reduce the high cost of gas separation to obtain the pure CO_2. When N_2 (which is more weakly adsorbed than CH_4) enters the natural fracture system, the partial pressure of CH_4 is reduced. As the partial pressure is reduced, the desorption rate of CH_4 increases. The CH_4 is then swept along with the N_2 through the natural fracture system to the production wells. Early N_2 breakthrough at the production well occurs, resulting in N_2 being produced with the CH_4, which must be separated from the CH_4 before sale in a commercial project. Combining CO_2 and N_2 as an injectant may improve CH_4 recovery performance, as less N_2 will be produced compared to a pure N_2 injection stream, with greater CH_4 productivity than during pure CO_2 injection.

A full understanding of all the complex mechanisms involved in the CO_2 storage process with pure CO_2 or flue gas injection is essential to have more confidence in numerical modeling results.

ALBERTA RESEARCH COUNCIL (ARC) CO_2-ECBM/STORAGE PROJECTS

To develop and improve CO_2-ECBM/storage technology, the Alberta Research Council (ARC) has performed two projects in Canada and China to study the technical and commercial viability of this process. The main objectives are:

- to reduce GHG emissions by subsurface injection of CO_2 into deep coalbeds; and
- to enhance CBM recovery and production rates.

ARC CO_2-ECBM/Storage Project in Canada

The project entitled "Sustainable Development of Coalbed Methane – A Life Cycle Approach to Production of Fossil Energy" [4] was split into five phases as follows:

- Phase I: A proof of concept study – initial assessment and feasibility of injecting CO_2, N_2 and flue gases (i.e., mixtures of CO_2 and N_2) into Mannville coals in Alberta, Canada.
- Phase II: The design and implementation of a CO_2 micro-pilot test.
- Phase III: The design and implementation of mixed gas micro-pilot tests.
- Phase IV: The matching of CO_2-rich sources (both existing and novel) with CBM in Alberta, Canada and the design of further field tests including micro-pilots and multi-well pilots.
- Phase V: Methanogenesis of coal gas.

The project was initiated in 1997 with the first three phases completed in 2001. Funding of the project is provided by government organizations in Canada (both Federal and Provincial), United States (U.S.A.), United Kingdom (U.K.), Australia, Japan and The Netherlands; major oil and gas companies; oil-field and gas-field service companies; electric power utilities, engineering and gas supply companies.

In Phases II and III, micro-pilot tests were carried out on two wells in the Mannville (Medicine River) coal seams at Fenn-Big Valley using a "huff and puff" scheme. The Fenn-Big Valley site located in Central Alberta, Canada (see Figure 2) was chosen because of prior work done by Gulf Canada (now ConocoPhillips): the reservoir properties were known, the productivity was in the range of a BP-Amoco ECBM micro-pilot [5], the permeability level is common in Alberta, and the area contains a huge resource of CBM if enhanced recovery is significant. The characteristics of the Mannville coal at the test site are listed in Table 1 and documented in greater detail elsewhere [8].

The "huff and puff" scheme involved injection of CO_2, N_2 or flue gas into a single well after establishing a primary production baseline, shutting in the well for a pre-determined period (i.e., pressure falloff period) and then monitoring the gas production from the well when it was re-opened. This huff and puff test was followed by a final shut-in test (i.e., pressure buildup period) to obtain pressure and permeability measurements after gas injection.

The fieldwork in Phase II began in 1998 by re-entering an old well, the ConocoPhillips FBV 4A-23-36-20 (FBV 4A) well (see Figure 3). The FBV 4A well was originally an oil production well but after the original completion was abandoned, the well was re-completed into two Mannville coal intervals. The upper coal interval was perforated from 1,259 to 1,263 m and the lower coal interval from 1,277 to 1,283 m. A bridge plug set above the lower coal interval isolated the upper interval, which was the target of the micro-pilot tests. Two micro-pilot tests were conducted in this well. The well was first produced followed by a shut-in test (i.e., pressure buildup) to obtain reservoir pressure (7,653 – 7,770 kPa), gas and water productivity data (15,900 to 2,970 m^3/D and 0.92 m^3/D, respectively) and produced gas samples for composition determination (91.2% CH_4, 1.8% C_2H_6, 0.3% C_3H_8, 1.6% CO_2, and 5.1% N_2). The pressure data collected during the shut-in period were used to estimate the permeability (3.65 md) before gas injection. This was followed by a CO_2 injection/falloff test using a 12-tonne slug of CO_2 to ensure that CO_2 injection was possible under fracturing pressure. A short production period followed. Since this test was successful, it allowed the testing program to continue with injection of twelve 15 tonne slugs of CO_2. Permeability and injectivity changes were calculated after the addition of each slug. Although CO_2 reduced the absolute permeability at original reservoir pressure, injectivity actually increased by 147%. This resulted as the gain in permeability due to the injection pressures during CO_2 injection was substantially greater than the loss due to coal swelling. The CO_2 was allowed to soak into the coal and the well was returned to production. After injection, during shut-in periods or subsequent production, the coal swelling effects reduced the permeability to levels below the original permeability (0.98 md). Fourteen months later in 2000 during Phase IIIA, the well was returned for a second production period to obtain gas composition data after the extended soak. The produced gas composition at the start of production was 65.1% CH_4, 1.4% C_2H_6, 29.8% CO_2, 3.2% N_2 and 0.5% C_{3+}, similar to that at the end of the first post-CO_2 production period. Then, a micro-pilot was conducted by injecting approximately 110 tonnes of flue gas (13% N_2 and 87% CO_2), which was generated from an under-balanced drilling system compressor, followed by a soak and a post injection-production test. Flue gas injection substantially increased the absolute permeability from the 0.98 md post-CO_2 injection to 23.7 md and 8.3 md in the near and far well regions, respectively. It is believed that this increase was due to stripping CO_2 and CH_4 from the coal matrix causing matrix shrinkage.

During Phase IIIA in 1999, a new well (FBV I#2) was drilled 493 m north of the existing FBV 4A well (see Figure 3). FBV I#2 was cored and logged for reservoir property data. The well was then cased and completed in the same upper Mannville coal seam (1,257.5 – 1,262.0 m) as the FBV 4A well. A combination of production tests and water-injection falloff tests were conducted to determine original reservoir pressure (7,901 kPa), permeability (1.18 md), and gas composition (94.42% CH_4, 1.53% C_2H_6, 0.26% CO_2, 3.46% N_2 and 0.33% C_{3+}). The first micro-pilot in the new well was a 100% N_2 injectivity test with an oxygen tracer. The oxygen was used to determine if the test was confined to the coal reservoir and was added in small amounts [5]. After approximately 10 tonnes of N_2 injection, a falloff test (soak period) was completed followed by a production period. During N_2 injection, the absolute permeability was estimated to be 13.8 md. The well was then shut in before the second micro-pilot test was started. The second micro-pilot was performed by injecting 120 tonnes of flue gas (53% N_2 and 47% CO_2) and followed the same steps as the 100% N_2 micro-pilot.

Details test procedures, quantitative data measured during this effort, and the interpretation thereof can be found in References [6-8].

ARC CO_2-ECBM/Storage Project in China

The micro-pilot in the project entitled "The development of China's Coalbed Methane Technology Carbon Dioxide Storage Project" was designed in six stages as follows:

- Stage 1: Inspection of wellhead equipment.
- Stage 2: Isolation of the #3 coal seam and installing additional downhole and surface equipment.
- Stage 3: Initial production testing to determine baseline reservoir properties.
- Stage 4: Intermittent injection of carbon dioxide for 30 days followed by a 30-day shut-in period.
- Stage 5: Production testing after the CO_2 injection period.
- Stage 6: Final shut-in test.

The Project was funded by the Canadian International Development Agency (CIDA) under the Canadian Climate Change Development Fund (CCCDF). The Canadian Consortium was led by the Alberta Research Council (ARC), Sproule International Ltd. (Sproule) and Computer Modelling Group Ltd.

(CMG), acting as the Canadian Executing Agency (CEA) for the Project. Other Canadian partners were Computalog Ltd., CalFrac Well Services, SNC-Lavalin and Porteous Engineering. China United Coalbed Methane Corporation Ltd. (CUCBM) was the Chinese Executing Agency (CHEA). The single well CO_2 injection micro-pilot test was conducted in the #3 coal seam (472 to 478 meters) in the Shanxi Formation at the TL-003 well located in the South Qinshui basin in the province of Shanxi (see Figure 4). Figure 5 illustrates the location of the test well. The characteristics of the Qinshui Basin coal at the test site are listed in Table 1.

TL-003 was the structurally highest well of all nine wells in the field and was the original well drilled and production tested before any of the other wells were drilled. Gas production of the TL-003 well began on March 16, 1998 and ended on March 25, 1999 after 374 days. Cumulative gas production during this period was 734,852 m^3 (25.95 MMscf) with a peak gas rate of 6,986 m^3/D (246.7 Mscf/D) and an average gas rate of 1,965 m^3/D (69.4 Mscf/D). Cumulative water production during this period was 13,195 m^3 (82,994 STB) with an average water rate of 35.28 m^3/D (222 STB/D). The well was then shut-in for a substantial amount of time from March 25, 1999 to December 3, 2000. The TL-003 well was placed back on production on December 3, 2000. The second production period was for 468 days and ended on March 16, 2002. Cumulative gas production during this period was 234,567 m^3 (8.28 MMscf) with a peak gas rate of 1,538 m^3/D (54.3 Mscf/D) and an average gas rate of 501.2 m^3/D (17.7 Mscf/D). Cumulative water production during this period was 21,698 m^3 (136,476 STB) with an average water rate of 46.36 m^3/D (291 STB/d).

Project activities commenced in March 2002. An initial workover of TL-003 was performed to isolate the lower water sand from the #3 coal seam and install down-hole pressure recording equipment. In Stage 3, a production test was initiated on October 28, 2003 to obtain baseline production information for the TL-003 well. Bottom-hole pressure and temperature data, gas composition, and daily fluid production rates were monitored from the start of production. The initial gas production period was 134 days with cumulative gas production and average gas rate of 74,632.5 m^3 (2.63 MMscf) and 556.9 m^3/D (19.7 Mscf/D), respectively. Cumulative water production during this period was 78.1 m^3 (491 STB) with an average water rate of 0.59 m^3/D (3.7 STB/D). Gas composition during initial production test was predominately methane (CH_4 - 97.52%) with a minor amount of nitrogen (N_2 - 2.42%) and traces of carbon dioxide (CO_2 - 0.04%), ethane (C_2H_6 - 0.01%) and other gases (C_3H_8, i-C_4H_{10}, n-C_4H_{10}, He, H_2, O_2 – 0.01%). Initial coal seam pressure before the micro-pilot test was estimated at 1,295.5 kPa.

The well was then shut-in on March 9, 2004 for a pressure build-up test. During the well shut-in, bottom-hole pressure data were collected for 5.8 days. The pressure data collected during the shut-in period were used to estimate the permeability (12.6 md) before gas injection. In Stage 4, which was started on April 6, 2004, intermittent injection of liquid carbon dioxide was performed over a 13-day period while monitoring surface and bottom-hole pressure and temperature. The Zhongyuan Oilfield trucked in liquid CO_2. Thirteen slugs of approximately 15 metric tonnes of CO_2 were injected in 13 days. Injection time per each slug ranged from 6.5 to 9.9 hours. Total CO_2 injection was 103,533 m^3 (192.8 metric tonnes or 3.66 MMscf) with an average CO_2 injection rate of 26,348.5 m^3/D (49 tonnes/D or 0.93 MMscf/D). The liquid CO_2 was injected at an injection pressure less than the fracturing pressure of approximately 7,557 kPa (1,096 psig). At the end of the 9th CO_2 slug, the near well permeability had reduced from the initial value of 12.6 md to 2.34 md, but no further reduction was observed after the 9th to the 13th CO_2 slugs. At the end of Stage 4, CO_2 could still be injected into the coal seam. Stage 4 was completed on April 30, 2004. After CO_2 injection, the well was shut-in for a pressure falloff test for 11.6 days.

During Stage 5, the well was placed on production from June 22 to August 2, 2004 with the well shut-in between July 2 to 12 for 11 days due to a pump problem. The total gas production period was 30 days with cumulative gas production and an average gas production rate of 30,522 m^3 (1.08 MMscf) and 1,017 m^3/D (35.9 Mscf/D), respectively. Cumulative water production was 46.6 m^3 (293 STB) with an average water production rate of 1.55 m^3/D (9.7 STB/D). Stage 6 was a final shut-in test to obtain estimates of reservoir properties and near-well conditions. After the post injection production test, the well was shut-in on August 2, 2004 for a pressure build-up test for 16 days. The final shut-in test was completed on August 18, 2004.

NUMERICAL HISTORY MATCH OF MICRO-PILOT TEST DATA

Numerical simulators have played a very important role in analyzing the micro-pilot test data and understanding the complex process mechanisms. The high quality database (i.e., bottom-hole pressure and temperature, gas/water production data, gas injection data and production gas composition) obtained from the successful micro-pilot tests was very useful in the validation of the numerical simulators. These data served as history matching files to test the accuracy of the numerical modeling methods.

Examples of the successful history match of the micro-pilot field data are given in Figures 6 and 7. Figure 6 illustrates very good match of bottom-hole pressure for the CO_2 injection micro-pilot test at well FBV 4A in the Fenn-Big Valley, Alberta, Canada while Figure 7 illustrates a very good match of bottom-hole pressure for the CO_2 injection micro-pilot test at well TL-003, in the Qinshui Basin.

NUMERICAL CHALLENGES

Seidle and Arri [9] have demonstrated that conventional oil and gas numerical simulators can be used for primary CBM recovery, provided that the diffusion of CH_4 from the primary porosity system into the natural fracture system of the coal is much faster than Darcy flow through the natural fractures into the production well. In this situation, numerical simulators with only a single porosity system can be used. Since then, many commercial and research numerical simulators have been developed to model primary CBM recovery process to take into account of many important phenomena such as:

- the dual porosity nature of the coalbed;
- Darcy flow of gas and water (i.e., multiphase flow) in the natural fracture system;
- diffusion of single component gas from the coal matrix to the natural fracture system;
- sorption/desorption of single component gas at the coal surface; and
- coal matrix shrinkage due to gas desorption.

However, process mechanisms become more complex during CO_2 or flue gas injection. Law et. al. [10] have suggested that in order to properly model the CO_2-ECBM recovery/storage process, improvements on the numerical simulators are needed to consider the additional features such as:

- capability to handle multiple gas components (e.g., three or more components: CH_4, CO_2 and N_2);
- coal matrix swelling due to CO_2 sorption;
- compaction/dilation of the natural fracture system due to stresses;
- diffusion of mixed gas through the coal matrix to the natural fracture system;
- sorption/desorption of mixed gas at the matrix-fracture surface;
- and non-isothermal effects due to differences in temperature between the coalbed and the injected gases

It has been demonstrated that many of these features are essential in the successful of history matching the aforementioned micro-pilot test data [11]. Although not necessary for the pilot test evaluation, it is also possible that in some cases, water may be produced from the coal matrix as well, although commonly this effect is not included in the simulation models.

Alberta Research Council (ARC) Permeability Theory

One of the special numerical features used in the history matching efforts was the ARC Permeability Theory [12], a comprehensive theory that combined coal matrix shrinkage/swelling and net confining stress effects to predict porosity and permeability of the natural fracture system of the coal. The ARC Permeability Theory is an extension of the Palmer & Mansoori Theory [13], which is commonly used for the primary CBM recovery process (single-gas component system), to multi-gas component system. The Palmer and Mansoori Theory for porosity and permeability ratios are presented in the following equations:

$$\frac{\phi}{\phi_i} = 1 + \frac{p - p_i}{\phi_i M} + \frac{\varepsilon_L}{\phi_i}\left(1 - \frac{K}{M}\right)\left(\frac{p_i}{p_i + p_L} - \frac{p}{p + p_L}\right) \tag{1}$$

$$\frac{k}{k_i} = \left(\frac{\phi}{\phi_i}\right)^3 \tag{2}$$

The second term on the right-hand side of Equation (1) takes account of pressure strain and the last term on the right-hand side of Equation (1) takes account of sorption strain. The rock mechanical properties K and M in Equation (1) are defined by the following equations:

$$\frac{K}{M} = \frac{1}{3}\left(\frac{1 + \nu}{1 - \nu}\right) \tag{3}$$

$$M = E\frac{1 - \nu}{(1 + \nu)(1 - 2\nu)} \tag{4}$$

where:

E *Young's modulus, kPa or psia*
k *absolute permeability of natural fracture system, md*
k_i *initial absolute permeability of natural fracture system, md*
K *Bulk modulus, kPa or psia*
M *constrained axial modulus, kPa or psia*
p *pressure of natural fracture system, kPa or psia*
p_i *initial coalbed pressure, kPa or psia*
p_L *Langmuir pressure, kPa or psia*
ε_L *volumetric strain change due to adsorbed gas content, dimensionless*
ν *Poisson's ratio*
ϕ *porosity of natural fracture system, fraction*
ϕi *initial porosity of natural fracture system, fraction*

On the other hand, the ARC Permeability Theory for porosity and permeability ratios are presented in the following equations:

$$\frac{\phi}{\phi_{atm}} = 1 + \frac{p - p_{atm}}{\phi_{atm} M} + \frac{\varepsilon_{\infty}}{\phi_{atm}}\left(1 - \frac{K}{M}\right)\left(\frac{p_{atm}}{p_{atm} + p_{\varepsilon}} - \frac{p}{p + p_{\varepsilon}}\right) \tag{5}$$

$$\frac{k}{k_{atm}} = \left(\frac{\phi}{\phi_{atm}}\right)^{\alpha} \tag{6}$$

where:

k_{atm}	absolute permeability of natural fracture system at atmospheric pressure, md
p_{atm}	atmospheric pressure, kPa or psia
p_{ε}	pressure of natural fracture system at a strain of 0.5 ε_{∞}, kPa or psia
α	exponent for porosity/permeability ratio, dimensionless
ε	volumetric strain, dimensionless
ε_{∞}	volumetric strain at infinitive pressure, dimensionless
ϕ_{atm}	porosity of natural fracture system at atmospheric pressure, fraction

Atmospheric pressure is used as the reference conditions for the ARC Permeability Theory compared to the initial coalbed reservoir pressure used as the reference conditions for the Palmer & Mansoori Theory. This pressure was selected as the sorption effects are the same for all gases at this condition.

The Palmer & Mansoori Theory assumes that the strain parameters, ε_{∞} and p_{ε} are constant which implies that gas composition is constant. The ARC Permeability Theory is a more general theory derived by analogy to multi-gas component adsorption. Multi-gas component isotherm relationships are computed from single component data by Extended Langmuir Isotherm Theory [14]:

$$G_{si} = G_{sLi}\left[1 - \left(w_{a} + w_{we}\right)\right]\frac{\dfrac{p y_{i}}{p_{Li}}}{1 + p\sum_{j=1}^{nc}\dfrac{y_{j}}{p_{Lj}}} \tag{7}$$

where:

G_{si}	multi-gas component in-situ adsorbed gas content of gas component i, m^3/kg or scf/ton
G_{sLi}	single-gas component Langmuir DAF adsorbed gas content of gas component i, m^3/kg or scf/ton
w_a	in-situ ash content, weight fraction
w_{we}	in-situ moisture content, weight fraction
p_{Li}, p_{Lj}	single-gas component Langmuir pressure of gas component i or j, [kPa or psia]
y_i, y_j	mole fraction of gas component i or j in the free gas phase, dimensionless
nc	number of gas components

The total adsorbed gas content, G_s, is the sum of the adsorbed gas content of each component as shown in Equation (8), while the adsorbed gas composition is computed with Equation (9).

$$G_{s} = \sum_{i=1}^{nc} G_{si} \tag{8}$$

$$x_i = \frac{G_{si}}{G_s} \tag{9}$$

Strain-pressure relationships are assumed to fit a Langmuir-like equation, Equation (10):

$$\varepsilon = \varepsilon_\infty \frac{p}{p + p_\varepsilon} \tag{10}$$

By analogy with an extended Langmuir isotherm, the volumetric strain caused by sorption at any composition and pressure due to each gas species can be computed with a relationship of the form of Equation (11). This relationship is based upon the composition of the gas in the adsorbed phase within the coal matrix as sorption strain is due to adsorbed gas content.

$$\varepsilon_i = \varepsilon_{\infty i} \frac{\dfrac{p x_i}{p_{\varepsilon i}}}{1 + p \sum_{j=1}^{nc} \dfrac{x_j}{p_{\varepsilon j}}} \tag{11}$$

where:

$p_{\varepsilon i}$, $p_{\varepsilon j}$ single-gas component p_ε of gas component *i* or *j*, kPa or psia
x_i, x_j mole fraction of gas component *i* or *j* in the adsorbed phase, dimensionless
nc number of gas components
ε_i Multi-gas component volumetric strain of gas component *i*, dimensionless
$\varepsilon_{\infty i}$ single-gas component volumetric strain at infinitive pressure of gas component *i*, dimensionless

The total strain is determined by summing the strain due to each gas species as Equation (12) indicates.

$$\varepsilon = \sum_{i=1}^{nc} \varepsilon_i \tag{12}$$

The coal fracture porosity relative to the porosity at atmospheric pressure is given in Equation (13):

$$\frac{\phi}{\phi_{atm}} = 1 + \frac{p - p_{atm}}{\phi_{atm} M} + \frac{1}{\phi_{atm}}\left(1 - \frac{K}{M}\right)\left(\varepsilon_{atm} - \varepsilon\right) \tag{13}$$

The ARC Permeability Theory is used by specification of the free gas composition in the natural fracture system, which is used to compute the adsorbed gas composition (Equations (7)-(9)). The sorption strain is computed from the adsorbed gas composition and the individual gas species strain parameters (Equation (11)). The total strain (Equation (12)) at the natural fracture pressure of interest specifies the porosity ratio (Equation (13)). The permeability ratio at this gas composition and natural fracture pressure is then computed (Equation (6)).

The ARC Permeability Theory was incorporated into the numerical simulator, GEM, developed by the Computer Modelling Group (CMG). Parameters for the ARC Permeability Theory for different coals were determined by history matching portions of the micro-pilot test data. Figure 8 shows the estimated permeability ratio based on the theory for the Qinshui Basin coal.

COMPARISON OF NUMERICAL SIMULATORS

In the past few years, there have been significant improvements in the numerical simulation of the CO_2-ECBM/storage process. A simulator comparison study was conducted to evaluate these ECBM numerical simulators. The comparison study, started in 2000 and completed in 2003, was part of the GEO-SEQ Project, funded by the National Energy Technology Laboratory (NETL) of the U.S. Department of Energy (DOE). The participants included: (1) GEM, Computer Modelling Group (CMG) Ltd., Canada; (2) ECLIPSE, Schlumberger Information Solutions (SIS), U.K.; (3) COMET, Advanced Resources International (ARI), U.S.A.; (4) SIMED II, Netherlands Institute of Applied Geoscience TNO, The Netherlands/Commonwealth Scientific and Industrial Research Organization (CSIRO), Australia; (5) GCOMP, BP-Amoco, U.S.A.; (6) METSIM2, Imperial College, U.K.; (7) COALCOMP, Pennsylvania State University (PSU), U.S.A. and (8) MoReS, Shell International, The Netherlands.

Alberta Research Council (ARC) organized and managed the simulator comparison study; facilitated the development and selection of appropriate test problems; distributed them to participants; and solicited, collected, reconciled, and documented solutions. Development and selection of sample test problems was made on the basis of major mechanisms expected to occur in the CO_2-ECBM/storage process, taking into account the existing simulation capabilities and future needs. The initial test problems emphasized the comparison of the performance of existing CBM simulators, which may not have had all the features required to properly model the CO_2-ECBM/storage process. Later test problems addressed more complex process mechanisms. At the early stage, improvements on existing numerical simulators were necessary. Finally, performance of numerical simulators was compared to determine their capability to history match the micro-pilot test data provided by ARC. Table 2 lists a description of the problem sets.

Part I of this study [15,16], which included problem sets 1 and 2 that dealt with a single well test (i.e., "micro-pilot test") and a CO_2-ECBM/storage process in an inverted five-spot pattern with pure CO_2 injection, respectively, compared the performance of CBM simulators that only had the features to model the primary CBM recovery process. In Parts II [17] and III [18], additional problem sets were developed that addressed more complex process mechanisms. In these cases, improvement of some of the existing CBM simulators was necessary by incorporating additional features for the CO_2-ECBM process. Part II was the extension of problem sets 1 and 2 in Part I, but with CO_2-enriched flue gas injection (i.e., 50% CO_2 & 50% N_2). Part III, with more complex problem sets 3 and 4, was the extension of the problem set 1 to include the CO_2-ECBM/storage process with pure CO_2 injection in an inverted five-spot pattern. Problem set 3 investigated gas diffusion between the coal matrix and the natural fracture system and problem set 4 investigated the effect of natural fracture permeability changes as a function of natural fracture pressure and adsorbed gas content (i.e., coal shrinkage/swelling). In Part IV [19], performance of CBM simulators was compared for their ability to history match field test data collected by ARC during micro-pilot tests with pure CO_2 and flue gas injection in the Fenn-Big Valley project. Detailed descriptions of the problem sets can be found in the password protected website: http://www.arc.ab.ca/extranet/ecbm.

In general, there was very good agreement between the results in Parts I - III predicted by the different CBM simulators. The differences between the predictions from different simulators resulted for a variety of reasons including:

- possible different initialization procedure (e.g., initial gas-in-place);
- possible different dual porosity approaches in the simulators;
- possible different diffusion models in the simulators;
- handling of wells (e.g., ¼ well in 5-spot pattern);
- tolerance on the convergence of iterations; and
- selection of numerical control parameters.

Figures 9 to 11 illustrate a few examples of the comparison results. Figure 9 illustrates CH_4 production rates for the ECBM process in an inverted 5-spot pattern with continuous pure CO_2 injection from problem set 2 in Part I. All simulators predicted enhanced CH_4 production caused by CO_2 injection when compared to the primary production case. Figure 10 illustrates the CO_2 distribution (CO_2 mole fraction in the gas phase in the natural fracture system). CO_2 breakthrough occurred after approximately 70 days of CO_2 injection which corresponded to the sharp decrease of CH_4 production rate (see Figure 9). Figure 11 illustrates CH_4 production rates for changes in the desorption time constant (t_{des} = 38.6 days) from the baseline case (t_{des} = 0 day) for continuous pure CO_2 injection. The baseline represented instantaneous sorption equilibrium at the coal natural fracture system pressure. The greater CH_4 desorption time resulted in lower initial CH_4 production rates because more time was required for CH_4 to desorb and flow to the natural fracture system.

Five numerical simulators participated in history matching the ARC micro-pilot test data in Part IV: (1) CMG's GEM; (2) ARI's COMET; (3) TNO/CSIRO's SIMED II; (4) Imperial College's METSIM2 and (5) PSU's COALCOMP. All five simulators were capable of history matching the field data. History match results comparing bottom-hole pressures and produced gas compositions are illustrated in Figure 12 for the pure CO_2 injection micro-pilot test. The methodologies used by different participants in the history match were inconsistent because the special features (e.g., stress and multi-component gas adsorbed dependent fracture permeability variation) needed to history match the field data were formulated quite differently in the CBM simulators. Nevertheless, all history matches indicated the significant effects of permeability variation due to stress and adsorbed gas content and mixed gas diffusion between the coal matrix and the natural fracture system during the micro-pilot tests. For example, during the gas injection stage, coal permeability increased due to stress (i.e., dilation of the natural fractures) even though swelling of the coal matrix was believed to occur due to CO_2 adsorption. On the other hand, during the pressure falloff stage, significant reduction of permeability was observed which was mainly due to swelling of the coal matrix and compaction of the natural fractures.

Global and/or local variations of permeability were necessary to match the bottom-hole pressure data. Local variation of permeability near the well were achieved dynamically by using theoretical models describing the dependence of coal porosity and permeability upon gas composition and pressure (e.g., ARC Permeability Theory in CMG's GEM and others in COMET, SIMEDII and METSIM2). These dynamic models, which were developed based on extension of the Palmer and Mansoori Theory [13] to mixed gas systems, not only minimized the adjustment of other parameters in the history match but also allowed the CBM models to be used as predictive tools after calibration by the history match. Without using a dynamic model (e.g., COALCOMP), it was necessary to adjust both the global permeability and the relative permeability relationships to obtain a history match. To match the produced gas composition, mixed gas diffusion between the coal matrix and the natural fractures was essential. In some cases, using different desorption time constants for the different gases was necessary. Desorption time constants from very small (instantaneous) to on the order of days were used with N_2 being the slowest diffusing gas and CO_2 being the fastest diffusing gas.

Detailed descriptions of all the comparison results can be found in the aforementioned website.

CONCLUSIONS

- Participation in the CBM simulator comparison study both from the commercial software developers (i.e., CMG, SIS, ARI and TNO/CSIRO) and non-commercial software developers (i.e., BP-Amoco, Imperial College, PSU and Shell International) indicated their commitment to improve their models for CO_2-ECBM/storage process simulation.

- All participating CBM numerical simulators in the comparison study have been improved significantly during the past few years to model the CO_2-ECBM/storage processes.

- Dynamic theories describing the dependence of coal porosity and permeability upon gas composition and pressure are useful predictive tools after their parameters have been determined by history matching the available field data. However, these theories are not perfect and continuous improvement of these theories is necessary.

REFERENCES

1. Gunter, W.D., Gentzis, T., Rottenfusser, B.A. and Richardson, R.J.H., (1997), *Deep Coalbed Methane in Alberta, Canada: A Fuel Resource with The Potential of Zero Greenhouse Gas Emissions,* Energy Convers. Mgmt, Volume 38, Suppl., P. S217-S222.
2. Wong, S., Gunter, W.D. and Mavor, M.J., (2000), *Economics of CO_2 Sequestration in Coalbed Methane Reservoirs*, Proceedings of SPE/CERI Gas Technology Symposium 2000, Paper No. SPE 59785, Calgary, Alberta, Canada, April 3-5, P. 631-638.
3. Wong, S., Gunter, W.D., Law, D.H.-S. and Mavor, M.J., (2001), *Economics of Flue Gas Injection and CO_2 Sequestration in Coalbed Methane Reservoirs,* Proceedings of the 5th International Conference on Greenhouse Gas Control Technologies (GHGT-5), Cairns, Australia, August 13-16, P. 543-548.
4. Wong, S. and Gunter, W.D. (1999), *Testing CO_2 Enhanced Coalbed Methane Recovery,* Greenhouse Issues, IEA Greenhouse Gas R&D Programme, No. 45, November, P. 1-3.
5. Puri, R., Voltz, R. and Duhrkopf, D., (1995), *A Micro-Pilot Approach to Coalbed Methane Reservoir Assessment,* Proceedings of Intergas 95, University of Alabama/Tuscaloosa, Paper 9556, Tuscaloosa, Alabama, U.S.A., May 15-19, P. 265-274.
6. Gunter, W.D. and Mavor, M.J., (2002), *Enhanced Coalbed Methane Recovery and CO_2 Storage: Micro-Pilot Testing at Fenn-Big Valley, Alberta Canada,* Proceedings of the JCOAL & JCOSCs' International Workshop on the "Current Status and Perspective of CO_2 Sequestration in Coal Seam," Tokyo, Japan, September 5. P. 101-116.
7. Gunter, W.D., Mavor, M.J. and Robinson, J.R., (2004), *CO_2 Storage and Enhanced Methane Production: Field Testing at Fenn-Big Valley, Alberta, Canada, with Application,* paper presented at the 7th International Conference on Greenhouse Gas Control Technologies (GHGT-7), Vancouver, British Columbia, Canada, September 5-9, 9p.
8. Mavor, M.J., Gunter, W.D. and Robinson, J.R., (2004), *Alberta Multiwell Micro-Pilot Testing for CBM Properties, Enhanced Methane Recovery and CO_2 Storage Potential,* SPE Paper No. 90256 presented at the SPE Annual Technical Conference, Houston, Texas, U.S.A., September 26-29, 14p.
9. Seidle, J.P. and Arri, L.E., (1990), *Use of Conventional Reservoir Models for Coalbed Methane Simulation*, Paper No. CIM/SPE 90-118, presented at The International Technical Meeting, Calgary, Alberta, Canada, June 10-13.
10. Law, D.H.-S., van der Meer, L.G.H. and Gunter, W.D., (2001), *Modelling of Carbon Dioxide Sequestration in Coalbeds: A Numerical Challenge*, Proceedings of the 5th International Conference on Greenhouse Gas Control Technologies (GHGT-5), Cairns, Australia, August 13-16, P 537-542..

11. Law, D.H.-S. and Gunter, W.D. (2003) *History Matching of Enhanced Coalbed Methane (CBM) Production Field Data*, Proceedings of the JCOAL, KANSO & JCOSCs' 2nd International Workshop on the "Research Relevant to CO_2 Sequestration in Coal Seam," Tokyo, Japan, September 25, P. 67-77.
12. Mavor, M.J. and Gunter, W.D. (2004) *Secondary Porosity and Permeability of Coal vs. Gas Composition and Pressure, SPE Paper No. 90255* presented at the SPE Annual Technical Conference, Houston, Texas, U.S.A., September 26-29, 15p.
13. Palmer I. and Mansoori, J., (1996), *How Permeability Depends on Stress and Pore Pressure in Coalbeds: A New Model,* SPE Paper No. 36737 presented at The SPE Annual Technical Conference, Denver, Colorado, U.S.A., October 6-9.
14. Arri, L.E., Yee., D., Morgan, W.D. and Jeansonne, N.W., (1992), *Modeling Coalbed Methane Production with Binary Gas Sorption*, Paper No. SPE 24363, presented at The SPE Rocky Mountain Regional Meeting, Casper, Wyoming, U.S.A., May 18-21.
15. Law, D.H.-S., van der Meer, L.G.H. and Gunter, W.D., (2001), *Comparison of Numerical Simulators for Greenhouse Gas Storage in Coal Beds, Part I: Pure Carbon Dioxide Injection,* Paper presented at The 1st National Conference on Carbon Sequestration, Washington, D.C., U.S.A., May 14-17, 15p.
16. Law, D.H.-S., van der Meer, L.G.H. and Gunter, W.D., (2002), *Numerical Simulator Comparison Study for Enhanced Coalbed Methane Recovery Processes, Part I: Pure Carbon Dioxide Injection,* SPE Paper No. 75669 presented at The SPE Gas Technology Symposium, Calgary, Alberta, Canada, April 30-May 2, 14p.
17. Law, D.H.-S., van der Meer, L.G.H. and Gunter, W.D., (2003), *Comparison of Numerical Simulators for Greenhouse Gas Storage in Coalbeds, Part II: Flue Gas Injection,* Proceedings of the 6th International Conference on Greenhouse Gas Control Technologies (GHGT-6), Kyoto, Japan, October 1-4, P. 563-568.
18. Law, D.H.-S., van der Meer, L.G.H. and Gunter, W.D., *(2003), Comparison of Numerical Simulators for Greenhouse Gas Sequestration in Coalbeds, Part III: More Complex Problems,* Paper presented at The 2nd Annual Conference on Carbon Sequestration, Alexandria, Virginia, U.S.A., May 5–8, 13p.
19. Law, D.H.-S., van der Meer, L.G.H. and Gunter, W.D., (2002), *Comparison of Numerical Simulators for Greenhouse Gas Storage in Coalbeds, Part IV: Numerical History Match of Field Micro-Pilot Test Data,* Paper presented at The 7th International Conference on Greenhouse Gas Control Technologies (GHGT-7), Vancouver, B.C., Canada, September, 4p.
20. Robinson, J., Kadatz, B., Wong, S., Gunter, W.D., Sangli, F. and Zhiqiang, F. (2004) *ECBM Micro-Pilot Test in the Anthracitic Coal's of the Qinshui Basin, China: Field Results and Preliminary Analysis*, Proceedings of the JCOSC, JCOAL & GETs' 3rd International Workshop on the "Prospective Roles of CO_2 Sequestration in Coal Seam," Hokkaido University, Japan, October 5, P. 33-44.

Table 1: Coal characteristics of ARC CO_2-ECBM/storage Projects

	Mannville Coal, Alberta, Canada	Qinshui Basin Coal, Shanxi, China
Age	Lower Cretaceous	Permian
Rank	High Volatile Bituminous B	Semi-Anthracite and Anthracite
Vitrinite Reflectance	0.51%	2.7 - 4.0%
Depth	1,250 m	472 m
Thickness	4 m	6.45 m (6.05 m net)

Table 2: Description of problem sets for the simulator comparison study ("X" means feature not tested).

Part	**I CO_2 Injection**		**II Flue Gas Injection**		**III CO_2 Injection**		**IV ARC Pilot Tests**
Problem Set	1	2	1	2	3	4	5
Model Attributes Tested	Single Well	5-Spot	Single Well	5-Spot	5-Spot	5-Spot	Field Tests
Two-Gas Components	√	√	√	√	√	√	√
Multi-Gas Components	X	X	√	√	X	X	√
Dual Porosity Approach	X	X	X	X	√	X	√
Mixed Gas Diffusion	X	X	X	X	√	X	√
Mixed Gas Sorption	√	√	√	√	√	√	√
Compaction/Dilation due to Stresses	X	X	X	X	X	√	√
Coal Shrinkage (Primary Process)	X	X	X	X	X	√	√
Coal Swelling (ECBM Process)	X	X	X	X	X	X	√

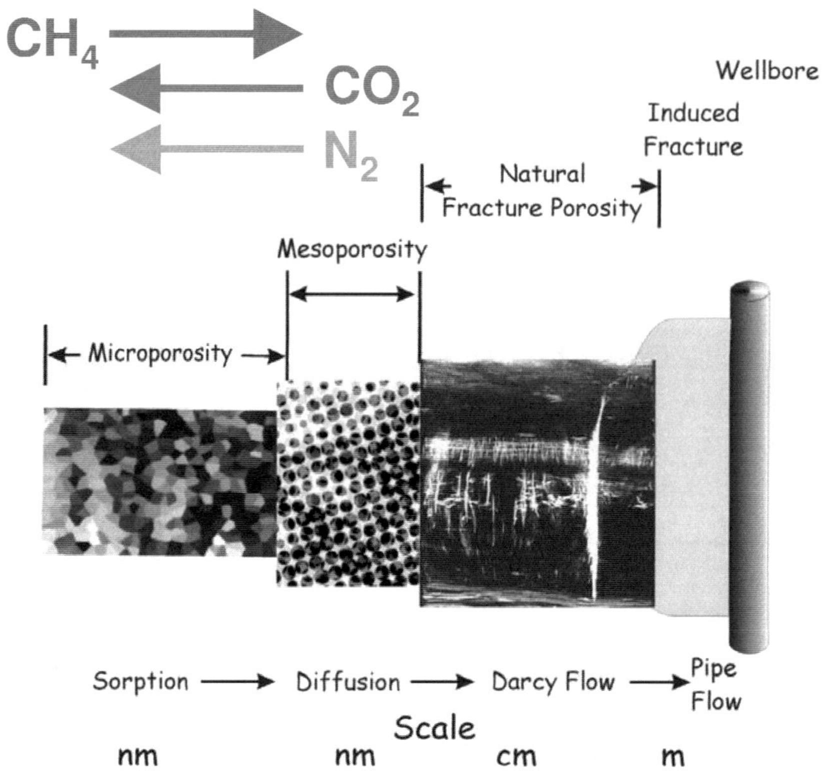

Figure 1: Overall process of gas storage and movement through coalbeds

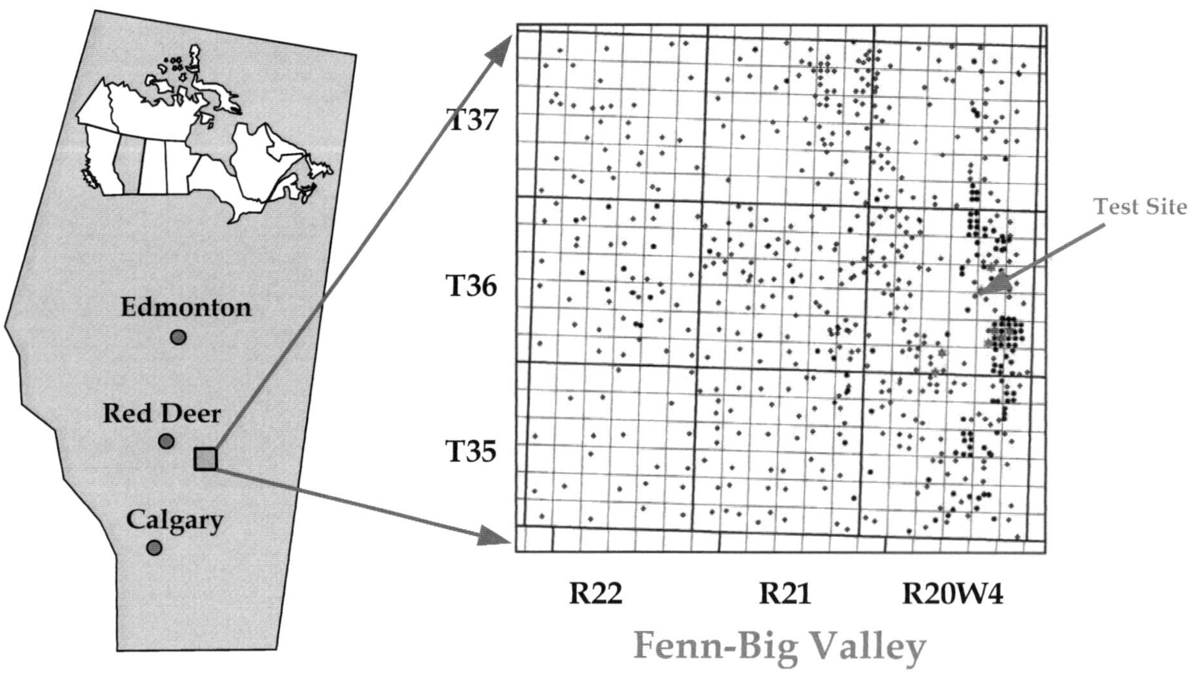

Figure 2: ARC micro-pilot test site in Fenn-Big Valley, Alberta, Canada

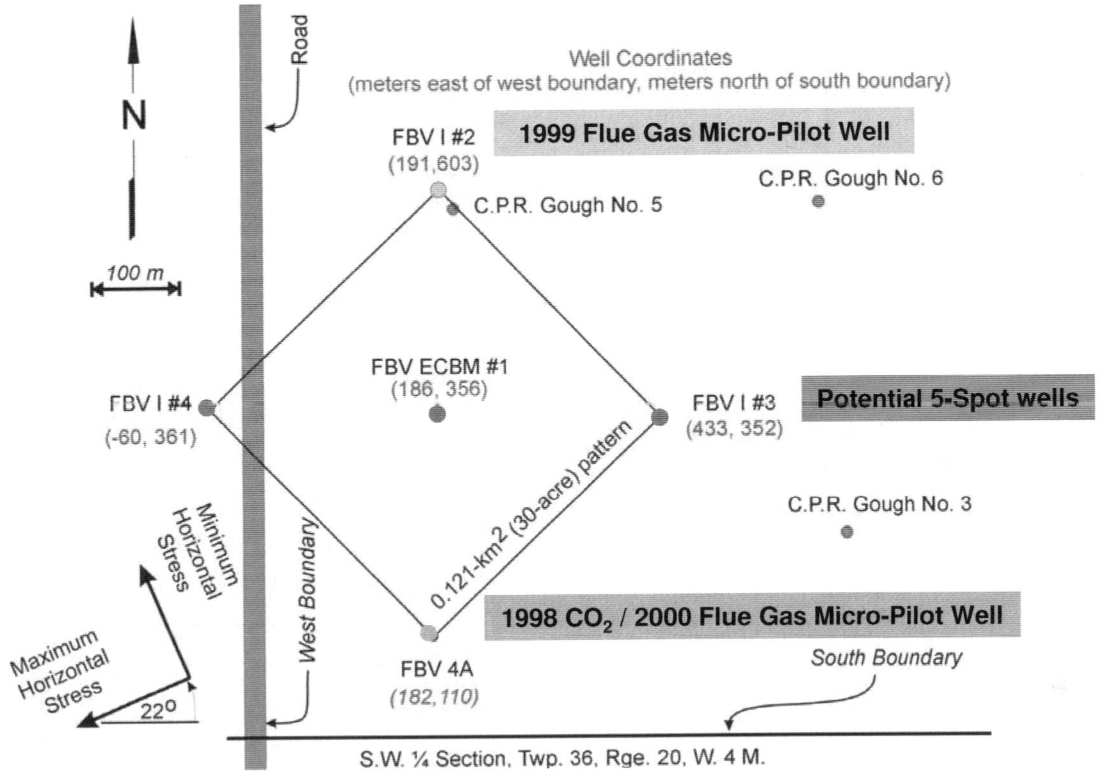

Figure 3: ARC micro-pilot test well locations in Fenn-Big Valley, Alberta, Canada

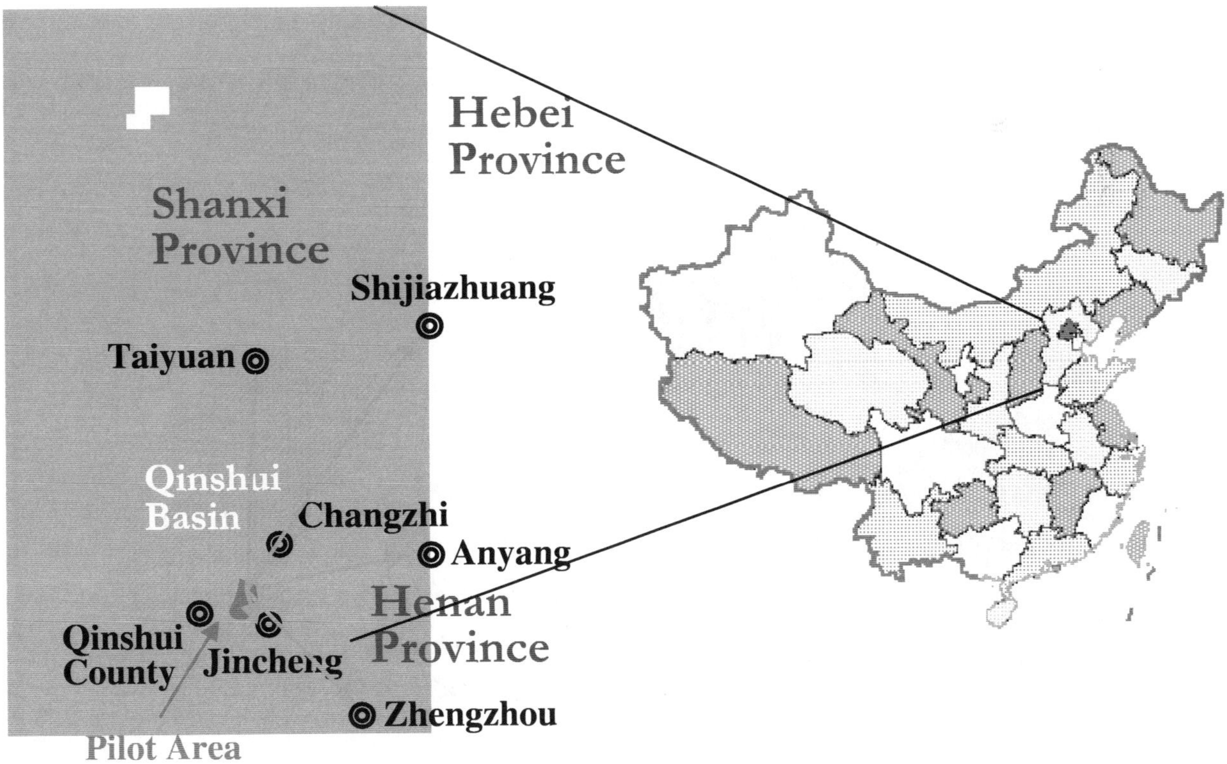

Figure 4: ARC micro-pilot test site location in Qinshui Basin, Shanxi, China

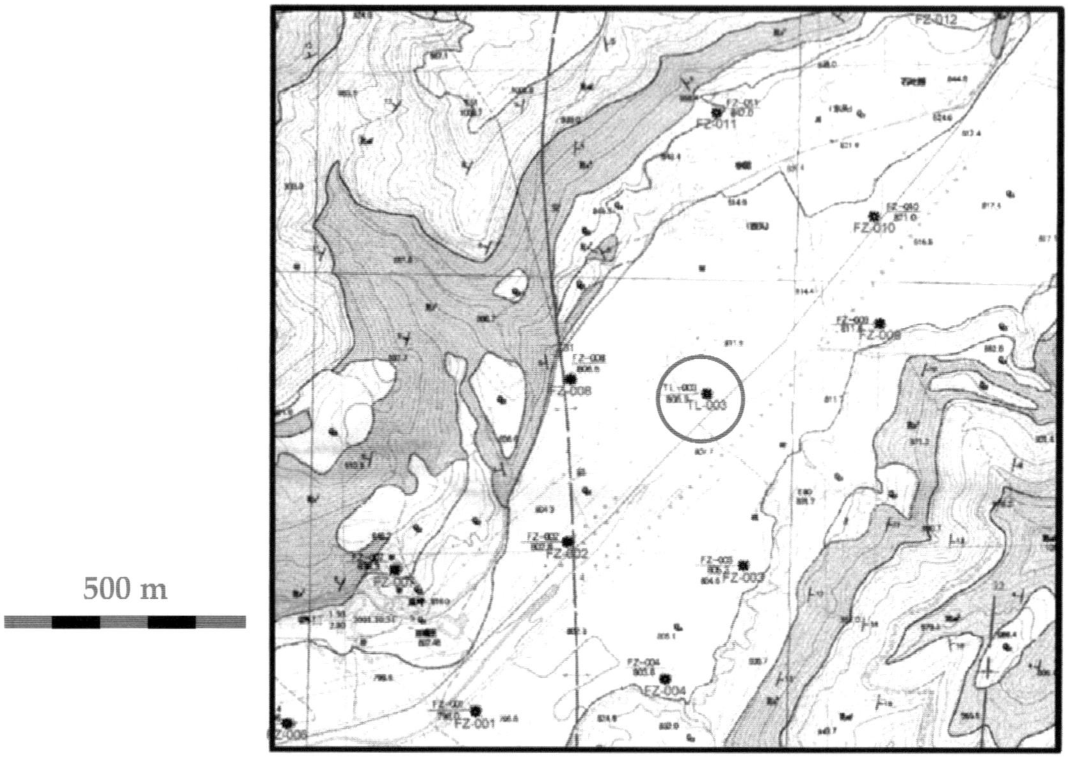

Figure 5: ARC micro-pilot test well location in Qinshui Basin, Shanxi, China

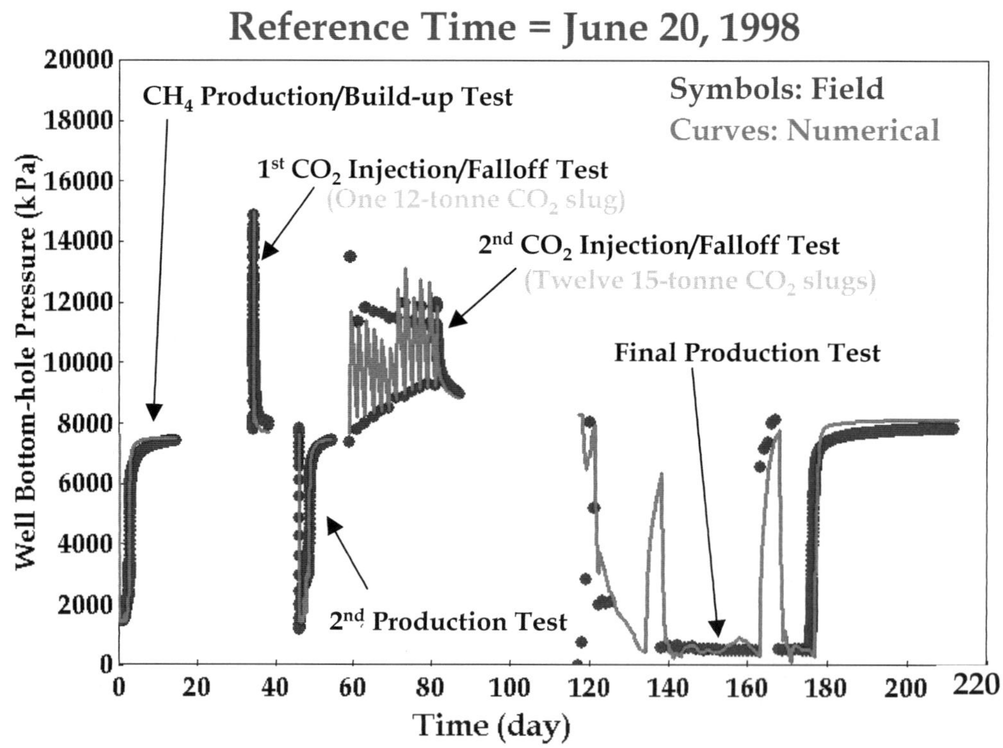

Figure 6: History match of bottom-hole pressure during CO_2 injection micro-pilot test at well FBV 4A in Fenn-Big Valley, Alberta, Canada (after [7])

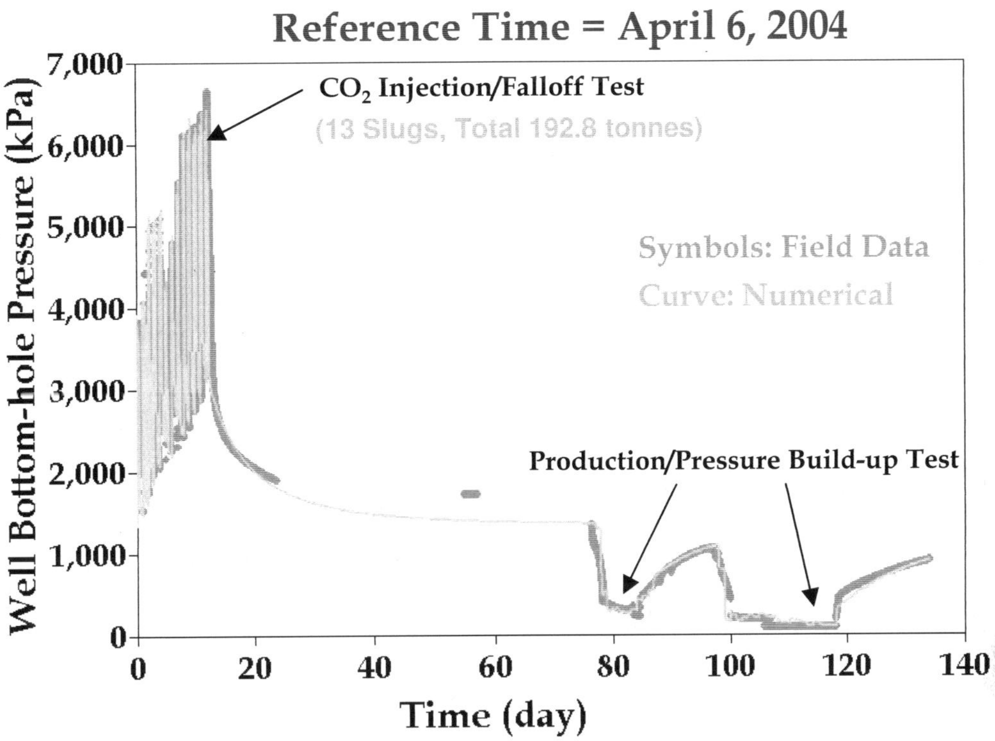

Figure 7: History match of bottom-hole pressure during CO_2 injection micro-pilot test at well TL-003 in Qinshui Basin, Shanxi, China (after [20])

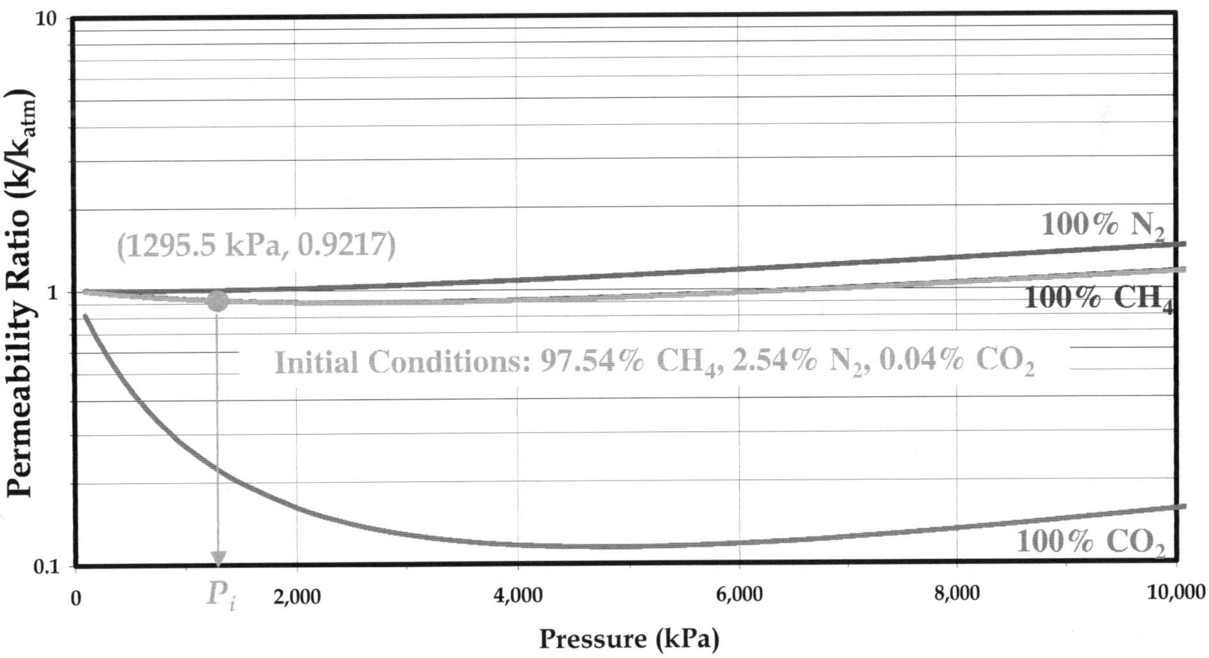

Figure 8: Permeability ratio based on the ARC Permeability Theory for Qinshui Basin coal in Shanxi, China

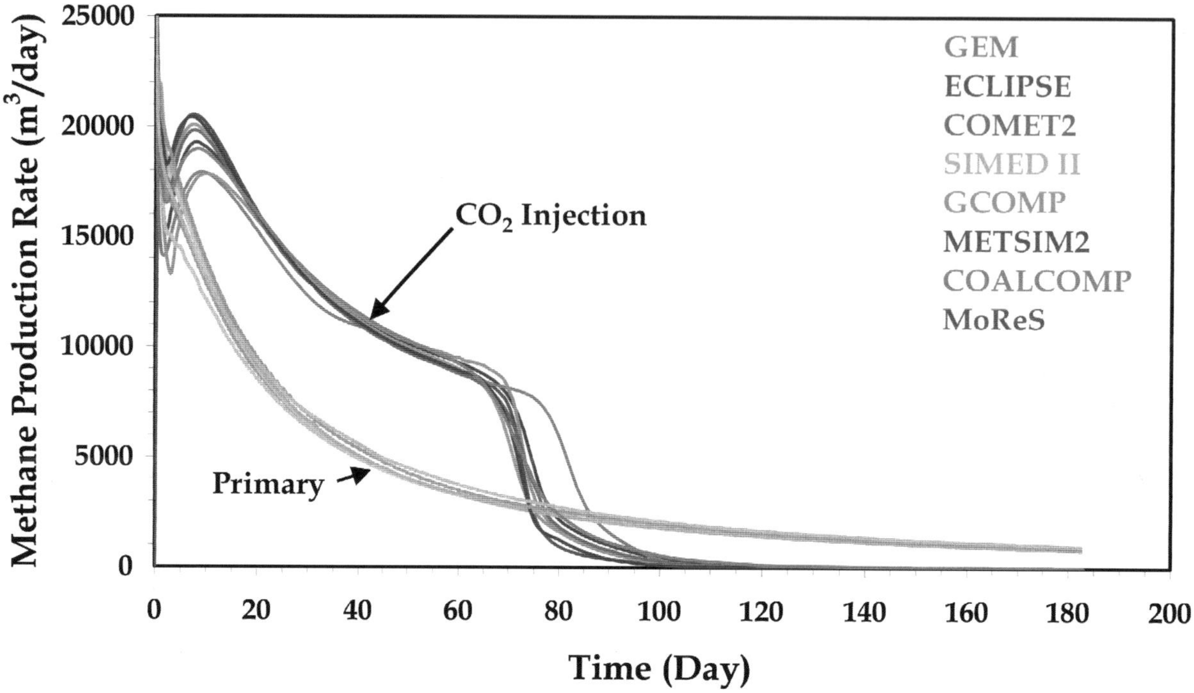

Figure 9: **Comparison of CH_4 production rate estimates for problem set 2 in Part I (ECBM in inverted 5-spot pattern with continuous pure CO_2 injection) (after [16])**

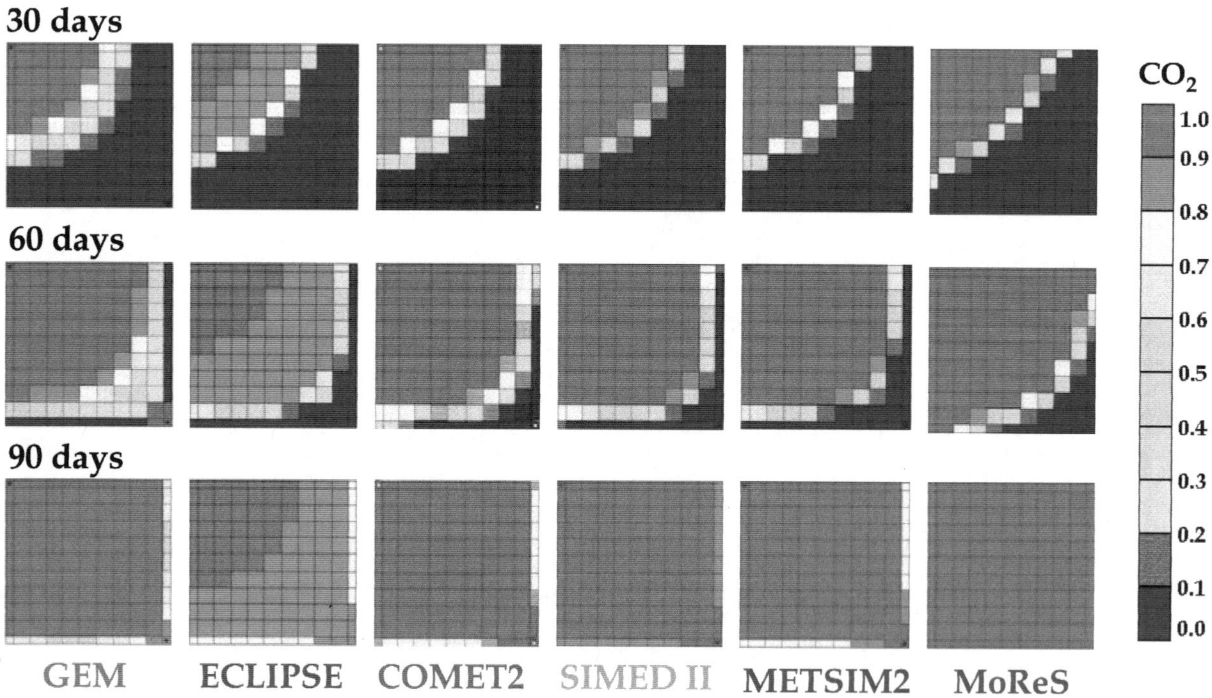

Figure 10: **Comparison of CO_2 mole fraction in natural fracture system for problem set 2 in Part I (ECBM in inverted 5-spot pattern with continuous pure CO_2 injection) (after [16])**

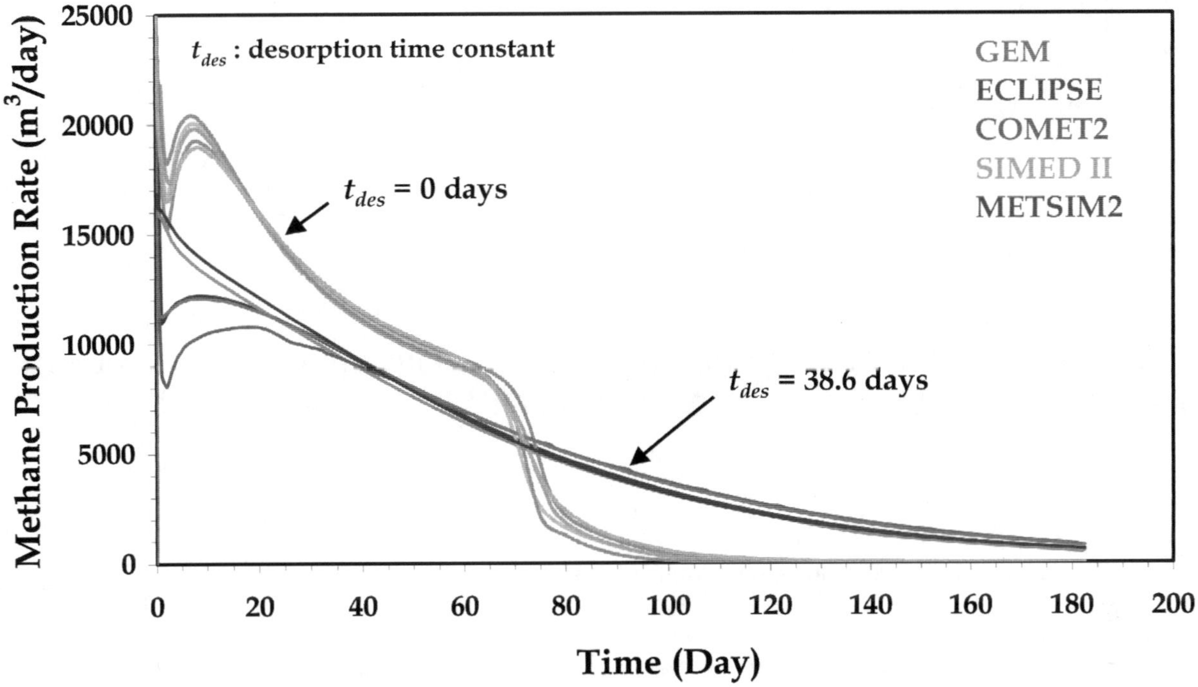

Figure 11: Comparison of CH_4 production rates for problem set 3 in Part III (ECBM in inverted 5-spot pattern with continuous pure CO_2 injection) (after [18])

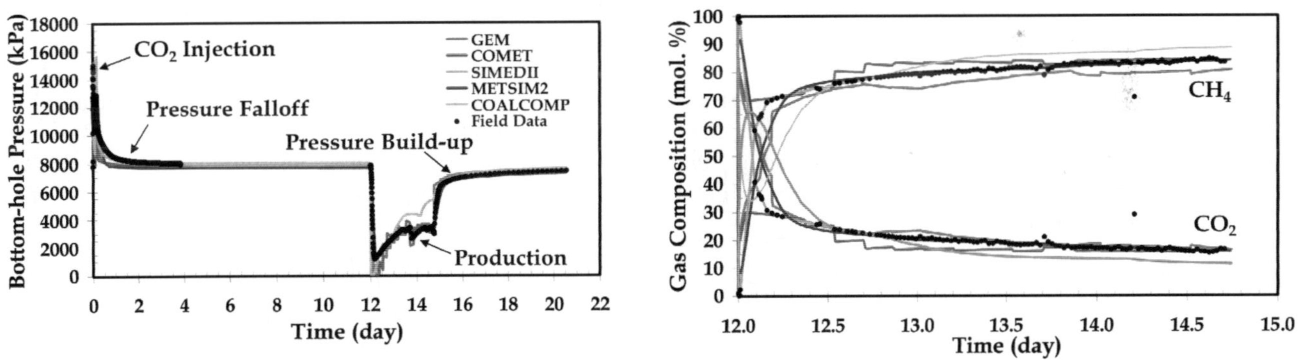

Figure 12: Comparison of bottom-hole pressures for problem set 5 in Part IV (ARC pure CO_2 injection micro-pilot test) (after [19])

Carbon Sequestration in Coal Seams: Defining the Nature of the Interactions Between CO_2 and Coal

A.L. Goodman and K.T. Schroeder, U.S. Department of Energy, National Energy Technology Laboratory; L.M. Campus, ORISE, Chemical Engineering, Carnegie Mellon University, Pittsburgh, and M.M. Hill, ORISE, Environmental Science, Indiana University

ABSTRACT

The CO_2 storage capacity in coal seams is typically estimated from isotherm measurements obtained from manometric techniques. In the calculation of the isotherm, two major parameters must be estimated. First, the compressibility factor must be calculated from an equation of state to account for the non-ideality of CO_2 at elevated pressures. Second, the volume change associated with the volume occupied by the sorbed CO_2 must be estimated. These two parameters can dramatically affect the shape of the CO_2-coal isotherm. Of the few papers that published CO_2–coal isotherms at high pressures, a variety of curve shapes have been reported. This lack of agreement reduces the confidence in the accuracy of CO_2 storage capacities estimated from volumetric isotherm measurements. In this study, the direct interaction between CO_2 and two Argonne premium coals [Pocahontas #3 (low volatile bituminous) and Beulah Zap (lignite)] was probed using Attenuated Total Reflectance – Fourier Transform Infrared (ATR-FTIR) spectroscopy at 328 K and pressures up to 8.0 MPa. Sorbed CO_2 on Argonne coals was detected at 2335 cm^{-1} for Buelah Zap coal and 2332 cm^{-1} for Pocahontas #3 coal. The energy of adsorption (18.8 - 20.5 kJ/mol), estimated using the Langmuir equation, was consistent with physisorption. The spectral data indicated that only one type of site was available for sorption. No evidence could be found for specific interactions between CO_2 and oxygen functional groups in the coals. The CO_2-coal sorption isotherm was derived without estimating the CO_2 compressibility and adsorbed layer density, both of which are needed in manometric techniques. The ATR-FTIR isotherms (units of net absorbance) and manometric isotherms (units mmol/g) compare well below the critical temperature providing some confidence in the values selected for the gas phase density and adsorbed layer density that were used to calculate absolute adsorption from the manometric data. In summary, we find that sorption of CO_2 is energetically similar for the two coal types, is due only to London forces and quadrupole interactions, and occurs preferentially on a hydrocarbon site.

INTRODUCTION

Storing CO_2 in coal seams has been proposed as a strategy to mitigate increasing atmospheric CO_2 levels.[1-5] An understanding of the mechanism by which CO_2 is stored in coals would provide a scientific foundation on which to base predictions of long-term CO_2 storage stability. Little information is available in the literature regarding the specific interactions between CO_2 and Coal. Although one report in the literature has implicated specific interactions between CO_2 and oxygen functional groups in coals,[6] most of the literature considers gases in coals to be stored as adsorbed species on the coal surface, both external and within the pore structure, and absorbed within the organic matrix of the coals.[1]

Also of interest is the CO_2 coal seam storage capacity. Typically the storage capacity is estimated from the CO_2-coal isotherm measurements by volumetric or manometric techniques. Calculation of absolute adsorption requires two major parameters to be estimated. First, to account for the non-ideality of CO_2 at elevated pressures, the compressibility factor must be calculated from an equation of state (EOS). The Span and Wagner[7] EOS is the most recent and most widely accepted but others have been used. Second and perhaps more importantly, the change associated with the volume occupied by the surface adsorbed CO_2 must be estimated from the CO_2 adsorbed layer density. A number of different values have been proposed for sub-critical and supercritical conditions.[8] There appears to be little experimental justification for choosing one value over another beyond better isotherm curve fits.

In this study, we report the interactions between CO_2 with the Argonne Premium coals using Attenuated Total Reflectance – Fourier Transform Infrared (ATR-FTIR) spectroscopy. Upon exposure of Argonne coals to CO_2, a positive absorption band due to sorbed CO_2 appeared in the spectrum. As the coals were exposed to increasing CO_2 pressure, the CO_2 absorption band grew in intensity. Because this technique allowed for the direct observation of sorbed CO_2, both the chemical interaction between CO_2 and coals and the isotherm shape were determined. This method of determining the adsorption isotherm shape avoided assumptions about the compressibility and the adsorbed layer density needed in manometric techniques.

ATR-FTIR SPECTROSCOPY

The interaction of CO_2 with coals was examined using ATR-FTIR spectroscopy, a technique which is well suited to study high-pressure gas systems[9-12] and coals.[13] All experiments were conducted using a modified ATR CIRCLE cell from Spectra Tech. The modified cell assembly consisted of two stainless steel high-pressure cells each containing a ZnSe ATR crystal. One cell contained a ZnSe crystal with no coal present (blank) and the other cell contained a ZnSe crystal coated with the coal (sample). The two cells were connected in tandem via 1/16 inch stainless steel tubing so that each cell experienced the same CO_2 pressure. The entire ATR cell assembly was enclosed in a heating jacket.

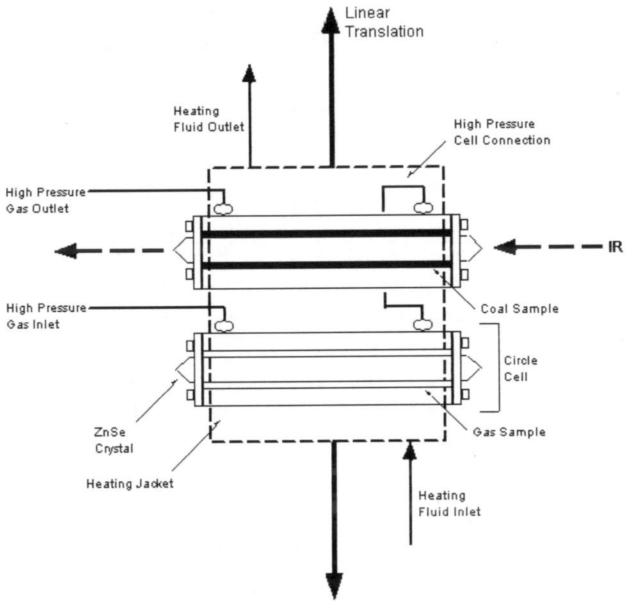

High pressure experiments were performed by connecting the ATR cell assembly to a gas handling system consisting of a syringe pump (ISCO model 260D), a port for gas introduction, and a pressure transducer (OmegaDyne Inc PX01K1-5KGV). The total volume of the system was estimated to be 1 ml. Carbon dioxide (99.998% SFC) and UHP nitrogen were used as supplied from Butler Gas Company. Infrared spectra were recorded with a single beam FTIR spectrometer (Thermo Electron Nexus 670 FTIR ESP) equipped with a wide-band MCT detector. Unless otherwise noted, 500 scans were collected with an instrument resolution of 2 cm^{-1} in the spectral range extending from 4000 to 650 cm^{-1}. Spectra were distorted below 650 cm^{-1} because the ZnSe crystal became opaque.

The ATR cell assembly was mounted on a linear translator inside the FT-IR spectrometer. The translator allowed each cell of the assembly, the blank side for CO_2 measurements and the sample side for CO_2-coal measurements, to be scanned by moving the cell of interest into the IR beam path. In the ATR-FTIR experiments, reference spectra were collected of the blank and sample cells. Then a known pressure of CO_2 was applied to both cells. After the pressure of the gas reached equilibrium, two spectra were recorded: one from the cell containing both CO_2 and coal (Spectrum a) and one from the cell containing CO_2 (Spectrum b). By calculating the difference between the two spectra, an absorption spectrum of CO_2 sorbed on the coal was obtained. The spectral subtraction result for CO_2 sorption on Pocahontas #3 coal at 328 K and 3.5 MPa is illustrated as Spectrum c. CO_2 pressures up to 8.0 MPa were investigated using this technique. At pressures above 8.0 MPa, CO_2

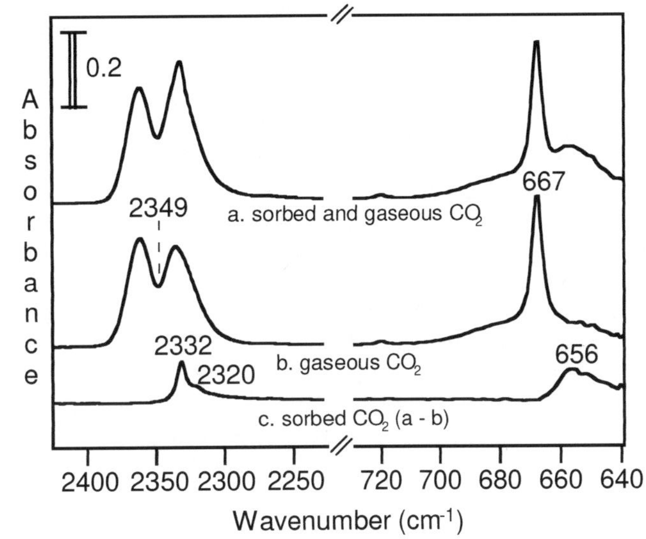

absorptions were too intense and saturated the wide-band MCT detector.

SAMPLES

Coals from the Argonne Premium Coal Sample Program was selected for this study. The Argonne coals provide the research community with the highest quality samples of coals for basic research.[14] The premium coal samples are chemically and physically as identical as possible, have well-characterized chemical and physical properties, and have been handled and packaged to be stable over long periods. The samples had a maximum particle diameter of 150 μm (-100 mesh) and were used as-received. In those cases where the amount of sample used was less than a full vial, the aliquot of coal was taken after mixing according to the Argonne instructions. An example of the ATR-FTIR spectra of Pocahontas # 3 and Buelah Zap coals are shown.

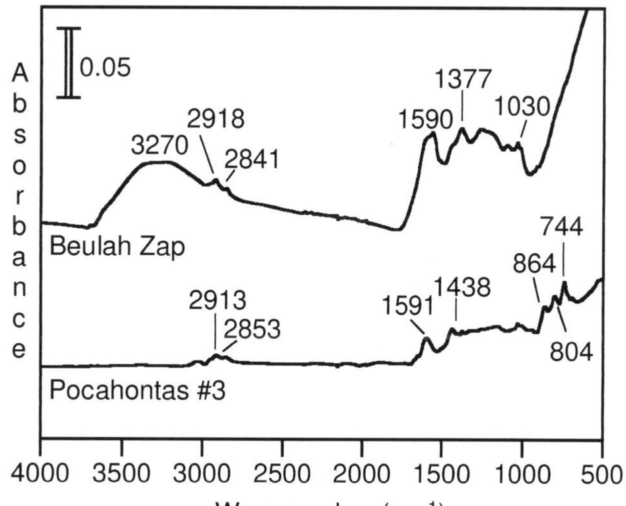

In these experiments, coals were prepared in a nitrogen-filled glove bag to avoid surface oxidation. The coals were immersed in nitrogen-sparged water to form a slurry. A ZnSe ATR crystal was then coated with a thin layer of the slurry (~10 mg) and allowed to dry at 295 K in the nitrogen-filled glove bag for 20 minutes. The crystal was loaded into the ATR cell and sealed under nitrogen. The ATR-FTIR spectra of the interaction of CO_2 with the coals was measured under one of two conditions: the coals were equilibrated under flowing N_2 at 328 K and then exposed to CO_2 pressure or the coals were additionally dried at 353 K for 36 hours under flowing N_2, equilibrated at 328, and then exposed to CO_2. The moisture contents of the nitrogen dried coals used in this study are listed in Table 1.

THE NATURE OF THE CO_2-COAL INTERACTION

Infrared spectroscopy was used as a tool to determine whether CO_2 interacted coals via specific or nonspecific interation. Is the interaction non-specific surface and pore filling where the CO_2 is adsorbed on the coal surface and absorbed within the coal organic matrix or is there specific interaction between CO_2 and coal functional groups such as carboxyl, carbonyl or hydroxyl groups? The spectral data indicate that there is no specific CO_2 – carbonyl nor CO_2 – carboxyl interactions. Upon exposure of Beulah Zap coal (0.2% moisture) to CO_2, two major positive absorption bands at 2335 and 654 cm^{-1} appeared in the difference spectrum (Spectrum a). A similar spectrum was obtained when a Buelah Zap sample containing more moisture (8.0% moisture) was exposed to CO_2 (Spectrum b). The higher moisture content of the coal did not result in new or shifted absorption bands. Similar spectra were obtained when Pocahontas #3 coals of different moisture contents were exposed to CO_2 (Spectra c-d). The absorption band frequencies, centered at 2332 and 656 cm^{-1}, were slightly shifted from those obtained with Beulah Zap samples. After desorption in flowing nitrogen, spectral absorption bands were not detected indicating that the interaction between CO_2 and the coals was reversible. The observed absorptions bands between 2335-2332 cm^{-1} (ν_3 anti-symmetric stretching mode) and 654-656 cm^{-1} (ν_2 bending mode) detected upon exposure of the coals to CO_2 were consistent with physisorption.[15-17]

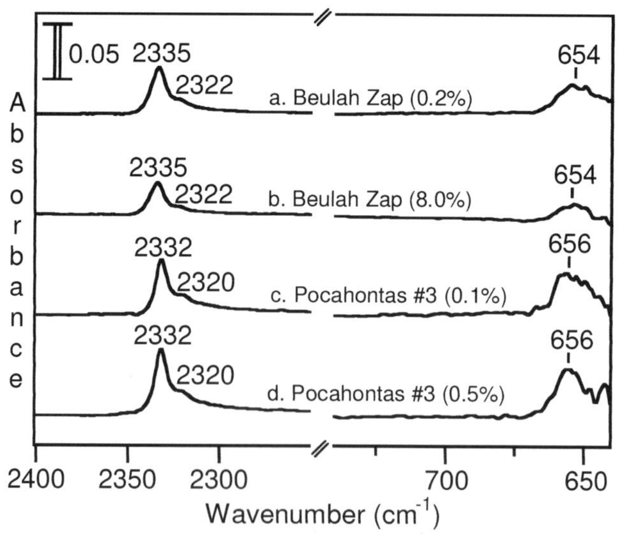

The CO_2 absorption band frequency and shape can help elucidate the chemical interaction between CO_2 and coal. That fact that there was one distinct absorption band suggests that there is one type of sorption site for the CO_2. If more than one distinct type of site were available for CO_2 to sorb, one would expect several distinct absorption bands corresponding to each type of sorption site.[17-18] The presence of inherent moisture does not appear to change the fundamental chemistry of the CO_2-coal interaction because intentional removal of essentially all of the inherent moisture does not affect the absorption band position, although it does appear to affect the adsorption capacity (*vide infra*). In his survey of the interaction of CO_2 with a number of polymers, Kazarian reported that the wavenumber corresponding to the anti-symmetric stretch of sorbed CO_2 was between 2338.0 cm^{-1} and 2339.5 cm^{-1} when the polymer contained carbonyl or carboxyl functionality.[9] In the absence of these functionalities, the wavenumber was between 2334.6 cm^{-1} and 2335.3 cm^{-1}. In this study, the wavenumber values of 2332 cm^{-1} for the low volatile bituminous coal and 2335 cm^{-1} for the lignite coal tend to indicate that specific interactions between CO_2 and oxygen functional groups were not responsible for the sorption phenomenon.

The ν_2 bending mode of sorbed CO_2 has also been used to provide evidence of specific interactions between CO_2 and polymer functional groups.[9,19] Upon sorption of gaseous CO_2, the degeneracy of the ν_2 bending mode is lost and the sorbed CO_2 absorption splits into two bands because of a change in symmetry. Kazarian and Nelson showed that the ν_2 split was large when CO_2 interacted with carbonyl functionalities (~ 6 cm^{-1}) found in polymers but that the ν_2 split was small when CO_2 interacted with aromatic rings (~1 cm^{-1}). They concluded that CO_2 interaction with carbonyl groups was favored over aromatic structures in the polymers they examined. The ν_2 CO_2 absorption bands we observed were broad. If the absorption band was splitting, the two bands were not clearly separated. According to Nelson, the presence of several surface sites can make the observed splitting of the bending mode of CO_2 complicated.[19] In our case, the heterogeneous nature of the coals may mask the ν_2 splitting. Although there is some ambiguity in the ν_2 region for coals, the ν_3 and ν_2 absorption bands taken together tend to support surface sorption without specific interactions with the oxygen functional groups in the coal.

DETERMINING THE CO_2-COAL ADSORPTION ISOTHERM QUANTITATIVELY

The interaction of the Argonne coals with increasing CO_2 pressure was also examined to determine the shape of the

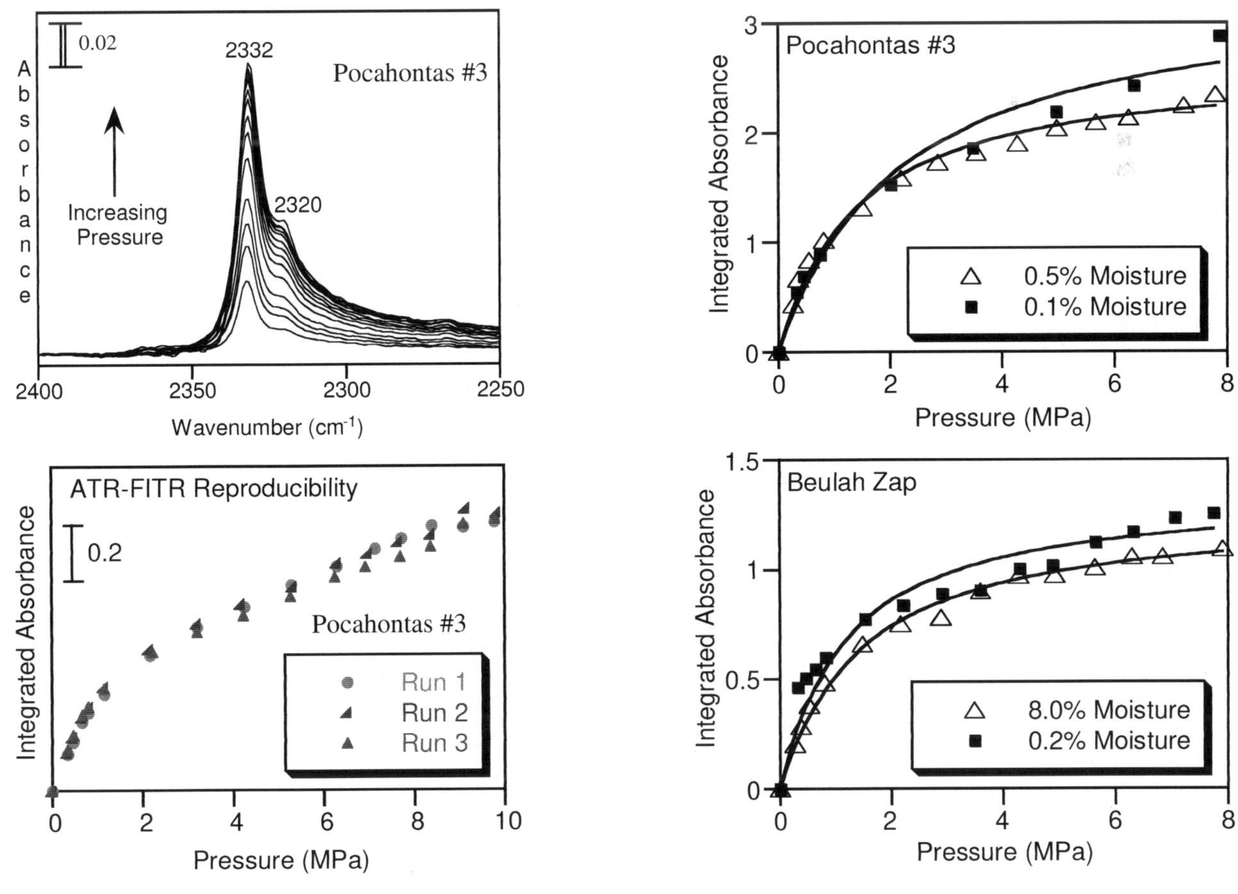

CO_2 sorption isotherms. Pressures up to 8.0 MPa were investigated at 328 K. Absorption spectra of CO_2 sorbed on Pocahontas #3 coal are shown in the figure as a function of increasing CO_2 pressure. As the CO_2 pressure increased, the intensity of the absorption band at 2332 cm^{-1} due to sorbed CO_2 increased. The Beulah-Zap coal produced similar sorbed-CO_2 spectra as a function of pressure (spectra not shown). Because ATR-FTIR allows for the direct observation of sorbed CO_2, the isotherm shape can be derived by plotting the net spectral absorbance of sorbed CO_2 versus pressure. The ν_3 stretching mode of sorbed CO_2 was integrated from 2360 and 2300 cm^{-1} and the area was plotted as a function of pressure.[11] This procedure was repeated three times for the Pocahontas #3 coal and reproducibility was found to be very good as shown in the figure. The sorption isotherms generated for Pocahontas #3 and Buelah Zap coals of two moisture contents are shown in the Figure. The plots show the shape of the CO_2 isotherm curve as a function of pressure using integrated spectral area units. For both Pocahontas #3 and Beulah Zap coals, more CO_2 was sorbed on the drier coal. Although this did not become obvious for the high rank coal until pressures over 3.5 MPa were attained. These results are consistent with the reports that moisture plays a role in determining the CO_2 adsorption capacity of coals.[20-22] If the sorption sites for the coals with different moisture contents are the same, water may compete for adsorption sites resulting in less CO_2 sorption.

Typically, volumetric or manometric techniques are used measure the CO_2 adsorption isotherm curve for coal samples.[23] The amount of CO_2 adsorbed on the coal sample (moles of CO_2 adsorbed per unit mass of coal), also known as the Gibbs excess adsorption,[8] is calculated by the real gas law shown in equation (1)

$$\Delta n^{ex} = \left(\frac{1}{RT_{iso}m_c}\right) * \left(V_R\left(\frac{P_{ri}}{Z_{ri}} - \frac{P_{rf}}{Z_{rf}}\right) - V_V\left(\frac{P_{sf}}{Z_{sf}} - \frac{P_{si}}{Z_{si}}\right)\right) \qquad (1)$$

where R is the molar gas constant, T_{iso} is the isothermal temperature, m_c is the mass of the coal, V_R is the volume of the reference cell, V_v is the void volume, Z is the compressibility factor, and P is the pressure of the corresponding cells (*ri* is the reference initial, *rf* is the reference final, *sf* is the sample final, and *si* is the sample initial). Equation (1) expresses the difference between the moles of CO_2 transferred from the reference cell and the moles of CO_2 present in the previously determined void volume of the sample cell. The coal volume, and thus the void volume of the sample chamber, is assumed to be constant throughout the adsorption measurement. The estimate of the total amount of Gibbs excess adsorbed gas, n^{ex}, at the end of the ith step is the sum of the incremental adsorptions obtained at each step (equation (2)).

$$n^{ex} = \Delta n_1^{ex} + \Delta n_2^{ex} + \Delta n_3^{ex} + \ldots + \Delta n_i^{ex} \qquad (2)$$

The isotherm curve is created by plotting the total amount of Gibbs excess adsorbed CO_2 as a function of the measured equilibrium pressure.

When calculating the Gibbs excess adsorption as shown in Equations (1) and (2), the gas compressibility coefficient Z at each pressure is required to determine the mass balance. This parameter describing the non-ideality of the gas should be known with the highest degree of precision because the shape of the CO_2 adsorption isotherm and the calculated amount of excess adsorption are affected by the choice of the compressibility factor. The most recent and widely-accepted method for computing the compressibility coefficients of carbon dioxide is the formulation published by Span and Wagner.[7] The figure shows the isotherm calculated for Buelah Zap coal using three different Z values available in the literature.[24]

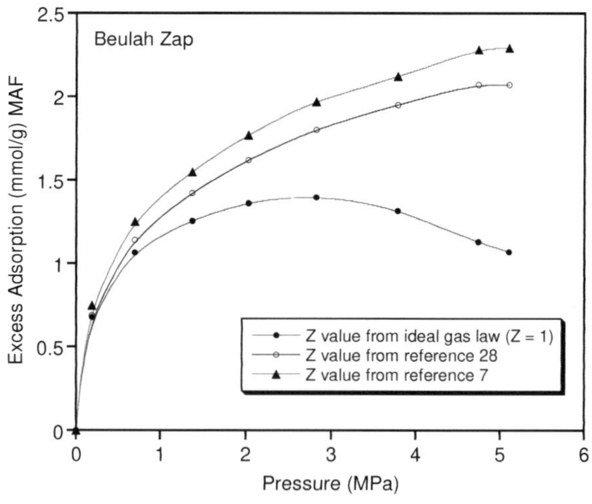

The Gibbs excess adsorption works well as long as the assumption that the void volume remains constant throughout the adsorption measurement. As higher pressures are reached, the volume occupied by adsorbed CO_2 increases, thus decreasing the void volume. The correction to convert Gibbs excess adsorption to absolute adsorption is shown in equation (3).

$$n^{ex} = \left[1 - \left(\rho_g / \rho_a\right)\right] * n^{abs} \qquad (3)$$

where n_{abs} is the absolute adsorption, ρ_g is the density of the gas phase, ρ_a is the density of the adsorbed phase.[8] ρ_g is calculated from an equation of state, typically from Span and Wagner.[7] ρ_a is assumed to be constant and different authors use various values. The inverse VanDerWalls volume (ρ_a = 23.45 mmol/cc) is often used for supercritical CO_2. For subcritical CO_2 the Dubinin Approx is often used

$$\rho_a - \rho_b - \left(\rho_b - \rho_{VDW}\right) * \left(T - T_b / T_c - T_b\right) \qquad (4)$$

where ρ_b equals the liquid CO_2 density at the boiling point, ρ_{VDW} is the VanDerWaals volume (ρ_a = 23.45 mmol/cc), T is the temperature of the measurement, T_b is the CO_2 boiling point, and T_c is the CO_2 critical temperature. The figure shows the absolute absorption calculated with four different values of ρ_a.[8]

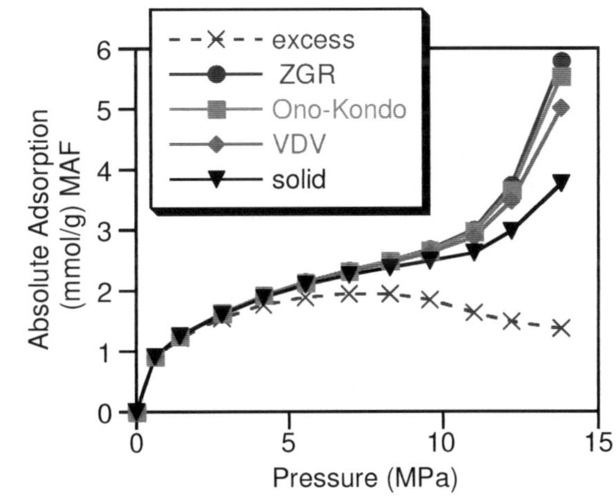

In this study, the spectral absorption intensity of the sorbed CO_2 is directly proportional to the absolute amount of CO_2 sorbed. Therefore, neither a compressibility factor nor an adsorbed layer density is needed to determine the shape of the absolute sorption isotherm. While it would be preferable to be able to determine the molar absolute sorption (mmol/g) with ATR-FTIR spectroscopy, the current experimental apparatus does not provide all of the parameters needed to do this. However, knowledge of the shape of the absolute sorption isotherm provides useful information.

The figure compares the CO_2–coal sorption isotherms generated using a manometric technique[24] and ATR-FTIR spectroscopy. The data were fit to the Langmuir equation [eq (5)] (solid lines),

$$\theta = \frac{bP}{1 + bP} \qquad (5)$$

where b is the Langmuir constant, P is pressure, and θ is the fractional sorption.[25] The unit-less parameter, θ, was used to directly compare the two techniques because the ATR-FTIR data is in units of integrated absorbance which cannot be converted to mmol/g without additional information. The similarity of the curves provides confidence in the values selected for the gas phase density (Span and Wagner EOS)[7] and adsorbed layer density (23.45 mmol/cc) that were used to calculate the absolute adsorption from the manometric data.

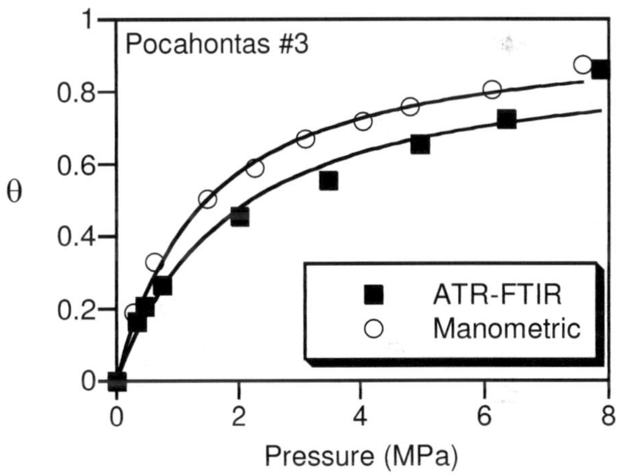

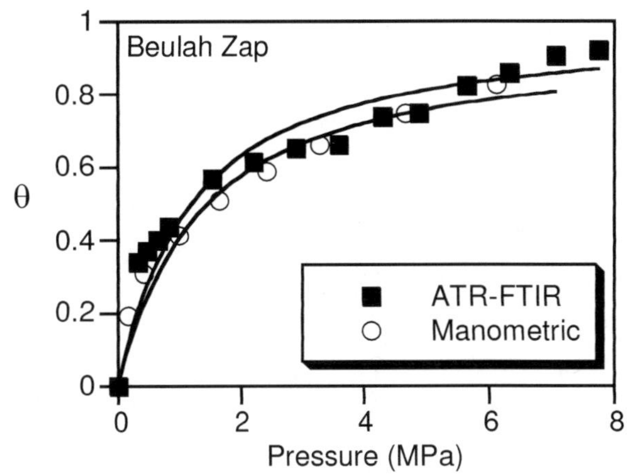

The energy of adsorption (Q) was estimated from the Langmuir constant b in equation 5 using equation (6),

$$b = b_o e^{Q/RT} \qquad (6)$$

where R is the gas constant, T is the temperature, and b_o is a constant defined as,

$$b_o = \frac{N\sigma\tau_o}{(2\pi MRT)^{1/2}} \qquad (7)$$

where N is Avogadro's constant, σ is the molecular area[26] of CO_2 (0.22 nm^2), τ_o equals $1x10^{-13}$ sec, and M is the molecular weight of CO_2. The energy of adsorption values are listed in Table 1. Nearly identical values were obtained regardless of coal type, moisture content, or technique. The average value of 19.8 kJ/mol obtained using the ATR-FTIR technique compared favorably with the 20.0 kJ/mol average obtained for the volumetric technique.[27] These values are consistent with CO_2 physisorption.

SUMMARY

The direct interaction between CO_2 and Argonne premium coals was probed using ATR-FTIR spectroscopy at 328 K. The sorbed CO_2 absorption bands at 2335 cm^{-1} (Buelah Zap) and 2332 cm^{-1} (Pocahontas #3) and an energy of adsorption of ~20 kJ/mol were consistent with a mechanism involving physical adsorption. The spectral data for CO_2 sorbed on the coals indicated that there was one type of sorption site for CO_2. No evidence for specific interactions between CO_2 and oxygen functionalities was found. The pressure dependence of the IR data was used to calculate the CO_2-coal adsorption isotherm while avoiding assumptions about the gas compressibility and the adsorbed layer density needed in manometric techniques.

TABLE 1

Comparison of the Langmuir Constants and Energies of Adsorption Obtained Using ATR-FTIR to Those Obtained Using the Manometric Technique.

Technique	Coal	% Moisture	Q (kJ/mol)	b (1/Mpa)
ATR-FTIR	Beulah Zap	8.0[a]	19.9	0.711
ATR-FTIR	Beulah Zap	0.2[b]	20.5	0.878
Manometric	Beulah Zap	4.0[c]	20.2	0.787
ATR-FTIR	Pocahontas #3	0.5[a]	20.0	0.706
ATR-FTIR	Pocahontas #3	0.1[b]	18.8	0.471
Manometric	Pocahontas #3	0.5[c]	19.9	0.708

[a]Nitrogen dried at 295 K as determined by TGA
[b]Nitrogen dried at 353 K as determined by TGA
[c]Dried at 353 K under vacuum as determined by TGA
Q = energy of adsorption
b = Langmuir constant

REFERENCES

(1) White, C. W.; Strazisar, B. R.; Granite, E. J.; Hoffman, J. S.; Pennline, H. W. *J. the Air and Waste Manage. Assoc.* **2003**, *52*, 645-715.

(2) Reeves, Scott R. Geological Sequestration of CO2 in Deep, Unmineable Coalbeds: An Integrated Research and Commerical- Scale Field Demonstration Project. 2001. SPE paper 71749 presented at the Annual Technical Conference and Exhibition, New Orleans, Louisiana, September 30-October 3, 2001.

(3) Richardson, R. J.; Hughes, J. D.; Rottenfusser, B. A.; Gentzis, T.; Gunter, W. D.; and Bachu, S. Coalbed methane and CO_2 disposal ; the answer for Alberta? 28[7], 41. 1996. United States, Geological Society of America (GSA), Boulder, CO, United States. Geological Society of America, 28th annual meeting. 10-28-1996.

(4) Gentzis, T. *Int. J. Coal Geol.* **2000,** *43,* 287-305.

(5) Allis, R.; Chidsey, T.; Gwynn, W.; Morgan, C.; White, S.; Adams, M.; Moore, J. Natural CO_2 Reservoirs on the Colorado Plateau and Southern Rocky Mountains:
Candidates for CO_2 Sequestration. DOE/NETL-2001/1144 - CDROM, 1-19. 2001. NETL Publications. 5-14-2001.

(6) Nishino, J. *Fuel* **2001,** *80,* 757-764.

(7) Span, R.; Wagner, W.; *J. Phys.Chem.* **1996**, *25*, 1509-1596.

(8) Sudibandriyo, M.; Pan, Z.; Fitzgerald, J. E.; Robinson, R. L.; Gasem, K. A. M. *Langmuir* **2003,** *19,* 5323-5331.

(9) Kazarian, S. G.; Vincent, M. F.; Bright, F. V.; Liotta, C. L.; Eckert, C. A. *J. Am. Chem. Soc.* **1996,** *118,* 1729-1736.

(10) Kazarian, S. G. *Macromol. Symp.* **2002,** *184,* 215-228.

(11) Yarwood, J. *Analytical Proceedings* **1993,** *30,* 13-18.

(12) Yokoyama, C.; Kanno, Y.; Takahashi, M.; Ohtake, K.; Takahashi, S. *Rev. Sci. Instrum.* **1993,** *64,* 1369-1370.

(13) Thomasson, J.; Coin, C.; Kahraman, H.; Fredericks, P. M. *Fuel* **2000,** *79,* 685-691.

(14) Vorres, K. S. *Energy & Fuels* **1990**, *4*, 420-426.

(15) Freund, H. J.; Roberts, M. W. *Surface Science Reports* **1996,** *25,* 225-27.

(16) Fastow, M.; Kozirovski, Y.; Folman, M. *Journal of Electron Spectroscopy and Related Phenomena* **1993,** *64,* 843-848.

(17) Mawhinney, D. B.; Rossin, J. A.; Gerhart, K.; Yates, J. T. Jr. *Langmuir* **1999,** *15,* 4617-4621.

(18) Gallei, E.; Stumpf, G. *Journal of Colloid and Interface Science* **1976,** *55,* 415-420.

(19) Nelson, M. R.; Borkman, R. F. *J. Phys. Chem* **1998,** *102,* 7860-7863.

(20) Clarkson, C. R.; Bustin, R. M. *Int. J. Coal Geol.* **2000,** *42,* 241-271.

(21) Ozdemir, E.; Schroeder, K.; Morsi, B. I.; White, C. W. Mechanism of Carbon Dioxide (CO2) Adsorption on Moist Coals. Fuel Chemistry Division Preprints 47[1]. 2003.

(22) Krooss, B. M.; Gensterblum, Y.; Siemons, N.; van Bergen, F.; Pagnier, H. J. M.; David, P. High-Pressure Methane and Carbon Dioxide Adsorption on Dry and Moisture-Equilibrated Carboniferous Coals. 2001. Proceedings: International Coalbed Methane Symposium, Univ of AL, Tuscaloosa, AL, May 14-18, pp 177-191.

(23) Mavor, M. J.; Owen, L. B.; Pratt, T. J. Measurement and Evaluation of Coal Sorption Isotherm Data. 1990. SPE paper 20728 presented at the 65th Annual Technical Conference and Exhibition, Society of Petroleum Engineers, New Orleans, LA, September 23-26, pp. 157-70.

(24) Goodman, A. L., Busch, A., Duffy, G. J., Fitzgerald, J. E., Gasem, K. A. M, Gensterblum, Y, Krooss, B. M., Levy, J., Ozdemir, E., Pan, Z., Robinson, R. L., Schroeder, K., Sudibandriyo, M., and White, C. M. *Energy & Fuels* **2004**, *18*, 1175-1182.

(25) Adamson, A. W. Physical Chemistry of Surfaces. John Wiley & Sons Inc.: New York, 1990; pp 662.

(26) McClellan, A. L.; Harnsberger, H. F. *J. Colloid and Interface Science* **1967,** *23,* 577-599.

(27) A. L. Goodman, L. M. Campus, K. T. Schroeder *Energy & Fuels* **2005**, *19*, 471-476.

(28) *Gas Encyclopaedia, L'Air Liquide;* Elsevier: Amsterdam, 1976.

0509

The Representation of Multi-component Adsorption in Reservoir Simulation of CO_2 Sequestration in Coal and Enhanced Coalbed Methane Recovery

Zhejun Pan and LD Connell
CSIRO Petroleum, Private Bag 10, Clayton South, Victoria, Australia, 3169

ABSTRACT

CO_2 sequestration in coal is a potential management option for greenhouse gas emissions. An attractive aspect to this process is that the CO_2 is adsorbed to the coal and therefore, for static reservoir pressure conditions, is immobile, reducing the risk of CO_2 migration to surface. Another aspect to this is that the injected CO_2 can displace adsorbed methane leading to a region within the coal seam of mixed gas. Therefore, in order to understand gas migration within the reservoir multi-component adsorption models are required. Langmuir based models, such as the extended Langmuir and Ideal Adsorbed Solution, are part of the reservoir simulators currently in use for modeling CO_2 sequestration in coal. However these models are inaccurate descriptions of the adsorption process particularly at the high pressures associated with sequestration in deep coals. Therefore new approaches are required that can be implemented in a computational efficient manner for reservoir simulation. This paper presents simulation work using several approaches to representing multi-component adsorption, implemented into the coal seam gas reservoir simulator SIMED II. The adsorption models are extended Langmuir, Ideal Adsorbed Solution (IAS) and 2D Equation of State (2D EOS). The 2D EOS is shown to be a more accurate description than the other models for experimental observations of adsorption on Fruitland coal. The effects of this greater accuracy on modeling the sequestration process is investigated using a hypothetical example. While simultion results for the extended Langmuir and IAS are very close, results calculated using 2D EOS are significantly different. The differences are greatest with the CO_2 injection rate. A conclusion from this work is that inaccurate isotherm models can lead to significant errors in simulation results and that the implementation of the 2D EOS model provides greater accuracy in the representation of the sequestration process.

INTRODUCTION

CO_2 sequestration in unminable coal is a potential management option for greenhouse gas emissions. The feasibility of this process will be determined by various factors including the coal capacity, the rate of injection and the long term behaviour of the sequestered CO_2. To investigate these requires representative models for the reservoir behaviour of the CO_2. One important process in determining the reservoir behaviour is adsorption and since migration will often involve the displacement of adsorbed methane with CO_2 and therefore a region of mixed gas within the seam, it is multi-component adsorption that needs to be represented. Currently, most of the reservoir simulators used for sequestration in coal implement the Langmuir and Extended Langmuir models (ELM) for pure and mixed gas adsorption, respectively. Since ELM is not thermodynamically correct [1] and subject to large errors for mixed gas adsorption, Ideal Adsorbed Solution (IAS) theory was implemented in some simulators. However, IAS is for mixture adsorption only and requires a pure gas adsorption model, and normally the Langmuir model is used. Thus, the inaccuracy of the pure gas adsorption propagates to the mixed gas adsorption calculation.

Recent work has proposed new adsorption models for multi-component adsorption for the full range of pressures, such as the Two-Dimensional Equation of State (2D EOS) approach. This paper presents simulation work using several approaches to representing multi-component adsorption, implemented into the coal seam gas reservoir simulator SIMED II. The adsorption models are extended Langmuir, Ideal Adsorbed Solution and 2D EOS. The simulation studies investigate the sequestration process and compare the results using these different adsorption approaches.

ADSORPTION THEORY

Langmuir and Extended Langmuir Model

The Langmuir model assumes adsorption occurs on a flat surface as given by the kinetic theory. At equilibrium, a continual process of bombardment of molecules onto the surface and a corresponding evaporation of molecules from the surface maintain a zero rate of accumulation at the surface. The assumptions of the Langmuir model are [1]:

(1) The surface is homogeneous; that is, the adsorption energy is constant over all sites.
(2) Adsorption on the surface is localized, which means that the atoms or molecules are adsorbed at definite, localized sites.
(3) Each site can accommodate only one molecule or atom.

Based on the assumptions above, the Langmuir model can be derived as:

$$\theta = \frac{\omega}{L} = \frac{BP}{1+BP} \qquad (1)$$

Where θ is the fractional coverage, ω is the amount adsorbed, P is the pressure, B is the Langmuir constant, and L is maximum adsorption capacity. When extended to mixture adsorption assuming no interaction among the adsorbed molecules, the Langmuir model becomes:

$$\omega_i = \frac{L_i B_i P y_i}{1+\sum_j B_j P y_j} \qquad (2)$$

where subscript "i" represents component i, and y is the mole fraction in the gas phase.

Langmuir model is successful in modeling adsorption at low or mid pressure (typically up to 8 MPa for gas adsorption on coals). However, the difference between the model and the experimental data is large at high pressure ranges. Moreover, the extended Langmuir model (ELM) is not thermodynamically correct unless the maximum capacity L_i is the same for each component [1]. Consequently, large errors may occur for high pressure multi-component adsorption predictions, particularly when the difference of L_i is large.

Ideal Adsorbed Solution

The Ideal Adsorbed Solution (IAS) model is for mixture adsorption only, developed by Myers and Prausnitz (1965) [2]. The IAS is an adsorption analog to the Raoult's Law for vapor-liquid equilibrium. The assumptions of the IAS model are [3]:
(1) The adsorbed solutions and the gas phase are ideal.
(2) All activity coefficients in the adsorbed phase are unity.

The equilibrium condition for the adsorbed phase and the gas phase is:

$$Py_i\hat{\phi}_i^g = P_i^0 x_i \phi_i\left(P_i^0\right) \qquad i = 1...N \tag{3}$$

where P_i^0 is the gas pressure of component i adsorbed at the same spreading pressure as the mixture, $\hat{\phi}_i^g$ is the partial fugacity coefficient of component i in the gas phase, x_i is the adsorbed phase composition and ϕ_i is the pure gas fugacity coefficient. The spreading pressure π_i can be determined by the Gibbs adsorption isotherm providing that the adsorbed phase is ideal:

$$\pi_i = \frac{RT}{A}\int_0^{P_i^0} \omega d\ln P \tag{4}$$

where R is the gas constant, T is the temperature and A is the surface area of the adsorbent.

The IAS theory assumes that the spreading pressure of each component should be equivalent to the multicomponent adsorption. When assuming each component accesses the same surface area, the following relation applies:

$$\int_0^{P_1^0} \frac{\omega_1^0}{P} dP = \int_0^{P_2^0} \frac{\omega_2^0}{P} dP = ... = \int_0^{P_n^0} \frac{\omega_n^0}{P} dP \tag{5}$$

Meanwhile, the mole fraction constraint is:

$$\sum_{i=1}^{n} x_i = \sum_{i=1}^{n} y_i = 1 \tag{6}$$

The total adsorption, ω_T, is given by:

$$\frac{1}{\omega_T} = \sum_{i=1}^{n} \frac{x_i}{\omega_i^0\left(P_i^0\right)} \tag{7}$$

The component adsorption, ω_i, is given by:

$$\omega_i = \omega_T x_i \tag{8}$$

To perform mixed-gas adsorption calculations, a pure-component model is needed. Any pure-component model, for instance the Langmuir model, can be utilized in the IAS calculation. However, for many models, analytical integration is unavailable. Thus numerical integration may be used in Equation (5). For high pressure adsorption, fugacity should be used in the gas phase. However, assuming an ideal adsorbed phase may be invalid at high pressure leading to large errors for these conditions.

Two-Dimensional Equation of State

The general 2D analog can be written as follows with an additional parameter m for added model flexibility [4]:

$$\left[A\pi + \frac{\alpha\omega^2}{1 + U\beta\omega + W(\beta\omega)^2}\right]\left[1 - (\beta\omega)^m\right] = \omega RT \tag{9}$$

where A is the specific surface area, π is the spreading pressure, ω is the specific amount adsorbed, and α and β are model parameters. The model coefficients, U, W, and m must be specified to obtain a specific form of the 2D EOS for application. For example, an analog of the van der Waals (VDW) EOS is obtained by setting m = 1 and U = W = 0; similarly for the Soave-Redlich-Kwong (SRK) (m = U = 1 and W

= 0); the Peng-Robinson (PR) (m = 1, U = 2, and W = -1); the Eyring (m = 1/2 and U = W = 0) EOS; and the Zhou-Gasem-Robinson (ZGR) EOS (m = 1/3 and U = W = 0). [4]

One-fluid mixing rules were used in the 2D EOS to describe multi-component adsorption:

$$\alpha = \sum_i \sum_j x_i x_j \alpha_{ij} \text{ and } \beta = \sum_i \sum_j x_i x_j \beta_{ij} \tag{10}$$

The combination rules below are chosen because they work best for adsorbed phase [4]:

$$\alpha_{ij} = (\alpha_i + \alpha_j)(1 - C_{ij})/2 \tag{11}$$

$$\beta_{ij} = \sqrt{\beta_i \beta_j}(1 + D_{ij}) \tag{12}$$

where C_{ij} and D_{ij} are binary interaction parameters. To predict the mixture adsorption, C_{ij} and D_{ij} are set to be zero.

To relate the adsorbed phase with the gas phase, at equilibrium, the chemical potential of species i in the adsorbed phase, μ_i^a, is equal to that in the gas phase, μ_i^g:

$$\mu_i^a = \mu_i^g \text{ and } d\mu_i^a = d\mu_i^g \tag{13}$$

Thus,

$$\int_{\pi^*}^{\pi} d\ln \hat{f}_i^a = \int_{P^*}^{P} d\ln \hat{f}_i^g \tag{14}$$

Integrating Equation 14, it yields:

$$\ln \hat{f}_i^a(\pi) - \ln \hat{f}_i^a(\pi^*) = \ln \hat{f}_i^g(P) - \ln \hat{f}_i^g(P^*) \tag{15}$$

At very low pressure, $\hat{f}_i^a(\pi^*) = \pi_i^*$ and $\hat{f}_i^g(P^*) = P_i^*$

$$P_i^* \hat{f}_i^a(\pi) = \pi_i^* \hat{f}_i^g(P) \tag{16}$$

At very low pressure, 2D ideal gas law applies:

$$\pi_i^* A = \omega_i^* RT \tag{17}$$

Henry's constant is defined as:

$$k_i = \frac{\omega_i^*}{P_i^*} \tag{18}$$

So, substitution into Equation 16, we obtain:

$$A\hat{f}_i^a = Ax_i \pi \hat{\phi}_i^a = k_i RT \hat{f}_i^g \tag{19}$$

For pure-gas adsorption, Equation 19 becomes:

$$\omega Z_a \phi^a = k_i f^g \tag{20}$$

The fugacity for the 2D EOS is:

$$\ln \hat{\phi}_i^a = \int_0^{\omega} \left[\frac{1}{RT\omega} \left. \frac{\partial (A\pi)}{\partial \omega_i} \right|_{T,M_s,n_j} - \frac{1}{\omega} \right] d\omega - \ln Z_a \tag{21}$$

where ω is the amount adsorbed, Z_a is the 2D compressibility factor, ϕ^a is the fugacity coefficient using the 2D EOS, f^g is the fugacity for the gas phase. The 3D PR EOS is normally used to calculate the gas fugacity. Detailed derivation of Equation 21 was provided by Zhou [3].

ADSORPTION COMPARISON

Pure CH_4, CO_2 and their mixture adsorption on wet Fruitland coal [5] were used to evaluate the models (the CO_2 adsorption data up to 8 MPa were used). The measurements were performed at 46.1 °C with pressures up to 12.8 MPa. The feed molar compositions for the mixture adsorption were 80%/20%, 60%/40%, 40%/60% and 20%/80%.

The pure gas adsorption data are used to regress the model parameters. The objective function, S, is used to correlate data with the adsorption models. The function minimizes the sum of the squared-percentage deviations in predicted adsorption:

$$S = \sqrt{\frac{\sum_{i=1}^{NPTS}\left[\frac{\omega_i^c - \omega_i^e}{\omega_i^e}\right]^2}{NPTS}} \tag{22}$$

where ω^c and ω^e are the calculated and experimental adsorption amount, respectively. NPTS is the number of data points. The percentage average absolute deviation (%AAD) is used to evaluate the results:

$$\%AAD = \frac{100}{NPTS}\sum_{i=1}^{NPTS}\left|\frac{\omega_i^c \quad \omega_i^e}{\omega_i^e}\right| \tag{23}$$

The model evaluation results for pure gas adsorption are summarized in Table 1 and the parameters for the models are listed in Table 2. The model evaluation results for the mixture adsorption are also listed in Table 1.

Figure 1 shows the CH_4 adsorption results represented with different models. The Langmuir model underestimates the pure CH_4 adsorption at pressures higher than 7 MPa and slightly overestimates the pure CH_4 adsorption from 2 MPa to 6 MPa. As a result, both ELM and IAS overestimate CH_4 component adsorption below 6 MPa and slightly overestimate the adsorption at higher pressure. 2D EOS, on the other hand, represents the pure CH_4 adsorption more accurately as well as CH_4 component adsorption.

Figure 2 shows the CO_2 adsorption results using different models. In a similar fashion as with the pure CH_4 adsorption results, the Langmuir model underestimates the pure CO_2 adsorption at pressure higher than 6 MPa and slightly overestimates the pure CO_2 adsorption below 6 MPa. Since the pure CO_2 adsorption data up to 8 MPa were used in the model evaluation, the models' behavior at pressures higher than 8 MPa are not known. However, for the CO_2 component adsorption, 2D EOS predicts the adsorption better than ELM or IAS, especially in the high pressure region.

Figure 3 shows the total adsorption results using different models. Both ELM and IAS underestimate the adsorption for pressures higher than 7 MPa. Figure 4 shows the compositional results for mixture adsorption. 2D EOS predicts the adsorbed phase composition better than the other two models.

SIMULATION RESULTS

SIMED II is a coalbed methane simulator capable of compositional dual-porosity reservoir simulation. The adsorption models integrated in SIMED II were ELM and IAS. In this work, the 2D EOS was implemented into SIMED II. A hypothetical one-dimensional enhanced coal methane through CO_2 injection case was studied to illustrate how the different adsorption models affect the reservoir simulation results using the adsorption data presented above for Fruitland coal. The description of the reservoir properties is listed in Appendix A.

Figure 5 presents the gas rate calculated with SIMED using the different adsorption models. The CH_4 production rate decreases after reaching a maximum and the different adsorption models predict similar rates till around day 400. After that, the CH_4 production rate increases and 2D EOS predicts a lower rate than the two other models. Figure 6 presents the adsorbed concentrations for each gas and isotherm model. In Figure 6(a), at the early stages of the simulation, the two wells have no interaction, thus the CH_4 production well behaves as conventional primary coalbed methane recovery, where the methane is desorbed from the coal by lowering the seam pressure. Since the representation of pure CH_4 adsorption at pressures less than 5 MPa is almost identical for the 2D EOS and Langmuir models, the initial production rate results for the two models are also very similar. However for Figure 6(b) at 1 year the CH_4 displaced due to CO_2 injection has started to arrive at the production well and the production rate increases. The adsorbed concentrations of CH_4 are higher for the 2D EOS than the other two models reflecting the adsorption isotherm differences at these high pressures. As a result the free gas volumes are less and thus production rate lower. Figure 5(b) shows the gas rate results for a 16 year period. Gas production rate increases sharply after day 4100, indicating a CO_2 breakthrough. Gas produced is mainly CO_2 after day 5000. Figure 6(e) shows that the adsorbed CO_2 concentration becomes higher near the production well, indicating a coming CO_2 breakthrough. Figure 6(f) shows no CH_4 adsorbed in the reservoir and CO_2 being produced at day 5480.

Figure 5 also presents CO_2 injection rate. All adsorption models considered have a maximum injection rate at around day 100. However, the CO_2 injection rate predicted by 2D EOS is significantly higher than by either ELM or IAS till around day 1000. Then the injection rate predicted by 2D EOS is slightly higher than the other models. The lower CO_2 injection rate at early times by ELM and IAS is due that both models underestimate the CO_2 adsorption at high pressures.

Figure 7 shows the gas phase composition profile using 2D EOS and ELM. The results are similar to the adsorbed gas concentration because they are closely related to each other.

Figure 8 shows the cumulative CH_4 produced and CO_2 injected for 2 year period. The results are consistent with the gas rate results.

SUMMARY and CONCLUSIONS

This paper has considered three approaches to representing adsorption in reservoir simulation of enhanced coalbed methane through CO_2 sequestration. Two of the adsorption isotherm models, extended Langmuir and Ideal Adsorbed Solution, are popular choices for reservoir simulation. These provide computationally efficient calculations but can involve inaccuracies particularly for multi-component gas mixtures and single component CO_2. The 2D EOS isotherm model has been considered by several authors as an approach to representing adsorption. The 2D EOS isotherm model has been shown to provide a closer match to experimental observations on Fruitland coal for a wide pressure range and gas component mixture than the extended Langmuir and Ideal Adsorbed Solution isotherms.

The 2D EOS isotherm model was implemented in the compositional dual porosity simulator SIMED. A hypothetical case study was formulated to investigate the relative effects of the three isotherm models on reservoir simulation of CO_2 sequestration and enhanced coalbed methane recovery. In this case study a CO_2 injection well is separated from a CH_4 recovery well by 1000m for a coal seam at an initial reservoir pressure of 8.8 MPa, corresponding to roughly 880 m depth. While the injection well operates at pressures significantly higher than the initial reservoir pressure, the production well produces methane initially through pressure drawdown below the desorption pressure.

In the case study, injected CO_2 displaces adsorbed methane which then migrates away from the injection well and towards the production well in response to the pressure gradient within the seam. In terms of adsorbed gases, three regions are created within the seam; towards the injection well there is a region of pure CO_2, around the production well there is a region of pure CH_4. In between these two regions there is a zone of mixed CO_2-CH_4, a transition region. The size of this transition region is determined by the rate of diffusion into the coal matrix; characterized in the modelling by the desorption time constant. Within this transition region a mixed gas adsorption isotherm is required.

Initially the methane production rates are very similar for the three adsorption models. This is consistent with the accuracy with which the adsorption behaviour is represented; for this low pressure, single component methane regime all three isotherm models are good descriptions of the adsorption behaviour. However with time this good agreement deteriorates, with the methane rates calculated using the 2D EOS being significantly lower than those calculated using the other two adsorption models. This can be attributed to the interaction of free methane displaced by the CO_2 injection arriving at the production well. As discussed above the CO_2 injection creates a mixed gas region where multi-component adsorption occurs. The 2D EOS is a more accurate description of this process than the ELM and IAS models. In addition, the injection of CO_2 involves high pressures; adsorption of CO_2 for this single component high pressure regime is also more accurately described using the 2D EOS. The combination of these two effects leads to increasing differences in methane production rate with time as the interaction between the two wells becomes stronger.

The most significant differences associated with the isotherm models are seen in the CO_2 injection rate. The rate with 2D EOS based calculations is significantly higher than with the other two adsorption approaches, particularly at early times. This behaviour reflects the differences in the adsorption isotherms, where at high pressures the CO_2 adsorbed amount, as represented by the 2D EOS, is greater than that estimated by the ELM and IAS models. Since the pressure behaviour is very similar across the simulations, the injection rate is therefore higher to compensate for the greater amount adsorbed. After reaching a maximum the injection rate decreases with time, in response to the pressure behaviour within the reservoir.

This paper has demonstrated that improved models for adsorption can have significant impacts on reservoir simulation of CO_2 sequestration and enhanced coalbed methane. The magnitude of this impact is a function of pressure and the role of multi-component gas migration. Implementation of the 2D EOS adsorption isotherm model into SIMED has therefore improved the capability of this simulator to represent CO_2 sequestration in coal and enhanced coalbed methane production.

Appendix A. Description of reservoir properties used in simulation case study

Grid system
Cartesian 78 × 1 × 1

x direction: see Table 3

Reservoir temperature: 46.1 °C
Permeability: 2.5 md
Relative permeability: see Table 4
Gas desorption pressure: 5 MPa
Initial reservoir pressure: 8 MPa
Desorption time constant: 10 days for CO_2, 27.5 days for CH_4

Well locations:
  Injection well: Grid 1
  Production well: Grid 78
Well skin factor=0

Bottomhole pressure: 344.75 kPa
Injection pressure: 14 MPa

References

1. Do, D. D., 1998: Adsorption Analysis: Equilibria and Kinetics: London, Imperial College Press.

2. Myers, A. L., Prausnitz, J. M., 1965: "Thermodynamics of Mixed-Gas Adsorption", AIChE J., V. 11, p. 121-129.

3. Zhou, C., 1994: Modeling and Prediction of Pure and Multicomponent Gas Adsorption, Ph.D Dissertation: Oklahoma State University, Stillwater, Oklahoma.

4. Zhou, C., Hall, F., Gasem, K. A. M., Robinson, Jr., R. L., 1994: "Predicting Gas Adsorption Using Two-Dimensional Equations of State", I&EC Research, V. 33, p. 1280-1289.

5. Hall, F., Zhou, C., Gasem, K. A. M., Robinson, Jr., R. L., 1994: "Adsorption of Pure Methane, Nitrogen, and Carbon Dioxide and Their Binary Mixtures on Wet Fruitland Coal", presented at the Eastern Regional Conference & Exhibition, Charleston.

Tables

Table 1. Adsorption results - %AAD

Pure Gas Adsorption			
	Langmuir		2D EOS
CH_4	2.8		0.5
CO_2	3.6		2.3
Mixture Adsorption			
	ELM	IAS	2D EOS
CH_4	13.7	10.4	8.5
CO_2	7.3	7.2	4.8
Total	4.6	4.8	3.2

Table 2. Best fit parameter values for adsorption isotherms of Fruitland coal

Langmuir			
	L (m^3/ton)	B (MPa^{-1})	
CH_4	24.83	0.3112	
CO_2	35.62	0.6462	
2D EOS			
	α (bar cm^3 g/mmol / mol)	β $(mmol/g)^{-1}$	-ln(k) ln(mmol/g bar^{-1})
CH_4	-13863	0.3556	3.2367
CO_2	-18138	0.1335	1.8805

Table 3. Grid spacing used in simulation case study

Grid	1	2	3	4	5	6	7	8	9	10
Δx(m)	0.1000	0.1193	0.1422	0.1697	0.2023	0.2413	0.2878	0.3433	0.4094	0.4883
Grid	11	12	13	14	15	16	17	18	19	20
Δx(m)	0.5824	0.6946	0.8284	0.9880	1.1783	1.4053	1.6761	1.9990	2.3842	2.8435
Grid	21	22	23	24	25	26	27	28	29	30
Δx(m)	3.3913	4.0447	4.8240	5.7534	6.8619	8.1840	9.7607	11.641	13.884	16.559
Grid	31	32	33	34	35	36	37	38	39	
Δx(m)	19.750	23.555	28.093	33.505	39.961	47.660	56.842	67.793	80.855	

* Δx[grid]=Δx[79-grid] when grid is 40 to 78

Table 4. Relative permeability of water and gas used in simulation case study

Water Saturation	Water phase relative permeability	Gas phase relative permeability
0.00	0.000	1.000
0.40	0.000	1.000
0.45	0.000	0.700
0.50	0.005	0.520
0.55	0.020	0.400
0.60	0.045	0.300
0.65	0.075	0.230
0.70	0.120	0.175
0.75	0.190	0.120
0.80	0.300	0.080
0.85	0.450	0.050
0.90	0.630	0.030
0.95	0.850	0.015
1.00	1.000	0.000

Figures

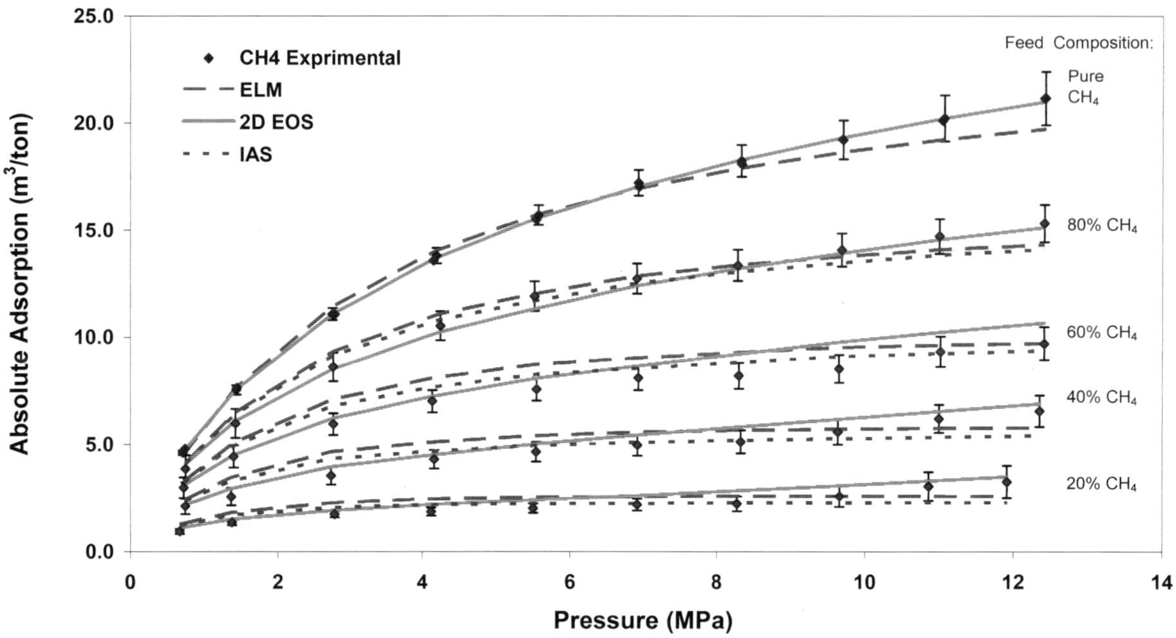

Figure 1. CH_4 adsorption on wet Fruitland coal at 46.1°C

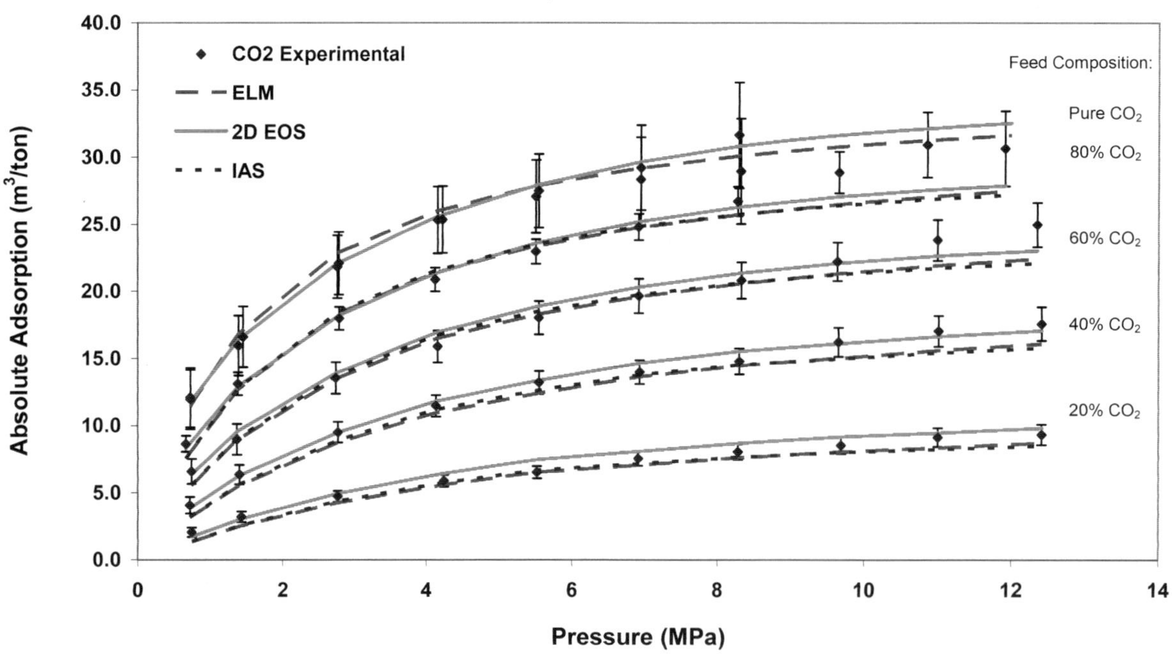

Figure 2. CO_2 adsorption on wet Fruitland coal at 46.1°C

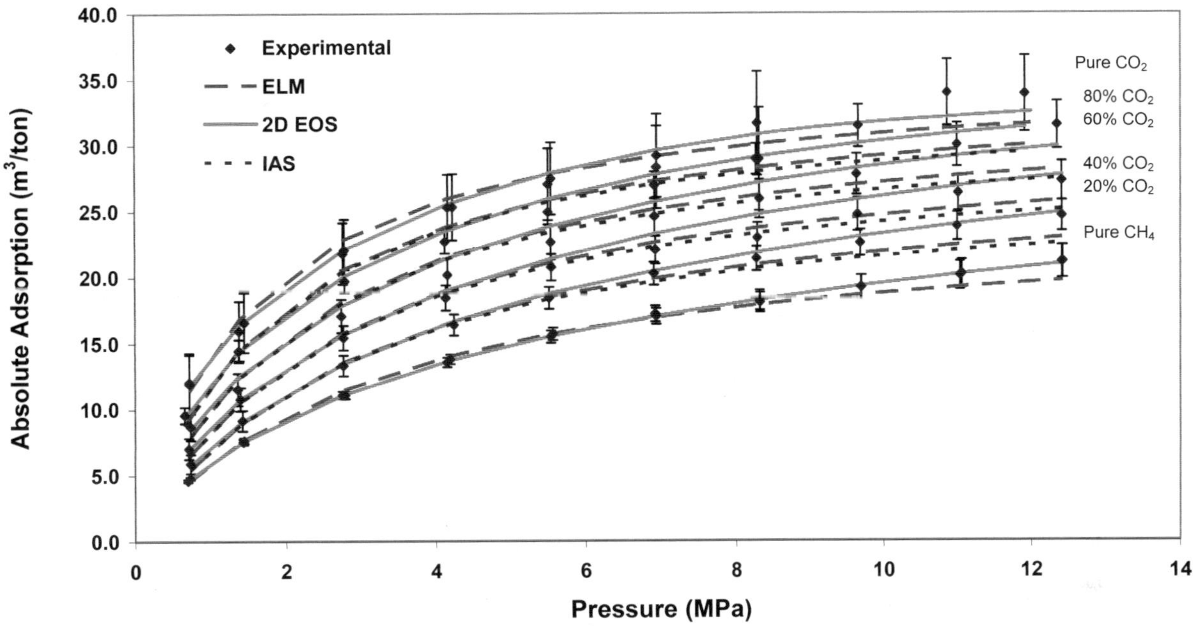

Figure 3. Total adsorption on wet Fruitland coal at 46.1°C

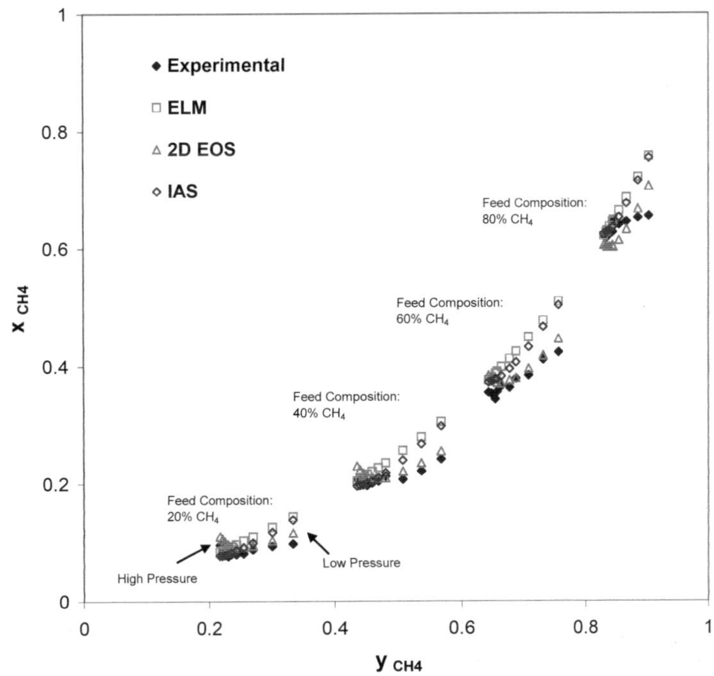

Figure 4. Composition for adsorption on wet Fruitland coal at 46.1°C

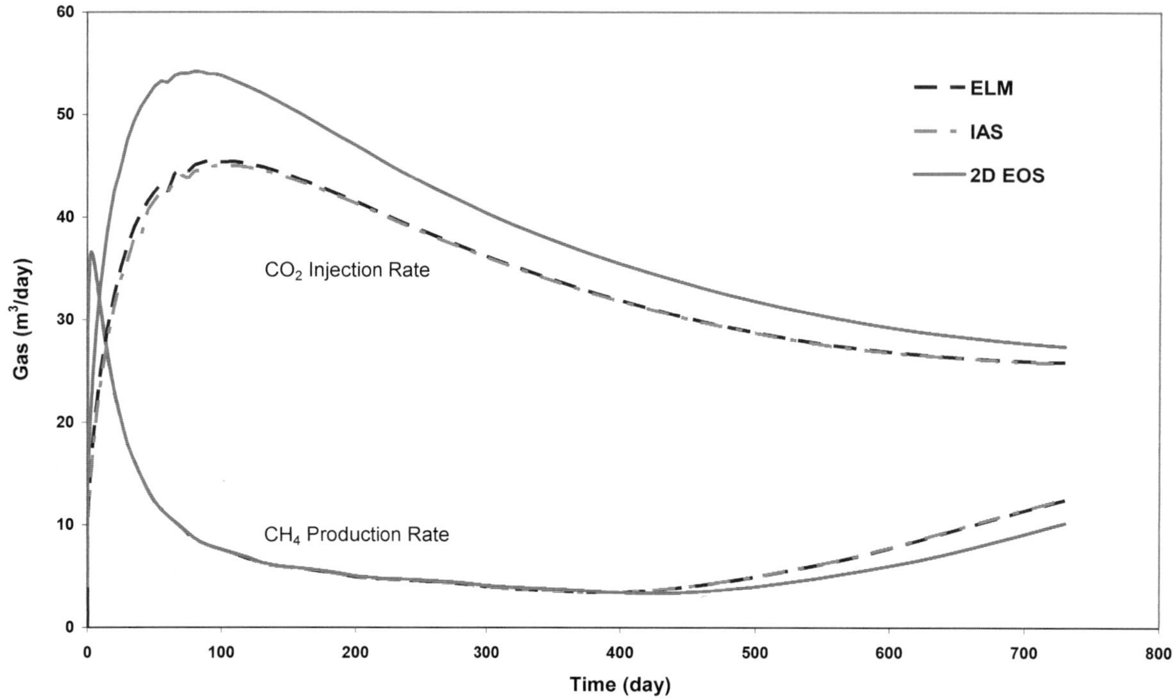

(a) 2 year (730 days) period

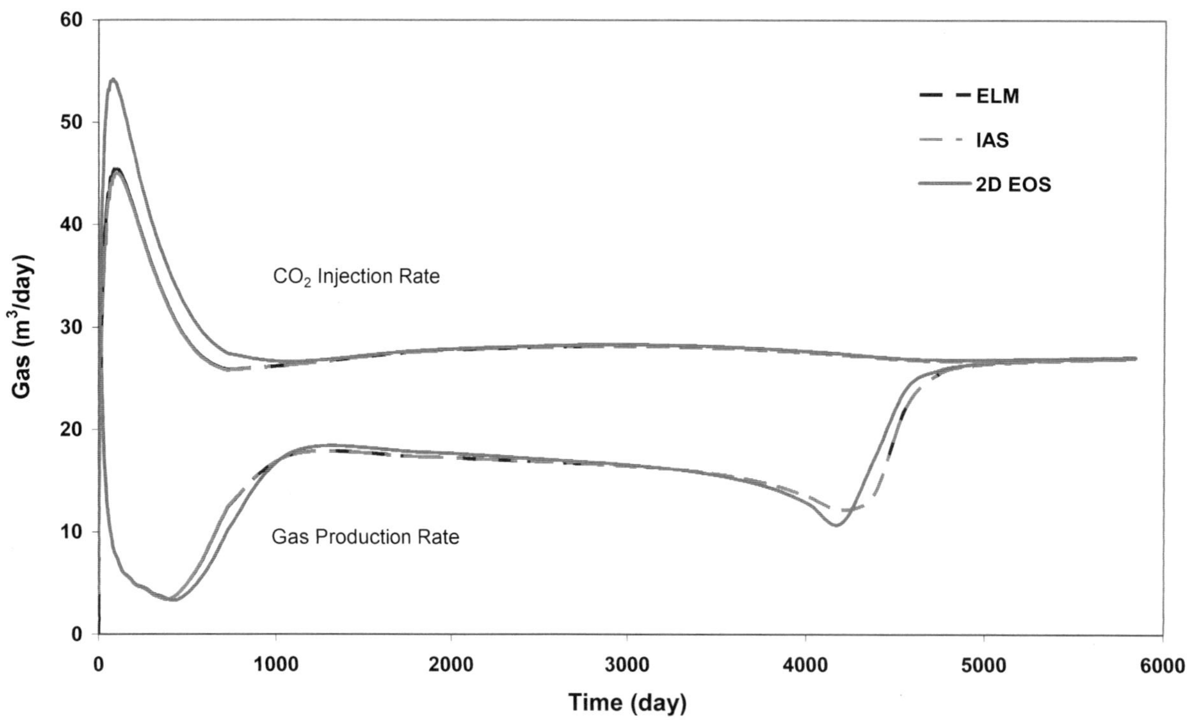

(b) 16 year (5840 days) period

Figure 5. Gas rates at injection and production wells for the hypothetical simulation example, calculated using SIMED, for the three adsorption isotherms

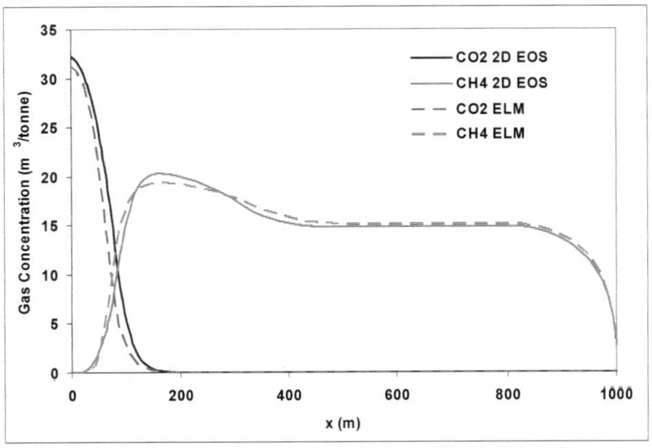

(a)

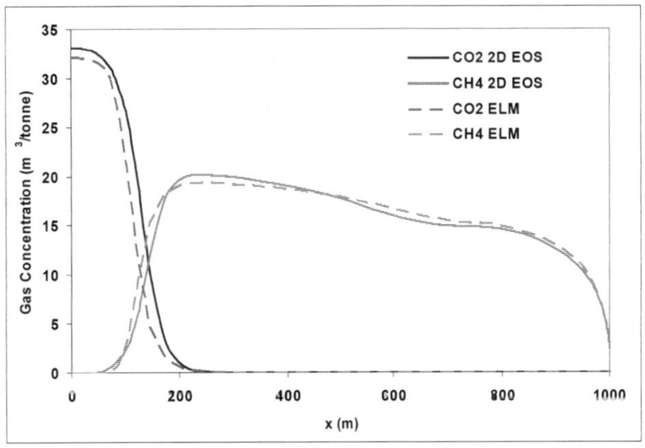

(b)

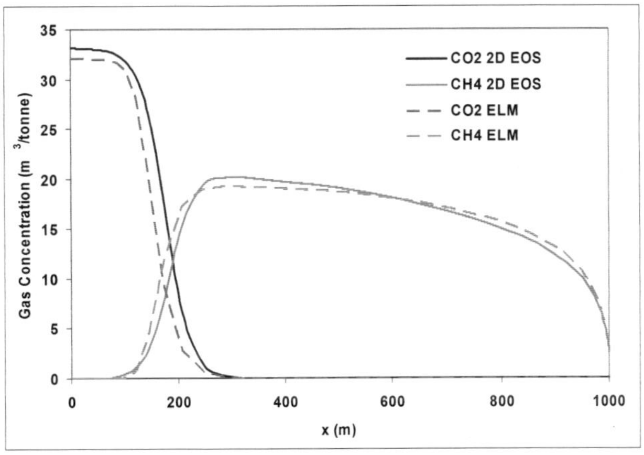

(c)

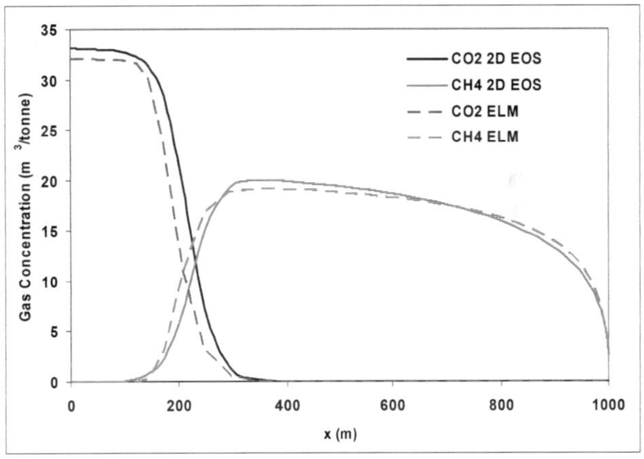

(d)

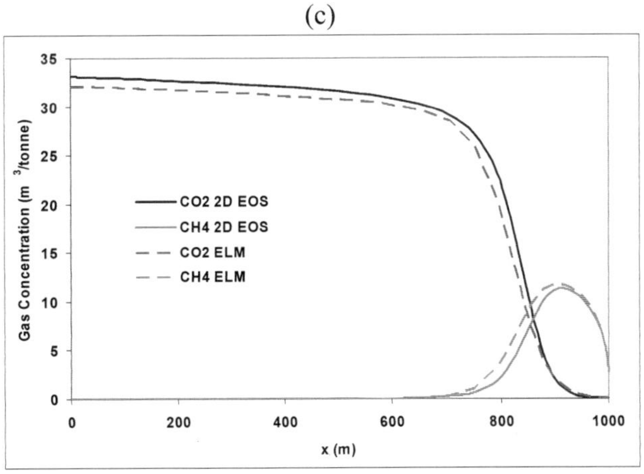

(e)

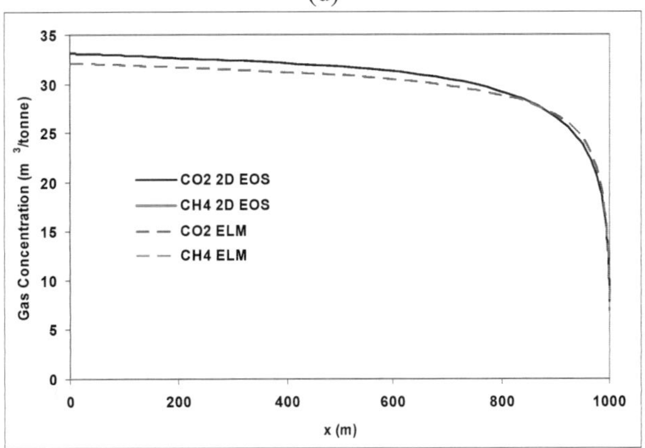

(f)

Figure 6. Adsorbed gas concentrations for SIMED simulation case study at (a) Day 182, (b) Day 365, (c) Day 547, (d) Day 730, (e) Day 3650, (f) Day 5480

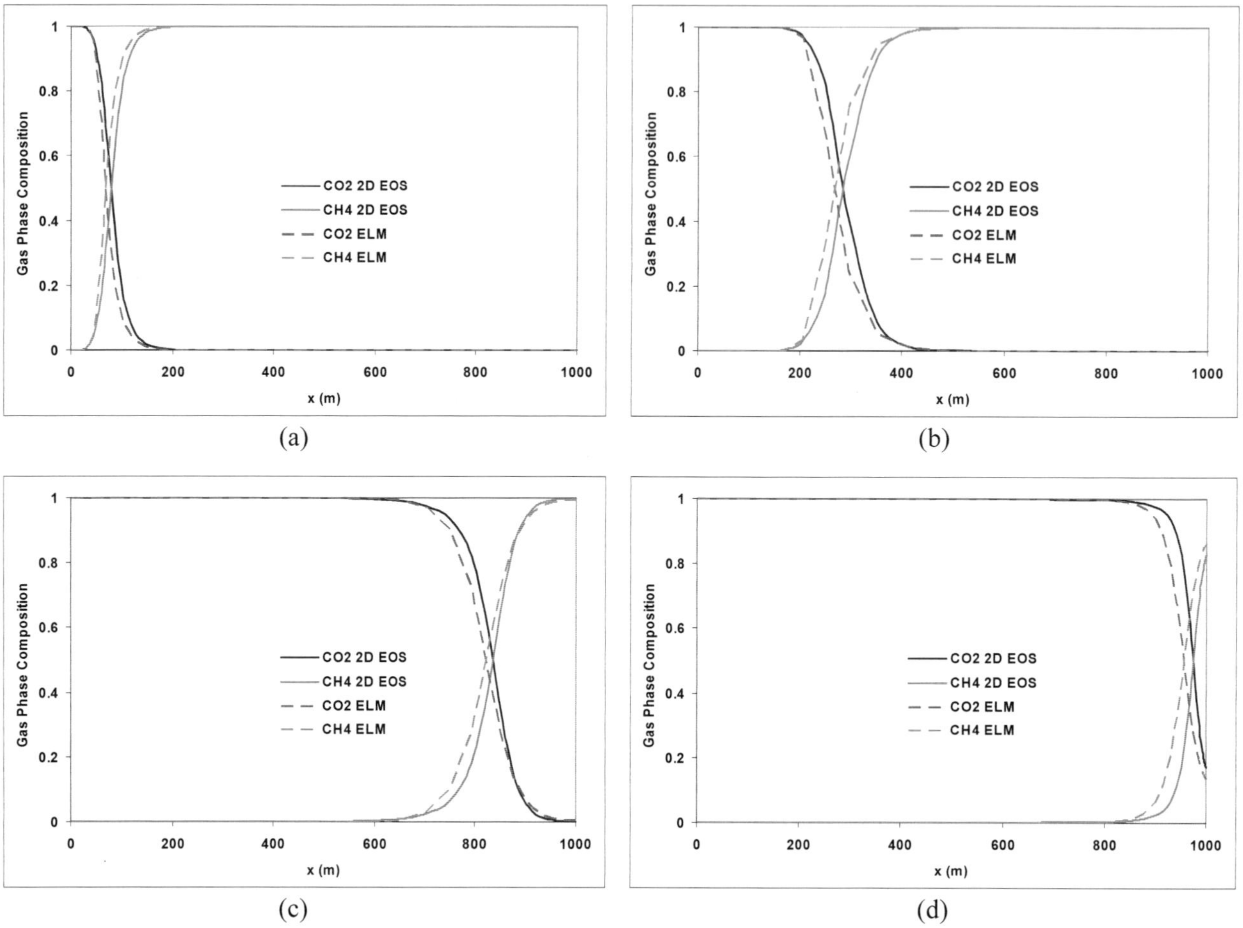

Figure 7. Gas phase composition for SIMED simulation case study at (a) Day 182, (b) Day 1095, (c) Day 3650, (d) Day 4197

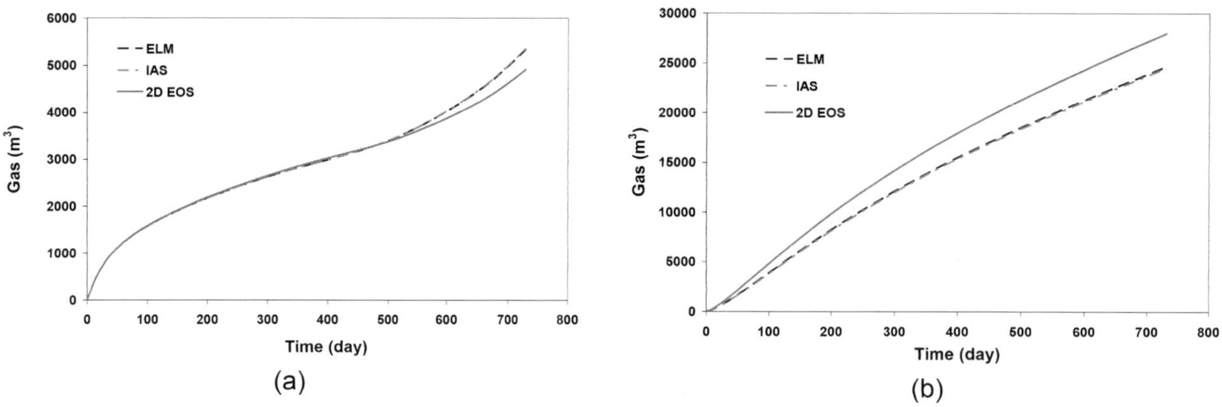

Figure 8. Cumulative gas production for simulation case study (a) CH_4 (b) CO_2

Development of a Neuro-simulation Tool for Coalbed Methane Recovery and CO_2 Sequestration

F. Burcu Gorucu, National Energy Technology Laboratory, U.S. Department of Energy and Pennsylvania State University
Turgay Ertekin, Pennsylvania State University
Grant S. Bromhal, National Energy Technology Laboratory, U.S. Department of Energy
Duane H. Smith, National Energy Technology Laboratory, U.S. Department of Energy
W. Neal Sams, National Energy Technology Laboratory, EG&G
Sinisha Jikich, National Energy Technology Laboratory, Parsons

ABSTRACT

The increase of atmospheric CO_2 levels due to releases from anthropogenic sources is an important environmental issue. One of the most promising technologies to mitigate emissions of this greenhouse gas is to sequester the CO_2 in geological formations for long time periods. For this purpose, coal seams are considered to be very attractive geological horizons because sequestration costs can be partially offset by methane production.

However, optimization of a coal seam sequestration project is a complicated process, requiring accurate and efficient methods and tools for predicting methane recoveries and CO_2 storage capacities. Because of the complex nature of coal seams, fluid flow through them, and other coal/fluid interactions, many parameters affect the outcome of a sequestration project. Our previous studies have indicated that at least fourteen parameters (*e.g.*, face cleat and butt cleat permeabilities, Langmuir sorption constants of carbon dioxide and methane, sorption time constant) are important. Well patterns and injection and production pressures are parameters that can be controlled in attempts to optimize rates and amounts of carbon dioxide sequestration and methane production. Because of the large number of parameters involved, optimization of a field project design can require thousands of reservoir engineering simulations. This paper outlines the application of a neuro-simulation methodology to predict the important performance indicators of a coal seam sequestration project.

Although coal-seam sequestration projects are now being started in the United States, Japan, Poland, and elsewhere, very few data from mature field projects exist in the public domain. Therefore, a compositional coalbed methane simulator, was used to model the enhanced recovery of methane by CO_2 injection. Results from the simulator were used to train artificial neural networks, utilizing the back-propagation learning algorithm. Neural networks were trained to predict the following production performance indicators: breakthrough time, cumulative CO_2 injection, and cumulative CH_4 production. Both the ability of the neural networks to fit the data sets used to train the neural network and the ability of the trained networks to predict other data sets were examined. The accuracy of the network's predictions was thus successfully established.

The networks trained in this work are capable of predicting project performance both quickly and accurately. They allow us to find the results of any of thousands of simulations that have been performed, and to rapidly predict results from combinations of parameter values that have never been simulated, without performing additional reservoir engineering computations.

INTRODUCTION

One of the most promising technologies to mitigate emissions of CO_2 is to sequester this greenhouse gas in geological formations for long time periods. Combining CO_2 sequestration with enhanced coalbed methane recovery has several potential environmental and energy advantages:

1. After conventional depletion, reservoir pressure can be recovered by CO_2 injection, which should contribute to additional gas recovery.
2. CO_2 storage costs can be partially offset because of the simultaneous methane production.
3. CO_2 transportation costs can be reduced due to the fact that coal fields exist in the near proximity of the major power plant sites.
4. CO_2 compression costs can be reduced relative to some forms of geologic sequestration, because the CO_2 sorption allows storage of large volumes of CO_2 at relatively low pressures.
5. Because of sorption of CO_2 in the coal and diffusion of CO_2 into the cleats, only a small portion of the CO_2 is sequestered as free gas, and any accidental release is retarded by the slow diffusion step. This provides an additional level of safety against any possibility of significant release.

Numerical reservoir simulation is the process of establishing relationships between geology, chemistry, petrophysics, reservoir engineering, and production operations. It is the only analysis method which integrates coal properties, operating protocols, and production characteristics. However, as a simulator is designed to become more accurate and reliable by inclusion of more and more effects, the number of input parameters that are included becomes increasingly large. In the present study, 14 input parameters were included. With this large number of parameters sensitivity studies can become very complex, as the number of possible combinations of input values becomes very large. Moreover, there is no practical way to fit a polynomial or other phenomenological equation to describe how any output variable depends on so many input variables.

Artificial neural networks (ANN's) have been developed to address problems where conventional computer systems cannot be utilized in a feasible manner. These problems can be classified in several ways; some are referred to as non-polynomial problems where no polynomial-time algorithm is available. There are many different types of neural networks. Each has a different data processing and learning method. According to the training method, neural networks can be classified under two broad groups: supervised and unsupervised. In the unsupervised neural networks, the aim is to classify the input vectors into groups and clusters. No feedback is provided to the network. These types of networks have been used to interpret well logs and identify lithology (Mohaghegh, 2000).

On the other hand, in the supervised training algorithm, both the inputs and outputs are shown to the network to permit learning on a feedback basis. This is the most commonly used network algorithm in many areas of the oil and gas industry; such as pressure-transit analysis and reservoir characterization. For this work, the multi-layer feed forward network which is trained by the error back propagation (BPN) algorithm is used. Supervised training was used for the work reported in this study.

A structure (network) is composed of a number of interconnected units (artificial neurons). Each unit has an input/output (I/O) characteristic and implements a local computation or function (Hagan, 1997; Schalkoff, 1997). The overall computing model consists of a reconfigurable interconnection of simple elements, or units. Figure 1 depicts a sample network, where units are denoted by circles and interconnections are shown as lines. The design of an ANN topology is complex, and usually it is an iterative and interactive task (Du et al, 2003). It is impossible to provide an all-inclusive algorithmic procedure. But there is a common sense that with the increasing complexity of the problem, the number of neurons should increase. The local functions that are being used in the neurons depend on the type of the problem. Finally, as a rule of thumb, the number of layers usually is not to be more than three. The BPN algorithm is simply the mapping of the input pattern space to the output pattern space by minimization of the error between the target output and the output produced by the network. The input (training data) is shown to the network to initiate the learning process. That input pattern is propagated through the entire network, until the output pattern is produced. Then, the error is calculated for the current pattern contributed by the every unit in the pattern. Finally the error is back-propagated so that each unit can modify its input connection weights.

In the present work, two different methodologies are integrated to develop a screening tool that is capable of predicting key project performance indicators. In other words, for a given set of coal seam properties and design parameters, the expert system developed can predict the performance of a CO_2 sequestration project. The production/injection performance indicators that we selected are the following:

- Breakthrough time: The time when the injected carbon dioxide appears at the production wells (i.e., the time when the concentration of carbon dioxide in the produced gas reaches 10%).
- Cumulative CO_2 injected: The total amount of CO_2 injected by the conclusion of the project.
- Cumulative CH_4 produced: The total amount of CH_4 produced when CO_2 injection is terminated.

METHODOLOGY

Reservoir Simulator

The reservoir simulator used, PSU-COALCOMP, is a three-dimensional, two-phase, dual-porosity, fully implicit, compositional reservoir simulator for primary coalbed methane recovery or injection of carbon dioxide [or other gas(es)] for sequestration and enhanced methane recovery. The complex physico-chemical processes that take place in coal seams are modeled in the simulator by coupling the multicomponent gas sorption formulation and the compositional fluid flow formulation. The ideal adsorbed solution model, with correction factors for real gases, is implemented to predict the adsorption capacity of multi-component gases. Two-phase flow equations are used to model the flow through the cleat network. Details of the mathematical formulations can be found in Manik et al (2003).

Coal Seam Properties and Operational Parameters

Neural networks were developed for 5-spot configurations of vertical wells on 20-acre and 80-acre drainage areas, respectively. Because of the complex transport and storage mechanisms of coal seams, 14 properties (Table 1A) were chosen to describe the coal seams. The range of each property considered was chosen to be very large, so that the majority of coals would fall into ranges over which the neural networks were trained. These ranges were based on available field data, previous simulation studies, and laboratory experiments (Bromhal et al, 2005; Smith et al, 2005).

Three operational parameters were varied in the simulations: skin factor, injection pressure, and production pressure. Their ranges are listed in Table 1B.

Table 1A: Coal seam properties and their ranges used in the simulations and training of the neural networks.

Parameter	Range
Porosity (%)	0.1 – 3
Permeability (md)	0.1 – 100
Permeability anisotropy	1:1 – 17:1
Initial water saturation (%)	20 – 100
Sorption time constant (days)	0 – 900
Langmuir volume constant of CH_4 (scf/ton)	100 – 800
Langmuir volume constant of CO_2 (scf/ton)	200 – 1600
Langmuir pressure constant of CH_4 (psia)	15 – 2000
Langmuir pressure constant of CO_2 (psia)	10 – 1000
Initial pressure (psia)	100 – 3500
Thickness (ft)	4 – 30
Temperature (^{0}F)	75 – 150
Irreducible water saturation (%)	0 – 30
Relative permeability to gas at Sg = 1 - Swc	0.05 – 1

Table 1B: Operational parameters and their ranges used in the simulations and training of the neural networks.

Operational Parameter	Range
Injection pressure (psia)	200 – 5000
Production pressure (psia)	15 – 3000
Skin factor	-4 – 4

Reservoir Simulations

The use of 17 variables (14 coal property and 3 operational), each with its own range of values, created a very large number of possible combinations of input values. The objective was to perform physically meaningful simulations that would cover or represent most of the scenarios that could possibly occur. Therefore, 335 simulations, each using a different combination of values for the 17 parameters, were performed for each pattern area. In these simulations the values of virtually all of the parameters were changed from one simulation to another. These 335 simulations were used during the network training phase of the study.

After the training phase had been completed, 50 more simulations were made for each pattern size. The results from these simulations were used to test the performance of the neural networks. For the test simulations only two or three input variables were changed from one simulation to another.

Neural Networks

In this stage of the work a different neural network was trained and tested for each pattern area, but the same network structure (Figure 1) was used for each pattern studied. To determine an appropriate network topology, various structures were considered. The search for an appropriate topology began with a simple structure, and then the architecture was progressively modified until a topology was found for which the trained network met the test criteria. The search for a suitable topology was an iterative and interactive procedure, during which the network properties were varied. The number of middle layers, number of neurons in the layers, and performance and transfer functions were varied during the search.

The mean-squared error with regularization performance function (msereg) was used as the performance indicator. For the network, this parameter measures network performance as the weighted sum of two factors: the mean-squared error and the mean-squared weights and biases. Transfer functions are used to calculate a layer's output from its net input.

Hyperbolic tangent (tansig) and logarithmic (logsig) sigmoid transfer functions were used in the middle layers. A linear transfer function (purelin) was chosen for the output layer. In the input layer, four functional links were included to improve the training capacity and to increase the weight effect of the input variables such as relative permeability, absolute permeability and pressure difference. These variables were selected to improve the errors in breakthrough time.

The structure of the final network is shown in Figure 1.

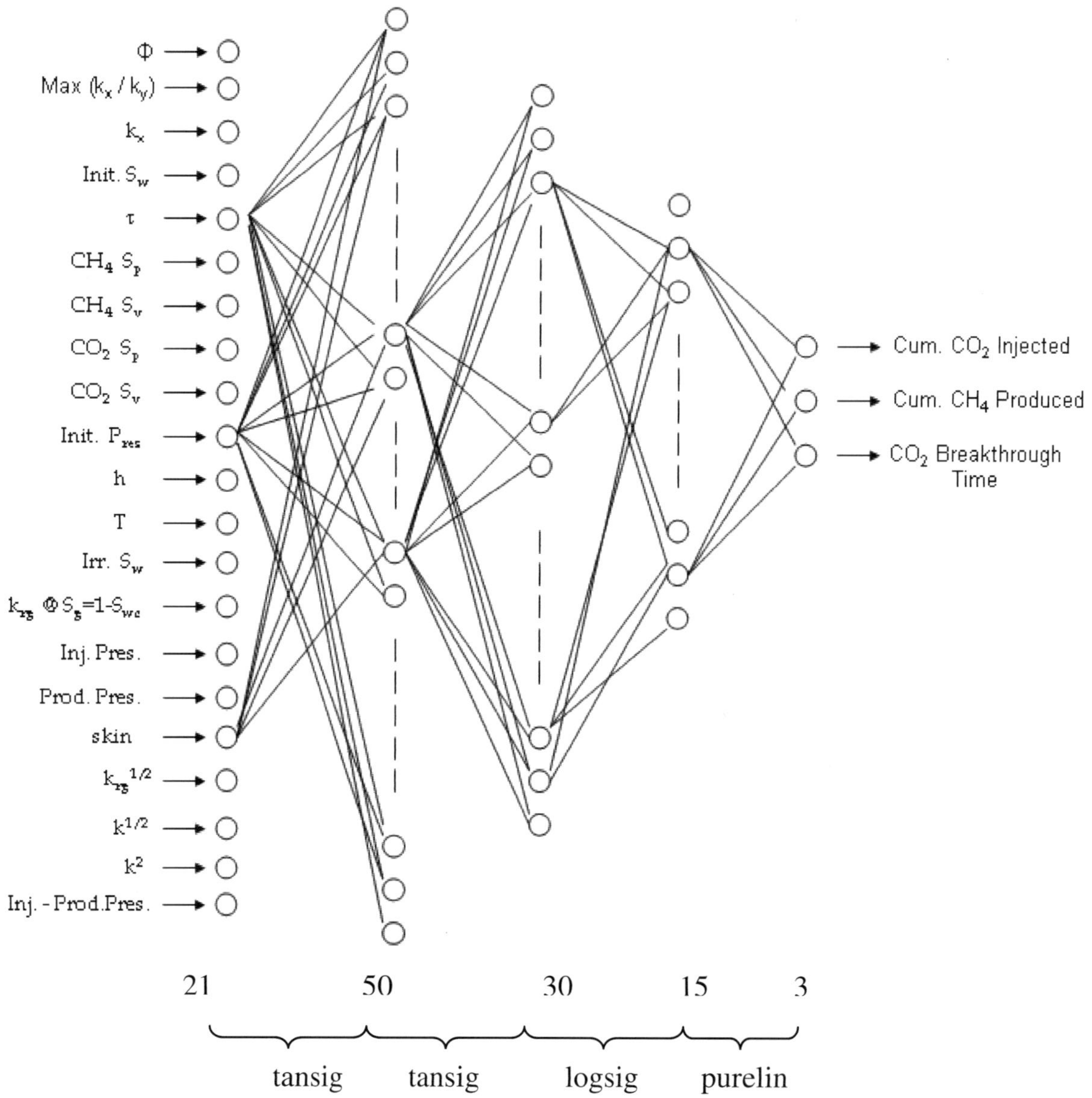

Figure 1: Network structure, showing the input variables, output variables, numbers of layers and neurons, and transfer functions.

The networks were tested in two steps. In the first step, the network was tested against the training data set. In other words, after a training session was completed, the network was asked to predict the performance for the same data set that was used in the training. This is one way of checking, whether the network has learned or not. During the second step, a new set of data which had not been exposed to the network was shown to the network. This is an important confidence check, as if it generates information on how well the network is trained and at what level of accuracy the network is trained. Sometimes, a "trained" network can generate outputs very close to the target outputs, but its predictions for totally new data sets can be poor. This may be an indication of "memorization" or "overtraining," as opposed to a trained network capable of accurate predictions for new conditions.

RESULTS AND DISSCUSSION

In this section, the training results as well as the results from the testing patterns are presented. For all networks, the errors observed during the testing of the training data were within the range of 1 to 2%. This is a good indication of the accomplishment of successful training. Moreover, the errors observed during the testing stage affirm that networks are capable of reasonably accurate predictions of the results that would be obtained from new simulations.

20-Acre Spacing

The first network was developed for 20-acre spacing. Figure 2 shows prediction results of the training data sets as generated by the network. The maximum errors involving the testing patterns (Figure 3, Figure 4, and Figure 5) for prediction of breakthrough time, cumulative methane produced, and cumulative carbon dioxide injected were 16.7%, 12.2%, and 9.0%, respectively. The observed average errors were 5.6%, 3.6%, and 3.5% for the same.

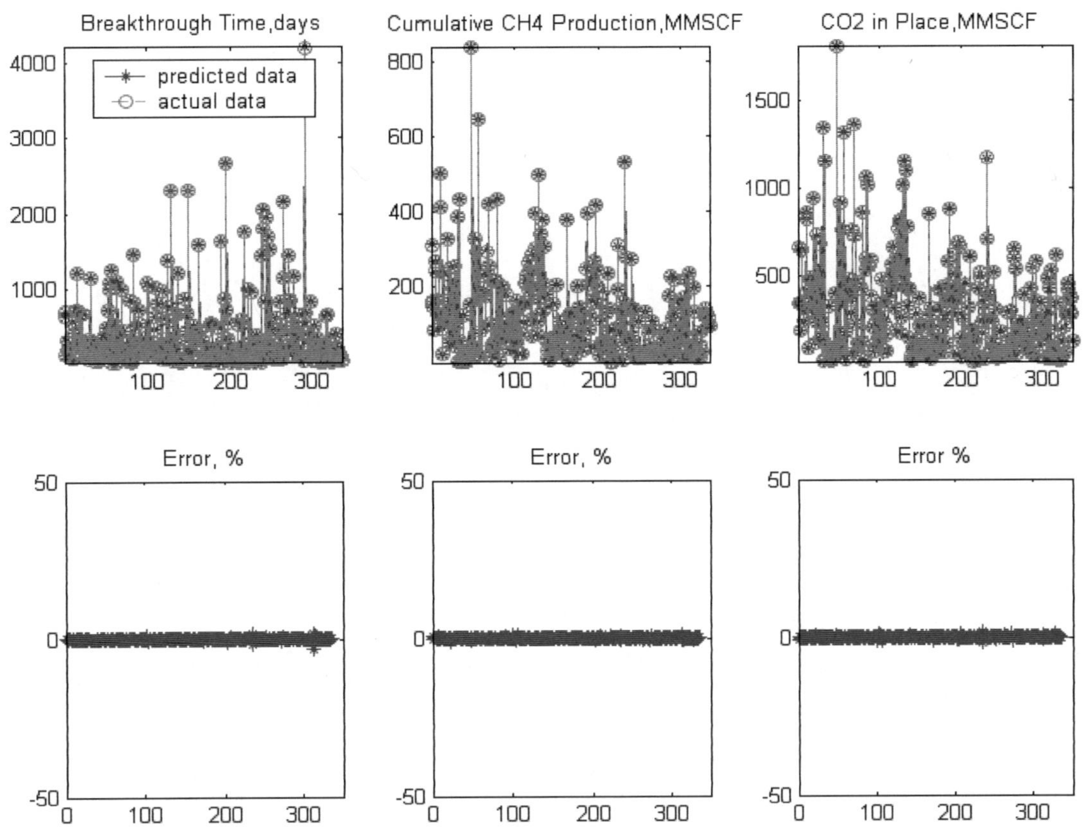

Figure 2: Training results for 20-acre spacing.

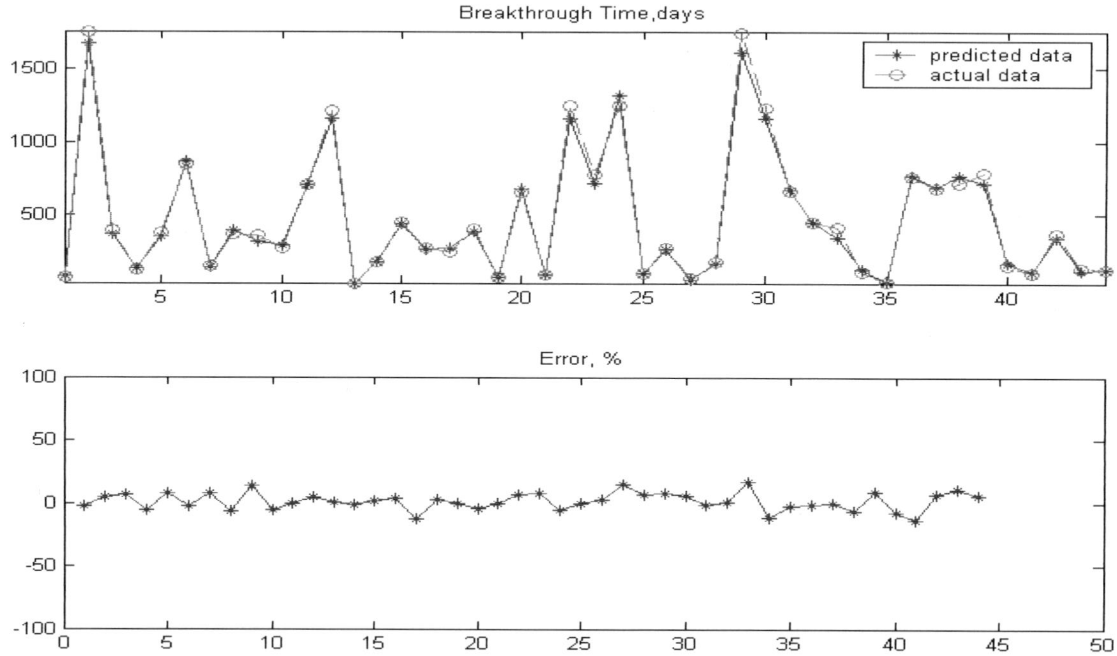

Figure 3: Testing results of breakthrough time for 20-acre spacing.

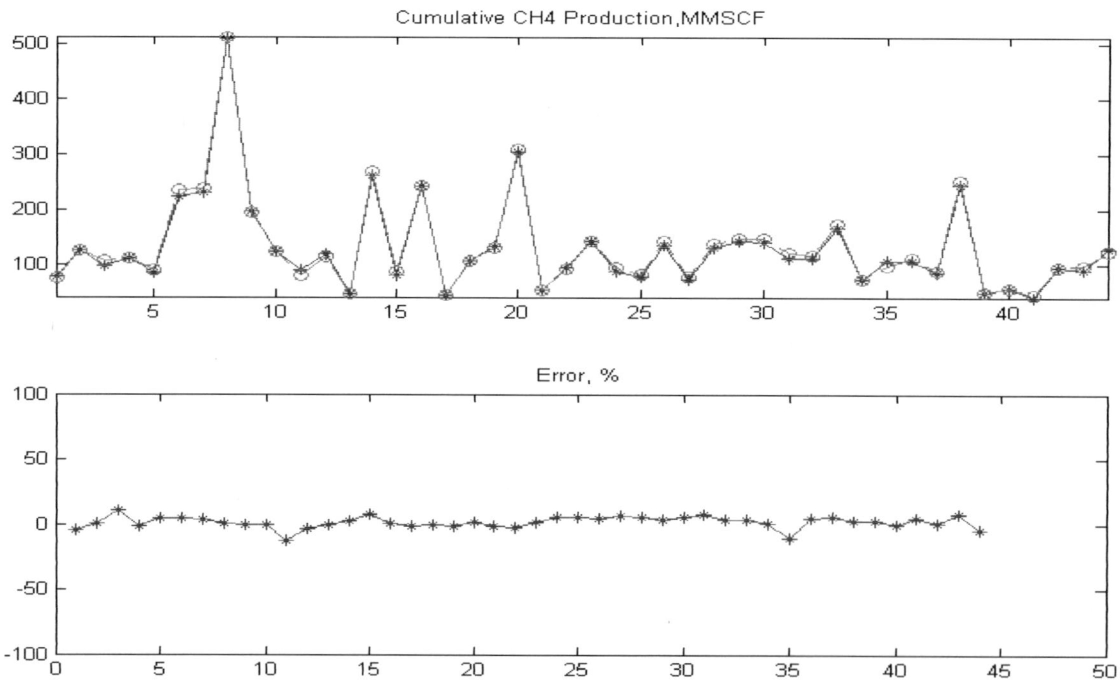

Figure 4: Testing results for cumulative methane production for 20-acre spacing.

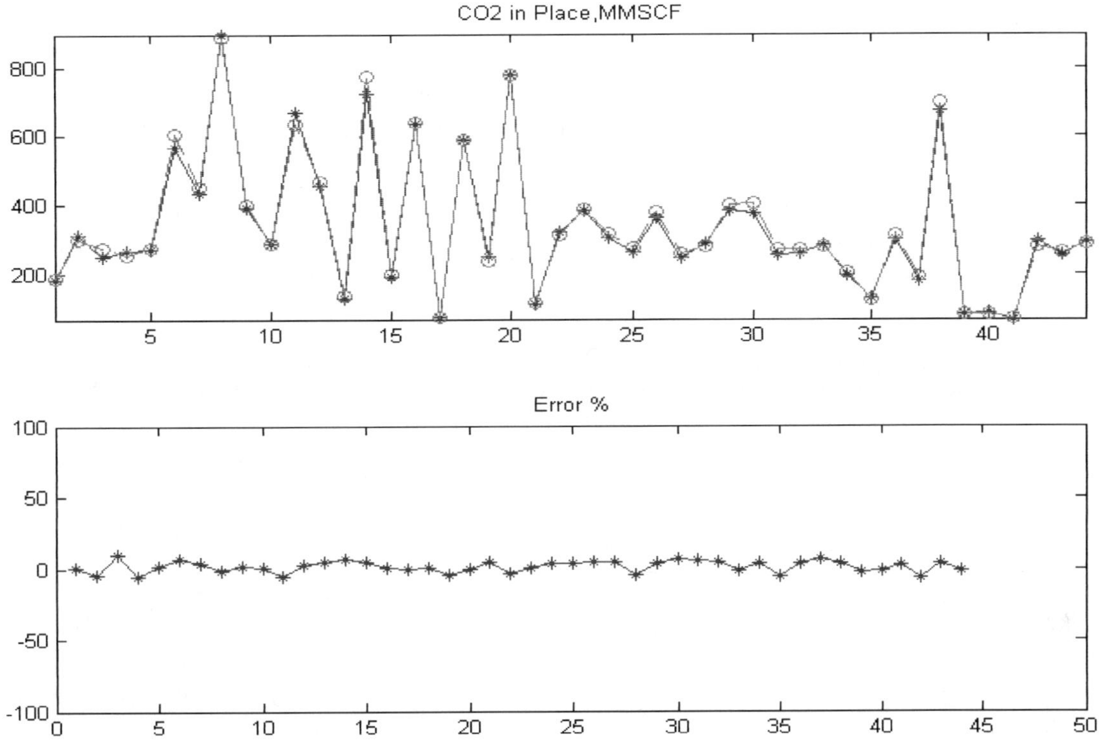

Figure 5: Testing results for cumulative carbon dioxide injection for 20-acre spacing.

80-Acre Spacing
The prediction results of the training data for 80-acre spacing can be seen in Figure 6. The maximum errors observed during the testing stage were 19.9%, 11.7%, and 8.8% for breakthrough time, cumulative methane produced, and cumulative CO_2 injected, respectively. The average errors observed at this stage were 3.9%, 2.3%, and 2.2%.

The errors of the testing results for prediction of breakthrough time, cumulative methane produced, and cumulative carbon dioxide injected are illustrated in Figure 7, Figure 8, and Figure 9, respectively.

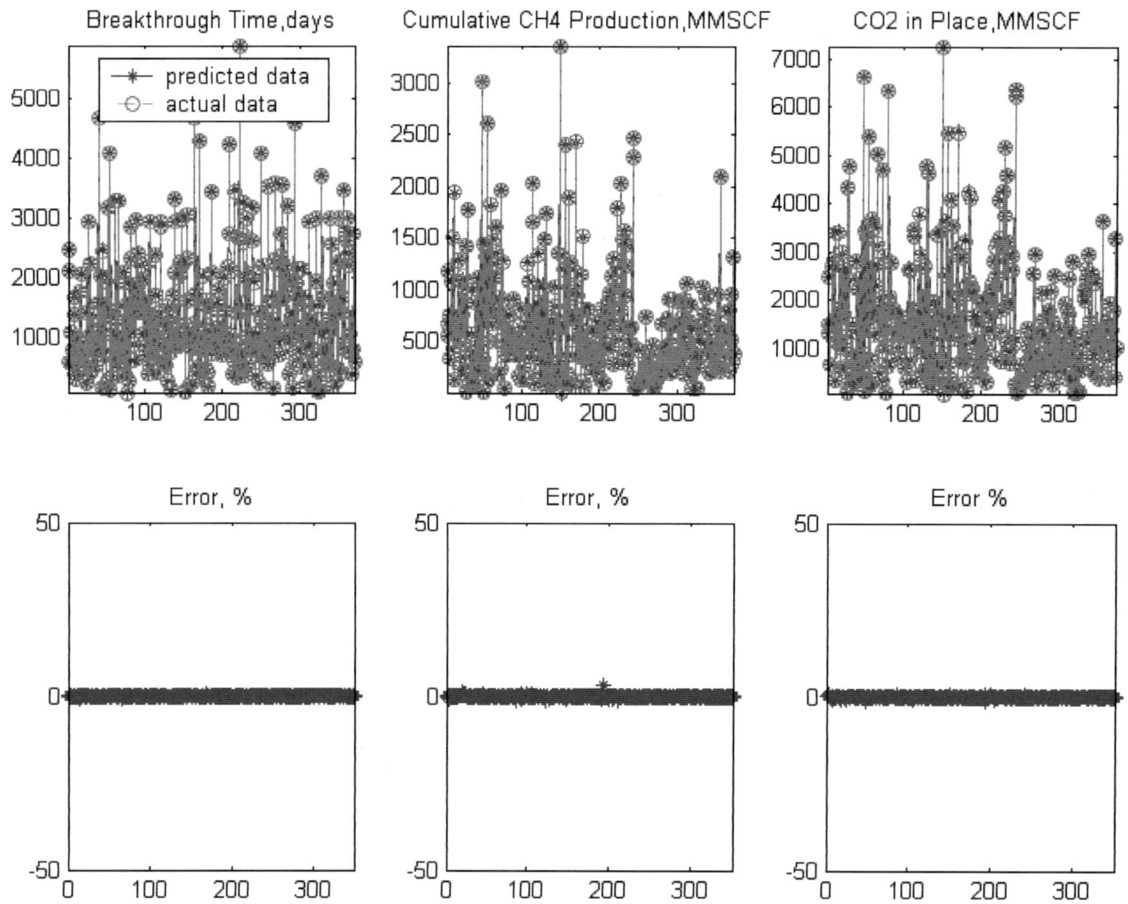

Figure 6: Training results for 80-acre spacing.

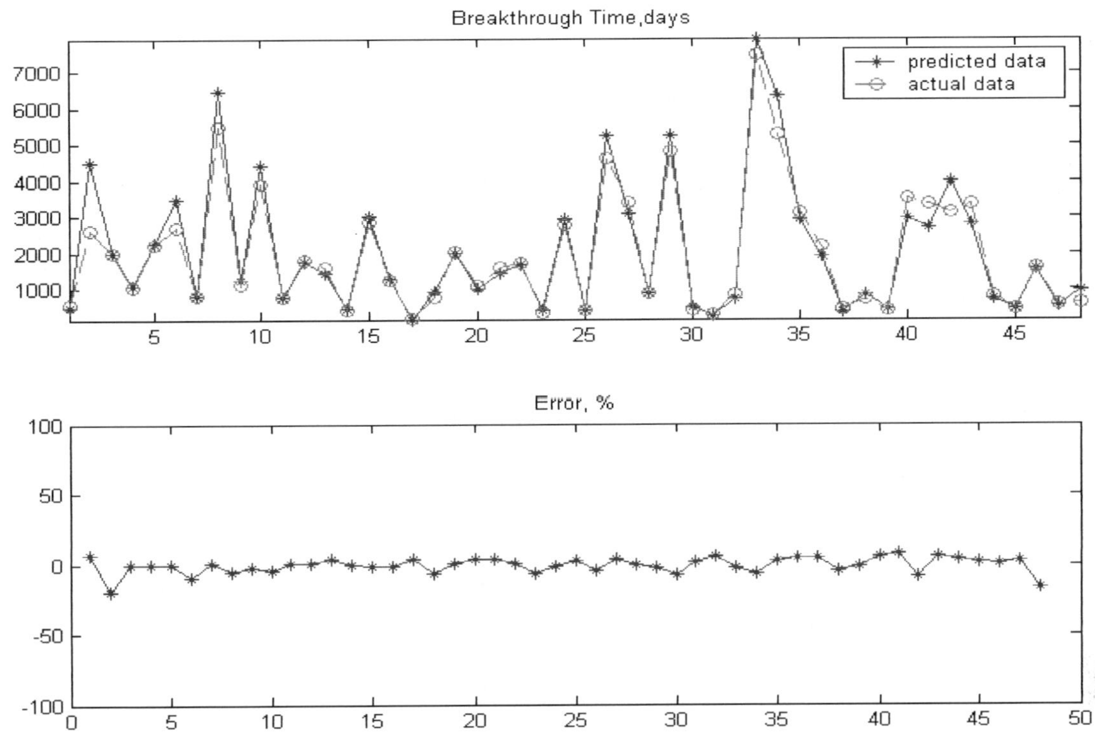

Figure 7: Testing results of breakthrough time for 80-acre spacing.

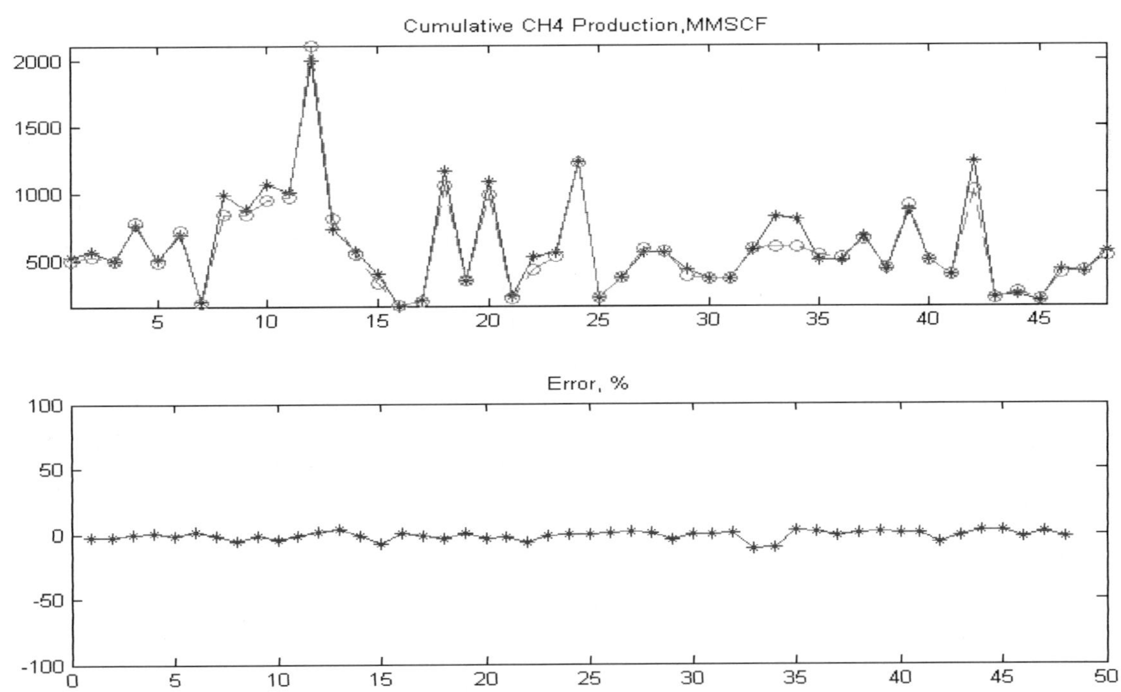

Figure 8: Testing results for cumulative methane production for 80-acre spacing.

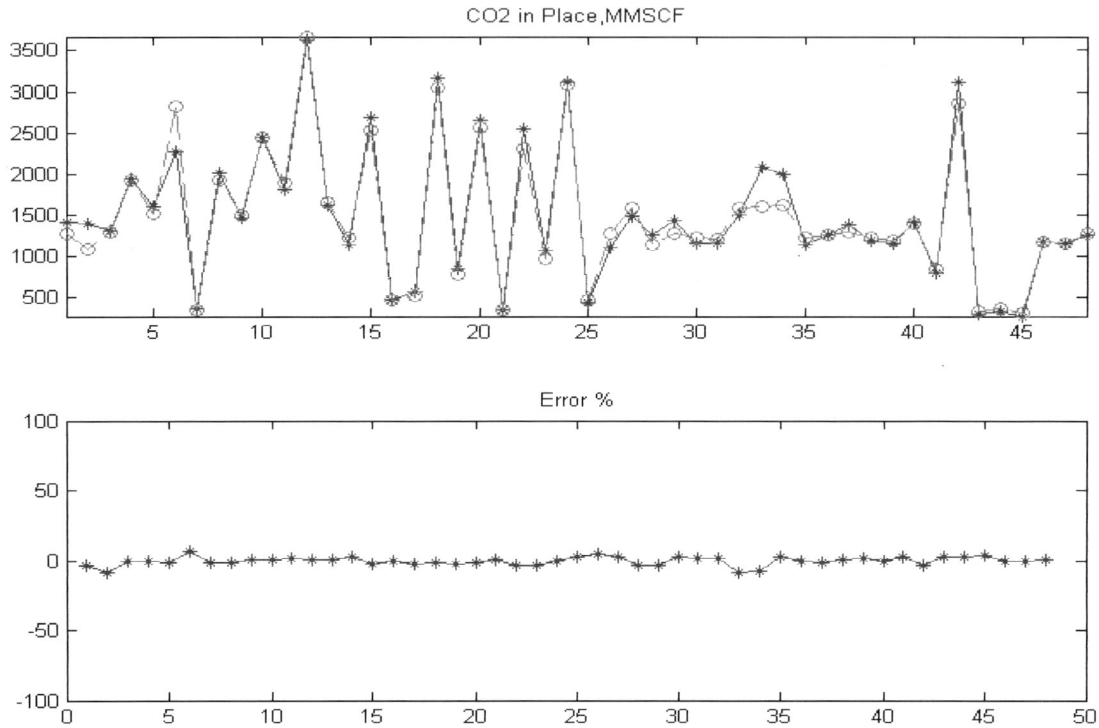

Figure 9: Testing results for cumulative carbon dioxide injected for 80-acre spacing.

CONCLUSIONS

For all of the cases that are systematically explored, the predicted cumulative carbon dioxide injected, cumulative methane produced, and carbon dioxide breakthrough time are smoothly and slowly varying functions of the values of the coal seam properties and operating parameters (Bromhal et al, 2005; Smith et al, 2005). The dependence of any one of these output variables as a function of any one of the input variables—with the values of all other variables held constant—could be fit by a simple phenomenological function, just as a plot of measured property y vs. independent variable x often be fit by a simple, non-theoretical equation. However, without neural networks, attempting to fit the dependence of any one of the output variables on all 14 input variables would be an enormous task. The results of this study indicate, however, that neural networks can be found and trained to provide such "fits" for the dependence of important field-performance variables on a complete set of many coal-seam and operational parameters. As a result, with the trained neural network sensitivity studies or Monte Carlo analyses can be quickly performed for any subset of the many input variables, without the need for hundreds of reservoir simulations that would otherwise be needed.

Moreover, one can quickly compare the results predicted for various candidate coal seams under consideration for possible coalbed methane and/or carbon sequestration projects, and thus use the trained neural networks to choose among these candidates. Of course, the predictions are subject to the limitations of the chemical, physical, flow, and numerical models of the simulator used to train the neural network. But for a state-of-the-art simulator, these limitations are simply the limitations of our collective knowledge of the technology of coal-seam sequestration.

ACKNOWLEDGMENTS

This work was performed at the National Energy Technology Laboratory and was funded by the Office of Fossil Energy, U.S. Department of Energy (DOE).

REFERENCES

Bromhal, Grant; Sams, W. Neal; Jikich, Sinisha; Smith, Duane H.: "Simulation of CO_2 Sequestration in Coal Beds: The Effects of Sorption Isotherms," *Chem. Geol.*, in press.

Du, Y; Weiss, W. W.; Xu, J.; Balch, R.S.; Li, D.: "Obtain an Optimum Artificial Neural Network Model for Reservoir Studies," SPE paper 84445, SPE Ann. Tech. Conf., Denver, Oct. 5-8, 2003.

Hagan, M.T; Demuth, H.B.; Beale M.H.: *Neural Network Design.* Boston: PWS Publishing, 1997.

Manik, J.; Ertekin, T.; Kohler, T.E.: "Development and Validation of a Compositional Coalbed Simulator," *J. Canadian Petrol. Technol.*, Vol. 41, No. 4, April 2003, pp. 3-45.

Mohaghegh, S.: "Virtual-Intelligence Applications in Petroleum Engineering: Part 1-Artificial Neural Networks," SPE paper 58046, *J. Petrol. Technol.*, Vol. 52, No. 9, Sept. 2000, pp. 64-73.

Schalkoff, R.J.: *Artificial Neural Networks.* New York: McGraw-Hill, 1997.

Smith, Duane H.; Sams, W. Neal; Bromhal, Grant; Jikich, Sinisha; Ertekin, Turgay: "Simulating Carbon Dioxide Sequestration/ECBM Production in Coal Seams: Effects of Permeability Anisotropies and Diffusion Time Constant," *Soc. Petrol. Engrs. Reserv. Eng. Eval.* (April 2005), in press.

0512

Experimental Study on Gas Displacement in Enhanced Coalbed Methane Recovery

S. Shimada, H. Li and K. Adachi
Institute of Environmental Studies, Graduate School of Frontier Sciences, The University of Tokyo; and Y. Oshima
Environmental Science Center, The University of Tokyo

Abstract Gas adsorption isotherms of Akabira coals were established for pure carbon dioxide (CO2), methane (CH4), and nitrogen (N2). Experimental data were proved to fit very well with the Langmuir model. The ratio of sorption capacity of CO2, CH4 and N2 is 8.5: 3.5: 1 at a lower- pressure (1.2MPa) regime and it becomes 5.5:2:1 when the gas pressures increase to 6.0MPa. This sorption behavior was explained by the different raise of the density of gases with increasing pressure. For a partially saturated with CH4 at 2.4MPa adsorption pressure, the desorption behavior of CH4 by injecting pure CO2 (at 3.0MPa, 4.0MPa, 5.0MPa, and 6.0MPa), and by injecting the CO2-N2 mixture and pure N2 (at 3.0MPa, and 6.0MPa) were experimentally studied. The results indicated that the preferential sorption property of coal for CO2 is significantly higher than that for CH4 or N2, and CO2 can displace almost all of the CH4 adsorbed on coal. Finally, when modeling the CH4-CO2 binary, and CH4-CO2-N2 ternary adsorption system by using the Extended Langmuir (EL) equation, it was found that EL model always over-predict the sorbed CO2 amount, while under-predict the sorbed CH4 when injection pressure increase up to 6.0MPa. This behavior was explained by that a part of CO2 may dissolve into the solid organic structure of coal besides its competitive adsorption with other gases. According to this explaination, the Extended Langmuir coefficients of CO2 in EL equation were revised, and the revised EL model was proved to be very accurate in predicting sorbed ratio of multi-component gases on coals.
Key words: Extended Langmuir (EL) model, Adsorption, Desorption, Displacement, Akabira coal

1. INTRODUCTION

Along with the Kyoto Protocol being put into effect from Feb.16, 2005, many technologies are being developed to reduce the atmospheric CO2 concentration. As reservoirs of CO2 sequestration, coal seams have good potential because of large deposits all over the world, and their proximity to CO2 generating coal-fired power plants. Furthermore, CO2 sequestration in coal seams can simultaneously enhance the coalbed methane recovery, known as ECBM technique, thus has both environmental and economic benefits.

Previous modeling work indicates that the 'stripping' mechanism associated with N2 injection is significantly more effective for ECBM recovery than displacement by CO2 (Reeves and Stevens, 2000). To achieve the desired low net-cost CO2 sequestration, the injected gas must consist of both N2 and CO2 in optimized proportions. Obviously, this is attractive since power plant flue gas is comprised of these two components. Desorption behavior of CH4 adsorbed on coals by injecting single gas (CO2 or N2) has been laboratory studied, and the displacement results have been discussed by many researchers (Prusty and Harpalani, 2004; Reeves and others, 2000). They commonly concluded that the injection of a lower-adsorbing gas such as N2, is considered to reduce the partial pressure of CH4 thus causing it to desorb (stripping effect). On the other hand, the injection of a higher-adsorbing gas such as CO2 will preferentially sorbed thus displacing the CH4 from coal. In this process, the injected CO2 can thus adsorbed and stay in coal at the same time, getting sequestered permanently (Prusty and Harpalani, 2004). However, the displacement mechanism of CO2-CH4 binary system has not been fully understood, and the actual experimental results have not been coinciding with prediction using EL model. Moreover, little was reported with injecting multi-component gas, such as the mixture of CO2 and N2, into coal seams for enhancing the methane recovery. Ternary gas mixtures adsorption/desorption behavior has not been studied, although the binary gas mixture adsorption/desorption was reported by a few papers (Prusty and Harpalani, 2004).

This study mainly demonstrate the desorption behavior of CH4 adsorbed on coals by injecting pure gas (CO2 or N2) and the mixture of CO2 and N2. In order to validate and Extended previous ideas on the mechanism of displacement of methane adsorbed on the internal surface of coal by injecting CO2 (carbon dioxide), and the mixture of CO2 and N2 (nitrogen), the adsorption and displacement experiments were carried out. The reasons why EL model can not accurately predicate the actual experimental results were explained by the CO2 dissolution into coals. The error between the actual experimental value and the calculated value by using Extended Langmuir model was extremely discussed. According to the knowledge that a part of CO2 may also dissolved into coal (Larsen, 2004) besides competitively adsorbing with CH4 or N2, a suggestion for revising the EL coefficients of CO2 was tried and the new EL equation was proposed.

2. SAMPLES PREPARATION AND EXPERIMENTAL APPARATUS

There are seven coalfields in Hokkaido, where more than 50% of Japanese coals exist. Ishikari coalfield is located in the centre of Hokkaido. The coal samples for these experiments were collected from the Akabira coal mine (closed) in this coalfield. Powdered sample was prepared from well preserved bituminous coal. Prior to adsorption experiments, the coal samples were ground, and the resulting powdered coal was sieved to obtain the desired size of 45-63μm. The sample so prepared was dried in a vacuum oven at 107C for a period of 1.5 hours. Then the dried sample was installed into the sample cell and vacuuming for 12 hours.

The experimental apparatus (Fig. 1) consists of a reference cell with a standard volume, and a sample cell which holds the powdered coal. Two cells are separated by a valve, and a filter is placed between two cells to prevent the fine coal from being away from the sample cell to reference cell. To eliminate the effects of temperature change on the adsorption amount, two cells are placed in a constant temperature bath. A pressure transducer is attached to the reference cell for measuring the change of pressure inside the reference cell.

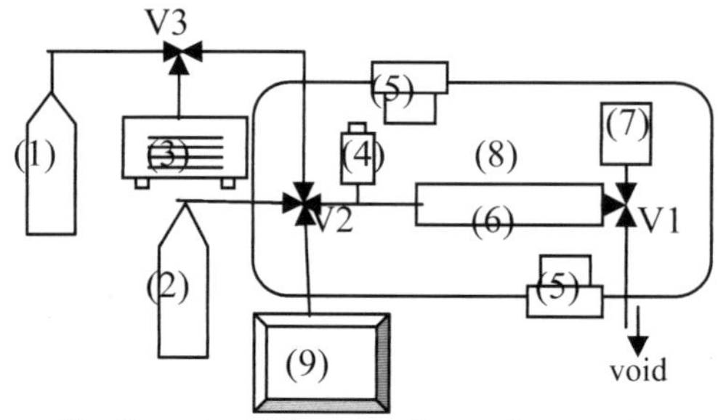

Fig. 1 Schematic diagram of the experimental apparatus.
(1) He container (2) CO2, CH4, N2 container (3) vacuum pump (4) pressure transducer (5)temperature controller (6) reference cell (7) sample cell (8) constant temperature bath (9) Gas Chromatograph (GC)

3. EXPERIMENTAL PROCEDURES

3.1 Adsorption isotherm of single (pure) gas (CH4, CO2, N2)

The adsorbed gas volume was measured using the Constant Volume method. Before the experiment, the volume of reference cell (RC) and sample cell (SC) was measured with helium gas (He), and $V_r=16.84cm^3$, $V_s=10.79cm^3$ was recorded. The procedures of the adsorption experiment were expressed as follow:

(1) After setting a prepared coal sample into sample cell, the cell was drawn by a pump at 35C for 12 hours to become a vacuum.

(2) The void volume of the sample cell was measured with helium gas, which can not be adsorbed on the sample.

(3) Single Gas (CO2, or CH4, or N2) was let into the reference cell and the equilibrium pressure p_1 was recorded.

(4) Opening the valve (V1) between RC and SC, and the single gas in RC was allowed to get into the SC.

(5) When the pressure change become less than 5×10^{-3}MPa/hour, the state was considered to be equilibrium and the pressure p_2 was recorded.

(6) By measuring the change of the pressure of reference cell, the amount of surface excess (Qs) of the gas on the coal was calculated with the real gas state equation. Using the adsorption density of gases (CO2=1.18g/cm, CH4= 0.421g/cm, N3=0.808g/cm), the absolute volume adsorbed (Q_{ad}) in coal was calculated with the equation (1).

$$Q_{ad} = \frac{Q_s}{1 - \frac{\rho_g}{\rho_{ad}}} \qquad (1)$$

where, Q_{ad} is the absolute volume adsorbed on coal, Q_s is the amount of surface excess of gas on coal, ρ_g is the density of gas phase, ρ_{ad} is the density of adsorbed phase.

(7) Steps (3)-(6) were repeated by increasing pressure up to 6.0MPa and thus can plot an adsorption isotherm, and get the Langmuir isotherms using Langmuir equation (2) within pressures 0-6.0MPa (Fig. 2).

$$Q = \frac{Q_{\max} KP}{1 + KP} \qquad (2)$$

where, Q is sorbed volume (mmoles/g) at pressure p (MPa), Qmax is saturated volume (mmoles/g), K is the Langmuir coefficient.

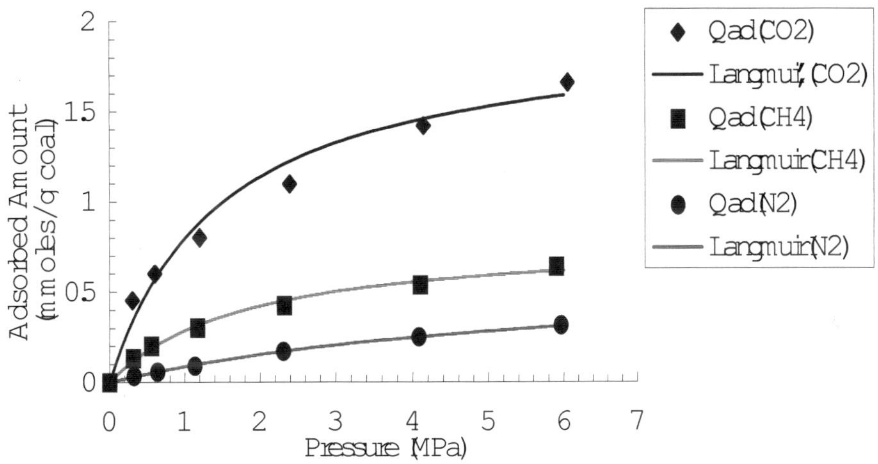

Fig. 2 Langmuir isotherms for CO2, CH4, and N2.

3.2 Displacement of CH4 adsorbed on coals by injecting CO2, CO2+N2 mixture, and N2

The same experiment apparatus as the adsorption experiments was used for displacement experiments. Experimental procedures were described as follows:

(1) For the displacement of CH4 by CO2 injection, the adsorption procedures were completed following the steps described above up to a pressure of 2.4MPa for pure methane.

(2) The RC was then evacuated and CO2 (or CO-N2 mixture, or N2) was filled into it at a pressure of 3.0MPa.

(3) The valve between the RC cell and SC cell was opened and the gas in two cells was allowed to mix.

(4) After equilibrium (it often takes 24 hours to get equilibrium), the final pressure was recorded.

(5) The valve between RC and SC was closed and a part of the mixed gas was passed through the Gas

Chromatograph (GC) to determine their molar compositions.

(6) Using the measured molar composition, the partial pressures of the two gases (or three gases) were calculated.

(7) The partial pressure of each gas was used to calculate the volume of each gas adsorption.

(8) Steps (2)-(7) were repeated for injecting CO2 with 4.0MPa, 5.0MPa, and 6.0MPa, or for injecting CO2-N2 mixture and single N2 respectively with 3.0MPa and 6.0MPa.

(9) Extended Langmuir (EL) equation (3) was used to predict adsorption equilibrium of multi-component gas.

$$Q_A = (Q_{\max})_A \frac{K_A p_A}{1 + \sum_{i=A}^{n} K_i p_i} \tag{3}$$

where, Q_A(mmoles/g) is the adsorbed volume of gas A, $(Q_{max})_A$ is the saturated volume (mmoles/g) of pure gas A, P_A (MPa) is the partial pressure of gas A ($P_A = PY_A$, Y_A is the gas phase mole fraction of the component A), K_A is the Langmuir coefficient of pure gas A,

4. RESULTS AND DISCUSSIONS

4.1 Adsorption experiments

The experimental adsorption data of three single gases was shown in Fig. 2. It can be seen from the isotherms that the CO2/CH4/N2 adsorption ratio is 8.5:3.5:1 in low pressure regimes (less than 1.2MPa). In the high pressure regimes (near 6.0MPa), the adsorption ratio of CO2/CH4/N2 becomes 5.5:2:1. This difference of adsorption ratio may relate with the different raise in the density of the three gases with the increasing pressure (Adachi, 2005).

Langmuir Equation was used to establish the adsorption isotherms of pure CH4, CO2 and N2 from the experimental data (Fig. 2). The figure clearly shows that the Langmuir model provides a very good fit to the experimental data. The coefficient of correlation (R^2 value) was 0.97 for methane, 0.95 for carbon dioxide, and 0.98 for nitrogen. It was also observed from the experimental data that adsorption/desorption of CH4 and N2 followed the same path, but there is a hysteresis between adsorption and desorption for CO2. This may be due to the relationship between the coal structure and the molecule size of CO2, which is smaller size than that of both CH4 and N2.

4.2 Displacement experiments

4.2.1 Single CO2 injection

In the first displacement measurement, CO2 injection was performed at pressure 3.0MPa, 4.0MPa, 5.0MPa and 6.0MPa. In the experiment, adsorption of methane was first completed to a pressure of 2.4 MPa. Then, CO2 was injected into the sample cell at a pressure of 3.0 MPa. After attaining equilibrium, the total pressure becomes 2.72 MPa, with CH4 partial pressure at 0.86MPa and CO2 partial pressure at 1.86MPa. The sorbed CO2 was 0.96mmol/g, and sorbed methane decreased to 0.07mmol/g.

By repeating above steps at CO2 injection pressure of 4.0MPa, 5.0MPa, 6.0MPa, the total pressure at equilibrium becomes 3.45MPa, 4.27MPa, 5.14MPa. The experimental data for the CO2 injection experiments are presented in Fig. 3, and the evaluated results using Extended Langmuir (EL) equation was expressed. This experimental data imply that the injection of CO2 can displace almost all of CH4 adsorbed on coals even at a low injection pressure.

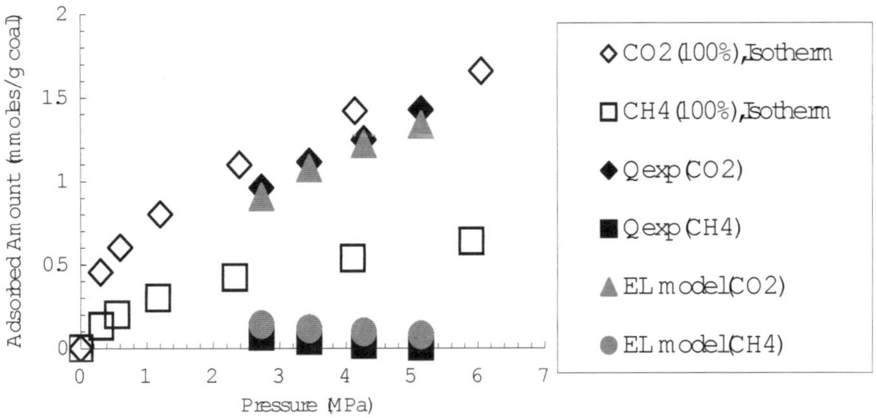

Fig. 3 Desorption of CH4 adsorbed on coal by CO2 injection with 3.0MPa, 4.0MPa, 5.0MPa, and 6.0MPa

4.2.2 CO2-N2 mixtures and single N2 injection

CO2-N2 mixtures with different ratio (35%CO2-65%N2, and 10%CO2-90%N2) were injected respectively, at pressure 3.0MPa and 6.0MPa to displace the CH4 adsorbed on coals. The experimental data and evaluated results using EL equation were shown in Fig. 4 and Fig. 5. Single N2 was also injected at 3.0MPa and 6.0MPa to displace the adsorbed CH4 from coal. The experimental data and the accuracy of EL equation are shown in Fig. 6.

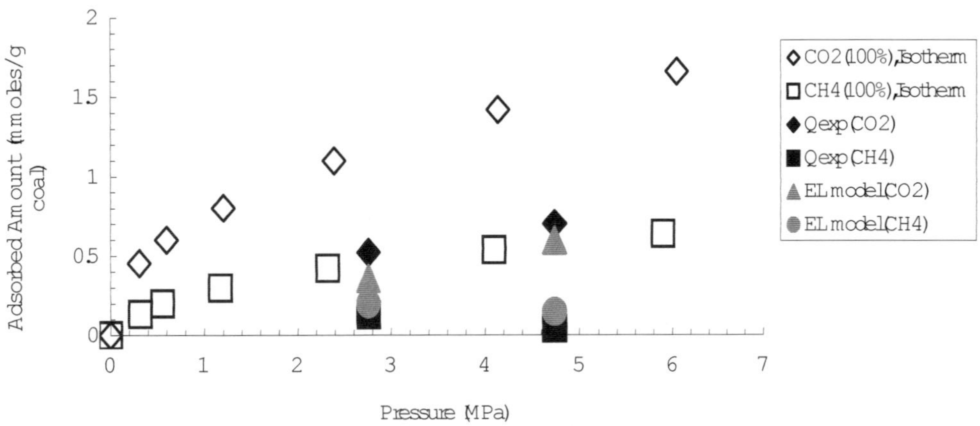

Fig. 4 Desorption of CH4 adsorbed on coal by mixed gas (35%CO2-65%N2) injection

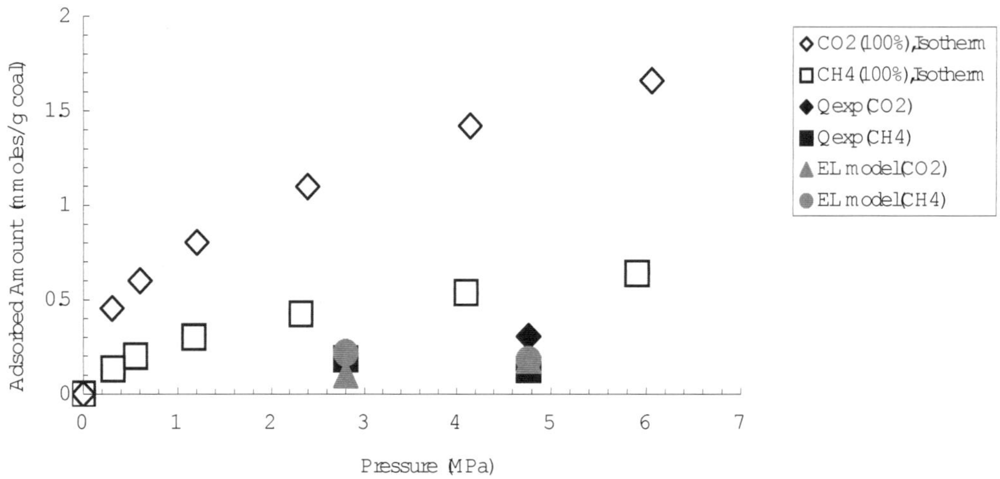

Fig. 5 Desorption of CH4 adsorbed on coal by mixed gas (10%CO2-90%N2) injection.

4.2.3 Evaluation of precision using EL equation

The Extendeded Langmuir (EL) equation was used to predict the adsorption equilibriums of multi-component gas of CH4/CO2, or CH4/CO2/N2 resulting from the injection of CO2, or CO2/N2 into the coal-methane system. The predicted values were compared with the actual experimental value.

Fig. 3 shows the sorption isotherm of methane along with the predicted adsorbed phase equilibriums of CH4 and CO2 mixture. And it was compared with the actual experimental results of the binary experiment. It was observed that EL model was under-predicting the CO2 sorbed value within 10% error, while over-predicting the sorbed CH4 value with very large errors of 95%, 150%, 380%, and 648% when injection pressure was 3.0MPa, 4.0MPa, 5.0MPa, and 6.0MPa.

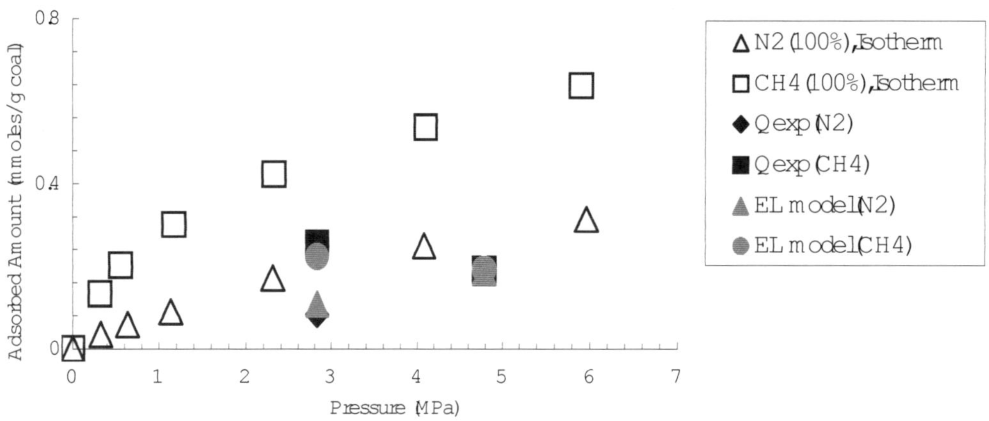

Fig. 6 Desorption of CH4 adsorbed on coal by N2 injection with 3.0MPa and 6.0MPa.

From Figs. 4-6, it was furthermore understood that:

(1) The higher the CO_2 ratio in mixed gas is, the lower the accuracy of predicted CH_4 adsorption becomes.

(2) The higher the pressure of injected CO_2 is, the lower the accuracy of predicted CH_4 adsorption becomes.

(3) The accuracy of predicted CH_4 adsorption is better in single N_2 than in single CO_2.

4.2.4 Discussions and suggestions

Similar studies as our experiments were performed by other researchers. Prusty and Harpalani (2004) conduced two similar experiments of CH_4 desorption by injecting CO_2 and concluded that the EL model is very accurate in predicting sorbed CO_2, while it is only moderately accurate in predicting sorbed CH_4. After performing a similar experiment, Reeves and Schoeling (2000) proposed that the EL model predicts more accurately for more strongly adsorbed component. However, the reasons why that states occur have not been explained by anyone yet. Here we will discuss the reasons and give a suggestion in order to improve the prediction precision.

The Extendeded Langmuir (EL) equation was commonly used to predict the adsorption equilibrium of mixed gases. The main assumption of this equation is stipulated that all gases competitively adsorbed on all sites of coal, and do not exist space where only entered by one gas. However, a lot of experiments proved that CO_2 behave preferentially adsorption as compared with CH_4 and N_2. Cui and others (2004) indicated that due to its relatively smaller kinetic diameter, CO_2 can permeate not only macropores but also ultra micropores, which likely block CH_4 and N_2 that have slightly larger kinetic diameters. This may result in one or two order of magnitude higher diffusivity of CO_2 than those of CH_4 and N_2 in the coal matrix.

The difference in the sorption mechanism of CO_2 and CH_4 was proposed (Mastalerz and others, 2004) that adsorption into the micropores being the major mechanism in the case of CH_4, whereas both adsorption into the micropores and absorption into the solid/elastic organic structure of coal being important in the case of CO_2. Based on the modeling study, Milewska-Duda et al. (2000) concluded that for CO_2 the contribution of pure absorption is comparable to that of the pure adsorption, whereas for CH_4 the pure absorption is much lower. Reucroft and Sethuraman (1987) also suggest that up to half of the total uptake of CO_2 could be due to CO_2 dissolved into the coal structure, and the contribution of dissolution increases with pressure. After reviewing a lot of literatures, Larsen (2004) summarized that the significant dissolution of CO_2 can lower the coal's softening temperature, besides swell them slightly. They also suggested that in considering sequestration of CO_2 in coal seams, attention must be paid to the possible effects of dissolved CO_2 on the structure and behavior of the coal.

Based on all above analysis, we found that it is not appropriate to predict the amount of CO_2 adsorbed on coals by traditional method using EL model, because a number of CO_2 is also preferentially absorbed into coal, besides the part is adsorbed competitively with other gases. In order to predict more accurately the adsorption amount of CO_2, CH_4, and N_2 using the EL model, the CO_2 dissolved preferentially into coal structure must be considered. Therefore, the actual experimental value of CO_2 is the sum of the preferential adsorption and the competitive adsorption which can be calculated using the EL equation. From this new point of view, the EL equation of CO2 was reestablished in accordance with following assumptions in this study:

(1) The amount of CO2 adsorbed preferentially at its partial pressure in mixed gas environment is equal to the amount at the same pressure in single gas environment.

(2) In injecting CO2 with 3.0MPa, the actual experimental value of CH4 is assumed to be equal to the predicted value using EL model. On this assumption, the Langmuir coefficient K (CO2) in EL equation was given.

(3) In injecting CO2 with 3.0MPa, the actual experimental value of CO2 is assumed to be equal to the sum of preferentially absorbed value and calculated value using EL model. On this assumption, the amount of saturated adsorption of CO2 Q_{max} (CO2) in EL equation was obtained.

According to above assumption, the true Extended Langmuir coefficients of CO2 were calculated as K (CO2) =2.1, and Q_{max}(CO2) =0.8. Figure 7 shows the new type of adsorption isotherm of CO2, the part under the new isotherm was considered as the amount of CO2 adsorbed competitively with other gases, while the part between the new isotherms and isotherms of single CO2 was expressed as the amount of CO2 absorbed preferentially.

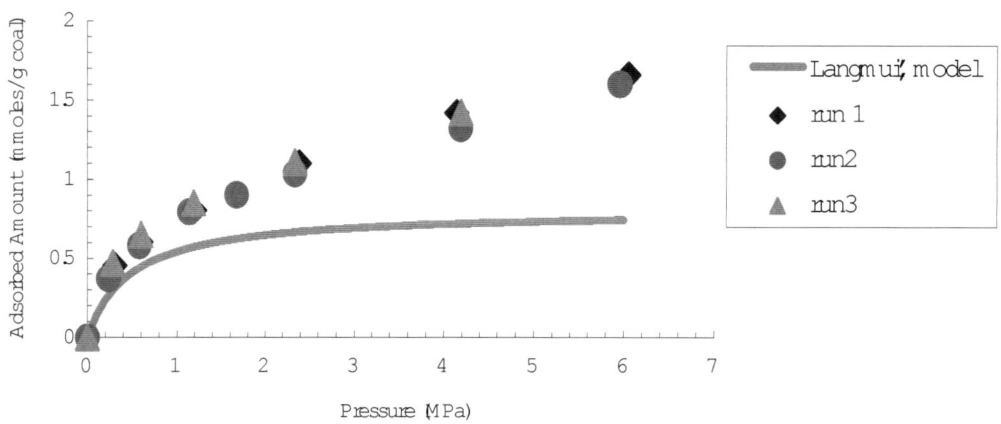

Fig. 7 A new Langmuir isotherm of CO2, which indicates only the part is adsorbed competitively with other gases.

Using the new adsorption isotherm of CO2, the experimental data of displacements were re-evaluated. It was found that the new evaluated results (Figures 8-10) have been considerably improved, that is, the new EL model can predict the actual experimental values accurately. The actual significance of this work lies in that more CO2 can be sequestrated in coal and more CH4 can be displaced from coal by comparing with the experimental data.

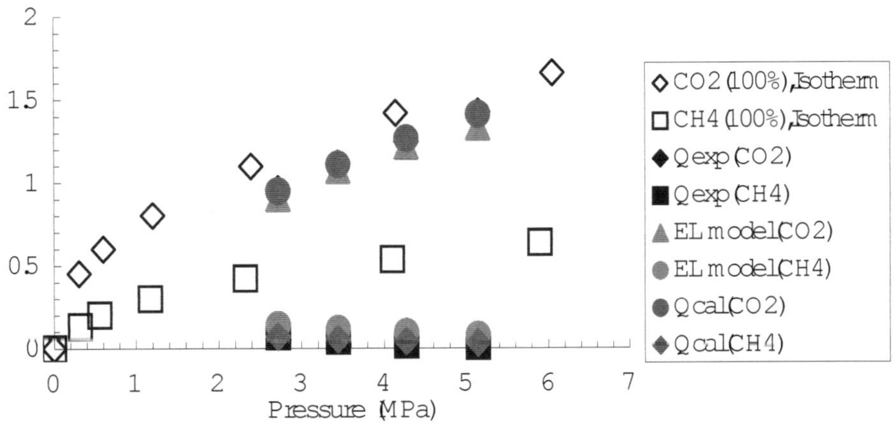

Fig.8 Revised model illustrating displacement of CH4 by CO2 injection at 3.0MPa, 4.0MPa, 5.0MPa, and 6.0MPa.

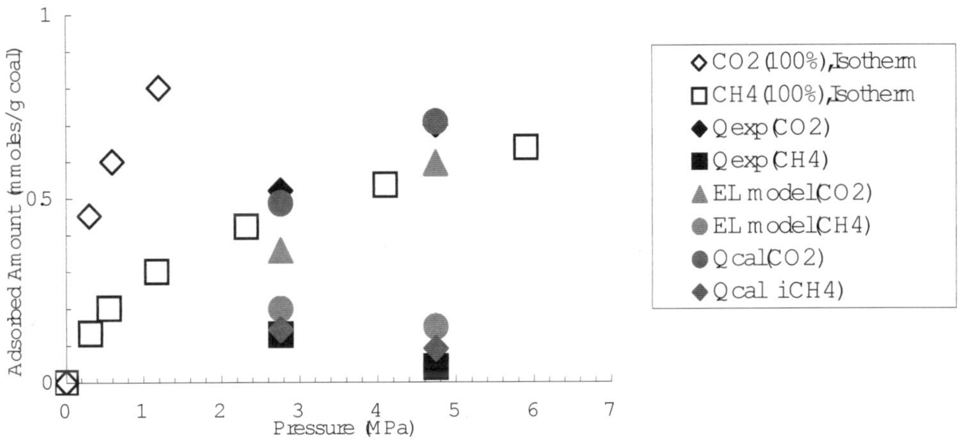

Fig. 9 Revised model illustrating displacement of CH4 by mixed gas (35%CO2-65%N2) injection

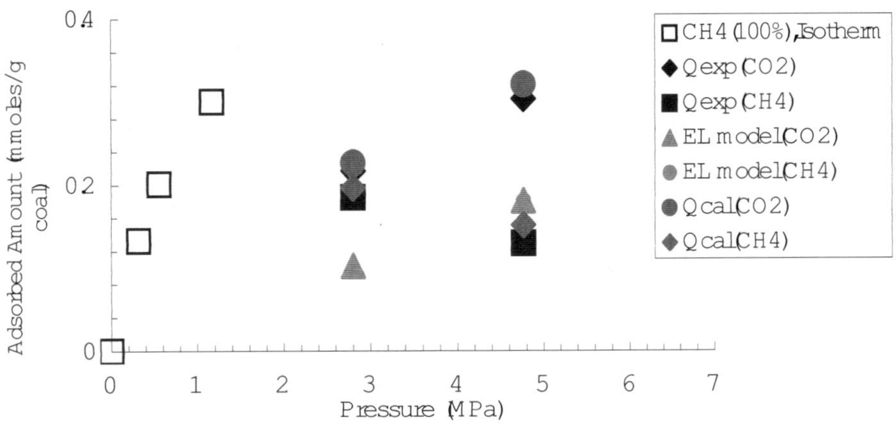

Fig. 10 Revised model illustrating displacement of CH4 by mixed gas (10%CO2-90%N2) injection

5. CONCLUSIONS

CO_2 has long been used to study coal surface areas by adsorption under a low-pressure regime. For the enhanced CBM recovery project, CO_2 will be injected coal seams with high-pressure regimes. The calculation of gas adsorptions on coal surface with EL model involves the assumption that CO_2 is only adsorbed and does not dissolve in coals. However, there are abundant data demonstrating that CO_2 dissolves into coals and swells them.

Displacement behavior of CH_4 adsorbed on coals by injecting single CO_2, mixed gas of CO_2 and N_2 with different ratio, and single N_2 was experimentally studied. This study indicated that the injection of CO_2 could effectively displace CH_4 even with a relatively low injection pressure. However, when comparing the actual experimental values with the predicted values using EL equation, it was found that EL model always under-predict the sorbed CO_2 value within 10% error, while over-predict the sorbed CH_4 value with a horrible error with the injection gas pressure up to 6.0MPa. The results remind us that the sorption mechanism of gas must be re-considered. Based on many reports that the gas of CO_2 not only adsorbed into the micropores of coal, but also absorbed (dissolved) into the solid organic structure of coal. We proposed an assumption that the actual experimental value of CO_2 sorbed on coals is the sum of the preferentially absorbed (dissolved) value and the value adsorbed competitively with other gases. The predicted value using EL model only involves the competitively adsorbed value, does not involve the preferentially absorbed (dissolved) value. From this assumption, the Extended Langmuir coefficients of CO_2 in EL equation were revised. With the revised EL equation of CO_2, the adsorption equilibrium of CO_2 and CH_4 were re-evaluated, and a higher calculation precision was achieved. This new evaluated results suggested that the prediction precision will be considerably improved by considering the preferentially absorbed (dissolved) CO_2.

REFERENCES

1. Prusty B.K. and Harpalani S. 2004, A Laboratory Study of Methane/CO2 Exchange in an Enhanced CBM Recovery Scenario, International Coalbed Methane Symposium, May, Alabama, USA
2. Larsen J.W. 2004, The Effects of Dissolved CO2 on Coal Structure and Properties. International Journal of Cola Geology, 57, 63-70
3. Reeves S.R. and Stevens S.H. 2000, CO2 Sequestration. World Coal, December 2000, CBM Review, 1-4
4. Adachi K. 2005, A Study of Displacement Behavior of CH4 adsorbed on coal by Injecting CO2-N2 mixture. Master Thesis, Institute of Environmental Studies, Graduate School of Frontier Sciences, The University of Tokyo, 20-50 (In Japanese)
5. Mavor M.J., Owen L.B. and Pratt T.J. 1990, Measurement and evaluation of isotherm data. Proceedings of the 65th Annual Technical Conference and Exibition of the Society of Petroleum Engineers, SPE 20728, New Orleans, LA, 157-170
6. Milewska-Duda J., Duda J.T., Nodenski A., L akatos, J. 2000 Absorption and Adsorption of methane

and carbon dioxide in hard coal and activated carbon. Langmuir 16, 5458-5466.

7. Reucroft P.J. and Sethuraman A.R. 1987 Effect of pressure on carbon dioxide induced coals swelling. Energy and fuels 1, 72-75
8. Cui X., Bustin R., Dipple G., 2004 Selective transport of CO2, CH4, and N2 in coals: insights from modeling of experimental gas adsorption data. Fuel 83, 293-303

0518

CHARACTERISTICS TESTS OF COAL OF CO2 INJECTION TEST FIELD IN JAPAN

Kotaro OHGA1) and Masaaki Fujioka2)

1) Hokkaido University, N13W8,Sapporo,060-8628,JAPAN

2) Japan Coal Research Center, Mita 3Chome, Minato-ku,Tokyo,JAPAN

ABSTRACT

CO_2 sequestration in coal seams project in Japan (JCOP) started at 2002. In the beginning two years, fundamental tests were carried out, micro-pilot test started from 2003 at Shuuparo Mine in Hokkaido, JAPAN.

The various tests, which were measurement of methane content of coal, adsorption tests and methane replacement test by carbon dioxide, were carried out before CO_2 injection.

From the results of measurement of methane content of coal and the results of adsorption tests, it is clear that the gas in place of this test field coal is not saturated and it is more than 20 m3/t. Adsorption volume of carbon dioxide was about 1.5 times as much as adsorption volume of methane.

In this paper, detail of these tests and results will be described.

INTRODUCTION

Recently, Carbon dioxide injection into coal seams has become a topic of major interest as one of carbon dioxide sequestration methods, because the sequestration into coal seam is more advantageous than other geo-sequestration methods in the following respects: 1) the principal of this sequestration is mainly relies on the mechanism of gas adsorption to the coal, therefore carbon dioxide injected into the coal seam is fixed firmly. 2) When carbon dioxide is injected into the coal seam, methane can be recovered as a by-product. This fact will reduce the cost of sequestration.

In the middle of 2002, the project of CO_2 injection into coal seams was started.

There are seven main coalfields in Japan. Most of the coalfields are in Kyushu and Hokkaido. More than 50% of coal is in Hokkaido and most gassy coalfield which is Ishikari coalfield is in the center of Hokkaido. The level of permeability of coal seams of southern part of Ishikari coal field is one of the highest among Japanese coal fields. Therefore, Yuubari area of Ishikari coal field in Hokkaido was selected as the most suitable pilot test field.

The injection well was drilled in the end of 2003. In order to obtain the values used for simulation. the various tests were carried out using the core samples collected from the injection well.

We also carried out the adsorption tests using carbon dioxide and methane, and methane replacement tests using carbon dioxide to clarify the mechanism of replacement of methane absorbed onto the internal surface of coal by carbon dioxide,.

Pilot Test Field

Injection well (IW-1)was drilled in the end of 2003, production well(PW-1) was drilled in the summer of 2004 as shown in Fig.1. The distance between injection well site and production well site was about 200m. The distance in the coal seam between the vertical injection well (IW-1) and the deviated production well (PW-1) was about 65m.

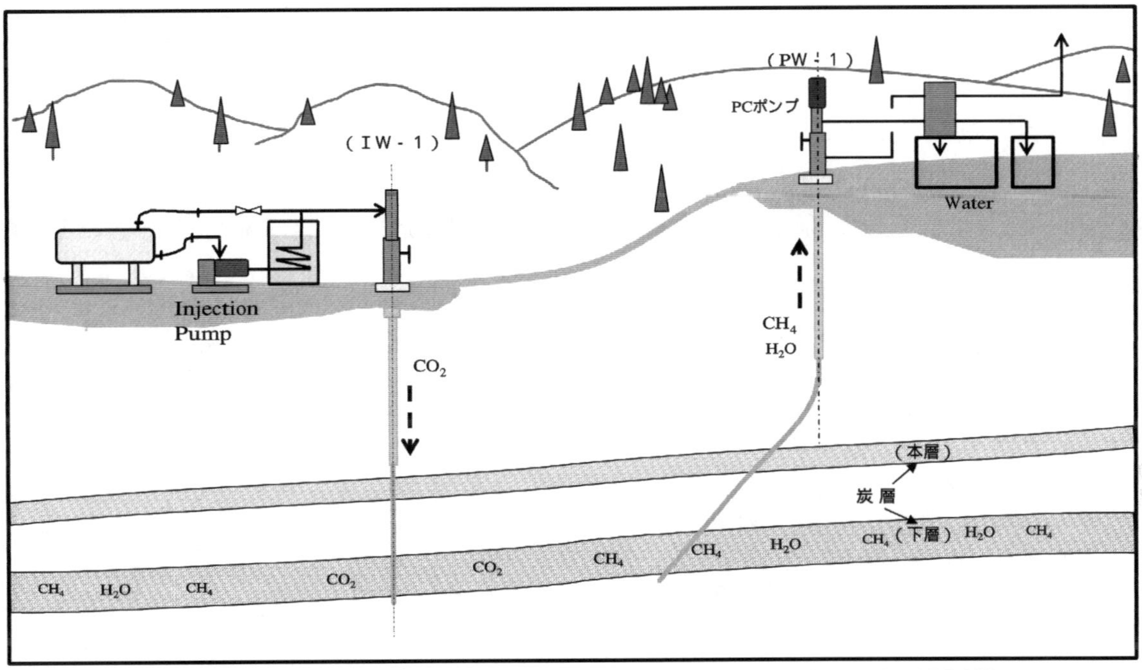

Fig.1 Outline of Pilot Test of CO2 Injection into Coal Seam in JAPAN

Gas Content of Coal

Gas content of coal was measured using coal core samples collected from the injection well by a direct method. Table 1 shows the results of measurement of gas content.

Sample No.	Sample Weight g	Emission Gas cc/g	Lost Gas cc/g	Residual Gas cc/g	Gas Content cc/g
1	895	13.03	4.52	2.36	19.91
2	1132	15.62	8.27	3.09	26.98
3	946	13.39	4.52	2.60	20.51
4	911	15.58	5.44	2.83	23.84
5	1211	12.09	4.68	2.61	19.38
6	978	10.98	7.51	2.68	21.17
7	897	16.16	3.88	2.70	22.74
8	958	16.29	3.60	3.39	23.29
Average	991.0	14.14	5.30	2.78	22.23

Table 1 Measurement Results of Gas Content of Coal

COAL SAMPLE FOR ADSORPTION TESTS

Coal core samples were collected during drilling of the injection well. A part of coal core samples were used as coal samples for the above tests. Most of the coal core samples could not be collected as solid core sample as shown in Fig.2. The space of cleats of coal was about from 2mm to 4mm. Therefore, we used coal samples whose size was between 4 mesh and 8 mesh.

Fig.2 Collected Coal Samples

EQUIPMENT

Fig.3 shows the equipment used for the adsorption tests and the replacement tests. The equipment consists of a reaction vessel (H) to absorb gas onto coal sample and a buffer tank (G) to measure the injection gas volume to the reactor vessel. Pressure sensors(D1,D2) are installed on the reaction vessel and the buffer tank. The injection gas volume into the reaction vessel and released gas are calculated from the change of pressure in the buffer tank and absorption volume into the coal sample is calculated from the change of the pressure in the reaction vessel. To reduce the effects of temperature change, whole of the equipment is submerged in a water bath (I) as shown in Fig.3

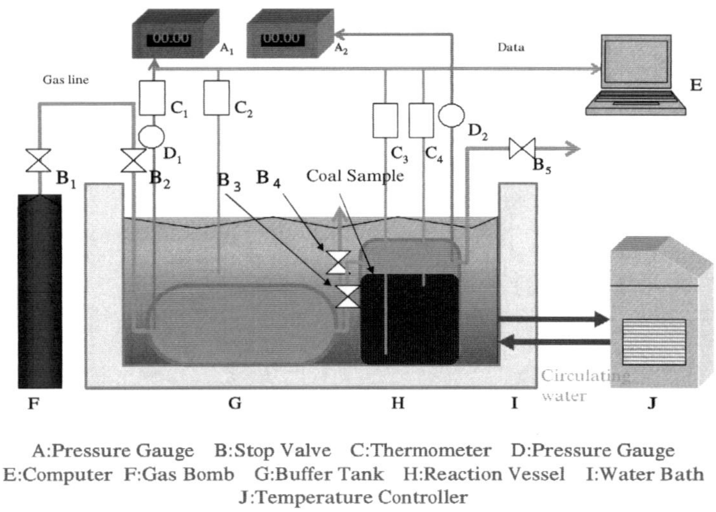

Equipment for Adsorption and Displacement Tests

Fig.3 Equipment for Adsorption and Replacement Tests

The temperature of the water is kept at a certain temperature by the temperature controller (J). About 50g of coal sample is put into the Reaction Vessel (H). Released gas is analyzed by gas-chromatography.

ADSORPTION TESTS

Adsorption tests by carbon dioxide and methane were carried out through the following procedures.

To remove the adsorbed gas from coal samples, gas in the reaction vessel is sucked by vacuum pomp at 40C , for 24hours. After that, the temperature of the reaction vessel decreased to 30C and gas is injected into the buffer tank and the reaction vessel until the pressure in the reaction vessel is at a certain pressure, and then adsorption gas volume at that pressure is measured. The adsorption volumes are measured under several pressures. Fig.4 shows the results of these tests by each gas. The open circles indicate an adsorption isotherm of carbon dioxide and closed circles indicate a desorption isotherm of carbon dioxide. The open and closed squares indicate adsorption and desorption isotherms of methane, respectively. From the results, the adsorption volume of carbon dioxide is 1.5 times as much as one of methane. There is a big difference between the methane adsorption isotherm and desorption isotherm. The hysterics of methane adsorption is clearly noticeable.

Generally, the hysterics of adsorption is often recognized in the case of carbon dioxide desorption, but it is not recognized in the case of methane desorption.

The methane adsorption volume is about 30 cc/g at 8.5 MPa and it is higher than the measurement results of gas content. Therefore, it is estimated that the coal around the injection well is not saturated by gas.

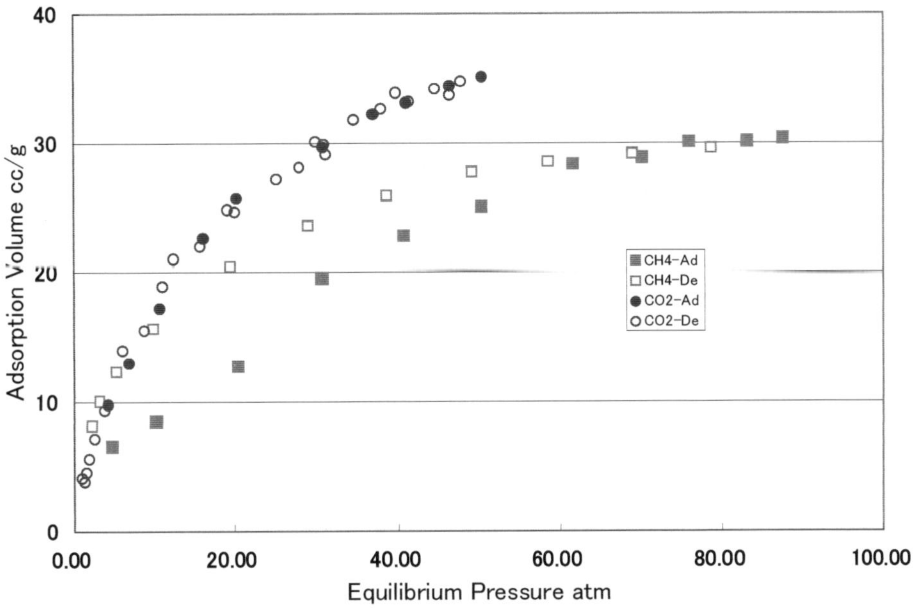

Fig.4 Results of Adsorption Tests

REPLACEMENT TESTS

Methane replacement tests by pure carbon dioxide are carried out using the following procedures. First, the adsorbed gas on the coal sample is removed by suction vacuum pump. Then, methane is injected into the reaction vessel to adsorb methane onto the coal sample at about 1Mpa. After the pressure in the reaction vessel reaches the equilibrium pressure, carbon dioxide is injected into the reaction vessel at a certain pressure. When the pressure in the reaction vessel reaches the equilibrium pressure, the gas is released and gas component is analyzed by gas-chromatography . Same procedures are repeated and each gas adsorption volume is measured at each pressure level. The results of the tests are shown in Fig.5.

ESTIMATION OF ADSORPTION VOLUM UNDER MIXTURE GASES

We tried to estimate the adsorption volume of each gas under the mixture of methane and carbon dioxide by the Markham-Benton equation extended Langmuir equation for mixed gases.
I t is expressed as follows:

$$Z_A = \frac{K_A y_A}{K_A y_A + K_B y_B}$$

where, Z_A expresses the molar fraction of gas A in the adsorption phase.
Y_A expresses the molar fraction of gas A in the gas phase.

K_A(0.0239,0.1256) and K_B(0.0639,0.0.715) were calculated from adsorption and desorption isotherms by gases. Fig.6 shows the estimation curves calculated from four combinations of K_A and K_B and the results of methane replacement tests .Most of the replacement results are around the curves which were calculated from the methane and carbon dioxide adsorption isotherms .

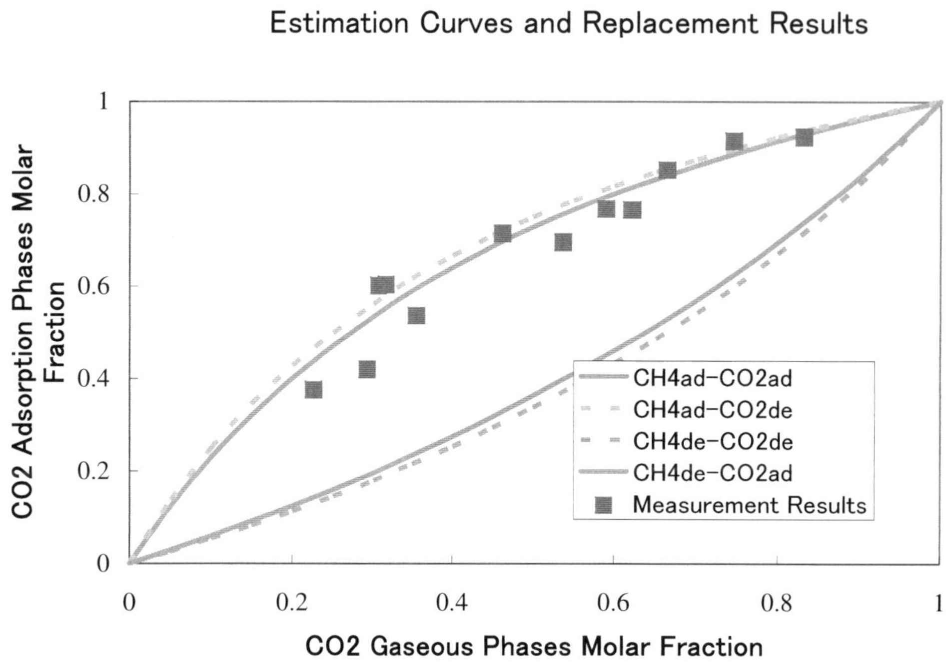

Fig.6 Estimation Curves and Replacement Results

CONCLUSIONS

From the tests we obtained the following results:

1) From the measurement of gas content, the gas content of the coal seam around the injection well is from 20 to 26 cc/g..
2) From the adsorption tests, the methane adsorption volume is about 30 cc/g. Therefore, it can be seen that coal seams around the injection well is not saturated by gas.
3) Estimation values by Markham-Benton equation are very much in agreement with the results of replacement tests. Therefore, we can use the Markham-Benton equation for estimation of methane replacement by carbon dioxide.

AKNOWLEDGEMENTS

This research was a part of the Project of Carbon Dioxide Sequestration Development Technology into the Coal Seams. It was funded by KANSO Technos CO.,LTD and METI(The Ministry of Economy Trade and Industry).
We thank to Ms.Sugiura and Mr. Nagai for measurement of adsorption tests..

0522

The Allison Unit CO_2-ECBM Pilot – A Reservoir and Economic Analysis

Scott Reeves, Advanced Resources International, Inc.
Anne Oudinot, Advanced Resources International, Inc.

ABSTRACT

In October, 2000, the U.S. Department of Energy, through contractor Advanced Resources International, launched a multi-year government-industry R&D collaboration called the Coal-Seq project. The Coal-Seq project investigated the feasibility of CO_2 sequestration in deep, unmineable coalseams by performing detailed reservoir studies of two enhanced coalbed methane (ECBM) recovery field projects in the San Juan basin. The two sites were the Allison Unit, operated by Burlington Resources, into which CO_2 was injected, and the Tiffany Unit, operating by BP America, into which N_2 was injected (the interest in understanding the N_2-ECBM process has important implications for CO_2 sequestration via flue-gas injection). The objectives of the field studies were to understand the reservoir mechanisms associated with CO_2 and N_2 injection into coalseams, demonstrate the effectiveness of the ECBM and sequestration processes, demonstrate an engineering capability to model them, and to evaluate ECBM/sequestration economics. In support of these efforts, laboratory and theoretical studies were also performed to understand multi-component isotherm behavior, and coal permeability changes due to swelling with CO_2 injection. This paper presents the results of The Allison Unit study, in which a detailed reservoir characterization of the field was developed, the field history was matched using the COMET2 reservoir simulator, future field performance was forecast under various operating conditions, and an economic analysis performed.

INTRODUCTION

In October, 2000, the U.S. Department of Energy (DOE), through contractor Advanced Resources International (ARI), launched a multi-year government-industry R&D collaboration called the Coal-Seq project[1]. The Coal-Seq project investigated the feasibility of CO_2 sequestration in deep, unmineable coalseams by performing detailed reservoir studies of two enhanced coalbed methane recovery (ECBM) field projects in the San Juan basin. The two sites were the Allison Unit, operated by Burlington Resources, into which CO_2 was injected, and the Tiffany Unit, operated by BP America, into which N_2 was injected (the interest in understanding the N_2-ECBM process has important implications for CO_2 sequestration via flue-gas injection). The objectives of the field studies were to understand the reservoir mechanisms of CO_2 and N_2 injection into coalseams, demonstrate the effectiveness of the ECBM and sequestration processes, demonstrate an engineering capability to model them, and to evaluate ECBM/sequestration economics. In support of these efforts, laboratory and theoretical studies were also performed to understand multi-component isotherm behavior, and coal permeability changes due to swelling with CO_2 injection. This paper presents the results of The Allison Unit study, in which a detailed reservoir characterization of the field was developed, the field history was matched using the COMET2 reservoir simulator, future field performance was forecast under various operating conditions, and an economic analysis performed.

SITE DESCRIPTION

The Allison Unit ECBM pilot is located in San Juan County, southern New Mexico, in close proximity to the border with Colorado (Figure 1). While the Unit consists of many wells, the pilot area for CO_2 injection consisted of 16 coalbed methane (CBM) producer wells, 4 CO_2 injectors, and one pressure observation well (POW #2). The study area well pattern is illustrated in Figure 2. At the center of the study area is a five-spot of CBM producers on nominal 320 acre spacing (wells 130, 114, 132 and 120 at the corners, and well 113 in the center), with the four CO_2 injectors roughly positioned on the sides of the five-spot between the corner producer wells (creating a nominal 160 acre spacing between injectors and producers). POW #2 is located on the eastern border of the central pattern, and the remaining CBM producers surround this central pattern.

The producing history for the study area is shown in Figure 3. The field originally began production in 1989, with CO_2 injection occurring between April, 1995 and August, 2001. Several points are worth making regarding the producing history:

- Upon commencement of the injection operations, the five producer wells in the central five-spot pattern were shut in. The purpose was to facilitate CH_4/CO_2 exchange in the reservoir. After about six months, CO_2 injection was suspended for about another six months, during which time the five shut-in producers were re-opened. These activities can be clearly identified in Figure 3; their impact on long-term production performance however, if any, is unclear.

- Shortly after CO_2 injection began, a program of production enhancement activities unrelated to the CO_2-ECBM pilot was implemented. Those activities included well recavitations, well reconfigurations (conversion from tubing/packer completions to annular flow with a pump installed for well dewatering), line pressure reductions due to centralized compression, and also the installation of on-site compression. These activities largely coincided with the dramatic increase in production observed beginning in mid-1998.

In addition, a plot of injection rate and pressure history for injector well # 143 is shown in Figure 4. Injection was performed at a constant surface pressure, and rate was allowed to vary. Note the reduction in injection rate during early time, presumably due to coal swelling and permeability reduction. The rebound in injectivity during later times is believed due to overall reservoir pressure reduction and resulting matrix shrinkage that occurred near the injector wells.

RESERVOIR DESCRIPTION

The Allison Unit wells produce from three Upper Cretaceous Fruitland Formation coal seams, named the Yellow, Blue and Purple (from shallowest to deepest) using Burlington Resources' terminology. A summary of basic coal depth, distribution, thickness, pressure, and temperature information is provided in Table 1.

Sorption isotherms for both CH_4 and CO_2 were measured for six coal samples taken from three wells within the study area. Average CH_4 and CO_2 isotherms based on these data for each coal interval, on a raw basis and at an average density of 1.5 grams per cubic centimeter (g/cc), are shown in Figures 5 and 6.

In May, 2000, pressure buildup tests were performed on 12 wells in the Allison Unit, eight of which were inside the study area. Analysis of these data provided estimates of effective gas permeability, skin factor, and reservoir pressure. Two adjustments of the results were made to 1) derive absolute permeability from the effective gas permeability results and 2) correct to initial conditions – accounting for both pressure-dependent permeability and matrix shrinkage. The resulting permeability map of the field is shown in Figure 7. Permeability values ranged from 30-150 millidarcies (md), with higher permeabilities concentrated within the central 5-spot pattern. No permeability anisotropy appeared to exist for the study area.

A novel technique was also used to estimate relative permeability and porosity for the study area based on historical gas and water production. This technique, described in a detailed report on the Allison Unit[2], provided average relative permeability curves for the study area, as well as a porosity map.

RESERVOIR MODEL CONSTRUCTION

The reservoir simulator used for the study was ARI's COMET2 (binary isotherm – CH_4 and CO_2) model. Details on the model theory are provided in the references[3,4].

A three-layer (Yellow, Blue, Purple), full-field model was constructed. The coal structure and thickness information for each layer was directly input per the maps generated. Coal permeability and porosity maps were similarly employed. Relative permeability curves from the analysis mentioned previously, as well as the laboratory isotherms, were also used.

Additionally, well completion and operating parameters were examined for input into the model, such as recavitations, well reconfigurations and producing pressure adjustments. This was particularly important given the complexity of the field history, and the desire to isolate and study the effects of CO_2 injection.

The model gridblock dimensions were 33 x 32 x 3 (approximately 3,200 total grid blocks, 2,600 of which were active), and covered an active area of about 7,100 acres (Figure 8). On average, the gridblock dimensions were 560 feet x 525 feet x 14 feet. The corners of the model were isolated using no-flow barriers to account for producing wells immediately adjacent to these portions of the study area.

HISTORY MATCH RESULTS

The independent parameter used for the simulator was gas production (and injection) rate to maintain material balance, and the dependent (history match) parameters were water production rate, flowing pressure (producing and injecting), and gas composition. Note that only some of these data were available for some periods for some wells; whatever was available was used. In addition, the pressure history at POW #2 was available.

All parameters were modified globally to obtain the best overall match for the field. The objective of the study was to understand the mechanisms of the CO_2-ECBM process by matching general trends, and not necessarily to make regional changes to the reservoir characterization to achieve matches on an individual well basis. While a large number of simulation trials were performed varying almost all significant reservoir parameters, it was ultimately found that the original reservoir characterization seemed to provide the best overall result.

A comparison of the actual versus simulated field gas rate is presented in Figure 9. The only conclusion that can be derived from this result, since the model was "driven" on gas rate, is that model (as constructed) was capable of delivering the gas volumes required.

The actual versus simulated pressure at POW#2 is presented in Figure 10. Actual pressure data is only available after the commencement of CO_2 injection. At that particular point in time (April, 1995), there appears to be excellent agreement between actual and predicted pressure, suggesting that material balance (at least during primary production) was achieved, and hence values for original gas/water storage capacities, as well as depletion characteristics, were reasonable. After that, however, there is considerable difference in pressure values. Of note is that the estimated pressure at the location of POW #2 based on the May, 2000 pressure transient analysis (PTA) is reasonably close to the simulated value. After considerable analysis of the discrepancy it is believed that the pressure data recorded at POW#2 may have been influenced by severe restrictions in wellbore-reservoir connectivity, and therefore may not have been valid.

Comparison plots of gas and water rates, flowing pressures, and produced gas compositions, for well 113 are presented in Figure 11. This well was selected because it was the central well of the 5-spot, it had data for comparison in all categories, and it had observable CO_2 breakthrough. In addition, this well typifies the differences in simulated versus actual results for the other wells. Several general comments can be made regarding the results:

- The quality of the water rate predictions varied, with some being too high and some too low. However, on balance the predictions were considered within reason (and that could be easily "fixed" with regional variations in porosity and/or water relative permeability).

- In all cases, the predicted bottomhole flowing pressures were higher than the measured values – which were actually surface casing pressure data – usually by 200-300 psi. While some difference might be expected due to the different types of data being compared (surface vs. downhole), the magnitude of the difference seems large. (The wells were believed to be pumped-off with little water head existing above the coal seam.) In most cases the predicted flowing pressures appear smooth through the period when the recavitation operations were performed. This result was per the model design.

- In general, the trend in gas composition was reasonably well replicated. In some cases (most noteworthy well #113), the increase in CO_2 content of the produced gas occurs more rapidly than that actually observed.

A comparison of actual to simulated bottomhole injection pressures for CO_2 injector well #142 is provided in Figure 12. Note that the results for the other three injector wells were very similar. The actual bottomhole pressure history data was computed using long-term surface pressure data, and flowing pressure gradients obtained during the August, 2001 injection/falloff tests. The simulated pressures are considerably lower than the actual values. While simulated bottomhole pressures could be increased substantially to better match the actual data by assuming lower initial permeability values for the injector well gridblocks, the objective was to see if the coal swelling formulation in the simulator could adequately account for sufficient permeability reduction to achieve the high injection pressures observed. The result suggests that the answer is negative. Therefore, coal swelling models with CO_2 injection may require further development to adequately replicate field data.

Pressure transient tests performed in August, 2001 in the CO_2 injection wells indicated near-well permeabilities of <1 md, considerably less than the estimated initial values. However, at most injectivity was only cut in half. The apparent discrepancy between the high permeability reduction and comparatively modest injectivity loss was investigated by examining the permeability profile that extended radially from one of the injector wells (#142) at about the time when injectivity was at its lowest value. The result is shown in Figure 13. This plot suggests that the permeability reduction effect is decreased radially from the well, and reached a distance of about 1000 feet. Simple analytic modeling confirmed that this type of permeability reduction profile would yield a reduction in injectivity by about a factor of two, all else being equal.

PERFORMANCE FORECASTS

In order to evaluate the long-term performance of the ECBM pilot, performance prediction cases were simulated using the history match result as the starting point. The specific cases evaluated were:

1. No CO_2 injection (i.e., primary production only).
2. Current conditions (i.e., CO_2 injection until August 2001).
3. Aggressive injection (i.e., CO_2 injection at four times actual rate until August, 2001)

For each ECBM forecast case, an economic limit of 50 Mcfd of methane per well and 50% CO_2 content per well was imposed; reaching those thresholds prompted the well in question to be shut-in in the model. Results of the forecast for the actual pilot conditions indicated that of the 6.4 Bcf of CO_2 injected in the pilot area, 1.6 would ultimately be reproduced. The incremental methane recovery was 1.6 Bcf, yielding a net CO_2/CH_4 ratio of 3.0. Figure 14 presents the simulated sweep of the CO_2 at the end of the forecast period for the pilot. Note that excellent sweep appears to have been achieved in the northern, western, and southern quadrants of the five-spot. However, due to the location of injection well #140, poor sweep was achieved in the eastern quadrant.

Since the model area was so large compared to the actual flooded area, the incremental recovery results were examined for each quadrant of the central 5-spot pattern. Methane recoveries with and without CO_2 injection were computed for each quadrant and are presented in Table 3. This analysis indicates that CO_2-ECBM was highly effective at recovering incremental methane, providing on the order of 17 – 18% of original-gas-in-place where the patterns were configured for effective sweep.

Two additional interesting observations were made regarding the modeling results:

- The stabilized CO_2/CH_4 ratio of about 3:1 is higher than normally cited for San Juan basin coals. However, if one examines the ratio as a function of pressure (based on the isotherms) the results are as expected (Figure 15). At an abandonment pressure of ~50 psi, the CO_2/CH_4 ratio is close to 3:1.

- It appears that some time was required after CO_2 injection ceased for the CO_2 to migrate through the reservoir and displace the "equilibrium" volume of methane. Figure 16 illustrates the CO_2/CH_4 ratio over time for the pilot. Note that the ratio increases during injection periods (and for some time afterwards), and then begins a gradual decline to the equilibrium value.

ECONOMIC ASSESSMENT

The final element of the study was to evaluate the economic performance of the pilot. The capital, operating and financial assumptions are presented in Table 3. Note that all economics were performed on an incremental basis (i.e., only the incremental production and costs were considered). Further, the effect of Section 29 tax credits was not considered.

The analysis first evaluated the performance of the existing pilot, with no future CO_2 injection considered. Note that the hot-tap and pipeline capital costs are included for this case, but only allocated at 25% of the total since the working assumption was that it would also be used for additional pilots and/or large-scale CO_2 flood implementation. The results are presented in Figure 17. There are several points worth making. First, at the prevailing gas price at the time of the pilot (~ $2.20/Mcf), the project had a negative net present value (NPV), not accounting for Section 29 tax credits. At $4.00/Mcf however, it would have yielded a peak NPV of $2 – 3 million. The breakeven gas price for the pilot was $2.57/Mcf.

Secondly, a peak in NPV occurs approximately five-years after CO_2 injection began. Examination of the incremental methane recovery profile provides insight into this finding, shown in Figure 18. The CO_2 injection resulted in some acceleration of methane recovery, and when the incremental methane rate became negative at later times, the NPV began to drop. This point in time also corresponds to the peak CO_2/CH_4 ratio in Figure 16. The implication is that there may be a fixed, optimum CO_2 volume that should be injected to a given pattern, probably corresponding to the volume of methane in place and the equilibrium CO_2/CH_4 ratio, and any further injection in addition to that volume merely represents additional cost without additional methane recovery. At the Allison Unit pilot, that optimum CO_2 injection volume appears to have been exceeded.

Finally, since the CO_2 injection rate was constrained by pressure limitations and coal swelling. The impact of a higher injection rate was examined. These cases, at $2.20/Mcf and $4.00/Mcf, are also shown on Figure 17. It is clear that higher injection rates substantially improves CO_2-ECBM economic performance. Therefore strategies for mitigating coal swelling and injectivity reduction should be a priority consideration for CO_2-ECBM projects.

CONCLUSIONS

Based on the results of this study, the following conclusions have been drawn:

- The injection of CO_2 at the Allison Unit has resulted in incremental methane recovery over estimated ultimate primary recovery, in approximately a proportion of one volume of methane for every three volumes of CO_2 injected. Methane recoveries of 17 - 18% of original-gas-in-place were estimated for effectively swept portions of the 5-spot.

- At the prevailing gas prices at the time the project was implemented (~$2.20/Mcf), and not considering any tax credit benefits, the pilot itself was uneconomic. However, with today's gas prices of ~$4.00/Mcf, CO_2-ECBM appears economically attractive. The breakeven gas price for the conditions at Allison was estimated to be ~ $2.60/Mcf.

- There appears to be clear evidence of significant coal permeability reduction with CO_2 injection. This permeability reduction, and the associated impact on CO_2 injectivity, compromised incremental methane recoveries and project economics. Finding ways to overcome and/or prevent this effect is therefore an important topic for future research.

ACKNOWLEDGEMENTS

This work was funded by the U.S. Department of Energy, National Energy Technology Laboratory, contract number DE-FC26-00NT40924. The DOE Project Manager was Charlie Byrer. The authors also wish to acknowledge the valuable contribution of Burlington Resources, which provided all the data for the study, invaluable insights, and permission for the results to be published.

REFERENCES

1) Reeves, S. R.: "Geologic Sequestration of CO_2 in Deep, Unmineable Coalbeds: An Integrated Research and Commercial-Scale Field Demonstration Project", SPE 71749, presented at the SPE Annual Technical Conference and Exhibition, New Orleans, September 30-October 3, 2001.

2) Reeves, S.R., Taillefert, A. Pekot, L., and Clarkson, C.: "The Allison Unit CO_2 – ECBM Pilot: A Reservoir Modeling Study", Topical Report, DOE Contract No. DE-FC26-00NT40924, February, 2003.

3) Reeves, S. and Pekot, L.: "Advanced Reservoir Modeling in Desorption-Controlled Reservoirs," SPE 71090, presented at the SPE Rocky Mountain Petroleum Technology Conference, Keystone, May 21-23, 2001.

4) Sawyer, W.K., Paul, G.W., Schraufnagel, R.A., "Development and Application of a 3D Coalbed Simulator," CIM/SPE 90-119, presented at the CIM/SPE International Technical Conference, Calgary, June 10-13, 1990.

Table 1: Basic Coal Reservoir Data, Allison Unit

Property	Value
Average Depth to Top Coal	3,100 feet
Number of Coal Intervals	3 (Yellow, Blue, Purple)
Average Total Net Thickness	43 feet Yellow - 22 ft Blue - 10 ft Purple - 11 ft
Initial Pressure	1,650 psi
Temperature	120º F

Table 2: Incremental Recovery by Quadrant, Case 2 vs. Case 1

Quadrant	Recovery (% OGIP)		
	w/o CO_2	w/ CO_2	Incremental
North	77%	94%	17%
West	77%	95%	18%
South	77%	95%	18%

Table 3: Economic Analysis Assumptions

Capex	
CO_2 Hot Tap:	$175,000 } Allocated @ 25%
36 mi (4 inch) Pipeline:	$3.5 million ($24,000/in-mi) } Allocated @ 25%
Field Distribution:	$80,000 ($20,000/in-mi)
Wells	$1.6 million ($400,000/ea; fully equipped)
Total	$5.355 million
Opex	
Injector Well Operating:	$1,000/mo (active only)
CO_2 Cost	$0.30/Mcf
Produced Gas Processing	$0.25/Mcf
Financial	
Gas Price:	$2.20/MMBTU (ex-field)
Methane BTU Content	1.04 MMBTU/Mcf
Net Revenue Interest:	87.5%
Production Taxes:	8%
Discount Rate:	12%

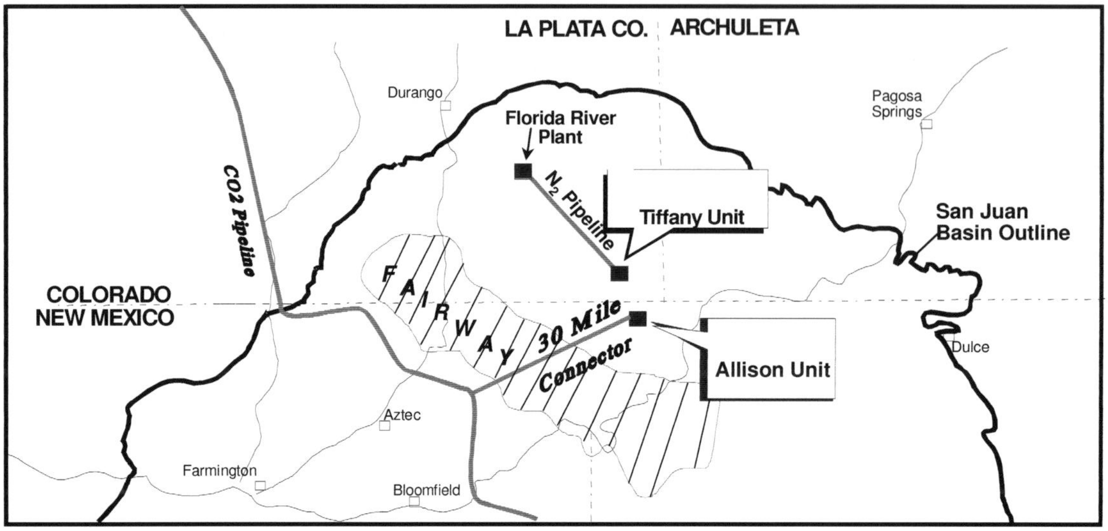

Figure 1: Location of the Allison Unit, San Juan Basin

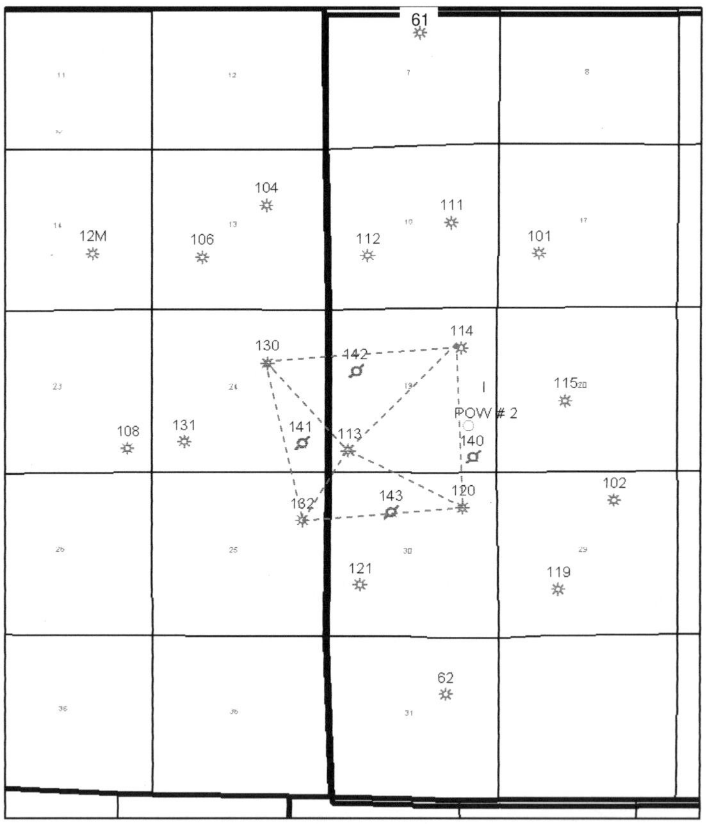

Figure 2: Producer/Injector Well Pattern, Allison Unit Study Area

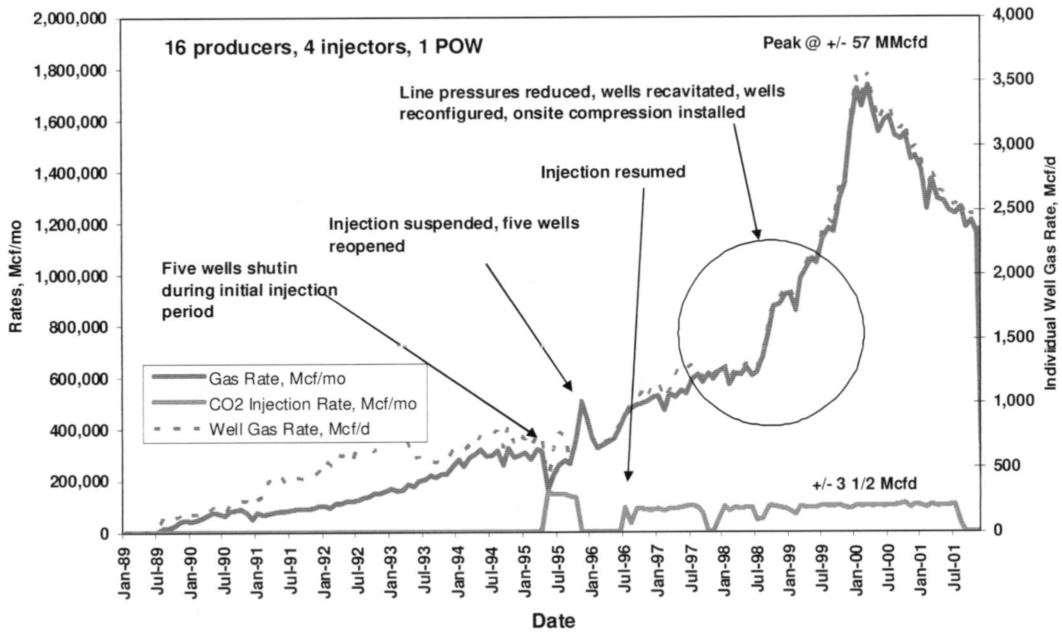

Figure 3: Producing History, Allison Unit Study Area

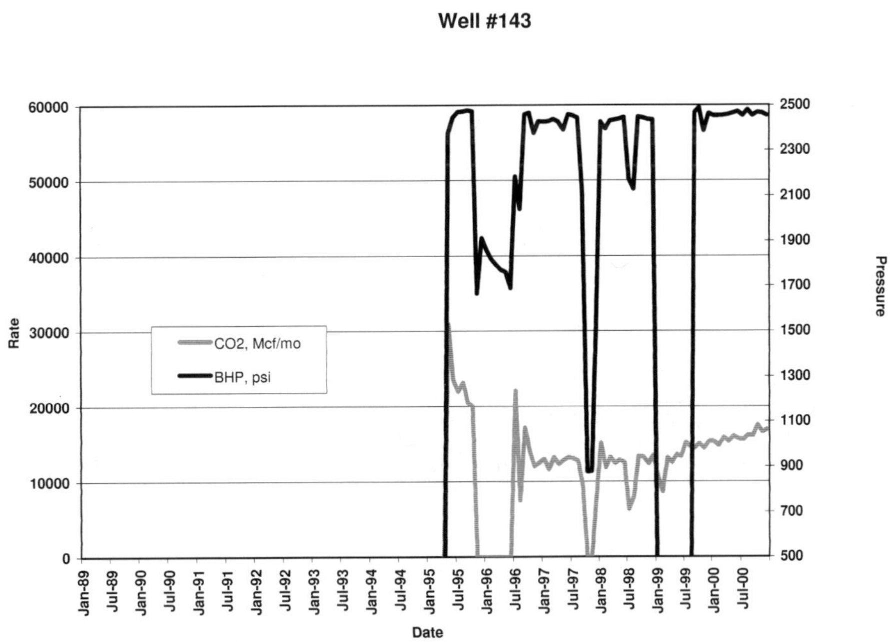

Figure 4: Injector Well # 143 Injection and Pressure History

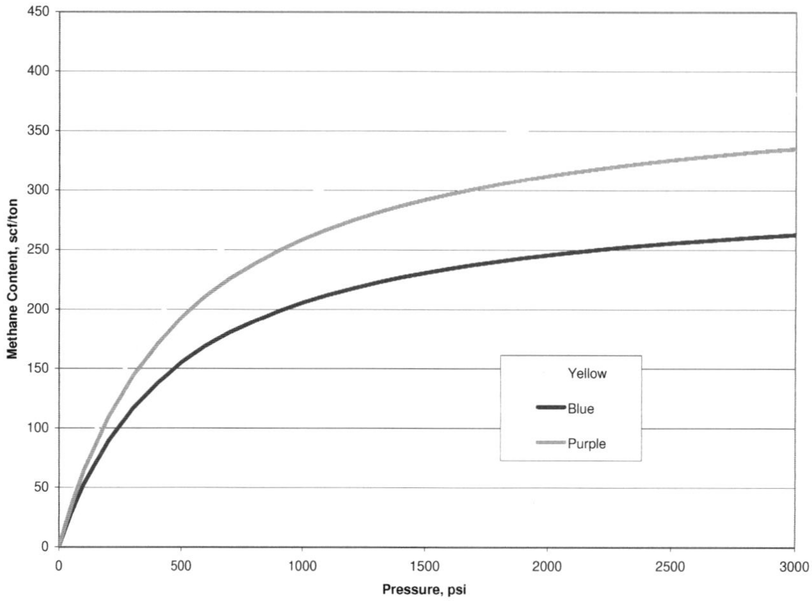

Figure 5: Methane Sorption Isotherms, Allison Unit Study Area

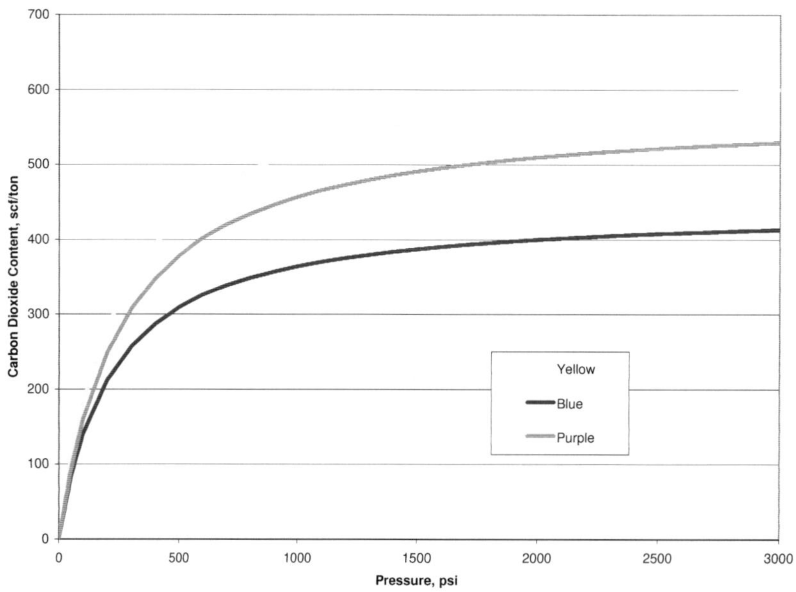

Figure 6: Carbon Dioxide Sorption Isotherms, Allison Unit Study Area

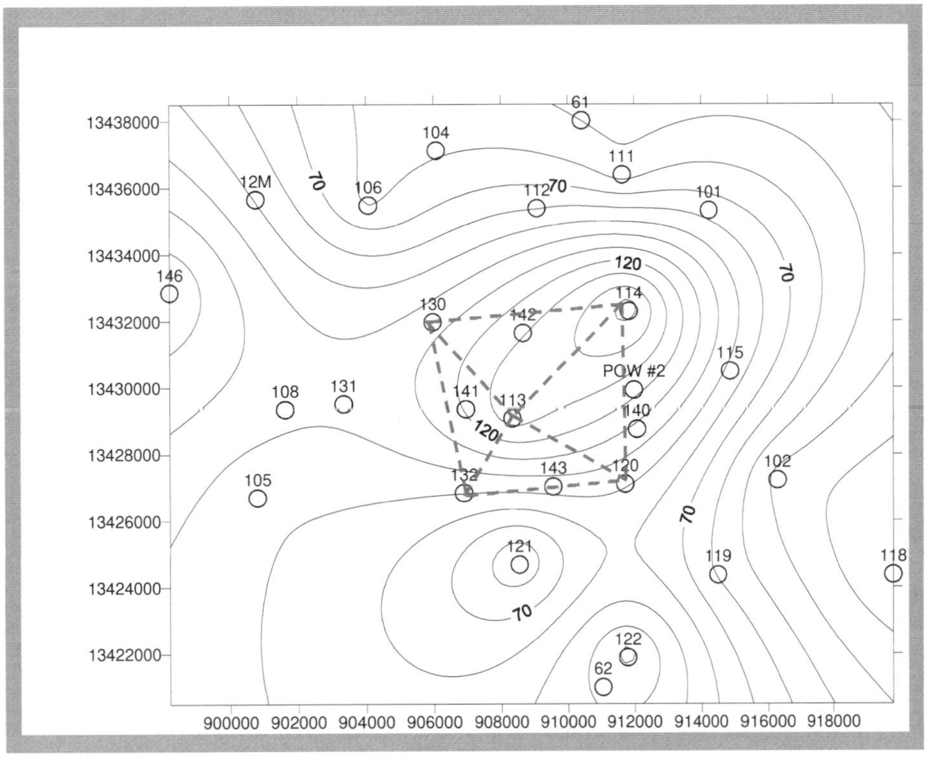

Figure 7: Permeability Map for Allison Unit Study Area

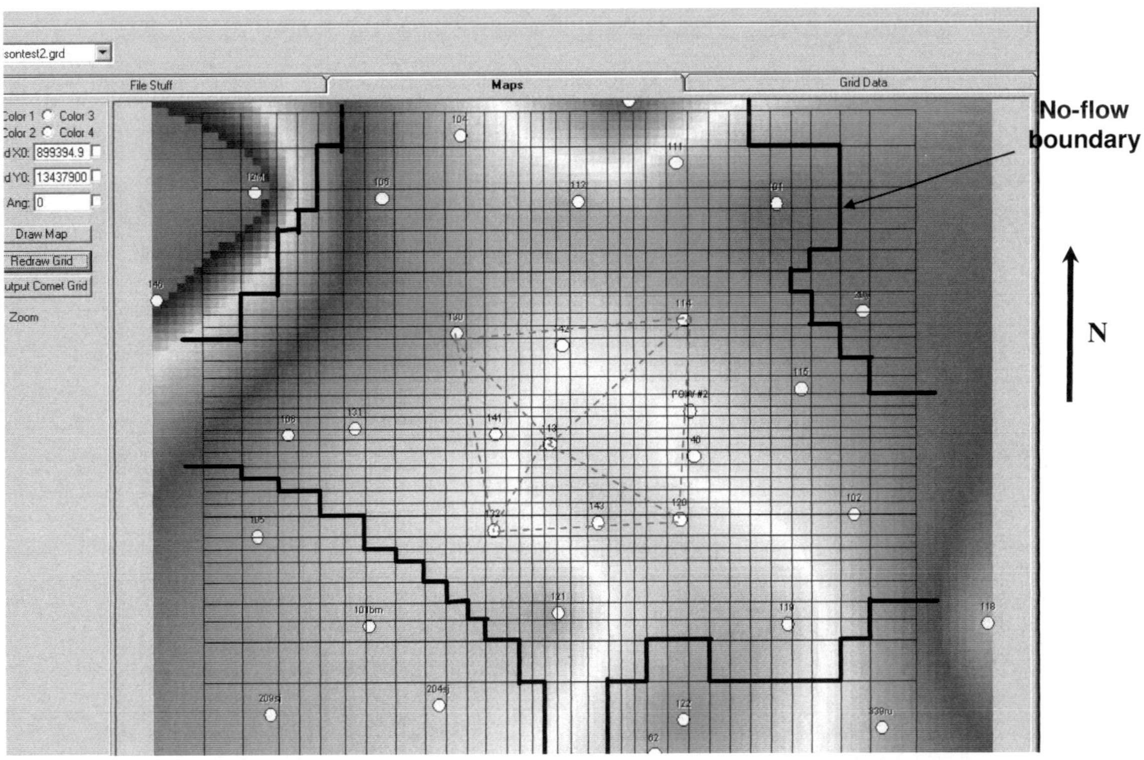

Figure 8: Map View of the Middle Layer of the Simulation Model

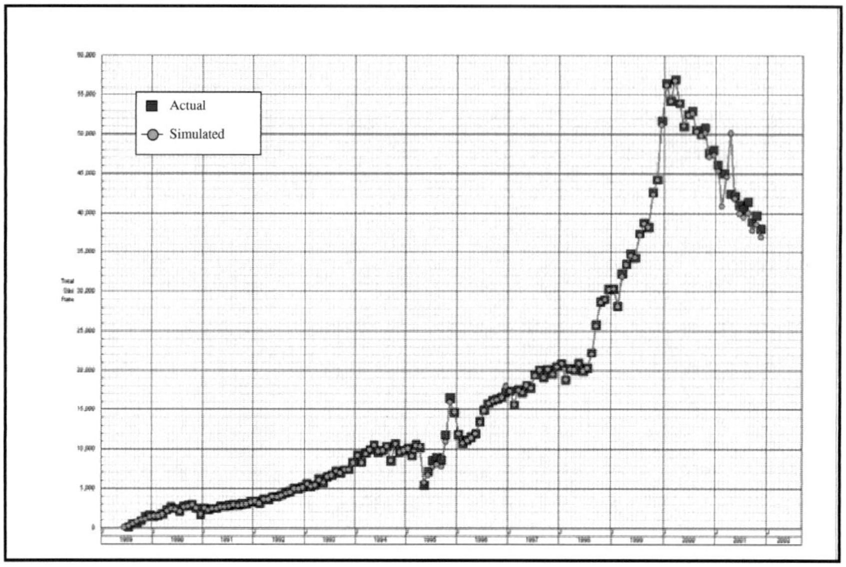

Figure 9: Actual versus Simulated Field Gas Rate, Allison

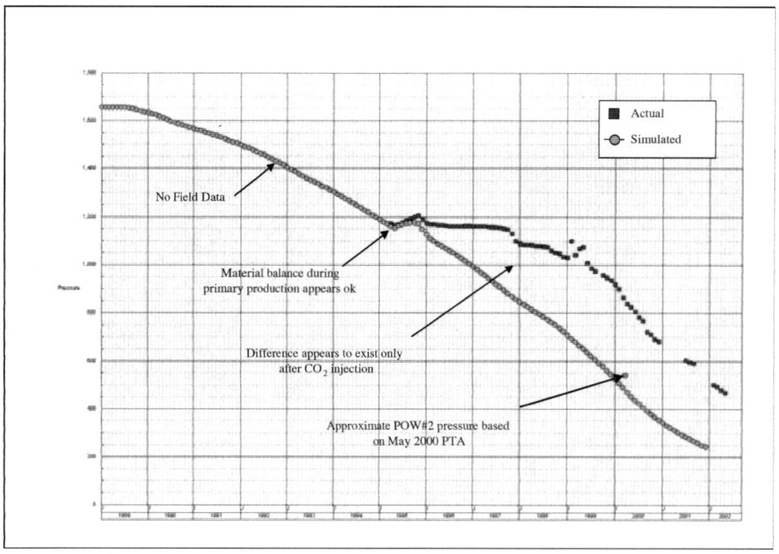

Figure 10: Actual versus Simulated Pressure at POW#2

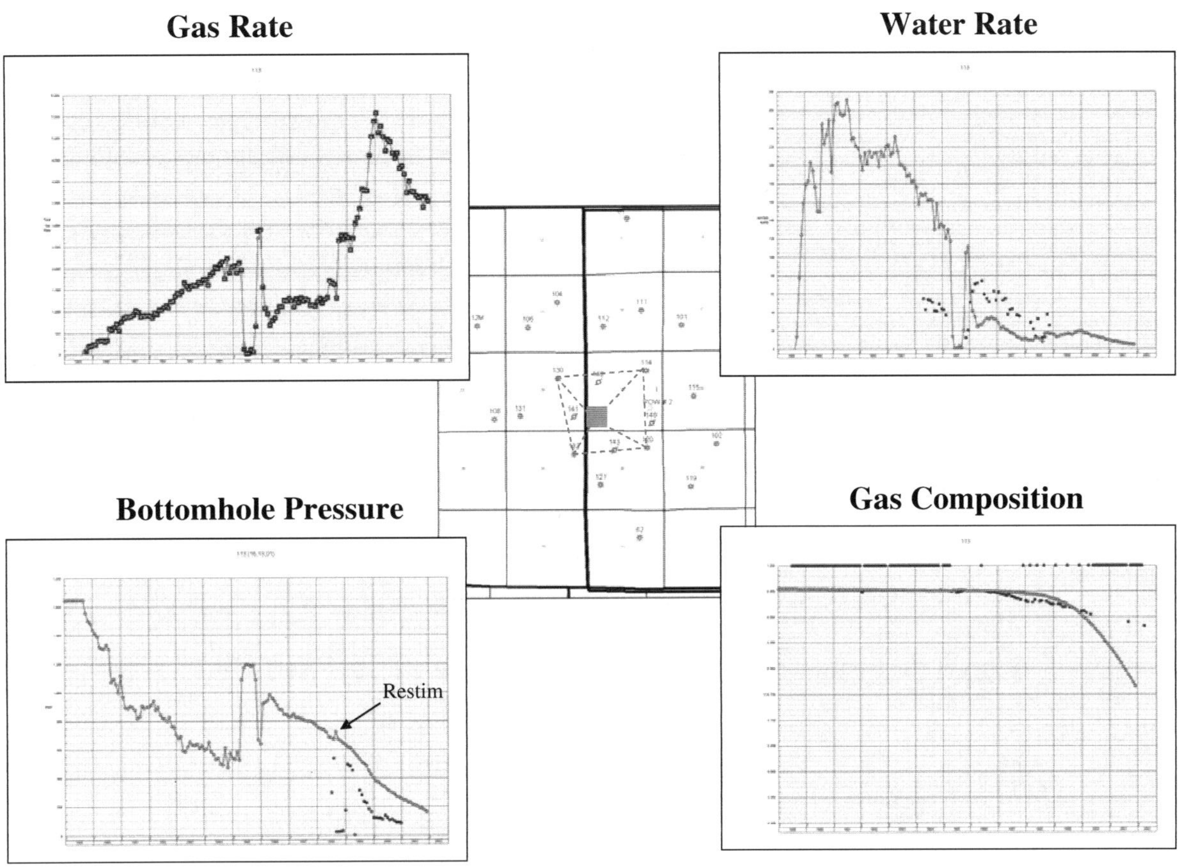

Figure 11: Comparison of Predicted to Actual Well Performance, Well 113

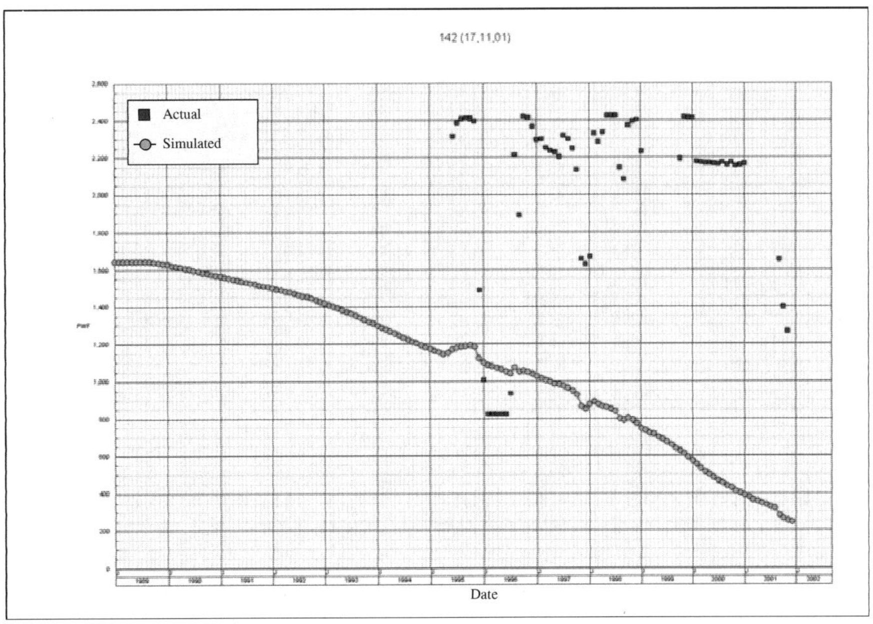

Figure 12: Comparison of Predicted to Actual Bottomhole Injection Pressures, Injection Well #142

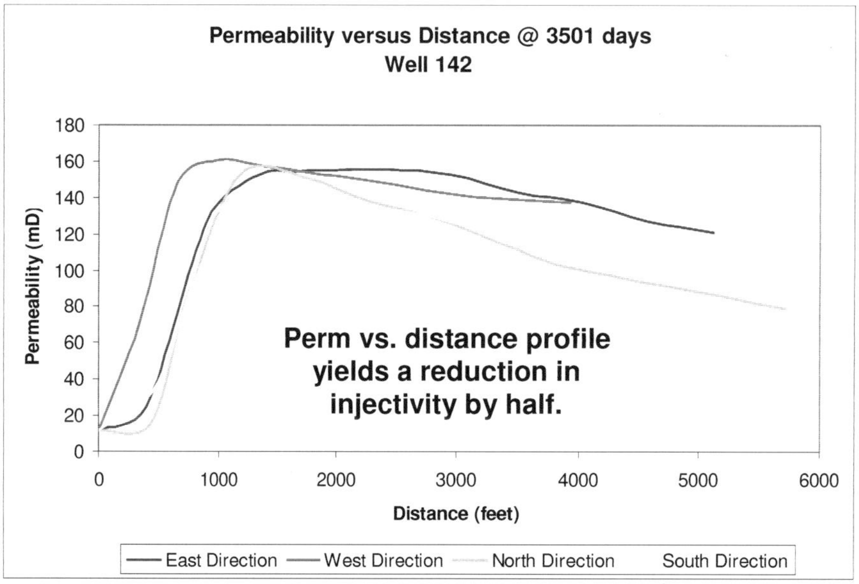

Figure 13: Simulated Permeability Profile from Injector Well #142

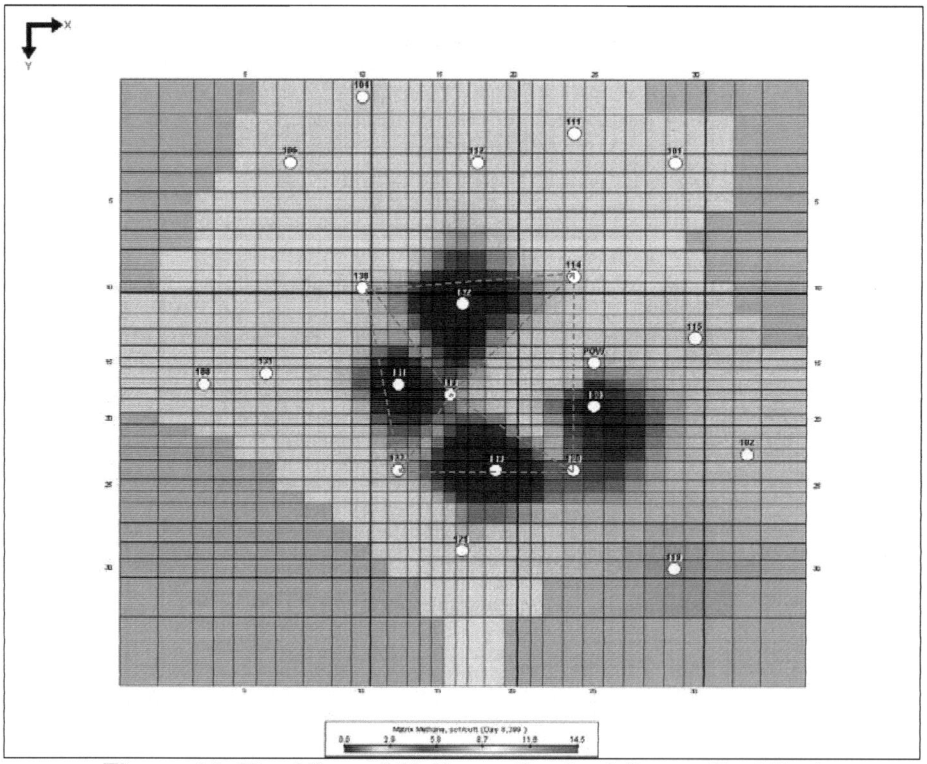

Figure 14: Map View of Methane Content (Layer 2) at End of Forecast Period (Case 2)

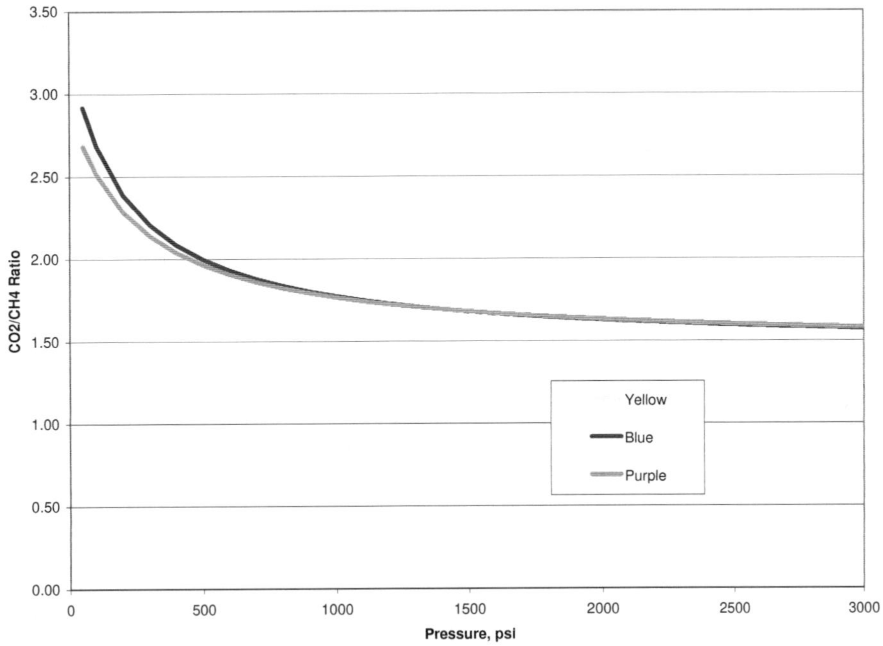

Figure 15: CO_2/CH_4 Ratio vs. Pressure

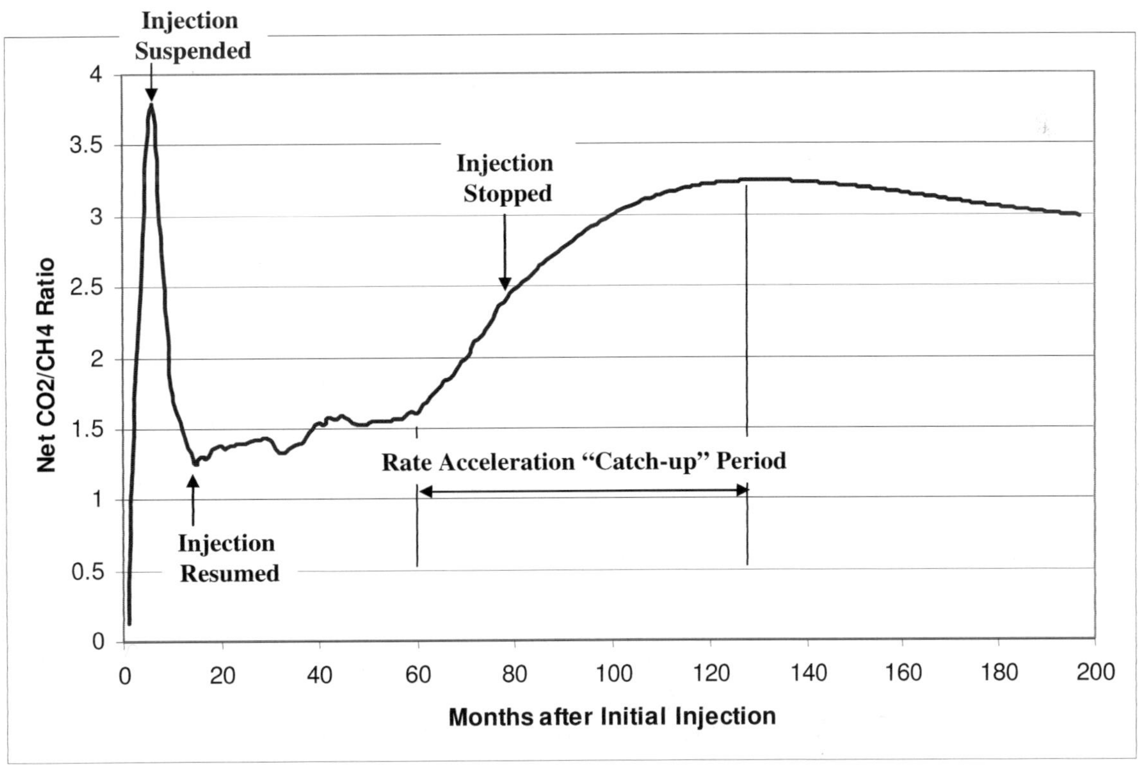

Figure 16: CO_2/CH_4 Ratio vs. Time, Allison Unit Pilot

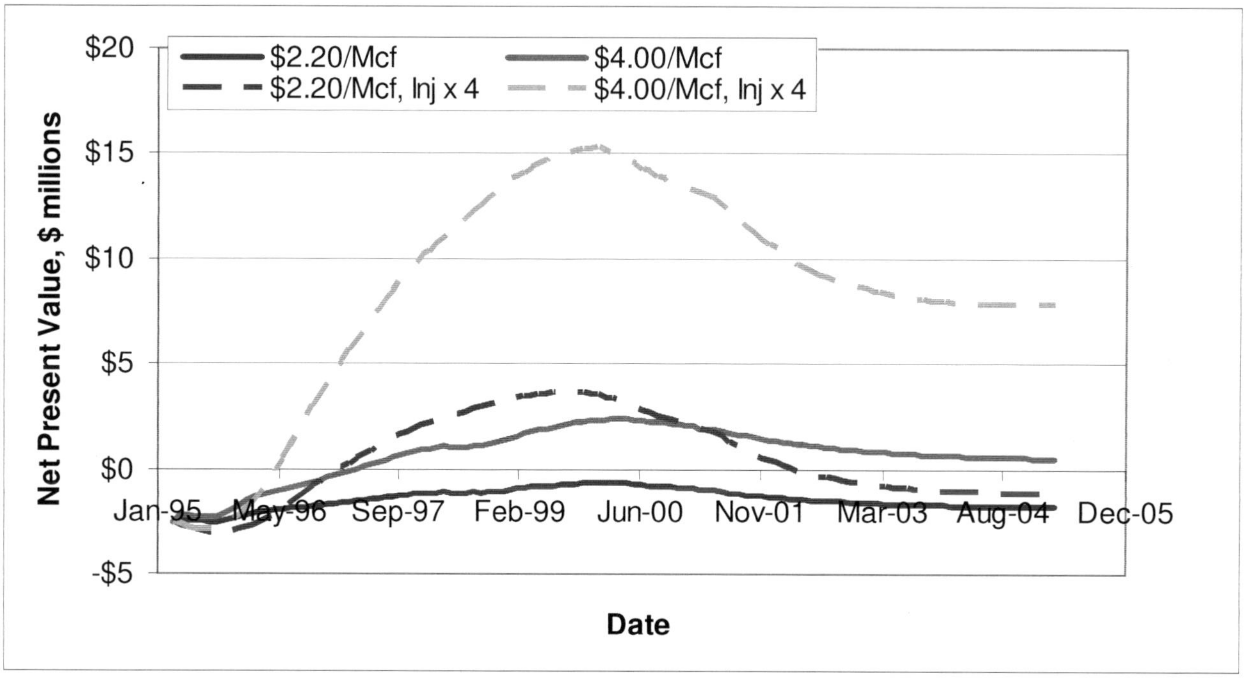

Figure 17: Economic Analysis Results, Case 2 vs. Case 1

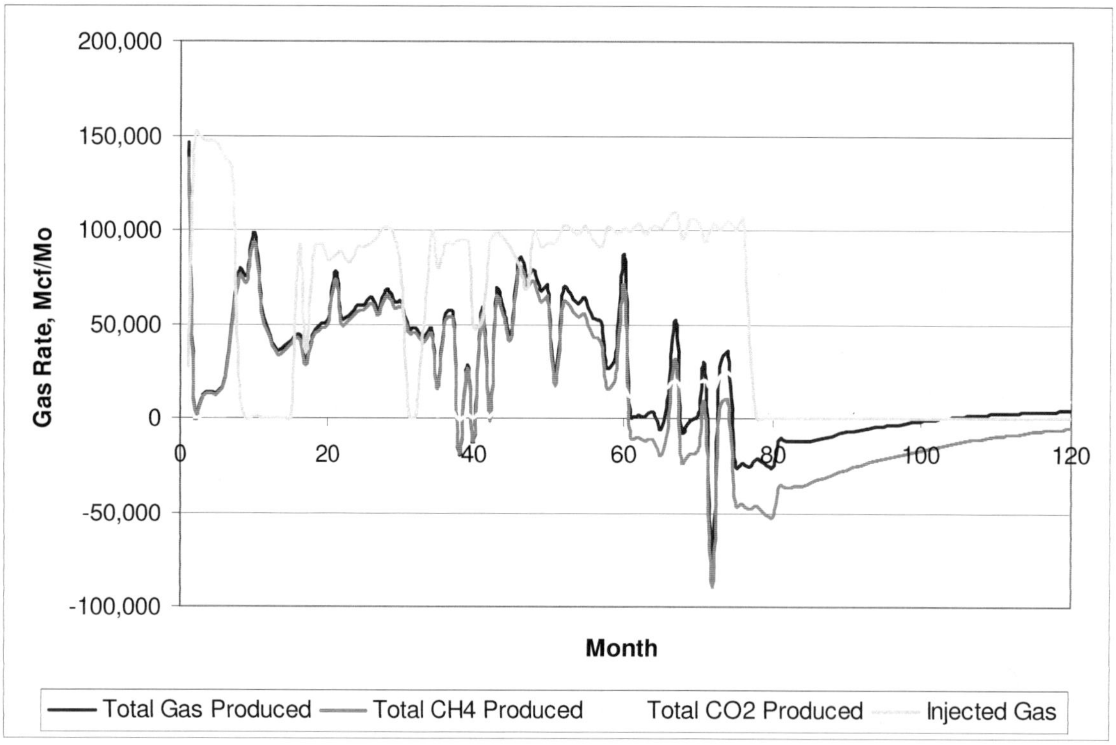

Figure 18: Incremental Gas Rates, Case 2 vs. Case 1

0523

The Tiffany Unit N_2-ECBM Pilot – A Reservoir and Economic Analysis

Scott Reeves, Advanced Resources International, Inc.
Anne Oudinot, Advanced Resources International, Inc.

ABSTRACT

In October, 2000, the U.S. Department of Energy, through contractor Advanced Resources International, launched a multi-year government-industry R&D collaboration called the Coal-Seq project. The Coal-Seq project investigated the feasibility of CO_2 sequestration in deep, unmineable coalseams by performing detailed reservoir studies of two enhanced coalbed methane recovery (ECBM) field projects in the San Juan basin. The two sites were the Allison Unit, operated by Burlington Resources, into which CO_2 was injected, and the Tiffany Unit, operating by BP America, into which N_2 was injected (the interest in understanding the N_2-ECBM process has important implications for CO_2 sequestration via flue-gas injection). The objectives of the field studies were to understand the reservoir mechanisms associated with CO_2 and N_2 injection into coalseams, demonstrate the effectiveness of the ECBM and sequestration processes, demonstrate an engineering capability to model them, and to evaluate ECBM/sequestration economics. In support of these efforts, laboratory and theoretical studies were also performed to understand multi-component isotherm behavior, and coal permeability changes due to swelling with CO_2 injection. This paper presents the results of the Tiffany Unit study, in which a detailed reservoir characterization of the field was developed, the field history was matched using the COMET3 reservoir simulator, future field performance was forecast under various operating conditions, and an economic analysis performed.

INTRODUCTION

In October, 2000, the U.S. Department of Energy (DOE), through contractor Advanced Resources International (ARI), launched a multi-year government-industry R&D collaboration called the Coal-Seq project[1]. The Coal-Seq project investigated the feasibility of CO_2 sequestration in deep, unmineable coalseams by performing detailed reservoir studies of two enhanced coalbed methane recovery (ECBM) field projects in the San Juan basin. The two sites were the Allison Unit, operated by Burlington Resources, into which CO_2 was injected, and the Tiffany Unit, operated by BP America (BP), into which N_2 was injected (the interest in understanding the N_2-ECBM process has important implications for CO_2 sequestration via flue-gas injection). The objectives of the field studies were to understand the reservoir mechanisms associated with CO_2 and N_2 injection into coalseams, demonstrate the effectiveness of the ECBM and sequestration processes, demonstrate an engineering capability to model them, and to evaluate ECBM/sequestration economics. In support of these efforts, laboratory and theoretical studies were also performed to understand multi-component isotherm behavior, and coal permeability changes due to swelling with CO_2 injection. This paper presents the results of the Tiffany Unit study, in which a detailed reservoir characterization of the field was developed, the field history was matched using the COMET3 reservoir simulator, future field performance was forecast under various operating conditions, and an economic analysis performed.

SITE DESCRIPTION

The Tiffany Unit ECBM pilot is located in La Plata County, northern Colorado, in close proximity to the border with New Mexico (Figure 1). While the Unit consists of many wells, the pilot area for N_2 injection consisted of 34 coalbed methane (CBM) producer wells and 12 N_2 injectors. The study area well pattern is illustrated in Figure 2. Note that the northwestern part of the study area was previously characterized and modeled by ARI as part of a Gas Research Institute effort to understand reservoir behavior in San Juan Basin coals[2].

The producing history for the study area is shown in Figure 3. The field originally began production in 1983, with N_2 injection beginning in January, 1998. Production just prior to nitrogen injection was about 5 MMcfd, or about 150 Mcfd per well. Injection was suspended in January 2002, after four years of intermittent N_2 injection, to evaluate the results. Several points are worth noting regarding the producing history:

- Nitrogen injection only occurred during the winter months, and was suspended during the summer months. The reason was that the nitrogen was sourced from a cryogenic air separation plant located at the Florida River gas processing facility, and the unit ran less efficiently at temperatures above 65 degrees Fahrenheit. Therefore nitrogen injection was only performed during the cooler winter months.

- The methane production response to N_2 injection was rapid and dramatic. During the initial injection period, total methane rate jumped from about 5 MMcfd to about 27 MMcfd, over a factor of 5. Production responses to subsequent shut-down and injection periods were also pronounced.

RESERVOIR DESCRIPTION

The Tiffany Unit wells produce from four Upper Cretaceous Fruitland Formation coal seams, named the B, C, D and E (from shallowest to deepest) using BP's terminology. A summary of basic coal depth, distribution, thickness, pressure and temperature information is provided in Table 1.

Sorption isotherms for CH_4, N_2 and CO_2 were measured for coal samples taken earlier (and preserved) from injection wells #1 and #10 (in the northwest and southeast portions of the field respectively). After careful quality control checking, the samples were mixed and single-component, binary and ternary isotherms measured[3]. The results for the pure component isotherms are shown in Figure 4, on an as-received basis.

In the previous reservoir study of the area[2], a permeability anisotropy of about 2.4 was determined to exist, with the maximum permeability in the northwest-southeast orientation. This coincides with the measured face-cleat orientation. The geometric average permeability from that study was also determined to be 1.6 md, and the average porosity 0.8%. These properties were initially retained for this study.

RESERVOIR MODEL CONSTRUCTION

The reservoir simulator used for the study was ARI's COMET3 (ternary isotherm – CH_4, N_2 and CO_2) model. Details on the model theory are provided in the references[4,5].

A four-layer (B, C, D, E), full-field model was constructed[6]. The coal structure and thickness information for each layer was directly input per the maps generated. Since information from BP and other sources suggested that the cleat orientations were approximately in the northwest-southeast (face) and northeast-southwest (butt), the model grid was so aligned. Permeability, relative permeability and porosity values were used per the prior study[2]. Finally, the isotherm properties as measured in the laboratory were used.

Additionally, well completion and operating parameters were examined for input into the model. Since the production wells had been restimulated in the mid-1990's, skin factors for these wells was set at –2. Since the N_2 injection wells were not stimulated, those skin factors were set at a value of 0.

The model gridblock dimensions were 73 x 37 x 4 (approximately 10,800 total gridblocks, about 7,800 of which were active), and covered an active area of about 16,400 acres (Figure 5). On average, the gridblock dimensions were 690 ft × 525 feet × 12 feet. The corners of the model were isolated using no-flow barriers to account for producing wells immediately adjacent to these portions of the study area.

HISTORY MATCH RESULTS

The independent parameter used to drive the simulator was gas production (and injection) rate to maintain material balance, and the dependent (history match) parameters were water production rate, flowing pressure (producing and injecting), and gas composition. Note that only some of these data were available for some periods for some wells; whatever was available was used.

The primary history match variables were permeability and porosity. These were modified globally to obtain the best overall match for the field. The objective of the study was to understand the mechanisms of the N_2-ECBM process by matching general trends, and not necessarily to make regional changes to the reservoir characterization to achieve matches on an individual well basis. Ultimately, it was found that a geometric average permeability of 13.4 md (retaining the permeability anisotropy of 2.4) and a porosity of 0.2% provided the best overall match.

A comparison of the actual versus simulated field gas rate is presented in Figure 6. The only conclusion that can be derived from this result, since the model was "driven" on gas rate, is that model (as constructed) was capable of delivering the gas volumes required.

Comparison plots of gas and water rates, flowing pressures, and produced gas compositions, for one of the production wells are presented in Figure 7. This was a typical result for many of the wells. Several general comments can be made regarding the results:

- The predicted water production rates were generally close to the actual rates, particularly in later times. BP noted that earlier water production data was suspect, whereas the latter data was more reliable.

- The predicted producing pressures were consistently and significantly higher than the actual values. This phenomenon was also observed in a separate, independent study of the field[7]. Changes in coal permeability and/or wellbore skin factors were unable to materially reduce this discrepancy. The cause for the discrepancy remains unclear.

- The predicted to actual comparisons of produced gas composition matches were of variable quality. In some cases, the predicted onset of gas breakthrough was earlier or later than actual, and increased either too quickly or slowly than actually observed in the field. In other cases however (such as that shown), they were quite good.

A plot of actual to predicted bottomhole injection pressures for N_2 injector well #2 is provided in Figure 8. The predicted injection pressures are in reasonable agreement with the actual values, suggesting the permeability and skin estimates for the injector wells were within reason.

PERFORMANCE FORECASTS

In order to evaluate the long-term performance of the ECBM pilot, under status quo conditions (i.e., no further N_2 injection) as well as under other "what if" future injection scenarios, performance prediction cases were simulated using the history match result as the starting point. The specific cases evaluated included:

1. No N_2 injection (i.e., primary production only).
2. Current conditions (i.e., intermittent N_2 injection until January 2002, and not resuming).
3. Intermittent future N_2 injection.
4. Continuous future N_2 injection.

For each forecast case, an economic limit of 50 Mcfd of methane and 50% N_2 content per well was imposed; reaching those thresholds prompted the well in question to be shut-in in the model. A summary of the results for each case are presented in Table 2. Since the total model area was so large compared to the actual flooded areas, the incremental recovery results were examined on two individual patterns, A & B (Figure 9). Incremental pattern recoveries for the actual pilot (Case 2) were in the 10 - 20% range of original gas-in-place (OGIP). Long-term N_2 injection would have added another 25 – 40% of OGIP to the total recovery, with continuous future injection providing more recovery than intermittent future injection. Thus, the N_2-ECBM was quite effective in enhancing methane recovery at the Tiffany Unit.

ECONOMIC ASSESSMENT

The final element of the study was to evaluate the economic performance of both the actual pilot, as well as the future injection scenarios. The capital, operating and financial assumptions utilized are provided in Table 3. Note that all economics were performed on an incremental basis (i.e., only the incremental production and costs were considered). Further, the effect of Section 29 tax credits was not considered.

Case 2 versus Case 1

This analysis evaluated the performance of the existing pilot, with no future N_2 injection considered. Note that the capital costs for a cryogenic air separation plant are included. It should also be noted that gas processing costs ($0.50/Mcf) have been included to account for costly separation of N_2 from the produced methane. In actuality, due to the small volume of N_2 relative to the total amount of natural gas processed at BP's the Florida River facility, the high N_2 gas was merely blended into the total facility product stream and no costs were actually incurred for separation. We have accounted for these costs in this analysis however to reflect what would be a more common economic reality.

The results are presented in Figure 10. The ultimate net present value (NPV) assuming $2.20/Mcf (at the time) was ($2.9 million). The breakeven gas price was $2.42/Mcf and the breakeven N_2 cost was $0.15/Mcf. This indicates the pilot was uneconomic under the assumed conditions (not accounting for Section 29 tax credits). Having said that however, an alternative scenario is presented that is more representative of today's environment: a more realistic gas price of $4.00/Mcf. Under this assumption, the pilot would have yielded an NPV of over $20 million. Thus N_2-ECBM appears commercially viable today, at least for the conditions that exist at Tiffany.

Cases 3 and 4 versus Case 2

For these analyses, no capital costs were included; they were considered sunk. Further, a gas price of $4.00/Mcf was used to reflect current economic conditions. The results of these analyses are presented in Table 4. Both cases are highly attractive economically, providing NPV's of $30-40 million.

CONCLUSIONS

Based on the results from this study, the following conclusions have been drawn:

- The injection of N_2 at the Tiffany Unit has resulted in incremental methane recovery over estimated primary recovery. In the swept areas, an incremental methane recovery of approximately 10 - 20% of original-gas-in-place resulted from N_2-ECBM operations. Future N_2 injection was forecast to add another to 25 – 40% of original-gas-in-place to the total recovery.

- At the prevailing gas prices at the time the project was implemented (~$2.20/Mcf), and not considering any tax credit benefits, the pilot itself was uneconomic. However, with today's gas prices of ~$4.00/Mcf, N_2-ECBM appears economically attractive. The breakeven gas price for the conditions at Tiffany was estimated to be ~ $2.40/Mcf.

- Future N_2 injection at Tiffany, assuming a gas price of $4.00/Mcf was also forecast to be economic.

ACKNOWLEDGEMENTS

This work was funded by the U.S. Department of Energy, National Energy Technology Laboratory, contract number DE-FC26-00NT40924. The DOE Project Manager was Charlie Byrer. The authors also wish to acknowledge the valuable contribution of BP America, which provided all the data for the study, invaluable insights, and permission for the results to be published.

REFERENCES

1) Reeves, S. R.: "Geologic Sequestration of CO_2 in Deep, Unmineable Coalbeds: An Integrated Research and Commercial-Scale Field Demonstration Project", SPE 71749, presented at the SPE Annual Technical Conference and Exhibition, New Orleans, September 30-October 3, 2001.

2) Paul, G.W. and Young, G.B.C.: "Characterization of Fruitland Coal through Reservoir Simulation," GRI Topical Report GRI-92/0461, November, 1993.

3) Gasem, K.A.M., Robinson, R.L., and Reeves, S.R.: "Adsorption of Pure Methane, Nitrogen, and Carbon Dioxide and Their Mixtures on San Juan Basin Coal", Topical Report, DOE Contract No. DE-FC26-00NT40924, May, 2002.

4) Reeves, S. and Pekot, L.: "Advanced Reservoir Modeling in Desorption-Controlled Reservoirs," SPE 71090, presented at the SPE Rocky Mountain Petroleum Technology Conference, Keystone, May 21-23, 2001.

5) Sawyer, W.K., Paul, G.W., Schraufnagel, R.A.: "Development and Application of a 3D Coalbed Simulator," CIM/SPE 90-119, presented at the CIM/SPE International Technical Conference, Calgary, June 10-13, 1990.

6) Reeves, S.R., Oudinot, A.Y. and Erickson, D.: "The Tiffany Unit N_2 – ECBM Pilot: A Reservoir Modeling Study", Topical Report, DOE Contract No. DE-FC26-00NT40924, May, 2004.

7) Liang, J., Raterman, K. T., and Robertson, E. P.: "A Mechanistic Model for CO_2 Sequestration in Tiffany Coal Bed Methane Field", 2003 International Coalbed Methane Symposium, Tuscaloosa, AL, May 7-8, 2003.

Table 1: Basic Coal Reservoir Data, Tiffany Unit

Property	Value
Average Depth to Top Coal (B)	3040 feet
No. Coal Intervals	7 total (A, A2, B, C, D, E, F) 4 main (B, C, D, E)
Average Total Net Thickness	47 feet B – 13 ft C – 11 ft D – 9 ft E – 14 ft
Coal Rank	Medium Volatile Bituminous
Initial Pressure	1600 psi
Temperature	120°F

Table 2: Summary of Model Forecast Results

Description	Case 1 No Injection	Case 2 Actual N_2 Injection	Case 3 Future Intermittent N_2 Injection	Case 4 Future Continuous N_2 Injection
Incremental CH_4 (Bcf) *	n/a	22.3	26.8	36.1
Total N_2 Injected (Bcf)	n/a	15.0	51.4	86.0
Total N_2 Produced (Bcf)	< 0.1	6.1	25.0	40.0
Net N_2/CH_4 Ratio	n/a	0.4	0.4	0.9
Incremental Pattern A Recovery (%OGIP)*	n/a	21.1%	27.2%	37.5%
Incremental Pattern B Recovery (% OGIP)*	n/a	9.4%	25.6%	36.2%

* Incremental recovery for Case 2 is relative to Case 1.
Incremental recoveries for Cases 3 & 4 are relative to Case 2.

Table 3: Economic Analysis Assumptions

Capex	**Value**	**Assumptions**
Cryogenic Air Separation Plant (includes compression)	\$ 7.5 million	\$250,000/MMcfd of capacity, 30 MMcfd Capacity
Pipeline	\$ 4.6 million	\$24,000/in-mi, 16 mi, 12-inch line
Field Distribution:	\$ 0.7 million	\$20,000/in-mi, avg 0.5 mi/well, 6 in lines, 12 wells
Wells	\$ 5.0 million	\$500,000/ea, fully equipped
Total	***\$ 17.8 million***	
Opex		
Injector Well Operating:	\$500/mo	Only when active
N_2 Cost	\$0.10/Mcf	
Produced Gas Processing	\$0.50/Mcf	
Financial		
Gas Price(Case 2 vs Case 1):	\$2.20/Mcf	Ex-Field
Gas Price (Cases 3 & 4 vs. Case 2)	\$4.00/Mcf	Ex-Field
Net Revenue Interest:	87.5%	
Production Taxes:	8%	
Discount Rate:	12%	

Table 4: Summary of Economic Results, Cases 3 and 4 (Incremental vs. Case 2)

	Case 3	**Case 4**
Assumed Gas Price (\$/Mcf)	\$4.00	\$4.00
Net Present Value (\$ millions)	\$32.0	\$42.2
Breakeven Gas Price (\$/Mcf)	\$1.49	\$1.59
Breakeven Injectant Cost (\$/Mcf)	\$2.12	\$1.89

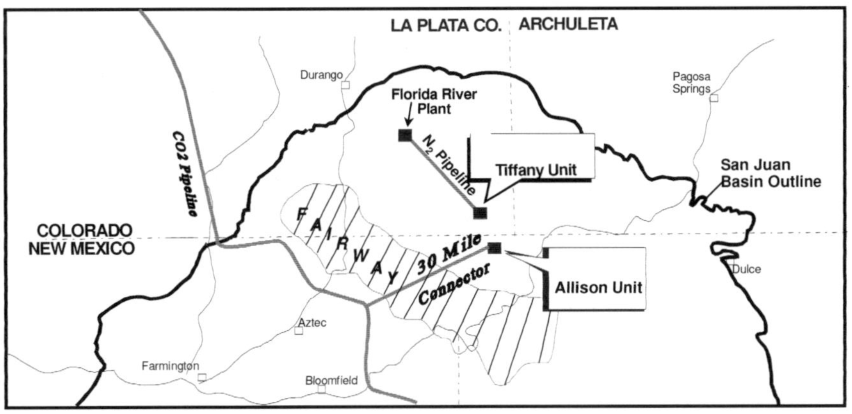

Figure 1: Location of the Tiffany Unit, San Juan Basin

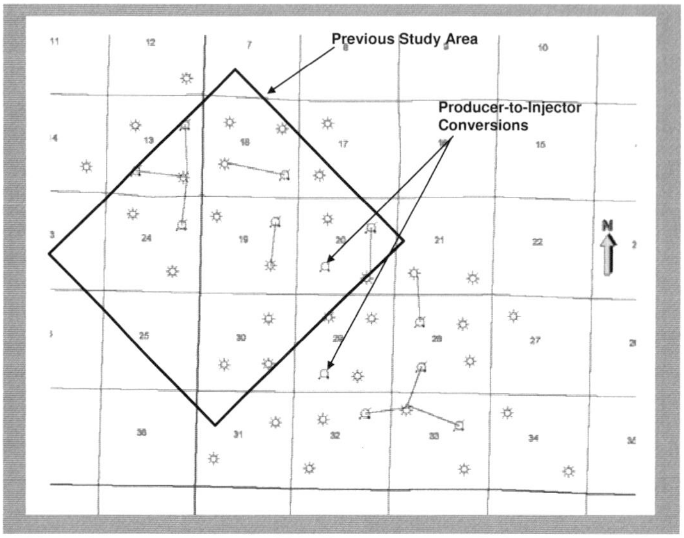

Figure 2: Producer/Injector Well Pattern, Tiffany Unit Study Area

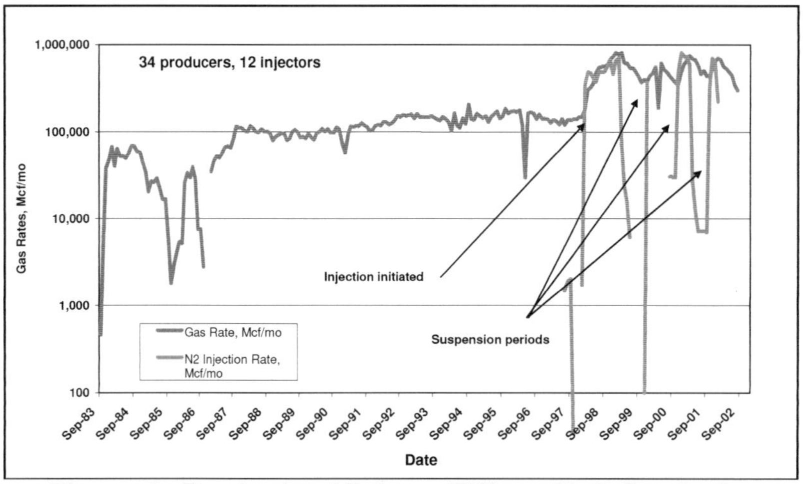

Figure 3: Producing History, Tiffany Unit Study Area

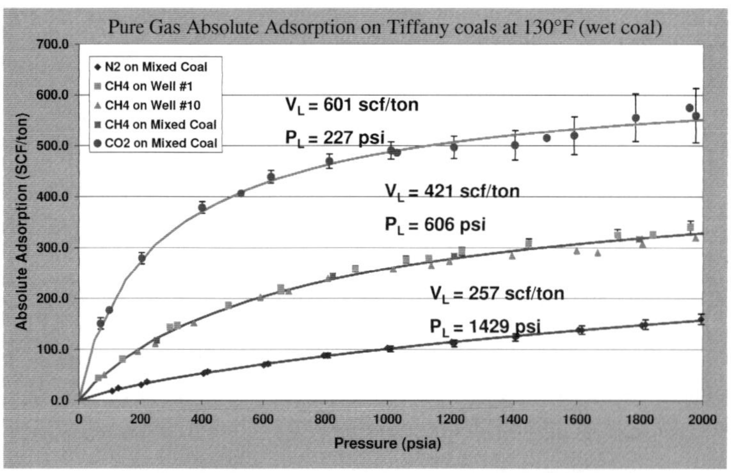

Figure 4: Carbon Dioxide, Methane and Nitrogen Isotherms for Tiffany Coal

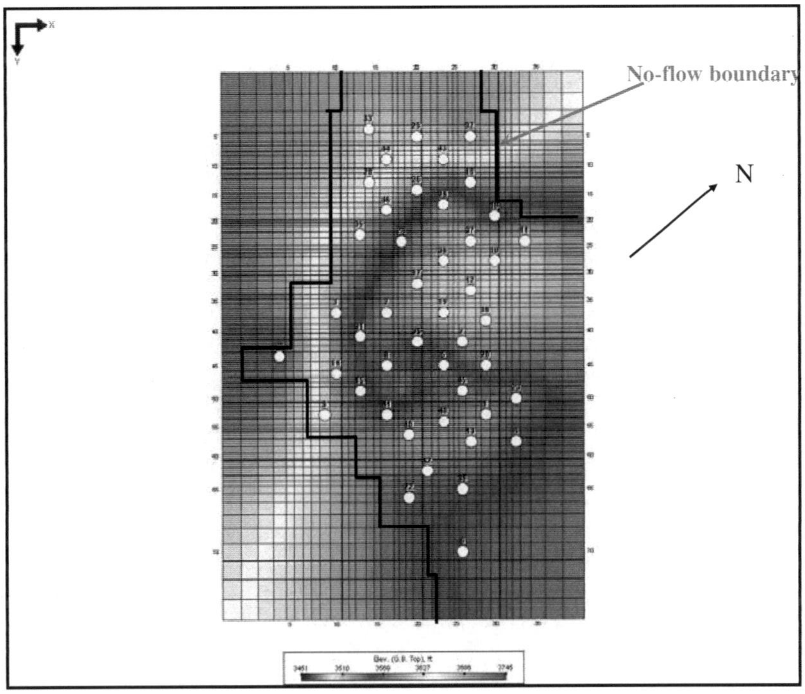

Figure 5: Map View of the Top Layer of the Simulation Model

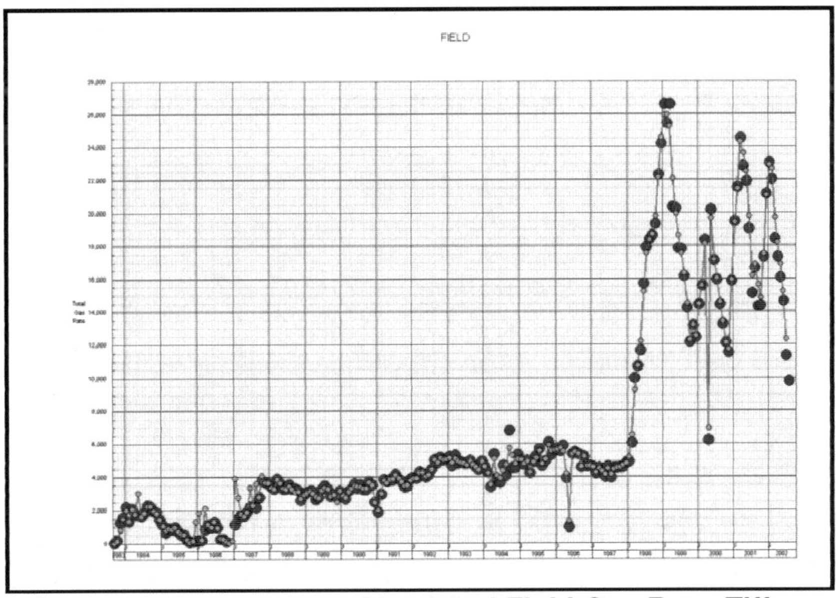

Figure 6: Actual versus Simulated Field Gas Rate, Tiffany

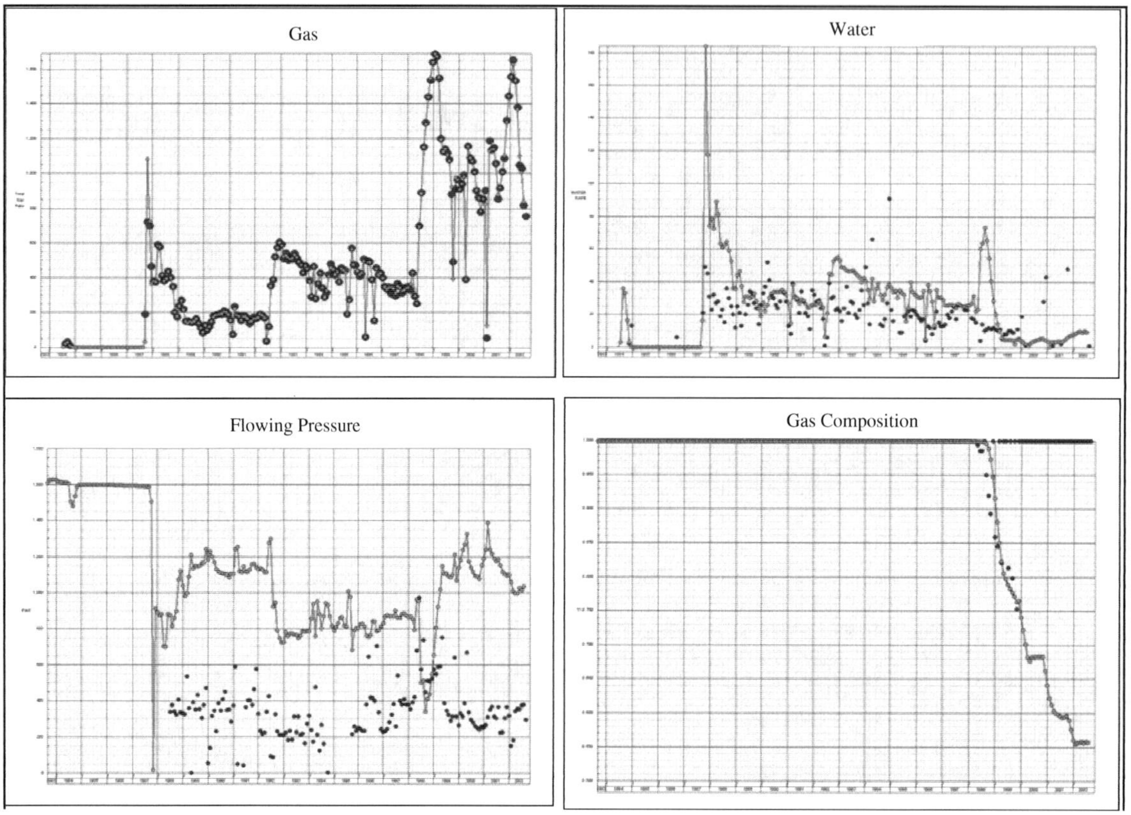

Figure 7: Comparison of Predicted to Actual Well Performance, Baird Gas Unit 18-01 No. 2

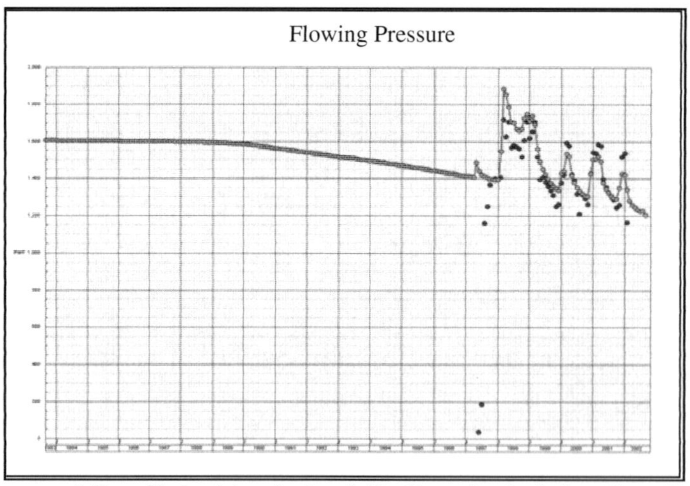

Figure 8: Comparison of Predicted to Actual Bottomhole Injection Pressures, Injection Well #2

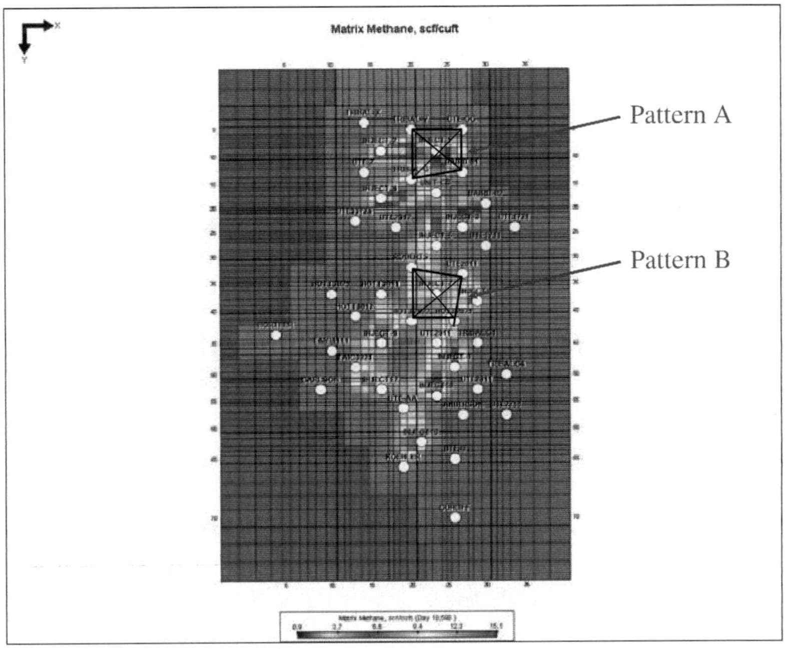

Figure 9: Locations of Patterns A & B, Tiffany Unit Study Area

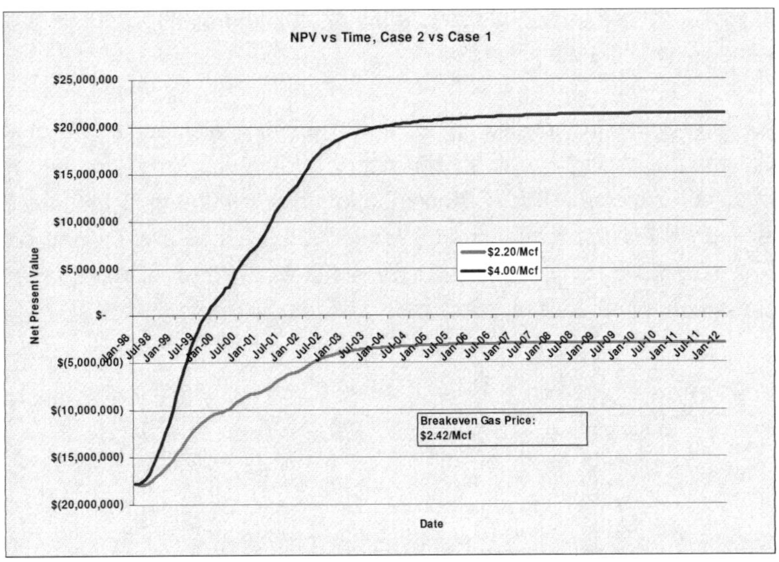

Figure 10: Economic Analysis Results, Case 2 versus Case 1

0525

CO_2 Leakage Risk Evaluation Using Enhanced Coalbed Methane Recovery Simulator

S. Shimada and Y. Funahashi, Institute of Environmental Studies,
Graduate School of Frontier Sciences, The University of Tokyo

INTRODUCTION

Coalbed methane (CBM) is explosive hazardous gas and dangerous for mining activity. On the other hand, CBM is considered to be unconventional natural gas when it is recovered with a certain concentration. Moreover, because of increased concern about anthropogenic CO_2 emissions, the geological sequestration of CO_2 is being examined as one means of reducing these emissions.

Enhanced coalbed methane recovery (ECBMR) is drawing attention as a production technology of CBM. ECBMR is the technology, which promotes the production of CH_4 by injecting CO_2 or N_2 gas. When CO_2 is injected into coal seam, CH_4 is replaced by CO_2 and CO_2 remains in coal seam (CO_2 sequestration). Therefore, ECBMR is considered to be one of the GHG emission reduction technologies. As an option to solve natural gas supply shortage and global warming simultaneously, ECBMR is a very promising method. Figure 1 shows the ECBMR scheme.

Coal as gas reservoirs are extremely complex, and the physical mechanism of fluid storage and transport in coal seams are strongly coupled and highly non-linear. Simulation is an essential tool for dealing with these complex mechanisms. CBM simulators are now being further developed and tested for the important environmental application of CO_2 sequestration in coal seams, and the enhancement of CH_4 production by CO_2 or flue gas injection. Reservoir simulation provides quantitative results for design, prediction and assessment of processes, production and leakage potential of CO_2.

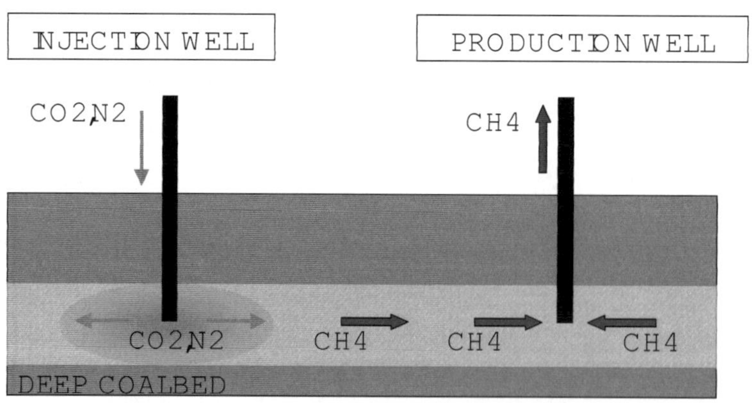

Figure 1. Coalbed methane recovery scheme

GAS STORAGE, TRANSPORT AND ENHANCED RECOVERY MECHANISMS AND MODELS

Coal reservoirs are systems of storage and transport mechanism that can be characterized using mathematical models [1]. In this section, gas storage, transport and enhanced CBM recovery mechanism and simplified equations are described.

Structure of coal

Coal contains a very fine micro-pore structure; pore diameters typically range from 5-10 . This micro-pore structure has a very low flow capacity (permeability in the microdarcy range), where as coal cleats have a much greater flow capacity (permeability in the millidarcy range). Thus, it is useful to conceptualize coal as a dual porosity system shown schematically in Figure 2. Water and gas flow to coalbed methane wells through the cleat and fracture systems. The cleats and fractures comprise the bulk permeability measured from well tests of coalbed methane wells.

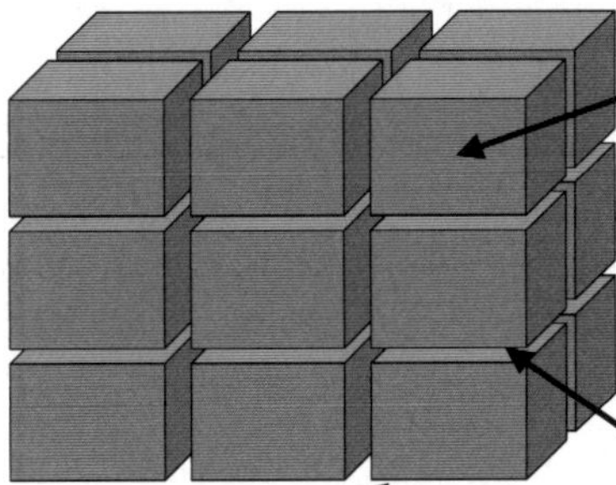

Figure 2. Dual porosity system

Gas storage mechanisms

Gas can exist in coal seam in two ways. It can be present as free gas within the natural porosity of the coal, and it can be present as an adsorbed layer on the internal surface of the coal. The fine micropore structure of coal has a very high storage capacity for methane. Methane molecules adsorb to the micropore walls. For most coal seams, the quantity of gas held in the coal is primarily a function of coal rank, ash content, moisture content, and the pressure of the coal seam. The adsorbed capacity of a coal seam varies non-linearly as a function of pressure. This relationship is best described using a Langmuir isotherm, which is generally expressed as,

$$C_m = V_L \frac{P}{P + P_L} \quad (1)$$

where C_m is the volume of gas at standard temperature and pressure per ton of coal, V_L is the Langmuir volume constant, P is the pressure in the coal cleat system and P_L is the Langmuir pressure constant. The physical model of adsorption described by the Langmuir isotherem has been extended to describe multi-component gas adsorption. This accounts for the effects of the presence of one adsorbable component on the other adsorbable components in the system, each component having its own characteristic affinity for the absorbent. The order of relative sorption strength is $CO_2 > CH_4 > N_2$.

Transport mechanisms

Most of the gas present in coal seam is physically adsorbed on the internal surface of the unfractured coal matrix. Gas production from coals occurs by a three-stage process in which gas: 1) diffuses through the coal matrix to the cleats, 2) desorbs from the cleat surfaces (natural fractures) 3) flows from natural fractures. This process is shown in Figure 3.

Diffusion process and desorption process in the micro pore system is described using Fick's law and sorption time, which are related to cleat spacing and diffusion coefficient. This relation is expressed as,

$$q_{gm} = \frac{V_m}{\tau}\left(C_m - C(P)\right) \quad (2)$$

where q_{gm} is gas production rate from the coal matrix, V_m is matrix volume, C_m is matrix gas concentration, C(p) is equilibrium concentration at matrix-cleat boundary and τ is sorption time.

Flow in the cleat system of coals can be described by Darcy's Law. Darcy's law considering the effective permeability of individual flowing phases is expressed as,

$$q_g = \frac{kk_{rg}}{\mu B_g}\nabla P_g \quad (3)$$

$$q_w = \frac{kk_{rw}}{\mu B_w}\nabla P_w \quad (4)$$

where q_g and q_w are flow rate of in cleat system , k is absolute permeability, k_{rg} and k_{rw} are relative permeability, $\mu$is dynamic viscosity, B_g and B_w are formation factor, P is pressure. Subscription g and w means gas and water, respectively.

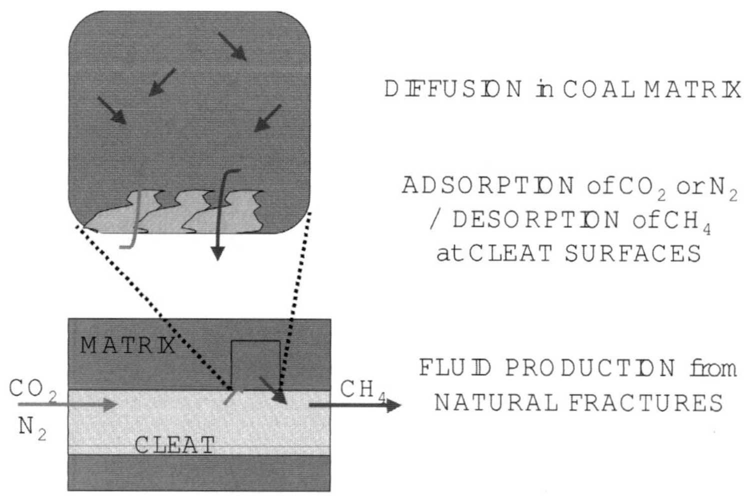

Figure 3. Gas transport scheme

Enhanced coalbed methane recovery (ECBMR)

The principle of ECBMR is that the amount of methane stored in coal is dependent on the pressure of methane rather than the total pressure. Because the partial pressure of methane is equal to the total pressure multiplied by the mole fraction of methane, N_2 can be injected to decrease the partial pressure of methane without reducing the total pressure. Moreover, High reservoir pressure also minimizes the impact of stress dependent permeability effects.

Lowering the partial pressure of methane by N_2 injection instead of pressure depletion should increase both the methane production rate and the recovery factor. By maintaining a higher reservoir pressure, greater driving force throughout the life of the reservoir. CO_2 injection is an alternative to N_2 injection and provides better methane recovery than N_2 injection. Not only does CO_2 reduce the partial pressure of methane but it also adsorbs in the coal and reduces the capacity of the coal to contain methane.

Though enhanced coalbed methane recovery is in its infancy, inert gas injection may prove to be very important to future coalbed methane production and CO_2 sequestration.

Features of Simulator

ECBMR simulator in this study (ECOMERS-UT) is a computational fluid dynamic (CFD) simulator. This simulator works by dividing the reservoir into an array of volume elements (1-3dimensional blocks). Each block is assigned reservoir properties such as porosity, permeability, phase saturation, pressure, gas molecular fraction and so on. The aggregate of volume elements represents the geologic model. The simulator also incorporates a well model, which uses fluid flow equation to represent extraction of fluids from the reservoir. As fluid is extracted from the well, the model uses equation of conservation of mass and the reservoir properties assigned to the volume elements to calculate the movement of fluids throughout the model.

ECOMERS-UT has following features:

- Dual porosity – matrix and fracture
- Two phase – gas and water
- Diffusion flow from matrix to fracture
- Darcy flow in fractures
- Stress sensitive cleat permeability
- Component dependent adsorption isotherm
- 1-D, 2-D, 3-D capability
- Numerical solution by implicit method
- Automatic time-step selector

SIMULATOR COMPARISON

This section describes the comparison of ECOMERS-UT with other CBM simulators under some conditions to confirm ECOMERS-UT to work correctly.

Problem:

1. Primary recovery (no injection)
2. Enhanced recovery using CO_2
3. Enhanced recovery using CO_2 and N_2 mixture
4. Enhanced recovery using CO_2 under long sorption time

Law et al. carried out simulator comparisons for ECBMR [2-4]. The problem sets dealt with CO_2 or N_2 injection/CBM production in an inverted five-spot pattern, and compared the basic features of numerical simulators for CBM. Reservoir and well properties in this section are the same as the problem set used by Law et al. Well position is 1/4 of 2.5-acre 5-spot pattern and basic reservoir parameters and operating conditions are shown in Table 1 and Table 2, respectively.

Table 1 Reservoir properties of three-dimensional quarter five-spot domain

Property	Value	Units
Quarter five spot size	50.294×50.294	m
Thickness	9	m
Porosity	0.001	-
Permeability (isotropic)	3.65	md
Initial water saturation	0.592	-
Temperature	45	
Initial pressure	7.65	MPa
Density	1.434	kg/m^3
Initial CH_4 fraction	1.0	-
Langmuir volume of CH_4	0.0152	m^3/kg
Langmuir volume of CO_2	0.0310	m^3/kg
Langmuir volume of N_2	0.0150	m^3/kg
Langmuir pressure of CH_4	4.6885	MPa
Langmuir pressure of CO_2	1.903	MPa
Langmuir pressure of N_2	27.241	MPa

Table 2 Operating conditions

Property	Value	Units
Wellbore radius	0.0365	m
Skin factor	0	-
Gas injection rate	28316.82	sm^3/d
	1×10^6	scf/d
Maximum bottom-hole pressure	15	MPa
Maximum gas production rate	100000	sm^3/d
	3.5315×10^6	scf/d
Bottom-hole pressure	0.275	MPa

Enhanced CBM recovery using CO_2

Figure 4 shows the methane production rate and mole fraction with injection of CO_2. Compared with primary recovery, the methane production rate with injection of CO_2 was larger in earlier period. Primary recovery is the no injecting production, in which natural reservoir energy, such as pressure and gravity gradient, displaces methane from the reservoir, into the wellbore and up to surface. The methane fraction of produced gas in primary recovery is 1.0. CO_2 mole fraction of produced gas with injection of CO_2 increased gradually. The mole fraction of methane is critical factor of the value of produced gas. The point that mole fraction of methane falls short of the lower limit is called "break through". Production rate and mole fraction pattern corresponded to the past study.

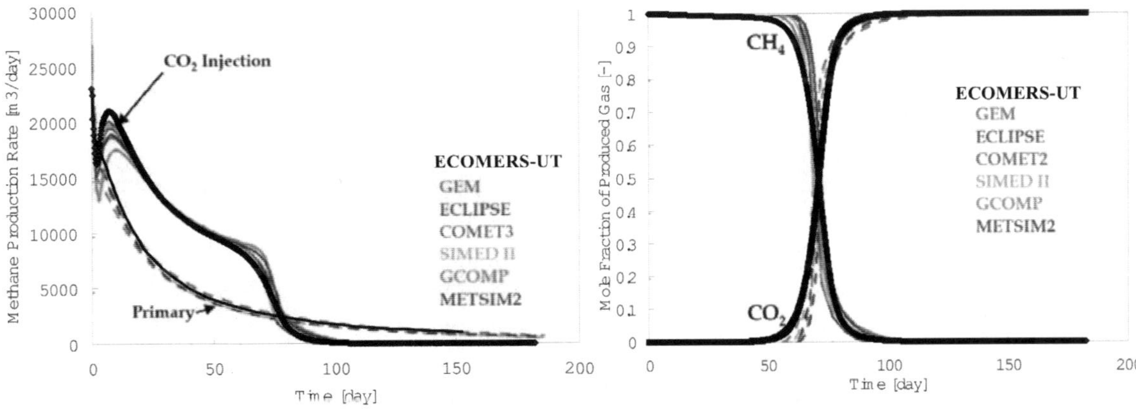

Figure 4. Simulator comparison (CO_2 injection)

Enhanced CBM recovery using mixture of CO_2 and N_2

N_2 is less adsorptive than CO_2 and CH_4. Thus N_2 injection leads to less methane mole fraction of pore. Figure 5 shows the effect on methane production rate and the mole fraction of produced gas with injecting CO_2 (50%) and N_2 (50%). Compared with primary recovery, the methane production rate was about twice larger in early period. Compared with pure CO_2 injection, the methane production rate in early period was larger. This phenomenon indicates that N_2 is better than CO_2 in a sense of enhancement. But N_2 breakthroughs earlier than CO_2 and N_2 injection lower the value of produced gas or the recovery factor.

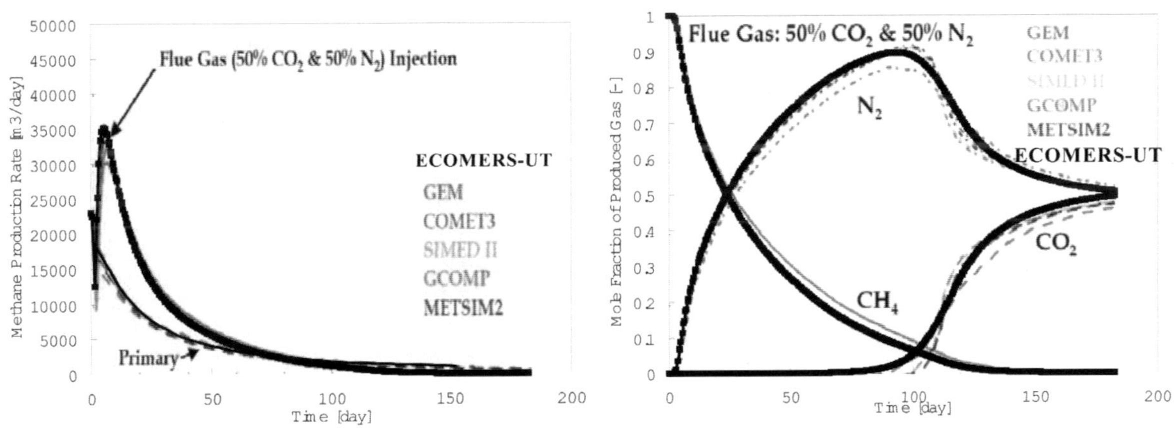

Figure 5. Simulator comparison (CO_2-50% and N_2-50% injection)

SENSITIVITY ANALYSIS

This section shows sensitivity analysis under some condition. Parameters changed are component of injection gas, absolute permeability of natural fracture and anisotropy of permeability. Generally, CBM reservoir properties like initial gas in place, coal thickness and permeability have great variety. Sometimes properties have great variety even in one coal seam. These varieties influence the productivity of methane and the efficiency of CO_2 sequestration. Methane recovery factor before breakthrough is adopted to assess the productivity. In this section, pattern half width was set to 400m and sorption time was set to 38.6days. Reservoir and operating conditions not mentioned are same as before.

Injection gas component

The above calculation showed that the difference of injection gas component has great influence on methane production rate and recovery factor. In this section, detailed study of gas component is discussed. Injection gas has three options;

1) High purity CO_2 separated from flue gas
2) Flue gas without separation
3) Intentional mixture of CO_2 and other gas.

CO_2 concentration in flue gas is 10-15% from thermal power plants, 25-30% from iron works or cement plants. Gas purification needs some cost but flue gas is easy to prepare. Concentration of CO_2 of flue gas exhausted from an oxygen-blown Integrated Gasification Combined Cycle (IGCC) plant may be more than 90%. Also intentional mixture of CO_2 and other gas (N_2 etc.) may be effective to accelerate methane recovery. Injection gas component was set to ([CO_2], [N_2])= (1, 0), (0.9, 0.1), (0.3, 0.7), (0.15, 0.85).

Figure 6 shows the cumulative CH_4 production comparison of various injection components. When N_2 is added to injection gas, the methane production rate is fostered and breakthrough is accelerated. The larger N_2 injection is, the faster production of methane is. But the later period and the larger CO_2 injection patterns exceed the larger N_2 injection patterns in cumulative CH_4 production. This phenomenon indicates that N_2 injection leaves more methane in coalbed than CO_2 injection. Figure 7 shows CH_4 fraction of produced gas comparison of various injection components. N_2 injection leads to early breakthrough and lower recovery factor.

Noted above, CO_2 and N_2 mixture injection has both good points and bad points. This section showed that the mole components of injection gas strongly affect the productive behavior. The best component depends on reservoir characteristics, required quality of produced gas and the economics of the project.

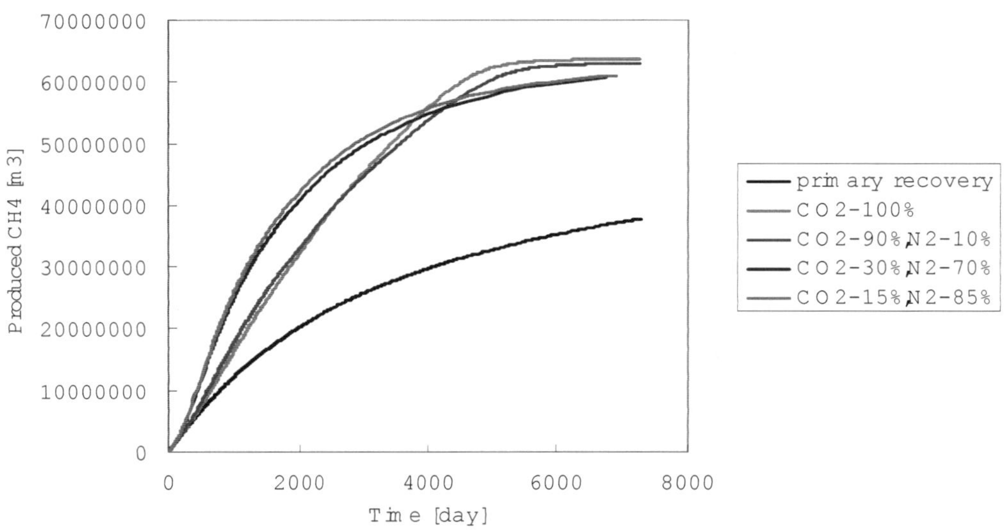

Figure 6. Cumulative CH_4 production comparison of various injection components

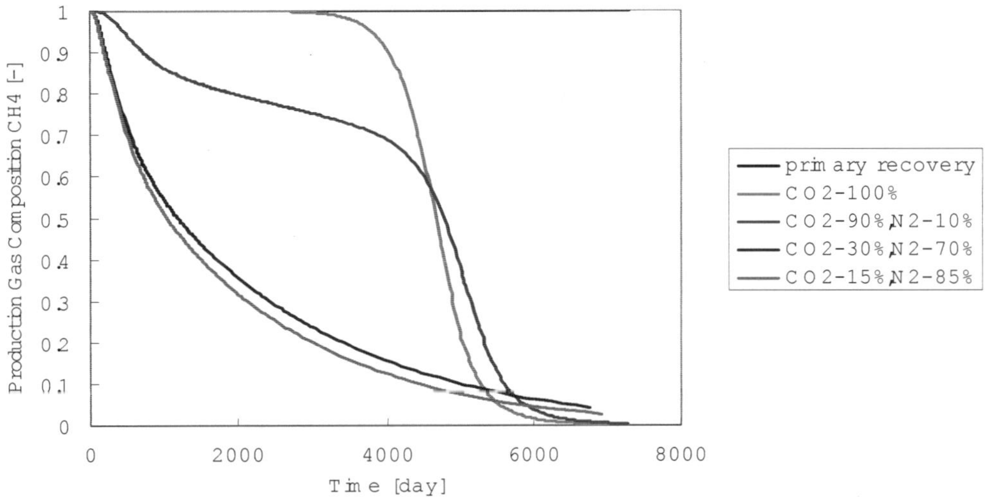

Figure 7. CH_4 fraction of produced gas comparison of various injection components

Absolute Permeability dependence

Absolute permeability is one of the most important parameters to production. As a preliminary study of concerning anisotropy in permeability, absolute permeability was set to 1md, 3.65md and 10md supposing isotropy.

Figure 8 shows the absolute permeability dependence of methane production rate. When permeability was set to 10md, the production rate of initial term was about twice larger than that set to 3.65md. High permeability leads to prompt pressure drop near production well, and fast pressure drop foster methane diffusion from matrix to fracture. When 1md, the production rate was one-forth that set to 3.65md. Low permeability leads to low flow rate and pressure gradient. Also, production rate pattern of 1md is not similar to others. This phenomenon is related to the maximum bottom-hole pressure of injection well. In this study, gas injection rate is set to be constant. Under low permeability, injection bottom-hole pressure increases to offset the low injection rate. However, the maximum bottom-hole pressure is set mechanically, in this case, the maximum bottom-hole pressure restricts the gas injection rate, or the enhancement effect is weaker than other cases.

Table 3 shows cumulative CH_4 production comparison of various absolute permeability cases at the time of breakthrough. In this study, the time of breakthrough is defined as the time when the mole fraction of produced gas lowers 0.9. The higher absolute permeability is the better methane recovery factor at the time of breakthrough and the difference between primary recovery and ECBMR is larger in lower absolute permeability case. This result indicates that the lower absolute permeability is the better enhancement effect is.

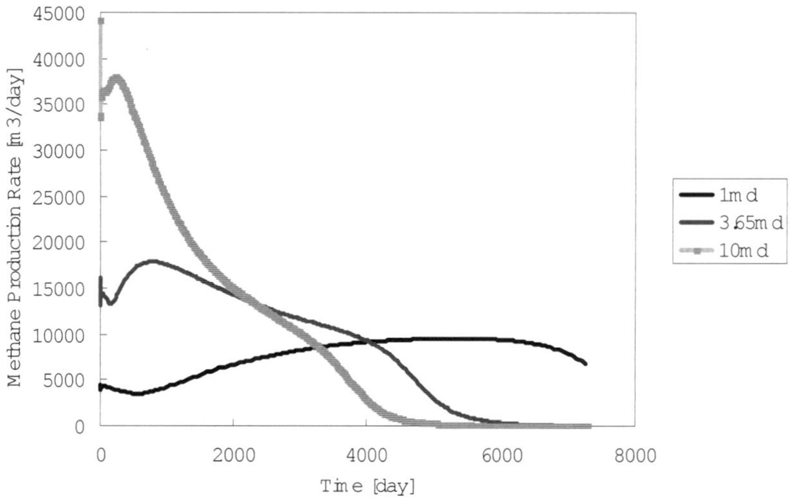

Figure 8. CH_4 production rate comparison of various absolute permeability cases

Table 3 Cumulative CH_4 production comparison of various absolute permeability cases at the time of breakthrough ($[CH_4] = 0.9$)

Permeability [md]	Production time [day]	A: Cumulative production [m^3]	B: Primary recovery [m^3]	A/B [-]
1	6648	5.03×10^7	1.93×10^7	2.6
3.65	4017	5.59×10^7	2.98×10^7	1.9
10	3178	6.62×10^7	4.51×10^7	1.5

EVALUATION OF CO_2 LEAKAGE

One of the risks associated with CO_2 sequestration in coal seam is leakage. CO_2 leakage leads to overestimate of sequestrable CO_2 amount and environmental impact [5]. This section shows evaluation of CO_2 leakage volume under some seal performance of rock over coal seam.

In this section, well pattern was set to be 400m and sorption time was set to be 38.6day. Reservoir and operating conditions not mentioned here are same as the former calculation. Rock strata lying over coal seam was considered. Rock is 108m in thickness and divided into 4 grids in vertical direction. Permeability of the coal was set to be 3.65md. Generally, permeability of shale or sand stone ranges $1x10^{-5}$ –0.1md or $1x10^{-2}$ –10md respectively. Rock permeability was set to be $1x10^{-3}$, $1x10^{-2}$, and 0.1. Gas flow from coal seam to rock was regarded as leakage.

Figure 9 shows the CO_2 leakage for some rock permeability. Higher permeability leads to earlier and larger leakage. At the time of breakthrough, production period was 4000-5000days and cumulative injected CO_2 was 1.2-1.4x10^8 m^3. At this time, in $1x10^{-2}$md case, CO_2 leakage volume is 6-9 x 10^5 m^3 and corresponds to about 0.6 % of injected CO_2. Rock smaller than $1x10^{-3}$md has enough seal performance. In both case, leakage volume is so small that production pattern is hardly affected. Leakage volume pattern after 5000days are explained as follows. Injected CO_2 flows in coal seam and is adsorbed in coal matrix. If coal matrix is saturated with CO_2, gas pressure in pore increases (larger pressure gradient) and CO_2 leaks to rock over coal seam. In 0.1md case, leaked CO_2 covered whole rock over coal seam in 6 years. Thus leakage potential is rather high.

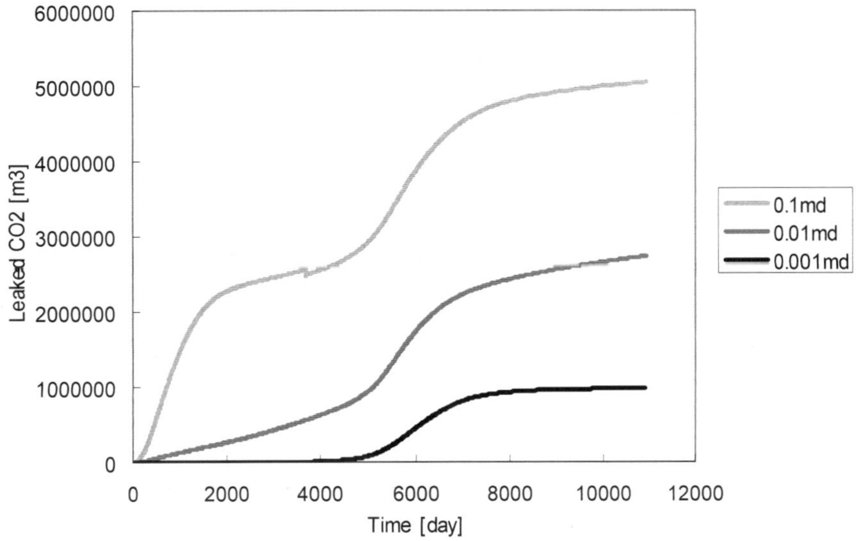

Figure 9. CO_2 leakage potential comparison considering rock permeability

CONCLUSION

The authors have developed the ECBMR simulator (ECOMERS-UT). Comparison study with other simulators confirmed that this simulator was valid. Sensitivity analyses showed that component of injection gas and absolute permeability are important parameters. Rock of low permeability (smaller than $1x10^{-3}$md, 108m in thickness) has enough seal performance. When rock over coal seam has higher permeability, leakage volume and risk need to be evaluated.

REFERENCES

1. Gas Research Institute, A guide to coalbed methane reservoir engineering, Gas Research Institute report No. GRI-94/0397, Chicago, Illinois (1994)

2. David H. -S. law, L.G.H.(Bert)van der Meer, W.D.(Bill) Gunter, Numerical Simulator Comparison Study for Enhanced Coalbed Methane Recovery Processes, Part : CO_2 Injection, SPE75669(2002)

3. David H. -S. law, L.G.H.(Bert) van der Meer, W.D.(Bill) Gunter, Comparison of Numerical Simulators for Green House Gas Storage in Coalbeds, Part :Flue Gas Injection, Proceedings of The 6th International Conference on Greenhouse Gas Control Technologies, Kyoto, Japan (2002),volume 563-568.

4. David H. -S. law, L.G.H.(Bert)van der Meer, W.D.(Bill) Gunter, Comparison of Numerical Simulators for Green House Gas Storage in Coalbeds, Part :More Complex Problems, http://www-esd.lbl.gov/GEOSEQ/papers.html

5. Sookyun Wang and Peter R. Jaffe, Dissolution of a mineral phase in portable aquifers due to CO2 release from deep formations; effect of dissolution kinetics, Energy Conversion and Management 45 (2004) p.2833-2848

0528

Development of a Probabilistic Forecasting and History Matching Model for Coalbed Methane Reservoirs

Anne Y. Oudinot, Advanced Resources International Inc.
George J. Koperna Jr., Advanced Resources International Inc.
Scott R. Reeves, Advanced Resources International Inc.

ABSTRACT

In recent years, coalbed methane (CBM) has become an essential component of the United States energy market. However, due to their heterogeneity, the productivity of CBM reservoirs depends upon an inter-related set of reservoir, completion and production parameters. Even though many powerful reservoir simulators are currently available to the industry, their sophistication coupled to the reservoir complexity tends to result in a large outlay of man-hours for a single reservoir study. Recognizing that stochastic (Monte Carlo) simulation techniques are well-known and have been used in a wide variety of applications to aid in the understanding of complex reservoir problems, particularly parametric studies, a stochastic modeling tool has been coupled to a reservoir simulator to address this workflow gap. Through this process, thousands of simulations can be automatically run to explore combinations of unknown parameters across their range of uncertainty. When historical data are available, a least square error function can be run between the simulation data and the historical data. In many cases, the results of this stochastic-based simulation methodology has provided more than one history match to a simulation problem, showing the non-uniqueness of the process and adding insight into the range of parameters that impact the production response of the CBM reservoir being investigated. Already applied to several projects, this approach has shown its efficiency and rapidity in solving complex reservoir simulation problems.

INTRODUCTION

Monte Carlo (or stochastic) modeling techniques have long been used for exploring the impact of uncertainty on oil and gas project performance worldwide. In its purest sense, Monte Carlo simulation employs a mathematical model that interjects randomness between limits to determine a probabilistic or likely outcome. Typically, this result is in the form of a probability distribution, the shape of which lends insight into what is likely to occur if the modeled course of action is pursued.

From a coalbed methane reservoir modeling standpoint, the implementation of a simulation project is in many ways similar to using Monte Carlo techniques, particularly when historical data is available for comparison to model output. Multiple executions of the simulation are typically required to converge on a solution that approximates reality (historical data). Depending on the complexity of the problem being modeled, the time to reach this convergence could range from hours to months of simulation time.

Many authors have seen the promise that these stochastic techniques hold with regard to simulation as well as its use in history match optimization[1-8]. In this work, a fractured reservoir simulator, *COMET3* [9-10], has been coupled to a commercially-available Monte Carlo simulation software, in order to automatically generate tens, hundreds or thousands of simulation cases for probabilistic forecasting and, if available, for comparison to historical data. While this number of simulation runs can generate voluminous data, the program has the capability to illustrate only those cases that meet pre-set statistical cut-offs. Through this process, the authors hope to show the utility of such a process as well as the functionality of automation, which ultimately provides the user a powerful complement to traditional reservoir simulation.

PROCEDURE

Creating the simulation input file

The first step of the process is to create the simulation input file. Once created, modifications to the reservoir properties, wells description, production and injection profiles are established.

Available options

The following are typical variable for the Monte-Carlo approach:

- For single well cases, well spacing can be adjusted at each iteration. Formulas have been implemented to compute the length of each grid block based on the length of the grid block in which the well is located (middle block) and an increment factor necessary to compute the length of each consecutive grid block.

- Another option allows varying the reservoir initial pressure versus depth. A very simple linear equation can be used as shown in equation (1) where P_i refers to the reservoir initial pressure in psia and P_{iFunct} is the pressure gradient in psi per foot. The model being highly flexible in that it may incorporate alternative pressure relationships. The range of possible values for the initial pressure corresponding to a specific probability distribution of this variable parameter is illustrated on Figure 1.

$$P_i = P_{iFunct} \times Depth \qquad (1)$$

- Desorption pressure can be computed versus initial reservoir pressure as shown in equation (2) where P_d refers to the desorption pressure in psia, P_{dFunct} is the variable parameter and P_i is the initial reservoir pressure in psia. In order to not violate the initial water saturation condition criteria, if the desorption pressure is less than the initial pressure (creating an undersaturated condition); the initial water saturation is automatically set to 100%. One may also establish a probability distribution around the initial gas saturation for cases where $P_d = P_i$

$$P_d = P_{dFunct} \times P_i \qquad (2)$$

- Absolute permeability is another parameter that can be varied versus depth. A preferred approach is to derive basin-specific permeability versus depth functions as shown on Figure 2. Equation (3) is an example of a possible function, where K refers to permeability.

$$K = 75{,}000 \times Depth^{-1.2} \qquad (3)$$

- Another option is also available regarding relative permeability curves. Two phase relative permeability curves can be approximated using the Corey equations. Equation (4) gives the relative permeability to water and equation (5) the relative permeability to gas where S*w* is the water saturation, S*wr* the irreducible water saturation, S*gr* the irreducible gas saturation, K*rwmax* the maximum water relative permeability, K*rgmax* the maximum gas relative permeability, m the Corey water exponent and n the Corey gas exponent.

$$Krw = K_{rw\max}\left(\frac{Sw - Swr}{1 - Swr - Sgr}\right)^{m} \qquad (4)$$

$$Krg = K_{rg\max}\left(\frac{1 - Sw - Sgr}{1 - Swr - Sgr}\right)^{n} \qquad (5)$$

These parameters have been added to the simulation input file so that a probability distribution can be defined, which generates new relative permeability curves at each iteration.

Defining uncertainties

- Uncertain parameters are then assigned distribution types and ranges. These functions simply represent a range of different possible values that a parameter could take instead of limiting it to just one value. Figure 1, for example depicts a normally distributed function. Other types that may be used are uniform, triangular, exponential, lognormal, Poisson and many more. There is no limitation to the number of parameters that can be varied.

- Most other simulator input parameters can also be probabilistically varied, such as porosity, gas content, isotherm properties, etc.

Defining outputs

Output parameters can also be defined, such as the cumulative gas production and/or the cumulative water production, for subsequent analysis. The program will automatically record the corresponding values from the output file back to the input file at the end of each simulation so that frequency and sensitivity charts (Tornado plots) can later be generated. The parameters that impact the production response of the reservoir investigated can therefore be determined.

Running the simulations

When all the assumptions have been made on the desired parameters and all outputs have been defined, the process is ready to be started. The workflow is outlined in Figure 3.

Following is a detailed description of the automated processes taking place after the number of simulations to be run has been input by the user:

- The Monte-Carlo simulator is started and generates random values for each uncertain parameter from the prescribed probability distribution functions.

- The *COMET3* input file is saved.

- The simulation run is executed.

- At the end of the run, all input/output data is saved.

- The Monte-Carlo simulator generates new random values and the next simulation starts.

Sensitivity charts

At the end of all the simulations, a sensitivity chart (also called Tornado plot) can be created that ranks the uncertain parameters according to the impact they have on the outputs. An example of a sensitivity chart is shown in Figure 4. Note that the most sensitive parameters, or variables impacting target output the most, are at the top of the chart. The shape of which is funnel-like, resulting in the name "Tornado chart". These tornado charts can be invaluable for identifying the impact of unknown or uncertain parameters upon the desired simulation output.

Analyzing the results: plots and error function

When historical data are available, a least square error function can be run between the simulation data and the historical data, bringing forth those simulation cases that yield the best statistical matches. The error function R^2 is computed using equation (6) where ActualData is the value at each time step of the historical data for the selected output parameter, SimData the value at the same time step of the simulation data and Mean the average value of the actual data.

$$R^2 = 1 - \frac{\sum (ActualData - SimData)^2}{\sum (SimData - Mean)^2} \qquad (6)$$

With

$$Mean = \frac{\sum ActualData}{NumberofData} \qquad (7)$$

A summary table is provided containing the input values for each parameter for each trial as well as the corresponding computed error function. The best acceptable match (or matches) can be identified (cases with the value of the error function R^2 closest to 1.0).

CASE STUDY 1

Problem Description

This analysis was performed for a property evaluation in Oklahoma. The work performed involved reservoir simulation history-matching of about 300 wells on 400,000 net acres. To simplify the work, three type wells were created based on geologic setting and production characteristics.

Set Up

The simulation input file for each type well was built based on existing reservoir parameters such as coal depth, coal thickness and isotherm data (Langmuir pressure, Langmuir volume and gas content). Likely ranges were established for 9 unknown parameters (permeability, porosity, pore compressibility, matrix compressibility, Corey water exponent, Corey gas exponent, initial water saturation, skin factor and desorption pressure function Pd/Pi). Table 1 summarizes for each uncertain parameter the type of probability distribution applied as well as the range selected. One thousand simulations were run using a producing bottomhole pressure constraint for each type well. The results were tabulated by computing the gas production rate error function.

Results

These one thousand simulations ran in approximately two hours, with an additional hour to compute the error function. A plot of the gas production rate for all the simulations for one type-well is shown on

Figure 5. After running the error function on the gas production rate and restricting the results to the simulation runs with $R^2>0.9$, only 15 out of the 1000 simulations were judged to be acceptable. The plot of gas production rate for those selected runs is shown on Figure 6.

The results of this stochastic-based simulation methodology applied to this particular case provided more than one reasonable (statistically) history match, showing the non-uniqueness of the process and adding insight into those parameters that impact the production response of the gas reservoir investigated. The final match parameters for the 15 highly ranked cases are provided in table 2. An average value and an acceptable range of values can be determined for each uncertain parameter.

Generating a Tornado plot on the cumulative gas production identified 6 performance drivers out of the 9 initially unknown parameters as shown on Figure 7. In order to improve this history match, the model could be run a second time, only defining probability distributions on the 6 identified parameters having the greatest impact on production and narrowing the range of uncertainty to the minimum and maximum values from the 15 acceptable simulation cases. However, this was not justified at the time due to the scoping-level nature of the exercise.

CASE STUDY 2

Problem Description

This analysis was performed in support of a possible asset acquisition in Texas. The work performed involved reservoir simulation history-matching of two existing production pilots, one with three wells and one with five wells. Each pilot had about three months of production data.

Set Up

Available reservoir data included coal depth, thickness and the Langmuir isotherm. A multi-well simulation model was built for each pilot, estimating bottom hole producing pressures over time for each well based on available surface pressure and fluid level data. Both models were run on pressure while computing gas and water production.

The variable parameters for history matching included permeability, relative permeability, compressibility (pore and matrix), initial water saturation, porosity, Pd/Pi (i.e., gas content), sorption time, cleat spacing, total model area (i.e., average well spacing), and skin factors for the wells.

Results

Even given the very short-term nature of the production data, reasonable history matches were achieved for each pilot. The gas and water rate matches for each pilot are provided in Figures 8 and 9. The final match parameters are provided in Table 3. The values determined for each unknown parameter were found to be very close between the two pilots.

DISCUSSION

What is clear from our experience with this approach is that simulation problems that would normally require considerable manpower to solve can now be accomplished much more effectively and efficiently by utilizing Monte-Carlo simulation, on automated procedure, and raw computing power. Further, the number and range of variables can be quite large, thus enabling the exploration of a large solution space. Sometimes this has yielded multiple acceptable solutions, demonstrating the non-uniqueness of each case. This has yielded results that would not have been obvious from the outset, and most likely overlooked in a normal process. Thus this approach, we believe, significantly reduces the time, improves the results, and expands the capability of CBM reservoir simulation.

CONCLUSIONS

- A means of probabilistic forecasting and automatic history matching has been created through the use of a coupled stochastic model and reservoir simulator for characterization of CBM reservoirs. These techniques provide the flexibility to run nearly unlimited simulation cases as specified by the user, whereupon completion, automatically high-grade cases of interest.

- Being an effectively automated model, the staff time devoted to the characterization is greatly minimized. Further, as the overall run results can be statistically ranked and compared to actual data, this one can focus only on those results of most interest.

- In several projects, the stochastic based simulation model provided more than one set of parameters judged to be favorably comparable to the historical data, showing the non-uniqueness of the process and adding insight into the range of parameters that impact the production response of the CBM reservoir investigated.

NOMENCLATURE

Depth = reservoir depth, ft
K = reservoir absolute permeability, md
K_{rg} = relative permeability to gas
K_{rgmax} = maximum relative permeability to gas @ $S_w = S_{wr}$
K_{rw} = relative permeability to water
K_{rwmax} = maximum relative permeability to water @ $S_w = 1 - S_{gr}$
m = Corey water relative permeability exponent
n = Corey gas relative permeability exponent
P_d = desorption pressure, psia
P_{dFunct} = desorption pressure over initial pressure, fraction
P_i = reservoir initial pressure, psia
P_{iFunct} = reservoir pressure gradient, psi/ft
R^2 = error function
S_{gr} = irreducible gas saturation
S_w = water saturation
S_{wr} = irreducible water saturation

REFERENCES

1. Nakayama, K.: "Estimation of Reservoir Properties by Monte Carlo Simulation," Paper SPE 59408 presented at the 2000 SPE Asia Pacific Conference on Integrated Modeling for Asset Management held in Yokohama, Japan, April 25-26, 2000.

2. Sawyer, W.K., Zuber, M.D. and Williamson, J.R.: "A Simulation-Based Spreadsheet Program for History Matching and Forecasting Shale Gas Production," Paper SPE 57439 presented at the 1999 SPE Eastern Regional Conference and Exhibition held in Charleston, West Virginia, Oct. 21-22, 1999.

3. MacMillan, D.J., Pletcher, J.L. and Bourgeois, S.A.: "Practical Tools to Assist History Matching," Paper SPE 51888 presented at the 1999 SPE Reservoir Simulation Symposium held in Houston, TX, Feb. 14-17, 1999.

4. Portella, R.C.M. and Prais, F. (Petrobras), "Use of Automatic History Matching and Geostatistical Simulation to Improve Production Forecast," SPE 53976 presented at the 1999 SPE Latin America and Caribbean Petroleum Engineering Conference, Caracas, Venezuela, 21-23 April.

5. Bennett, F, and Graf, T. (Schlumberger), "Use of Geostatistcal Modeling and Automatic History Matching to Estimate Production Forecast Uncertainty," SPE 74389 presented at the 2002 SPE International Petroleum Conference and Exhibition, Villahermosa, Mexico, 10-12 February.

6. Landa, J.L., and Guyaguler, B. (ChevronTexaco), "A Methodology for History Matching and the Assessment of Uncertainties Associated with Flow Prediction," SPE 84465, presented at the 2003 SPE Annual Technical Conference and Exhibition, Denver, Colorado, Oct. 5-8.

7. Gao, G., Gaoming, L., and Reynolds, A. (Univ. Tulsa), "A Stochastic Optimization Algorithm for Automatic History Matching," SPE 90065, presented at the 2004 SPE Annual Technical Conference and Exhibition, Houston, Texas, Sep. 26-29.

8. Roadifer, R. D., Farnan, R. A., Krabtree, B. J., Raterman, K. T., Moore, T. R., "History Matching (Reservoir Parameter Estimation) for Coalbed Methane Reservoirs via Monte Carlo Simulation," presented at the 2003 International Coalbed Methane Symposium , Tuscaloosa, Alabama, May 5-9.

9. Sawyer, W. K., Paul, G. W., and Schraufnagel, R. A., "Development and Application of a 3D Coalbed Simulator," paper CIM/SPE 90-119 presented at the 1990 CIM/SPE International Technical Conference, Calgary, June 10-13.

10. Paul, G. W., Sawyer, W. K., and Dean, R. H., "Validation 3D Coalbed Simulators," paper SPE 20733 presented at SPE Annual Technical Conference and Exhibition, Houston, Sept. 23-26, 1990.

Table 1: Case study 1 probability distribution functions.

Variables	Range	Distribution Type
Permeability	10 - 60 mD	Uniform
Porosity	0.5 - 3 %	Uniform
Pore Compressibility	100E-06 - 300E-06 psi^{-1}	Uniform
Matrix Compressibility	5E-07 – 2.5E-06 psi^{-1}	Uniform
Corey Water Exponent	1 - 6	Uniform
Corey Gas Exponent	1 - 6	Uniform
Acreage	80 - 160 acres	Uniform
Initial Water Saturation	0.6 – 1.0	Uniform
Desorption Pressure Function	0.6 – 1.0	Uniform

Table 2: Case study 1 final matches parameters.

Simulation	K	Phi	Pd/Pi	Swi	Skin	Cp	Cm	KrwExp	KrgExp	R2
320	25.5	0.018	0.99	1.00	0.68	1.85E-04	2.1E-06	6.70	1.06	0.95
445	35.5	0.019	0.86	1.00	-0.62	2.27E-04	1.9E-06	5.54	1.19	0.92
451	43.4	0.025	0.77	1.00	-1.70	2.49E-04	1.2E-06	5.27	1.46	0.94
547	66.8	0.030	0.70	1.00	-1.89	2.76E-04	7.5E-07	3.56	1.44	0.90
647	24.6	0.008	0.79	1.00	0.39	2.85E-04	2.0E-06	2.75	1.06	0.93
688	23.6	0.028	0.63	1.00	-2.84	1.31E-04	2.1E-06	1.39	1.10	0.90
733	38.4	0.022	0.62	1.00	-1.95	1.10E-04	7.6E-07	2.70	1.43	0.90
758	49.6	0.021	0.71	1.00	-1.38	1.93E-04	1.5E-06	4.10	1.54	0.92
769	30.6	0.007	0.72	1.00	0.85	2.02E-04	1.9E-06	2.86	1.47	0.93
785	31.4	0.027	0.82	1.00	-1.29	1.22E-04	1.4E-06	3.79	1.17	0.93
799	32.4	0.017	0.68	1.00	0.32	2.79E-04	6.7E-07	1.03	1.05	0.93
843	20.7	0.015	0.80	1.00	-2.42	1.23E-04	6.2E-07	4.74	1.31	0.91
888	34.8	0.023	0.71	1.00	-1.45	2.27E-04	1.6E-06	2.79	1.04	0.92
910	25.6	0.015	0.92	1.00	0.93	1.12E-04	1.2E-06	4.24	1.05	0.93
911	37.9	0.009	0.65	1.00	-0.93	1.58E-04	1.5E-06	5.32	1.06	0.91
970	28.7	0.012	0.71	1.00	-2.87	1.13E-04	1.3E-06	4.87	1.64	0.92
Min	20.7	0.007	0.62	1.00	-2.87	1.10E-04	6.2E-07	1.03	1.04	
Max	56.8	0.030	0.99	1.00	0.93	2.85E-04	2.1E-06	5.70	1.64	
Average	33.7	0.018	0.75	1.00	-1.01	1.87E-04	1.4E-06	3.79	1.26	

Table 3: Case study 2 final match parameters.

	Pilot 1	Pilot 2
Spacing	478 acres/well	420 acres/well
Permeability	30 mD	30 mD
Porosity	0.10%	0.10%
Tau	7 days	7.8 days
Fracture Spacing	2.7 in	2.9 in
Pd/Pi	0.66	0.67
Cp	250E-6 psi-1	250E-6 psi-1
Cm	1E-6 psi-1	1E-6 psi-1
Skin	-2.9	-2.9
Swi	1	1
Swirr	0.26	0.24
KrwExp	5.69	5.93
KrgExp	5.65	5.94
Alpha (Corey Exp)	0.91	0.95

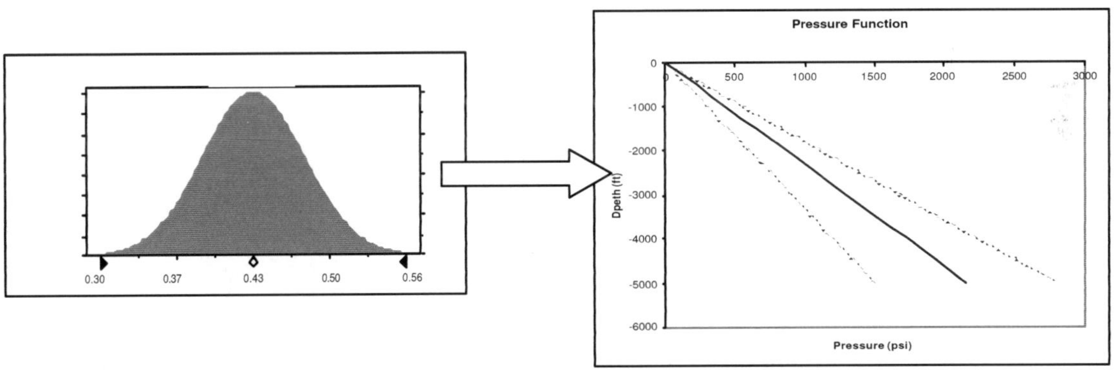

Figure 1: Possible range for the initial pressure given a specific probability distribution for the pressure gradient.

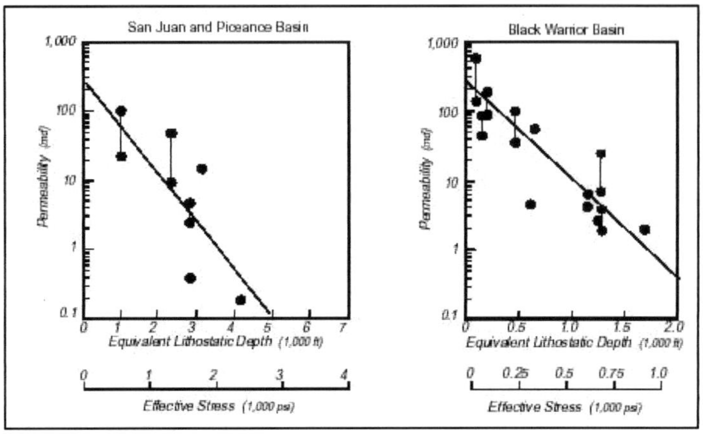

Figure 2: Typical permeability versus depth functions for several CBM basins.

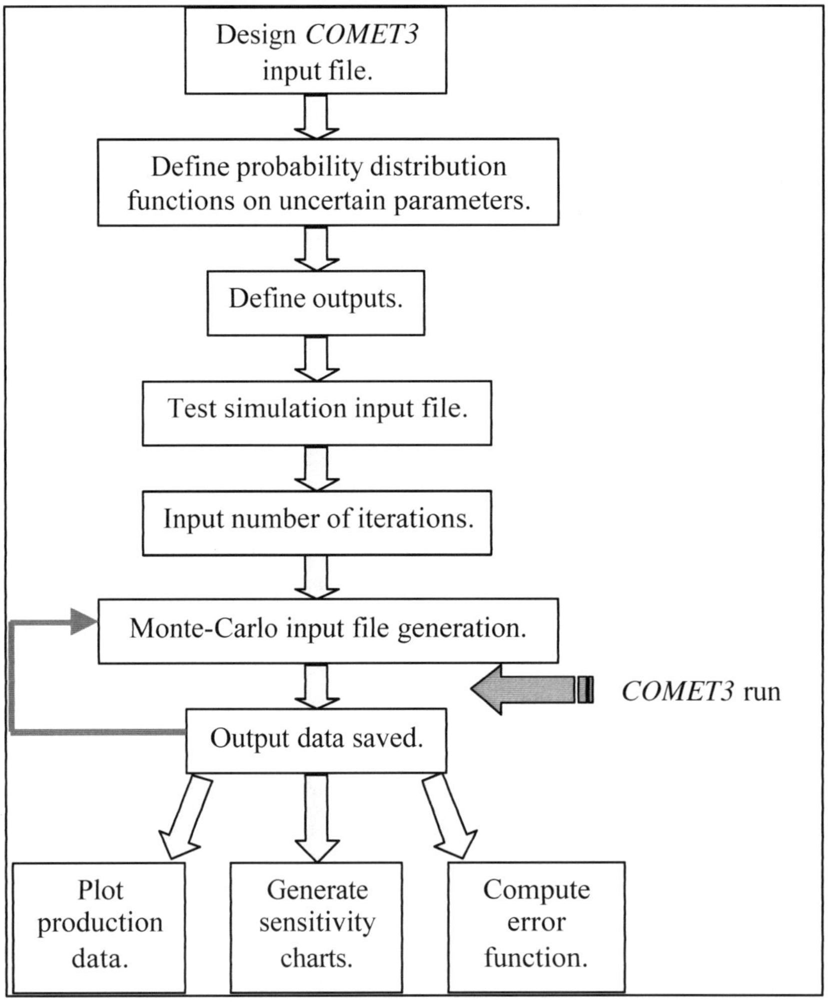

Figure 3: Model flow chart.

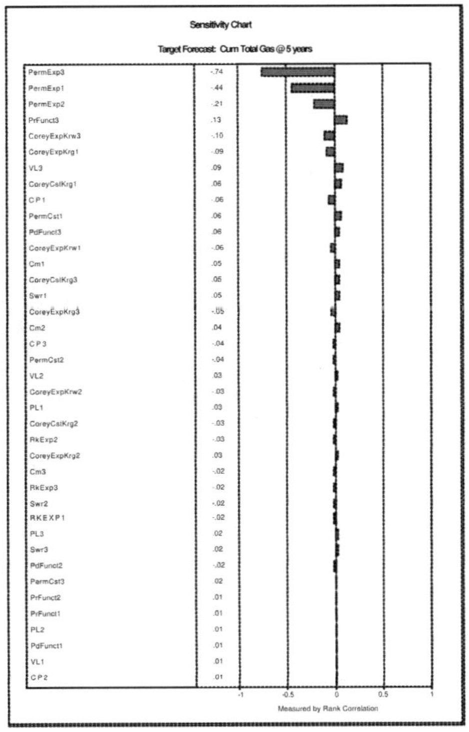

Figure 4: Example of a sensitivity chart.

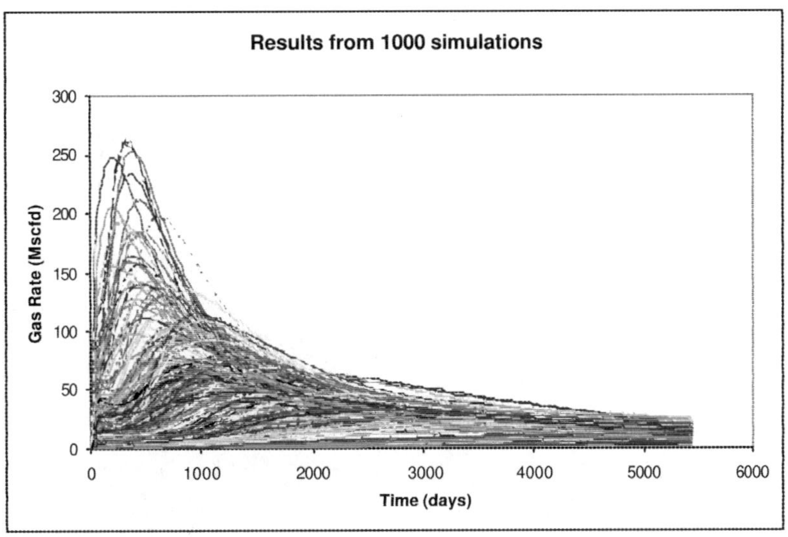

Figure 5: Gas production from 1000 simulations.

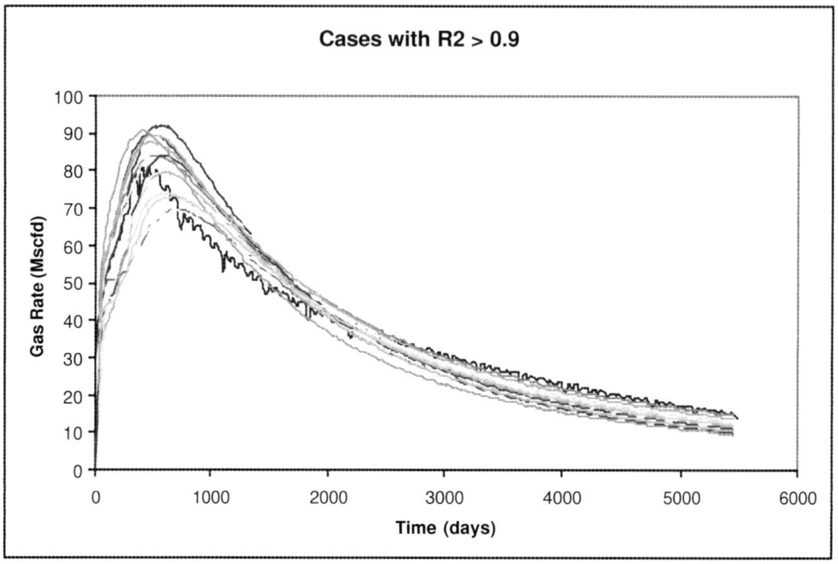

Figure 6: Gas production for simulations with R^2>0.9.

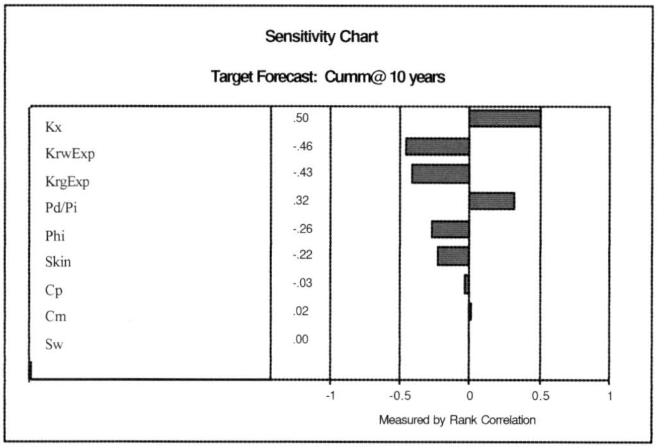

Figure 7: Tornado plot: performance drivers identified.

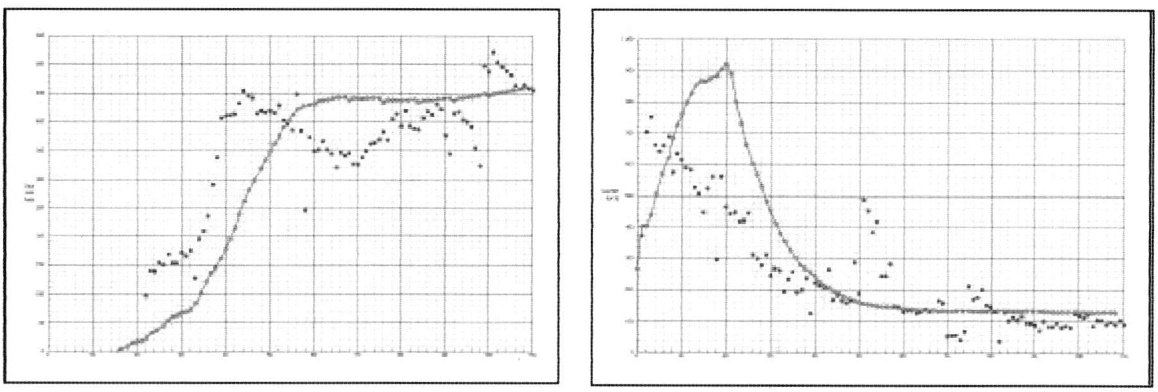

Figure 8: Gas and water production for pilot 1.

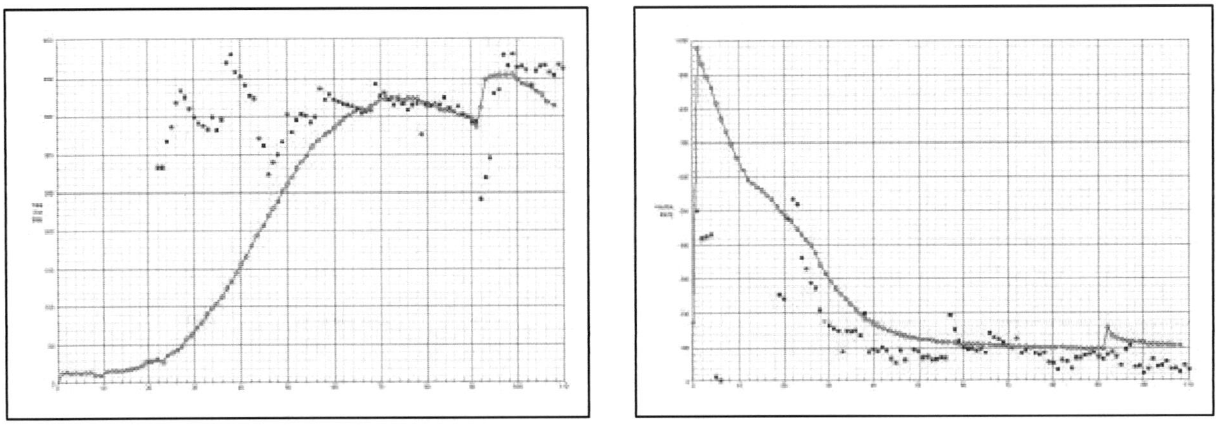

Figure 9: Gas and water production for pilot 2.

0531

The Development of Analytical Model for Hydraulic Treatment of Coal Bed and Determination of its Main Hydrodynamic Parameters

L.A. Puchcov, S.V. Slastunov, G.G. Karkashadze and K.S. Kolilov
(Moscow State Mining University, Russia)

ABSTRACT

We have developed an analytical model of Hydraulic treatment of coalbed. To solve the problem, we employed continuity equation and Darcy's law with variable coefficient of filtration. The coefficient of filtration has minimum value before treatment, however, under pressure, as a result of a crack opening, it increases. The decreasing of filtration coefficient with decreasing pressure is also taken into consideration, which simulates incomplete crack closure.

We have also developed and used numerical method of estimation using computer. It takes into consideration the influence of water flow and pressure per radius of hydraulic treatment area at different time periods. One problem is solved at present condition of water flow change in time at well's boundary. The other describes the process at present conditions of change in water pressure with time. A part from that we have developed an algorithm to estimate actual bed parameters as a result of correlation made between theoretical solution and actual experiment. The choice of most reliable correlation is made on basis of minimum squares method. We have provided some results of computer simulation, which allows predicting the radius of area to be treated.

INTRODUCTION

Hydraulic treatment involves water injection into the coal bed, crack opening and subsequent dewatering. In practice this technological operation is conducted in order to degas coal beds and ensure safety at working as well as to enhance methane recovery process [1]. One important task is to choose the best possible mode of water flow or pressure throughout the process of hydraulic treatment. The answer to this question provides an opportunity of optimizing technological modes as well as increasing area of hydraulic treatment with minimum material and energy costs.

Analytical model developed is based on the following assumptions:

- Coal bed has radial anisotropy: that is filtration coefficients along two mutually perpendicular directions are different.
- At the pressure the order of 10 MPa the dependence of water density ρ on pressure P takes the form:

$$\rho = \rho_0\left[1 + \beta_b(P - P')\right], \qquad (1)$$

where: ρ_0 – water density at atmospheric pressure, $P' = 10^5$ Pa;

β_b - contraction coefficient, $\beta_b = (0.27\ldots0.5)\cdot10^{-5}$ Pa^{-1}.

- Dependence of porosition pressure looks same:

$$n = n_0\left[1 + \beta_c(P - P')\right], \qquad (2)$$

where: n- porosity of coal at pressure P;

n_0 - porosity at atmospheric pressure P';

β_c – indicator of specific contraction of mineral's framework, $\beta_c=(0.8\ldots5)\cdot10^{-11}$ Pa^{-1}.

- Coal bed is constant thickness.
- Water filtration along the bed is described by linear Darcy's Law and continuity of flow equation.
- Filtration coefficient is dependent on water pressure.
- Roof and ground water drainage depend on water pressure in a coal bed.

The physical pattern of water spread along the bed is the following:

Under pressure water is spreading from a single source in radial direction along the coal bed. The first stage of process involves pores and hollows being filled with water in a low pressure mode. Once the pores have been filled by elementary volume, the second stage involves pressure mode filtration with water spreading throughout coalbed in horizontal direction.

Low pressure mode may take much time, therefore it must be considered. As for pressure mode filtration, it is being realized immediately after low pressure filtration and is characterized by crack opening resulted from water pressure and horizontal water flow along those cracks. Thus, hydro treatment mechanism is described by two similar but substantially different water modes (low pressure and pressure modes).

In case of low-pressure water filtration continuity looks like:

$$\frac{\partial(\rho V_x wH)}{\partial x} + \frac{\partial(\rho V_y wH)}{\partial y} = -\frac{\partial(n\rho wH)}{\partial t} . \quad (3)$$

This equation holds true if

$$w(x,y,t) \le 1 . \quad (4)$$

In equation (3):

V_x, V_y – water velocity in direction x, and in direction y, m/sec;

$w(x,y,t)$ – coefficient of pore filling in by water in a coalbed;

H - seam thickness, m;

t – time, sec.

The same equation in case of pressure filtration i.e. if $w(t)$=1 looks like:

$$\frac{\partial(\rho V_x)}{\partial x} + \frac{\partial(\rho V_y)}{\partial y} = -\frac{\partial(n\rho)}{\partial t} . \quad (5)$$

The description of water flow in our model employs linear Darcy's Law. In case of low pressure filtration we have:

$$V_x = -K_x K_0 \frac{\partial h}{\partial x} , \qquad V_y = -K_y K_0 \frac{\partial h}{\partial y} , \quad (6)$$

Where K_x, K_y – nondimensional coefficients of filtration anisotropy in directions x and y.

In case of pressure filtration of thrust water pressure P in a coalbed:

$$V_x = -\frac{K_x K(P)}{\rho g}\frac{\partial P}{\partial x} , \qquad V_y = -\frac{K_y K(P)}{\rho g}\frac{\partial P}{\partial y} , \quad (7)$$

where $K(P)$ – water dependent filtration coefficient, m/sec.

Darcy's Law being taken into account, continuity equation, in case of low pressure filtration, takes the form

$$\frac{\partial}{\partial x}\left(\frac{K_x K_0 (P-P_0)}{\rho g}\frac{\partial P}{\partial x}\right) + \frac{\partial}{\partial y}\left(\frac{K_y K_0 (P-P_0)}{\rho g}\frac{\partial P}{\partial y}\right) = n\frac{\partial P}{\partial t} . \quad (8)$$

In this case we take into account that for low pressure filtration

$$P - P_0 = \rho g w H . \quad (9)$$

Apart from that

$$n \approx n_0 = \text{const}, \quad (10)$$

i.e. in the process of low pressure filtration porosity does not depend on water pressure (due to low pressure).

In case of pressure filtration similar rearrangement gives us:

$$\frac{\partial}{\partial x}\left(K_x K(P)\frac{\partial P}{\partial x}\right) + \frac{\partial}{\partial y}\left(K_y K(P)\frac{\partial P}{\partial y}\right) = (\beta_c n_0 \rho g + \beta_b n \rho_0 g)\frac{\partial P}{\partial t} . \quad (11)$$

Equation (11) is derived from equation (5) if $w(t)=1$ and taking into account dependability of (1) and (2). In differentiation of the right side of equation (5) we have:

$$\frac{\partial(n\rho)}{\partial t}=\frac{\partial n}{\partial t}\rho+\frac{\partial \rho}{\partial t}n \quad . \quad (12)$$

Major parameter which characterizes the process of filtration is filtration coefficient $k(P)$. It depends on water pressure in coalbed. At pressure filtration and low pressure filtration coefficient may differ by a factor 10 or more, its maximum value being reached when the process of crack opening in coalbed has taken place. With subsequent drop in pressure so does the filtration coefficient, but its value never reaches it initial value that is hysteresis of filtration coefficient take place. These features of filtration coefficient change can be seen in fig. 1.

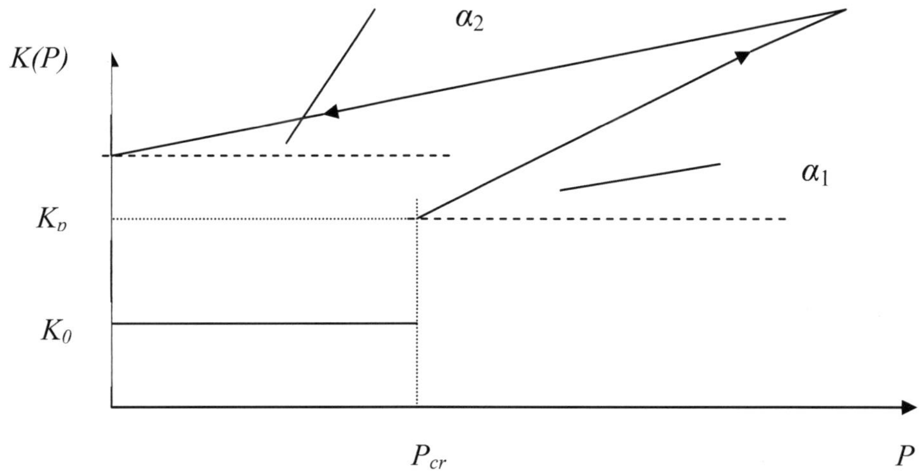

Fig.1 Pressure dependent filtration coefficient

In the process of water filtration along the coalbed part water must drain into roof or ground formations. The assumption that those losses $f(P)$ have linear of the process. Under this condition we get the following equation in Cartesian system of coordinates:

$$\frac{\partial}{\partial x}\left(K_x\cdot K(P)\frac{\partial P}{\partial x}\right)+\frac{\partial}{\partial y}\left(K_y\cdot K(P)\frac{\partial P}{\partial y}\right)+f(P)=\left(\beta_c n_0\rho g+\beta_b n\rho_0 g\right)\frac{\partial P}{\partial t}. \quad (13)$$

Here

$$f(P)=q\frac{g}{H}\frac{P-P_0}{P_1-P_0}. \quad (14)$$

Where: q – mass-transfer coefficient, considering water flow onto the roof, kg/(m^2·sec);
P_0 and P_1 – initial and average water pressure in coalbed. We can assume: P_1=4MPa.

In case of low pressure filtration ground water drainage is consider only. That is water loss is twice as less as that observed during pressure filtration, other conditions being equal. Thus differential equation looks like

$$\frac{\partial}{\partial x}\left(\frac{K_x K_0 P}{H\rho g}\frac{\partial P}{\partial x}\right)+\frac{\partial}{\partial y}\left(\frac{K_y K_0 P}{H\rho g}\frac{\partial P}{\partial y}\right)+0.5f(P)=\frac{n}{H}\frac{\partial P}{\partial t} \quad . \quad (15)$$

Of great interest is the discovery of relationships with present arbitrarily changing in time water flow rate – $Q(t)$ (dimensional representation – m^3/sec) at the well's boundary in coalbed. It is defined like this

$$-H\oint_L \vec{n}\cdot\vec{V}dl = Q(t)\,. \quad (16)$$

Here:

$\vec{n}$ - normal to well's boundary;

$\vec{V}$ - radial velocity of water filtration, m/sec;

L - horizontal contour of a well, m;

dl- element of horizontal contour of a well.

Taking Darcy's Law into account we rewrite equation (16) in following way

$$-\frac{H}{\rho g}\oint_L \vec{n}\cdot(K\cdot\nabla P)dl = Q(t)\,. \quad (17)$$

Here filtration coefficient is different in mutually perpendicular directions:

$$K = \begin{pmatrix} K_x\cdot K(P) & 0 \\ 0 & K_y\cdot K(P) \end{pmatrix}. \quad (18)$$

Besides, we suppose that at infinity and at initial time water pressure in coalbed is equal to initial P_0:

$$P\big|_{x=\infty} = P_0,\quad P\big|_{y=\infty} = P_0,\quad P(x,y,t)\big|_{t=0} = P_0. \quad (19)$$

Thus, combining both cases pressure and low pressure filtrations we can general equation of pressure-free and pressure filtration:

$$\begin{cases} \frac{\partial}{\partial x}\left(K_x\cdot K_2(P)\frac{\partial P}{\partial x}\right) + \frac{\partial}{\partial y}\left(K_y\cdot K_2(P)\frac{\partial P}{\partial y}\right) + F(P) = C(P)\frac{\partial P}{\partial t} \\ -\frac{H}{\rho g}\oint_L \vec{n}\cdot(K\cdot\nabla P)dl = Q(t) \\ P\big|_{x=\infty} = P_0 \\ P\big|_{y=\infty} = P_0 \\ P(x,y,t)\big|_{t=0} = P_0 \end{cases} \quad (20)$$

We have employed the following symbols:

$$K_2(P) = \begin{cases} \frac{K_0 P}{\rho g H}, & P - P_0 \le \rho g H \\ K(P), & P - P_0 > \rho g H \end{cases}, \quad (21)$$

$$F(P) = \begin{cases} 0.5f(P), & P - P_0 \le \rho g H \\ f(P), & P - P_0 > \rho g H \end{cases}, \quad (22)$$

$$C(P)=\begin{cases}\dfrac{n}{H}, & P-P_0\le\rho gH,\\ (\beta_c+\beta_b)n_0\rho g, & P-P_0>\rho gH\end{cases}. \quad (23)$$

The task in hand has a number of properties making numerical solution difficult to select. First of all, this is a two dimensional task, that means it will take a lot of computer time to solve it. Then, substantial non-linear character of filtration coefficient limits the choice of differential scheme to a great extent. However, considering non-substantial anisotropy of a coalbed we have managed to make the task one-dimensional problem, depending on radius only. We did it by changing the following:

$$x_1=\sqrt{\frac{K_y}{K_x}}x,\quad y_1=y. \quad (24)$$

At that we assume the boundary of well equation remains the same as before the change of variables. It is reasonable since the well's radius is much less than that of treatment area. And at great distances the influence of the well's cross-section form is not substation. However, the water flow rate at the compression of coordinates' system changes, as less volume will be filled with water. Thus, we assume

$$Q_1(t)=\sqrt{\frac{K_y}{K_x}}Q(t). \quad (25)$$

If the above-mentioned change is followed by polar system of coordinates then, we will have an axial-symmetric task, i.e. we face one dimensional non-stationary non-linear equation of the parabolic type. In polar system of coordinates the initial task of non-pressure filtration takes the form:

$$\boxed{\begin{cases}\dfrac{1}{r}\dfrac{\partial}{\partial r}\left(r\cdot K_2(P)\dfrac{\partial P}{\partial r}\right)+F(P)=C(P)\dfrac{\partial P}{\partial t}\\ -\dfrac{2\pi r_0 H}{\rho g}K(P)\left.\dfrac{\partial P}{\partial r}\right|_{r=r0}=Q_1(t)\\ P\big|_{r=\infty}=P_0\\ P(r,t)\big|_{t=0}=P_0\end{cases}} \quad (26)$$

With the task (26) to be solved numerically it is much easier to select differential scheme. Besides, we can substantially reduce the volume of calculations and we can also simulate hydraulic fracture process with the help of a common personal computer. In essence, the technological process of hydro treatment can be controlled by two ways:

- first, when pressure is purposefully changed in time, the water flow rate being the consequence of the above-mentioned controlling facts;
- second, when water flow rate is purposefully regulated. In this case the pressure is no longer the reason but the consequence.

We have also considered the task when pressure $P^*(t)$ variable in time, is present at the well's boundary with the radius to. At infinity the gradient of pressure change is equal to zero. Pressure in a coal bed at initial moment of time is constant and equal to bed pressure P_0. In this case boundary conditions look like:

$$\begin{cases}P\big|_{r=r_0}=P^*(t)\\ \left.\dfrac{\partial P}{\partial r}\right|_{r=\infty}=0\\ P(x,y,t)\big|_{t=0}=P_0\\ r=\sqrt{x^2+y^2}\end{cases} \quad (27)$$

Having gone though rearrangements, scale changes being the same as in the previous case, we arrive at the following initial system of equations in cylindrical system of coordinates:

$$\begin{cases} \frac{1}{r}\frac{\partial}{\partial r}\left(r\cdot K_2(P)\frac{\partial P}{\partial r}\right)+F(P)=C(P)\frac{\partial P}{\partial t} \\ P\big|_{r=r_0}=P^*(t) \\ \frac{\partial P}{\partial r}\bigg|_{r=\infty}=0 \\ P(r,t)\big|_{t=0}=P_0 \end{cases} \quad (28)$$

To arrive at numerical solution of quasi-linear filtration tasks (26), (28) presented here, we employed non-linear differential advance scheme, which is absolutely stable and monotonous at any stage. The procedure of numerical method is described in paper [2]. Thus, both tasks are realized using a computer which allows to make theoretical simulation of process.

Despite the multifactor reflection of actual process of hydro fracturing the above-mentioned approach in problem-solving created certain difficulties, in particular the right selection of K_0, P_{cr}, K_p, $tg\alpha_1$, $tg\alpha_2$, that is physical parameters given in figure 1, as well as roof and ground loss of water q factor, reflecting actual technological process. Under laboratory condition of simulation the above parameters are unlikely to be gained. Analytical way of gaining exact values due to a great number of unpredictable factors is problematic either. Thus, there is, perhaps only one preferable way left. That is estimation of analytical model and production experiment results. The search for the most reliable parameters is possible with the help of computer. For instance, the principle of minimum squares in correlation made between analytical and experimental relationship. We have solved such problem as well. The developed algorithm of search for the most correct physical parameters, determining bed hydro treatment mode involves the following:

- indication of actual intervals of parameters change known to the user from empirical observation or theoretical estimation as well as minimum stage of search for each parameter;
- the user also introduce, known to him from actual experiments, water flow rate which variable in time of the process studied;
- reliability criterion is the minimum of square deviation of estimated correlation $P(t)$ from those gained under real conditions.

For computer implementation of inverse problem, we have selected the quickest possible method. This is a modified method of coordinate by coordinate descent. As well as being easy it is fast and effective in operation. Thus, unknown hydrodynamic parameters of a coalbed are found.

Let us demonstrate some practical results gained by using the model. Water flow and water pressure dependence (in figure are not submitted), identified in hydro fracturing of coalbed $Д_6$ (well №12, Lenin mine in Karaganda basin). Besides, we have the following parameters known: coalbed depth 422.5 m; coal porosity n_0=0.022 m^3/m^3; coalbed thickness H= 6.3 m. Exact values of the other parameters are considered to be unknown. We only knew their range of variation on the basis of physical understanding. Unknown parameters include: P_{cr} – pressure at which initial crack growth begins; K_0 – initial coefficient of coal bed filtration; K_p – filtration coefficient at the start of hydro fracturing; q – dimension coefficient of water drainage into roof and ground; $tg\alpha_1$ and $tg\alpha_2$ tangent of angle of incline (fig. 1).

During simulation process by theory and practice comparison, we have discovered that the most reliable results of that comparison are gained under the following values:

P_{cr} = 9.9 MPa;
K_0 = $5.79\cdot10^{-6}$m/sec;
K_p = $34.7\cdot10^{-6}$m/sec;
q= - $4.5\cdot10^{-3}$ kg/(m^2·sec);
$tg\alpha_1$ = $6.37\cdot10^{-13}$ m/(sec·Pa);
$tg\alpha_2$ = $1.62\cdot10^{-13}$ m/(sec·Pa).

The results obtained are of practical importance since they have paved the way for treatment optimization. For example, we can solve the first task of optimization: the question of radius of hydraulic treatment area dependence on water injection time at constant water consumption. For example, $Q(t)$ =80 liter/sec=const.

We use equivalent radius of hydraulic treatment as an efficiency indicator, it is calculated:

$$R_{eq} = \sqrt{R_1 \cdot R_2}\ , (29)$$

where R_1 and R_2 – Simi axes of hydraulic treatment area, m.

Figure 2 shows the change of equivalent radius R_{eq} of hydraulic brake up area over time for above mentioned conditions. It is noticed that equivalent radius of hydraulic brake up area first grows up rapidly, and then very soon in about ten minutes its growth stops. This fact is accounted for by loss of water into the roof and ground. It is particularly felt the area of hydraulic treatment becomes bigger. Thus, as it is clear from calculation results analysis at fixed water consumption the area of treatment is limited in its dimensions and longer water injection time does not give substantial growth.

The second task of optimization goes like this: let's suppose injection of water is possible at rate of 60-140 liters per second, the question is: what can be best possible hydraulic treatment mode to implement maximum radius of crack extension, water volume being, for example $100m^3$?

Figure 3 shows predicted equivalent radius, the mentioned modes having been implemented. It is important to mention the linear radius of hydraulic treatment growth while the rate of water injection is being increased. It is explained by less water loss into the roof and ground in a process taking place rapidly.

Figure 4 shows the dependency of pressure developed by a pump over time at constant water consumption. We can see, first an abrupt pressure growth, then its stabilization which is in accord with the known practical results.

The model developed also allows calculating maximum pressure developed by a pump depending on water consumption. We have established linear connection between these parameters (fig.5).

Evidently, the choice of rational radius for hydraulic treatment of coal seam should be based on possible minimum of specific costs towards technological operation. In this case, such parameters as the radius of treatment, the volume of water consumption, time, etc. can be obtained duly as a result of a computer modulation process. Analytical model, as we see it, can also be used not only in projects for promising, as well as calculation of necessary coal seam parameters (based on reverse task solution) and finally, it helps in decision making about particular mode and duration of subsequent hydraulic treatment. The developed analytical model and computer software allow foreseeing the hydraulic treatment mode not only at water consumption being constant, but at any law of change in consumption over time.

The results obtained by the authors can be discussed over E-mail: karkash@msmu.ru.

Reference to an article:

1. Puchkov L.A., Slastunov S.V., Kolikov K.S. «Methane recovery from Coal Beds». Moscow, MSMU Publishing House, 2002, 383 p.
2. Kalitkin N.N. «Numerical Methods».-Moscow: «Science» Publishing House, 1978.

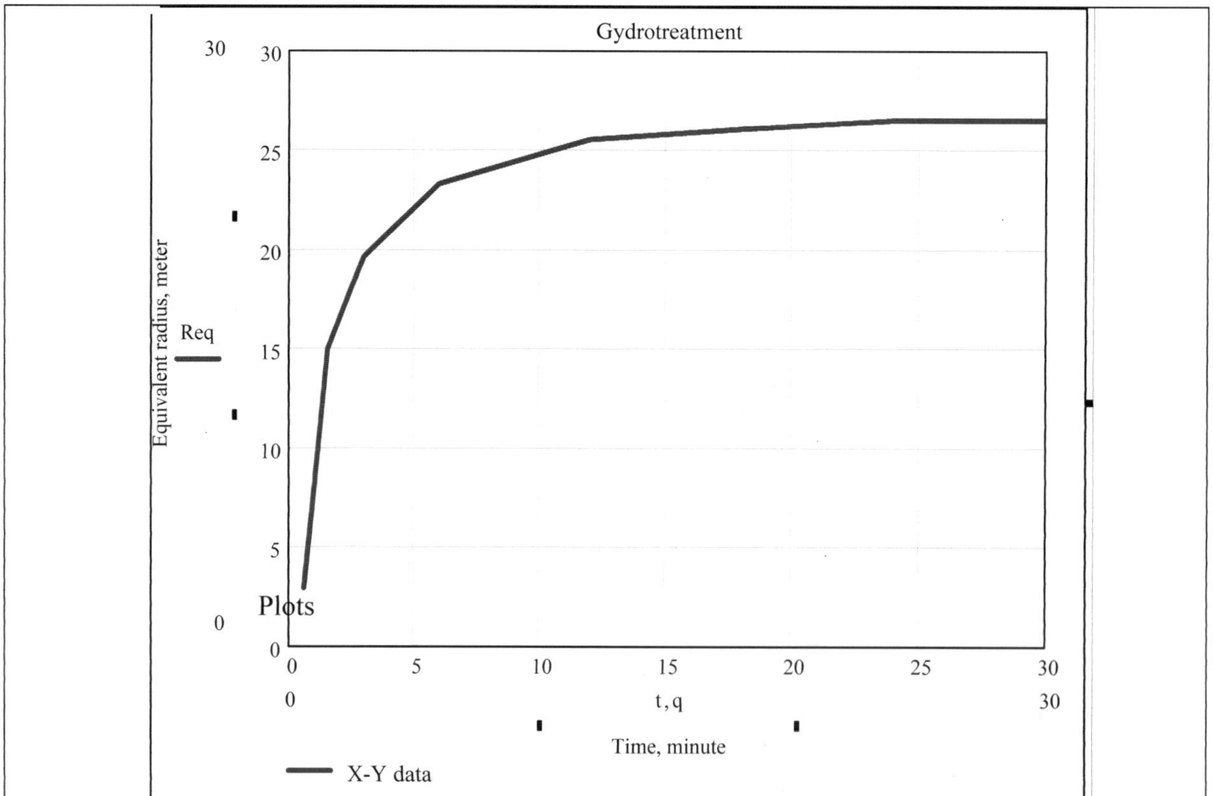

Fig.2 The change of equivalent radius of hydraulic treatment over time

Figure 3.

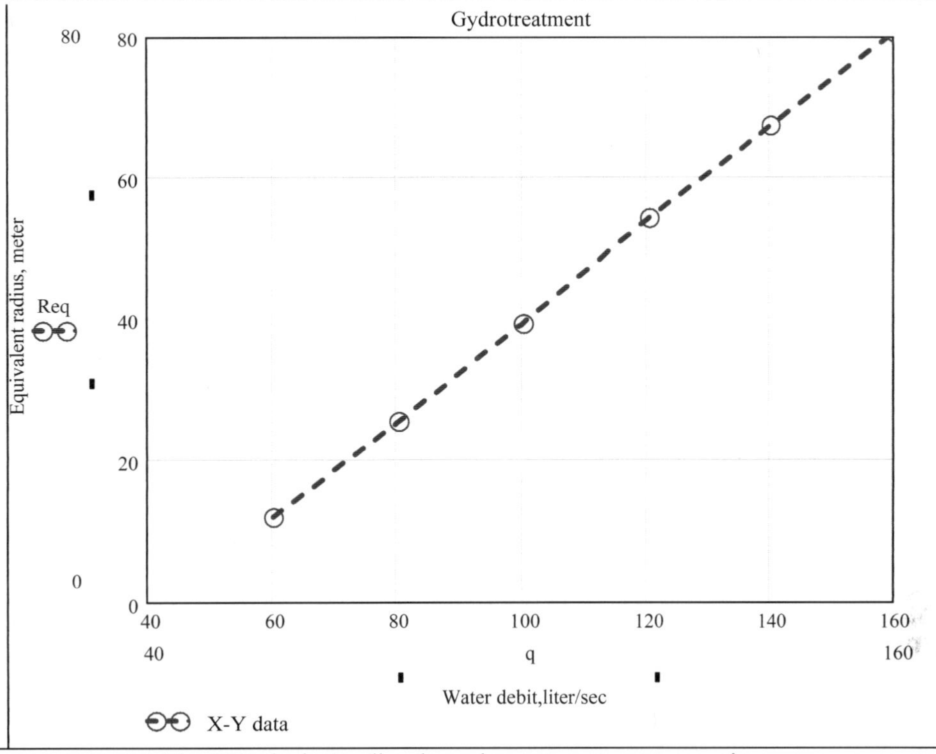

Fig. 3 Equivalent radius dependence on water consumption

Figure 4.

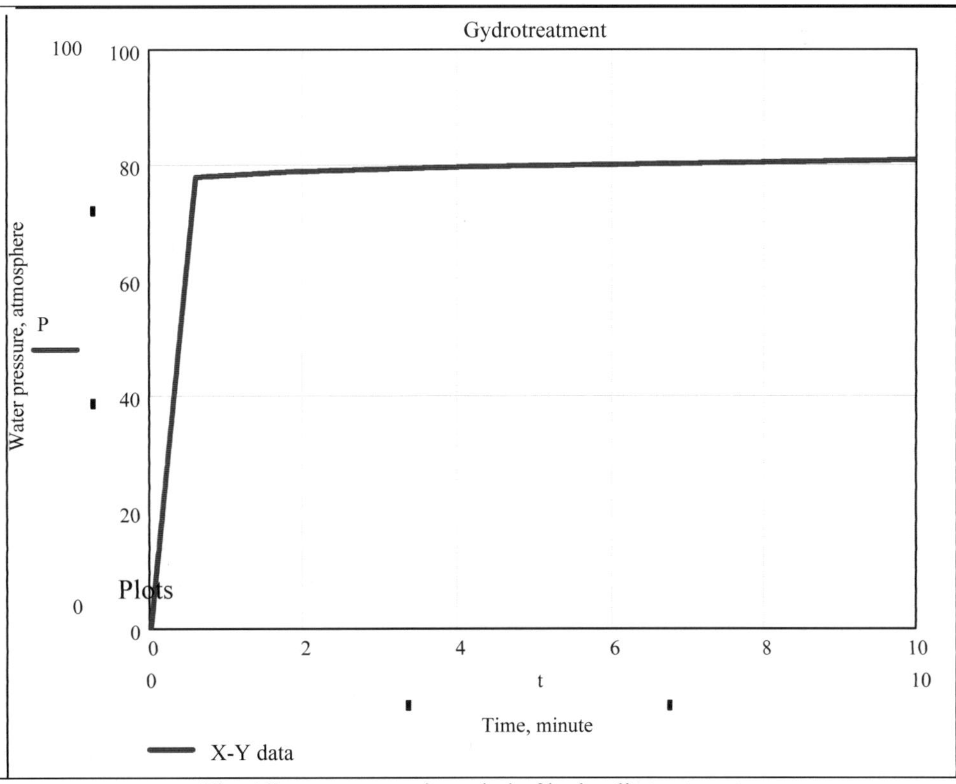

Fig. 4 Pump pressure in period of hydraulic treatment

Figure 5.

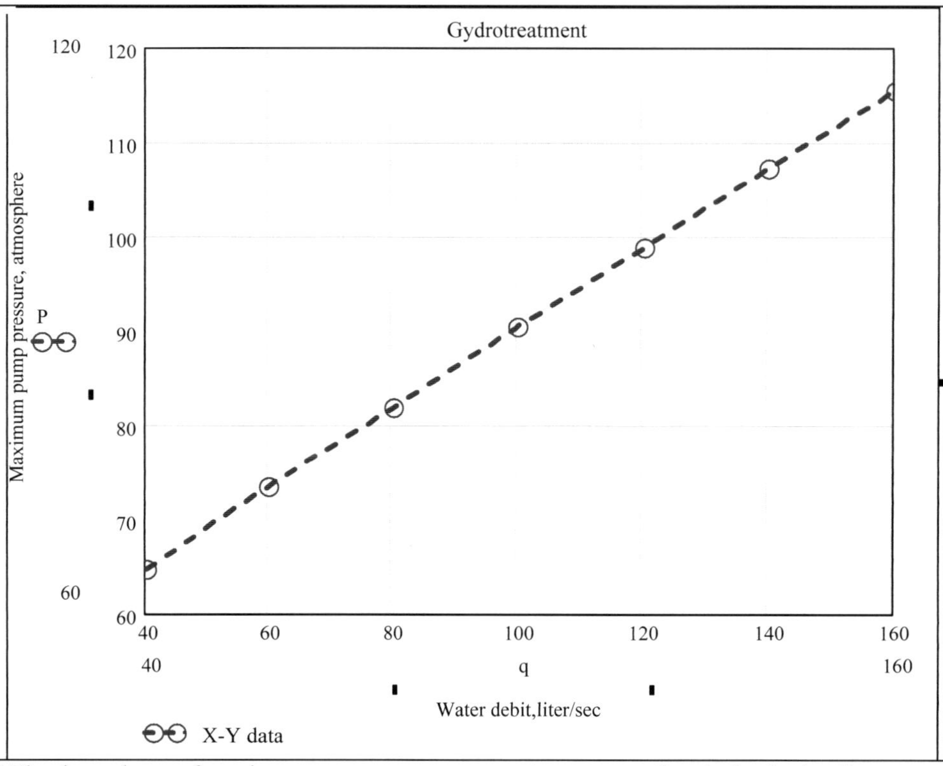

Fig.5 The dependence of maximum pump pressure on water consumption during hydraulic treatment

0532

Measurement of Sorption-Induced Strain

Eric P. Robertson, Idaho National Laboratory
and
Richard L. Christiansen, Colorado School of Mines

ABSTRACT

Strain caused by the adsorption of gases was measured in samples of subbituminous coal from the Powder River basin of Wyoming, U.S.A. and high-volatile bituminous coal from east-central Utah, U.S.A. using an apparatus developed jointly at the Idaho National Laboratory (Idaho Falls, Idaho, U.S.A.) and Colorado School of Mines (Golden, Colorado, U.S.A.). The apparatus can be used to measure strain on multiple small coal samples based on the optical detection of the longitudinal strain instead of the more common usage of strain gauges, which require larger samples and longer equilibration times. With this apparatus, we showed that the swelling and shrinkage processes were reversible and that accurate strain data could be obtained in a shortened amount of time. A suite of strain curves was generated for these coals using gases that included carbon dioxide, nitrogen, methane, helium, and various mixtures of these gases. A Langmuir-type equation was applied to satisfactorily model the strain data obtained for pure gases. The sorption-induced strain measured in the subbituminous coal was larger than the high-volatile bituminous coal for all gases tested over the range of pressures used in the experimentation, with the CO_2-induced strain for the subbituminous coal over twice as great at the bituminous coal.

INTRODUCTION

Coal seams have the capacity to adsorb large amounts of gases because of their typically large internal surface area. Some gases, such as carbon dioxide, have a higher affinity for the coal surfaces than others such as nitrogen. Two gases are of particular importance: methane, because of its inherent value as a fuel; and carbon dioxide, because of its purported effect on global warming.

Fluid Movement in Coal

As reservoir pressure is lowered, gas molecules are desorbed from the matrix and travel to the cleat (natural fracture) system where they are conveyed to producing wells. Fluid movement in coal is controlled by diffusion in the coal matrix and described by Darcy flow in the fracture (cleat) system. Because diffusion of gases through the matrix is a much slower process than Darcy flow through the fracture (cleat) system, coal seams are treated as fractured reservoirs with respect to fluid flow.

Sorption-Induced Strain and Relationship to Permeability

Coalbeds, however, are more complex than other fractured reservoirs because of their ability to adsorb (or desorb) large amounts of gas. Adsorption of gases by the internal surfaces of coal causes the coal matrix to swell and desorption of gases causes the coal matrix to shrink. The swelling or shrinkage of coal as gas is adsorbed or desorbed is referred to as sorption-induced strain. Sorption-induced strain of the coal matrix causes a change in the width of the cleats or fractures that must be accounted for when modeling permeability changes in the system. A number of permeability-change models for coal have been proposed that include the effect of sorption-induced strain [1,2,3,4,5,6]. Accurate measurement of coal strains induced by the sorption of gases becomes important when modeling the effect of gas sorption on coal permeability.

Previous Strain Measurements with Reported Difficulties

Until recently, sorption-induced strain data that can be incorporated into permeability models have been difficult and tedious to collect; however, recent advancements have allowed for the collection of strain data in a much more timely fashion. The measurement of sorption-induced strain in coal has been reported by a relative few researchers. Gray [1] reported that strain varied linearly with gas pressure for both CO_2 and CH_4, but did not include any details about how those values were obtained. Harpalani and Schraufnagel [7] showed that sorption-induced coal strain was not necessarily a linear function of gas pressure, but might be non-linear with decreasing gas pressure. For a coal sample 1.5 inches in diameter and 3 inches in length and measuring strain with strain gauges they found that the sorption process can be extremely slow requiring long stabilization times.

Seidle and Huitt [3] also used strain gauges to measure the volumetric strain of coal samples subjected to various gases under pressure and noted that the resulting strain tests yielded curves resembling sorption isotherm curves and opined that sorption-induced strain was correlated with sorbed gas content. They found that with the large samples required for measurements using strain gauges, the sorption process was slow, requiring nearly three months to stabilize.

Levine [8] presented both longitudinal and volumetric strain for high-volatile bituminous Illinois coal using both carbon dioxide and methane showing that volumetric strain is roughly three times the longitudinal strain (see also Appendix A for derivation relating longitudinal and volumetric strain). Resistance-type strain gauges also presented some problems that Levine noted: 1) they may not adhere properly to the coal, 2) they may not deform homogeneously with the coal, and 3) the length of time required for equilibration caused by restrictions of the coal sample size could be very long. He noted that some samples required exposure times as long as 200 hours (over 8 days), with larger samples requiring even longer equilibration times; and any measurement errors due to the lack of equilibration would result in measurements lower than reality.

Zutshi and Harpalani [9] presented coal volumetric strain data collected using strain gauges attached to coal samples of unspecified dimensions. The coal samples were placed in vessels and pressurized with various gases. Equilibration times for these tests were long; resulting in total time of about 220 days to collect five data points. Chikatamarla and Bustin [10] also recently reported strain measurements using strain gauges and although they do not mention the amount of time needed for equilibration, they did report problems with gas reacting with the strain gauges forcing an early termination of some experiments.

Table 1 summarizes some coal strain data collected using resistance-type strain gauges as reported by previous researchers.

Need for More Coal Strain Data

In light of the rapid development of coalbed methane plays and the potential for CO_2 storage in unmineable coal seams, very little sorption-induced strain data have been reported. Perhaps this is because of the difficulties associated with resistance-type strain gauges and the length of time required for equilibration of the samples. Clearly, if strain data are to be incorporated into the permeability equations that are employed in coalbed simulators, more sorption-induced strain data are needed. In order to avoid the difficulties encountered by previous researchers and to speed up the data collection process, we wanted to determine if strain could be accurately measured on smaller coal samples without the use of resistance-type strain gauges. The use of smaller samples would conceivably result in shorter equilibration times and faster data collection.

STRAIN MEASUREMENT APPARATUS

An earlier paper discusses the development of the strain measurement apparatus used in this research and more detail can be found there regarding the different ideas tried and how the current version of the apparatus was developed [11]. The apparatus used for measuring coal strain currently includes an optical microscope mounted to a pressure vessel with transparent view ports through which changes in coal length can be optically detected. A brief description of its components and operation follows.

Figure 1 is a photograph of the apparatus developed to measure the longitudinal strain of multiple samples under identical conditions of pressure, temperature, and gas composition. In the photograph, a microscope can be seen attached to the pressure vessel and it is through this microscope that the change in the length of the samples is monitored. The microscope that can be positioned anywhere along the outside of the vessel by turning the positioning screw handle. Two opposite sides of the pressure vessel are composed of thick, yet transparent glass through which the coal samples inside the pressure vessel can be seen with the aid of a backlight. The microscope currently being used has a magnification of 2.3X, which gives an object field of 4.8 mm. The coal samples used in these experiments are roughly 20 mm in length, 3 mm in width, and 3 mm in height (¾ in. by ⅛ in. by ⅛ in.). Because of the magnification of the microscope, only the ends of the samples can be seen. The apparatus is tilted about 45 degrees with respect to the horizontal plane so that the coal samples are at rest at one end of the bed. Monitoring changes in length of the samples' free ends allows the calculation of longitudinal strain. The strain is calculated by comparing the initial end position of each of the samples relative to a static mark with subsequent end positions relative to the same mark at different conditions.

The apparatus is capable of holding six samples that lie in individual beds machined into a stainless steel support rod. The ends of the rod are drilled through to allow gas to enter and exit the pressure vessel when desired. When the samples need to be changed, the entire rod is simply removed from the vessel and new samples are placed in the sample beds and reinserted into the chamber. The lower limit of accurate and repeatable measurements of changes in length is 0.001 mm, which translates into a strain of 0.005% for a 20 mm sample. The upper limit is 4 mm, yielding a strain of 20% for a 20 mm sample, which is an order of magnitude larger than typical coal strain induced by gas adsorption.

Volumetric Strain versus Longitudinal Strain

Volumetric strain is three times the measured longitudinal strain under isotropic conditions (see Appendix A for derivation of this relationship). The assumption of isotropic conditions is supported by the strain results reported by Levine [8]. In his experiments, he recorded strain in the direction of all three dimensions and plotted both longitudinal strain and calculated volumetric strain. The ratio of the longitudinal strain to the volumetric strain in Levine's data is 1:3, which supports the assumption that small samples of coal can be considered isotropic.

COAL DESCRIPTION

The small coal samples used for strain measurement were taken from larger blocks of coal that were collected from coal mines in Utah and Wyoming. The high-volatile bituminous Utah coal was collected from the Aberdeen seam, the Gilson seam, and the lower Sunnyside seam of the Book Cliffs coal field from underground mines near Price, Utah. Subbituminous, low-contaminant coal from the Powder River basin was collected from the Anderson seam and the Canyon seam from an open pit coal mine near Gillette, Wyoming. At the mine location, the Canyon and Anderson seams are each over 100 ft thick and separated by about 150 feet of non productive rock. Proximate, ultimate, and heating value analyses were subsequently done on samples of the collected coal. Results of these analyses are shown in Table 2. Literature values obtained for the both the Wyoming coal [12] and the Utah coal [13] are also shown for comparison. Additional coal information can be found in an Argonne National Laboratory study [14] using a Utah coal collected from a mine adjacent to the coal used in this study.

COAL COLLECTION AND SAMPLE PREPARATION

The Utah coal was taken from the conveyer belt carrying recently mined coal out of the mine as close to the active mine face as possible to limit oxygen exposure. Immediately after being taken from the conveyer, each sample was double wrapped in plastic bags and sealed by tape. Transporting the sample from the mine face to the surface took from 5 to 20 minutes depending on the collection site. Upon reaching the surface, the samples were removed from the bags and placed under de-ionized water inside containers for transport back to the Idaho National Laboratory (INL) in Idaho Falls, Idaho. One coal sample was taken from the Aberdeen mine (Aberdeen seam), one from the Centennial mine (Gilson seam), and two from the West Ridge mine (Sunnyside seam).

The Wyoming coal was collected from recently exposed walls from an open pit mine. Large boulders were broken to expose fresh coal inside and smaller samples (roughly one cubic foot) were then taken from this fresh area and immediately placed under water in sealed containers for transport to the INL. Large coal blocks were taken from the Anderson seam and also from the Canyon seam in this manner.

The large coal blocks were kept underwater at the INL laboratories until needed. The small samples used to measure strain were taken from the larger blocks by using a rock saw cooled by de-ionized water. The small samples were then dried on the lab bench using paper towels, measured, and placed into the strain measurement apparatus. We did not attempt to retain the moisture within the coal samples. The moisture within the samples was evaporated at 80° F inside the strain measurement apparatus by cycling between a vacuum and high pressure moisture- and oxygen-free gas until strain at the pressure extremes was constant.

EXPERIMENTAL PROCEDURE

All strain measurements were done at a constant temperature of 80° F. Initially, a hard vacuum was applied to the pressure chamber for 24 hours and then the initial location of each of the sample ends with respect to specific marks on the measurement standard was recorded. Gas was then introduced into the chamber and pressurized to a desired pressure and changes in the length of the samples were monitored over time as gas was adsorbed by the coal and the samples equilibrated to the new pressure state. The strain of five coal samples were averaged to arrive at an average strain value and one stainless steel sample was used as a control. This procedure was repeated for a number of increasingly higher pressures until the maximum desired pressure was achieved. At that point, a hard vacuum was applied to determine the reversibility of the strain as gas was desorbed. The process was repeated for other gases of interest.

RESULTS AND DISCUSSION

Figure 2 shows the average strain with respect to adsorption time of five Anderson coal samples under differing carbon dioxide pressures. We found that the increase in strain over time at a given pressure could be modeled very well using a Langmuir-type equation. This approach was helpful in determining what the strain would be at infinite time if the strain were not completely stabilized after 24 hours. Figure 3 shows the same strain data as in Figure 2, but modeled using a Langmuir-type equation of the following form:

$$S = S_{\max} \frac{t}{t_L + t}, \quad \text{Eq (1)}$$

where S is the measured strain, S_{max} is the maximum, equilibrated strain, t is time in hours, and t_L is a constant representing the time at which S equals ½ S_{max}. The constant t_L can be used as a measure of equilibration time; a small value for t_L means that the strain stabilized rapidly, while a large value for tL represents a long stabilization time. As can be seen in Figure 3, the CO_2 strain data stabilized very quickly at lower pressures (t_L = 0.032 hrs for 100 psia) and the equilibration rate became progressively longer at the higher pressures (t_L = 10.7 hrs at 800 psia). Similar plots showing the relationship between equilibration time and strain stabilization could be constructed for each gas and coal combination, but are not included in this paper due to space constraints.

The strain measurement procedure called for collecting data for 24 hours at each pressure point and then modeling the data using Eq (1) to extrapolate to infinite time. In retrospect, more time could have been allowed at the higher pressure regions of the tests. However, because the data is modeled very well by the Langmuir-type equation, any error associated with the extrapolation is expected to be small. Further testing should be done to determine the reason for the longer equilibration times in the higher pressure region.

The calculated value of S_{max} at the various pressures was then used to construct a strain versus pressure plot, which can be seen in Figure 4. In this figure the average longitudinal strain of the Anderson seam coal is plotted against pressure under an atmosphere of CO_2.

A Measure of the Associated Error

Figure 4 also shows the measured strain of a non-reactive stainless steel sample of the same size as the coal samples used as a control. Because the control is non-reactive, any measured strain in the control is equivalent to the error associated with the measured strain in the reactive coal samples. The standard deviation of the error (measured strain of control sample) is 0.0016%. The error associated with these measurements appears to be random error resulting from the ability to accurately and repeatedly determine the location of the ends of the samples.

Modeling Strain Data

The strain data in Figure 4 are modeled using a Langmuir-type equation of the following form:

$$S = S_{max} \frac{P}{P_L + P}, \qquad \text{Eq (2)}$$

where S is the measured strain, S_{max} is the maximum strain at infinite pressure, P is the gas pressure in psia inside the pressure chamber, and P_L is a constant representing the pressure at which S equals ½ S_{max}. Eq (2) is the same basic equation as Eq (1), but with different constants representing pressure instead of time.

Strain was measured using both the Gilson coal and the Anderson coal as a function of gas pressure for three pure gases: CO_2, CH_4, and N_2. Figure 5 shows the data resulting from these experiments along with the modeling of the data using Eq (2). Table 3 shows the Langmuir strain constants accompanying the model curves for the different coals and gases. The values of P_L are fairly constant for a given gas regardless of the coal rank as seen in Table 3. However, different sorbing gases result in different total strains (S_{max}). Data show that as the total strain decreases, the Langmuir pressure, P_L, increases; meaning that gases with large sorption-induced strains such as CO_2 approach their maximum strain at lower pressures than those with low sorption-induced strains such as N_2. Figure 6 plots the value of P_L, as a function of the average S_{max} for the Anderson and Gilson coals. This figure also shows data suggesting that the product of Langmuir strain constant and the Langmuir pressure ($S_{max} \cdot P_L$) decreases as S_{max} decreases; from 14 psia for CO_2 to 6 psia for N_2.

Shape of Sorption-Induced Strain Data

Seidle and Huitt reported pre-1990 coal strain data from several authors suggesting that sorption-induced coal strain varied linearly with pressure [3]. Some early results from Harpalani and Schraufnagel [7] appear to indicate that sorption-induced strain in coal could be linear with respect to pressure. However, our data (Figure 4, Figure 5, and Table 3) clearly indicate that sorption-induced strain is not a linear function of gas pressure, but can be very satisfactorily fit using a Langmuir-type equation. This should be expected because the principal cause of the change in coal dimensions (strain) is the sorption of gas, which is modeled by the Langmuir isotherm.

A secondary cause of strain in these experiments is the gas pressure acting to compress the coal samples. As gas pressure changes, the resulting compressive and sorptive strains are counter acting. The matrix compressive strain is small relative to the sorption-induced strain in the presence of carbon dioxide, but may become important when the sorbed gas is only slightly adsorptive such as nitrogen. The matrix compressibility can be determined by measuring the strain of a coal sample subjected to the pressure of a non-adsorbing gas such as helium. Figure 7 shows the strain induced by the exposure of the coal samples to helium at various pressures with temperature held constant at 80° F. These data were obtained after allowing sufficient equilibration time to arrive at constant values with respect to time. From the data shown in Figure 6, the longitudinal matrix compressive strain is calculated to be 0.84E-6 psi^{-1}, which translates into a volumetric matrix compressibility of 2.5E-6 psi^{-1} for Anderson (subbituminous) coal at 80° F. In the derivation of permeability models that will employ sorption-induced strain data, matrix compressibility is neglected because it is so much smaller than the fracture compressibility. Therefore, it would be appropriate to include the matrix compressibility portion of the total strain measured under gas pressure and it would not be appropriate to add it back into the measured sorption-induced strain data.

Both adsorption isotherms and strain isotherms can be modeled using forms of the Langmuir equation and it may be possible to predict one based on the character of the other. There is evidence that for most gases and coals, the relationship between strain and adsorption is linear [8,10]. However, Pekot and Reeves suggest that for high adsorbed gas concentration, the adsorption-strain relationship may become non-linear [15]. More work needs to be done in this area before any strong conclusions can be made as to the relationship between strain and adsorption curves, but there is strong evidence that there is a positive (and possibly linear) relationship between the sorption and strain isotherms.

Strain and Coal Rank

The data presented in Figure 5 show a marked difference in the sorption strain between the two ranks of coals studied here. The relationships between sorption-induced strain and coal rank for the three gases used in this study are shown in Figure 8. In this figure, S_{max} (strain at infinite gas pressure) is used as a measure of strain and vitrinite reflectance is used as a measure of coal rank. The CH_4 strain decreases only slightly with the change in coal rank, but the CO_2 and N_2 strain curves each decrease by a factor of about two as vitrinite reflectance increases from 0.24 to 0.53. One would expect the strain curves of all three gases to be affected in the same manner by the change in coals, and more work should be done to determine if the CH_4 data are truly unaffected by a change in coal rank or if it is simply an experimental anomaly.

Regardless of the amount of decrease in strain resulting from an increase in coal rank, the strain of all three gases did decline as rank increased. Other researchers, however, have shown different results. Bustin [16] compared the adsorption capacity of coals of different ranks and found a trend towards higher capacity with higher coal rank. Later, Chikatamarla and Bustin [10] measured sorption-induced strains of various coals and found that strain also increased with higher coal rank, which is contrary to the data reported here (see Figure 8). Based on these results, it may not be possible to derive a general relationship between coal strains (and by analogy, adsorbed gas) and coal rank without further testing on a much larger group of coals of various ranks.

Strain Ratios and Coal Rank

The ratio of strain induced by CO_2 adsorption to strain induced by CH_4 adsorption is of some importance when considering using coal seams as CO_2 sinks for sequestration of carbon or considering CO_2-enhanced CBM production (CO_2-ECBM). During either of these two processes, the strain in the coal increases as CO_2 displaces methane. If the increase in strain is large enough, this can cause a significant reduction in permeability. Figure 9 shows the ratio of CO_2 strain and CH_4 strain for the two coals tested to date. This figure shows that if all the methane in the coal were to be displaced by carbon dioxide, the expected increase in strain would be about twice as large in the subbituminous coal compared to the bituminous coal. Therefore, permeability reduction due to coal swelling during CO_2-ECBM or CO_2 sequestration in coal appears to be sensitive to the rank of the coal and may be more of a detriment in coals of lower rank. This knowledge should be considered during the field selection process for CO_2-ECBM or CO_2 sequestration planning stages.

Reeves [17] compiled some historical data showing a relationship between the CO_2/CH_4 adsorption ratio and coal rank and found that as coal rank increased from subbituminous to high volatile bituminous the CO_2/CH_4 adsorption ratio decreased. However, Chikatamarla and Bustin [10] published strain data showing very little trend in CO_2/CH_4 strain ratio with respect to coal rank. The data shown in Figure 9 tend to support the results presented by Reeves, but definite conclusions regarding the relationship between CO_2/CH_4 strain ratios and coal rank are difficult to make because data for only two coals have been collected to date, and further testing should be done.

CONCLUSIONS

A novel method to measure the longitudinal strain in coal induced by the sorption of gases has been developed that greatly reduces the amount of time required for sample equilibration. The total time needed to obtain data necessary to construct a strain-pressure plot for a given temperature has been reduced from over 100 days with traditional strain gauges to less than 10 days using this new technique.

As sorption pressure increased the equilibration time also increased for the samples tested. Further experimentation should be done to determine the cause for this behavior.

Sorption-induced strain data with respect to time (at constant pressure and temperature) can be modeled using a Langmuir-type equation, which allows the extrapolation of strain data to infinite time. Strain data with respect to adsorption pressure can also be modeled using a Langmuir-type equation and is not a linear function of pressure.

When comparing the strain curves for different gases, CO_2 adsorption caused the highest strain, followed by CH_4, and N_2 adsorption caused the lowest strain in both coals tested. As total strain decreased, the Langmuir pressure, P_L, increased. Gases with large sorption-induced strains such as CO_2 approach their maximum strain at lower pressures than those with low sorption-induced strains such as N_2.

Sorption-induced strain decreased as coal rank increased for all gases tested. The CO_2 and N_2 strains were about twice as large in the subbituminous coal as the high-volatile bituminous coal, while CH_4 strain was only 1.1 times larger in the subbituminous coal than the high-volatile bituminous coal. The CO_2/CH_4 strain ratio also decreased with an increase in coal rank.

Because sorption-induced strain is important to modeling reservoir performance during coalbed CO_2 sequestration, CO_2-ECBM, as well as CBM production by pressure depletion, more strain measurements are needed for a wide variety of coals and under different conditions.

APPENDIX A – DERIVATION OF RELATIONSHIP BETWEEN LONGITUDINAL AND VOLUMETRIC STRAIN

By definition, longitudinal strain,

$$S_L = \frac{\Delta L}{L}, \quad \text{Eq (3)}$$

where ΔL is the change in length of a dimension and L is the original length of that dimension. Also by definition, volumetric strain,

$$S_V = \frac{\Delta V}{V}, \quad \text{Eq (4)}$$

where ΔV is the change in volume of a body and V is the original volume of that body.

Assume we are given a solid, rectangular box with dimensions L_1, L_2, and L_3. By applying a strain-inducing force in all directions, the change in volume can be calculated by:

$$\Delta V = (L_1 + \Delta L_1)(L_2 + \Delta L_2)(L_3 + \Delta L_3) - L_1 L_2 L_3. \quad \text{Eq (5)}$$

Multiplying these terms and simplifying,

$$\Delta V = L_1L_2L_3 + L_2L_3\Delta L_1 + L_1L_3\Delta L_2 + L_1L_2\Delta L_3 + L_1\Delta L_2\Delta L_3 + L_2\Delta L_1\Delta L_3 + L_3\Delta L_1\Delta L_2 + \Delta L_1\Delta L_2\Delta L_3 - L_1L_2L_3; \quad \text{Eq (6)}$$

and

$$\Delta V = L_2L_3\Delta L_1 + L_1L_3\Delta L_2 + L_1L_2\Delta L_3 + L_1\Delta L_2\Delta L_3 + L_2\Delta L_1\Delta L_3 + L_3\Delta L_1\Delta L_2 + \Delta L_1\Delta L_2\Delta L_3. \quad \text{Eq (7)}$$

Volumetric strain then becomes:

$$S_V = \frac{\Delta V}{V} = \frac{L_2L_3\Delta L_1 + L_1L_3\Delta L_2 + L_1L_2\Delta L_3 + L_1\Delta L_2\Delta L_3 + L_2\Delta L_1\Delta L_3 + L_3\Delta L_1\Delta L_2 + \Delta L_1\Delta L_2\Delta L_3}{L_1L_2L_3}. \quad \text{Eq (8)}$$

Simplifying,

$$S_V = \frac{\Delta V}{V} = \frac{L_2L_3\Delta L_1}{L_1L_2L_3} + \frac{L_1L_3\Delta L_2}{L_1L_2L_3} + \frac{L_1L_2\Delta L_3}{L_1L_2L_3} + \frac{L_1\Delta L_2\Delta L_3}{L_1L_2L_3} + \frac{L_2\Delta L_1\Delta L_3}{L_1L_2L_3} + \frac{L_3\Delta L_1\Delta L_2}{L_1L_2L_3} + \frac{\Delta L_1\Delta L_2\Delta L_3}{L_1L_2L_3}. \quad \text{Eq (9)}$$

Canceling terms,

$$S_V = \frac{\Delta V}{V} = \frac{\Delta L_1}{L_1} + \frac{\Delta L_2}{L_2} + \frac{\Delta L_3}{L_3} + \frac{\Delta L_2 \Delta L_3}{L_2 L_3} + \frac{\Delta L_1 \Delta L_3}{L_1 L_3} + \frac{\Delta L_1 \Delta L_2}{L_1 L_2} + \frac{\Delta L_1 \Delta L_2 \Delta L_3}{L_1 L_2 L_3}. \quad \text{Eq (10)}$$

Substituting definition of longitudinal strain, Eq (3), into Eq (10),

$$S_V = S_{L_1} + S_{L_2} + S_{L_3} + S_{L_2} S_{L_3} + S_{L_1} S_{L_3} + S_{L_1} S_{L_2} + S_{L_1} S_{L_2} S_{L_3}, \quad \text{Eq (11)}$$

and assuming the body is isotropic (strain equal in all directions), then volumetric strain becomes:

$$S_V = 3S_L + 3S_L^{\,2} + S_L^{\,3}. \quad \text{Eq (12)}$$

If strain is small, then last two terms can be neglected and volumetric strain is approximately equal to three times the linear strain,

$$S_V \approx 3S_L. \quad \text{Eq (13)}$$

The error associated with Eq (13) is roughly equivalent to the linear strain. So that with a linear strain of 1%, the actual strain is about 1% more than that given by the above equation; with a linear strain of 10%, the actual strain is about 10% greater; and so on. However, there is no error associated with Eq (12) if the material is isotropic with respect to strain.

APPENDIX B – SI METRIC CONVERSION FACTORS

psi	X	6.894 757	E + 00	= kPa
in.	X	2.54*	E + 00	= cm
Btu	X	1.055 056	E + 00	= kJ
ft	X	3.048*	E – 01	= m
ft3	X	2.831 685	E – 02	= m3
lbm	X	4.535 924	E – 01	= kg
°F	(°F – 32)/1.8			= °C
ton	X	9.071 847	E – 01	= Mg

*Conversion factor is exact.

NOMENCLATURE

- L = Length, cm
- P = Pressure, psia
- P_L = Langmuir pressure constant, psia
- S = Strain, dimensionless
- S_L = Longitudinal strain, dimensionless
- S_{max} = Langmuir strain constant; strain at infinite pressure or time, dimensionless
- S_v = Volumetric strain, dimensionless
- t = time, hours
- t_L = Langmuir time constant, hours
- V = Volume, cm^3
- ΔL = Change in length, cm
- ΔV = Change in volume, cm^3

ACKNOWLEDGMENTS

This work has been funded by Battelle Energy Alliance, the prime contractor at the Idaho National Laboratory. The authors wish to express their thanks to the management of the Idaho National Laboratory for funding this work and allowing its publication. We also express our gratitude to Mike Glasson of Andalex Resources and Greg Gannon of Kennecott for help in the collection of the coal samples.

REFERENCES CITED

1. Gray, I., 1987: "Reservoir Engineering in Coal Seams: Part 1 – The Physical Process of Gas Storage and Movement in Coal Seams," paper SPE 12514, *SPE Reservoir Engineering* (February 1987) 28-34.

2. Sawyer, W.K. and G.W. Paul, 1990: "Development and Application of a 3D Coalbed Simulator," paper CIM/SPE 90-119, presented at the 1990 International Technical Meeting hosted jointly by the Petroleum Society of CIM and the Society of Petroleum Engineers, Calgary, Alberta, Canada, 10-13 June.

3. Seidle, J.P. and L.G. Huitt, 1995: "Experimental Measurement of Coal Matrix Shrinkage Due to Gas Desorption and Implications for Cleat Permeability Increases," paper SPE 30010, presented at the 1995 International Meeting on Petroleum Engineering, Beijing, China, 14-17 November.

4. Palmer, I. and J. Mansoori, 1998: "How Permeability Depends on Stress and Pore Pressure in Coalbeds: A New Model," paper SPE 52607, *SPE Reservoir Evaluation & Engineering* (December 1998) 539-544.

5. Pekot, L.J. and S.R. Reeves, 2003: "Modeling the Effects of Matrix Shrinkage and Differential Swelling on Coalbed Methane Recovery and Carbon Sequestration, paper 0328 presented at the 2003 International Coalbed Methane Symposium, University of Alabama, Tuscaloosa, Alabama, May 2003.

6. Shi, J.Q. and S. Durucan, 2003: "Changes in Permeability of Coalbeds During Primary Recovery – Part 1: Model Formulation and Analysis," paper 0341 presented at the 2003 International Coalbed Methane Symposium, University of Alabama, Tuscaloosa, Alabama, May 2003.

7. Harpalani, S. and R.A. Schraufnagel, 1990: "Influence of Matrix Shrinkage and Compressibility on Gas Production From Coalbed Methane Reservoirs," paper SPE 20729, presented at the 1990 SPE Annual Technical Conference and Exhibition, New Orleans, Louisiana, 23-26 September.

8. Levine, J.R., 1996: "Model study of the influence of matrix shrinkage on absolute permeability of coal bed reservoirs," *Coalbed Methane and Coal Geology*, R Gayer and I. Harris (eds.) 1996, Geological Society Special Publication No. 109, 197-212.

9. Zutshi A. and S. Harpalani, 2004: "Matrix Swelling with CO_2 Injection in a CBM Reservoir and Its Impact on Permeability of Coal," paper 0425 presented at the 2004 International Coalbed Methane Symposium, University of Alabama, Tuscaloosa, Alabama, May 2004.

10. Chikatamarla, L. and M.R. Bustin, 2004: "Implications of Volumetric Swelling/Shrinkage of Coal in Sequestration of Acid Gases," paper 0435, proceedings of the 2004 International Coalbed Methane Symposium, paper 0360, University of Alabama, Tuscaloosa, Alabama, May 2004.

11. Robertson, E.P. and R.L. Christiansen, 2004: "Optically-based Strain Measurement of Coal Swelling and Shrinkage," paper 0417 presented at the 2004 International Coalbed Methane Symposium, University of Alabama, Tuscaloosa, Alabama, May 2004.

12. Stricker, G.D. and M.S. Ellis, 1999: "Coal Quality and Geochemistry, Powder River Basin, Wyoming and Montana," 1999 Resource Assessment Of Selected Tertiary Coal Beds And Zones In The Northern Rocky Mountains And Great Plains Region, Part I: Powder River Basin, *in* U.S. Geological Survey Professional Paper 1625-A.

13. *Annual Review and Forecast of Utah Coal, Production and Distribution – 2003*, prepared by the Utah Energy Office, Department of Natural Resources, printed October 2004.

14. Vorres, K.S., 1990: *Users Handbook for The Argonne Premium Coal Sample Program*, www.anl.gov/PCS/report/part1.html.

15. Pekot, L.J. and S.R. Reeves, 2002: "Modeling Coal Matrix Shrinkage and Differential Swelling with CO_2 Injection for Enhanced Coalbed Methane Recovery and Carbon Sequestration Applications," topical report, Contract No. DE-FC26-00NT40924, U.S. DOE, Washington, DC (November 2002).

16. Bustin, M., 2002: "Research Activities on CO_2, H_2S, and SO_2 Sequestration at UBC," Coal-Seq I Forum, Houston Texas, 14-15 March.

17. Reeves, S.R., 2003: "Enhanced CBM recovery, coalbed CO_2 sequestration assessed," *Oil & Gas Journal* (July 14) 49-53.

TABLES

Table 1. Coal strain values as reported by previous researchers.

Author	Gas	Strain Type	Strain Value, percent	Pressure, psi
Gray, 1987	CO_2	Longitudinal	1.0%	800
	CH_4	Longitudinal	0.06%	800
Harpalani and Schraufnagel, 1990	CH_4	Volumetric	0.6%	1000
Seidle and Huitt, 1995	CO_2	Volumetric	0.8%	800
Levine, 1996	CO_2	Longitudinal	0.5%	750
	CH_4	Longitudinal	0.2%	1000
Zutshi and Harpalani, 2004	CO_2	Volumetric	1.1%	750
	CH_4	Volumetric	0.5%	1000
Chikatamarla and Bustin, 2004	CO_2	Volumetric	2.41%	800
	CH_4	Volumetric	0.49%	1000

Table 2. Properties of coal samples collected and used in this research as ascertained from various analyses on an "as received" basis.

	Powder River basin, subbituminous		Eastern Utah, high-volatile bituminous		
	Anderson	Canyon	Gilson	Sunnyside	Aberdeen
Proximate Analysis wt%:					
Moisture	26.60	20.36	7.52	4.61	3.71
Ash	6.18	24.50	2.99	19.30	3.38
Volatile Matter	30.99	24.46	37.42	31.14	41.49
Fixed Carbon	36.23	30.68	52.07	44.95	51.42
Total	100.00	100.00	100.00	100.00	100.00
Ultimate Analysis wt%:					
Moisture	26.60	20.36	7.52	4.61	3.71
Hydrogen	2.08	1.83	3.86	3.68	4.56
Carbon	50.57	41.96	71.66	62.38	75.74
Nitrogen	0.43	0.34	1.36	0.80	1.60
Sulfur	0.27	0.54	0.49	1.37	0.59
Oxygen	13.87	10.47	12.12	7.86	10.42
Ash	6.18	24.50	2.99	19.30	3.38
Total	100.00	100.00	100.00	100.00	100.00
Heating Value, Btu/lb					
Measured	8,514	6,939	12,437	10,788	13,685
Literature values	8,220 [12]	—	12,000 [13]	—	12,300 [13]
Vitrinite Reflectance	0.24	0.28	0.53	0.62	0.54

Table 3. Langmuir strain constants for sorption-induced strain for Anderson and Gilson coals at 80° F.

Gas	Coal	Langmuir strain constants		R^2 value for curve fit
		S_{max}	P_L	
CO_2	Anderson	0.03447	529.19	0.9985
	Gilson	0.01596	581.32	0.9972
CH_4	Anderson	0.00822	818.98	0.9966
	Gilson	0.00749	853.06	0.9962
N_2	Anderson	0.00429	1891.44	0.9989
	Gilson	0.00251	1648.41	0.8229

FIGURES

Figure 1. Photograph of the apparatus developed to measure the strain of multiple samples without the use of strain gauges.

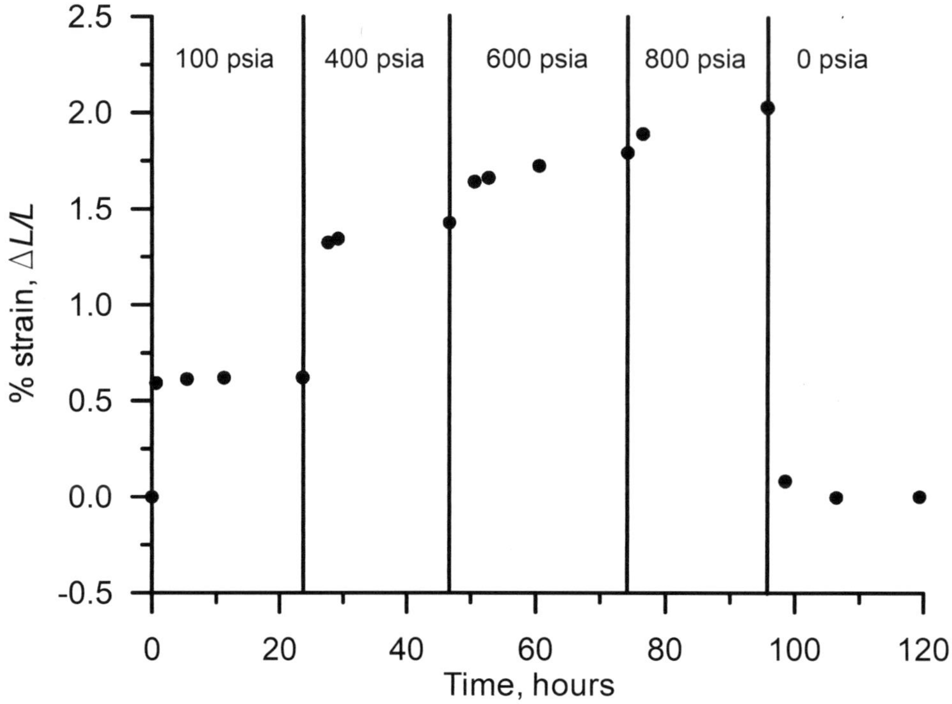

Figure 2. Graph of Anderson seam coal strain with respect to time under different carbon dioxide pressures.

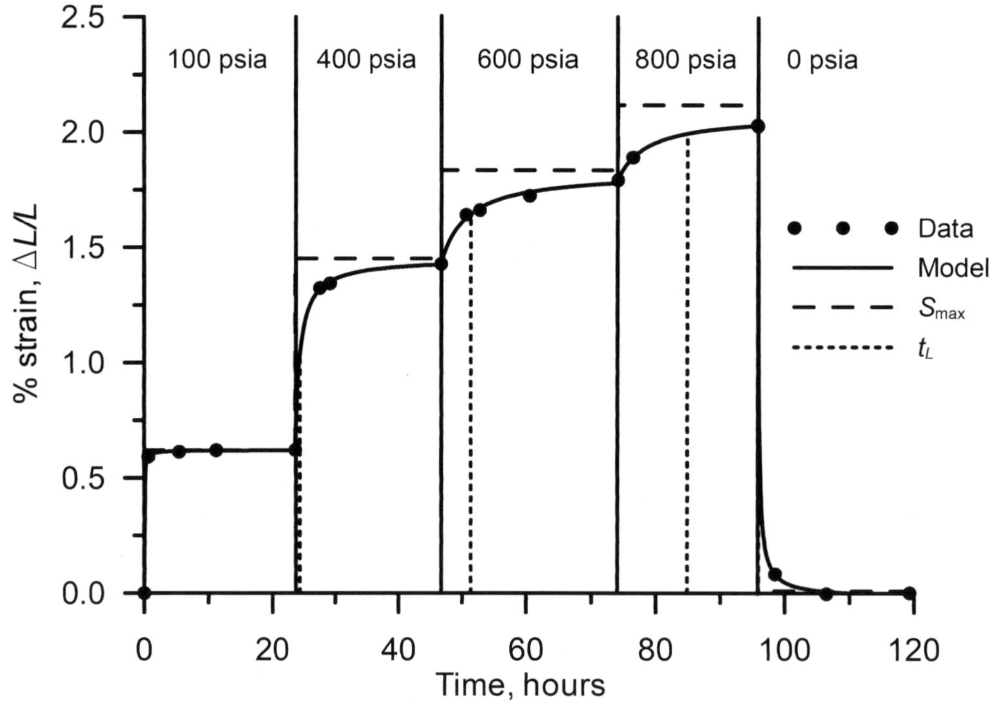

Figure 3. Graph of Anderson seam coal strain with respect to equilibration time under different carbon dioxide pressures. Also shown are the Langmuir-type model with corresponding constants.

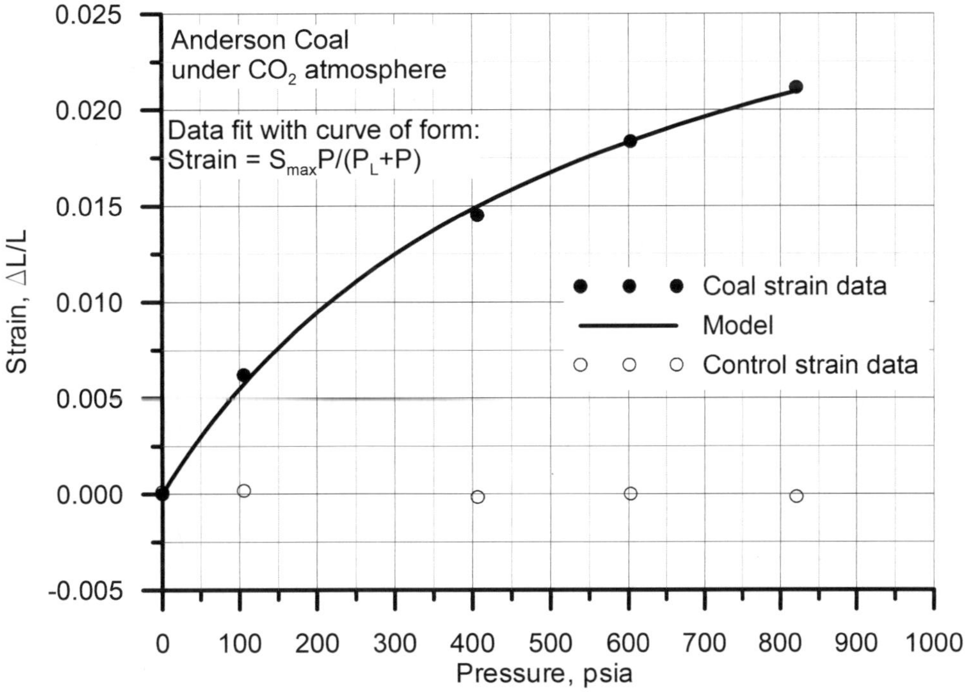

Figure 4. Measured Longitudinal strain using coal from the Anderson seam, Powder River basin of Wyoming under various CO_2 pressures. Also shown is the measured strain of a stainless steel sample used as a non-reactive control.

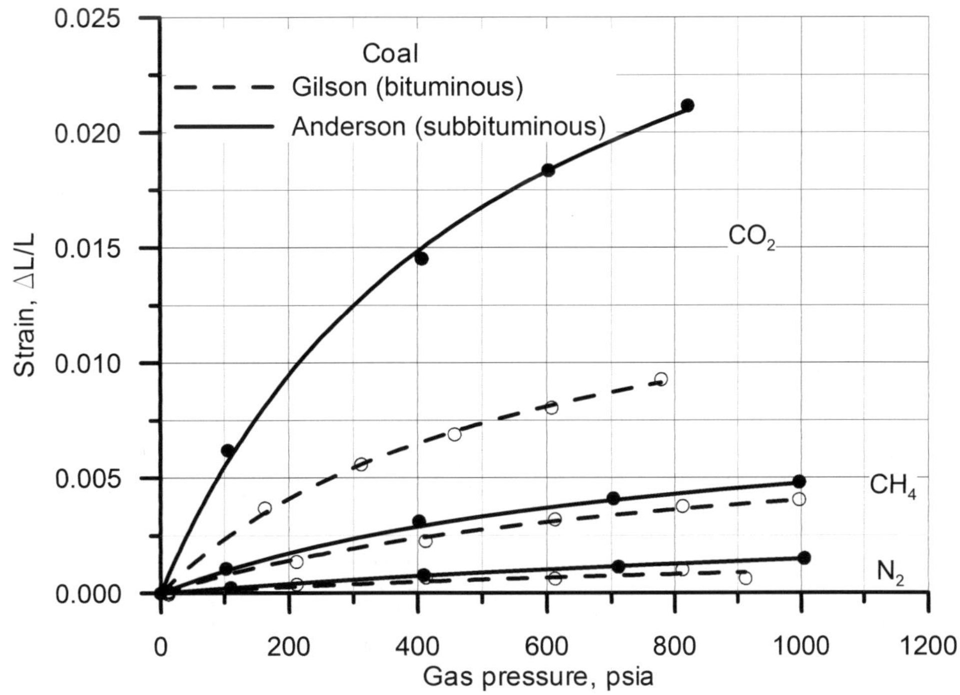

Figure 5. Strain curves for two different coals subjected to three different pure gases at various pressures.

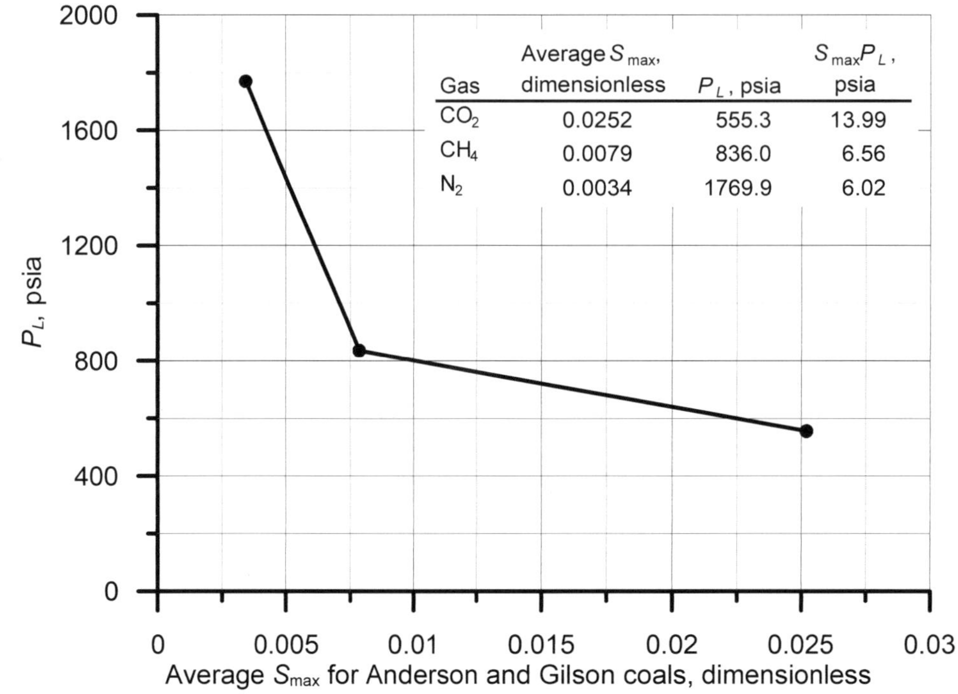

Gas	Average S_{max}, dimensionless	P_L, psia	$S_{max}P_L$, psia
CO_2	0.0252	555.3	13.99
CH_4	0.0079	836.0	6.56
N_2	0.0034	1769.9	6.02

Figure 6. The relationship between the average Langmuir strain, pl, and the average maximum strain for two different coals using three gases.

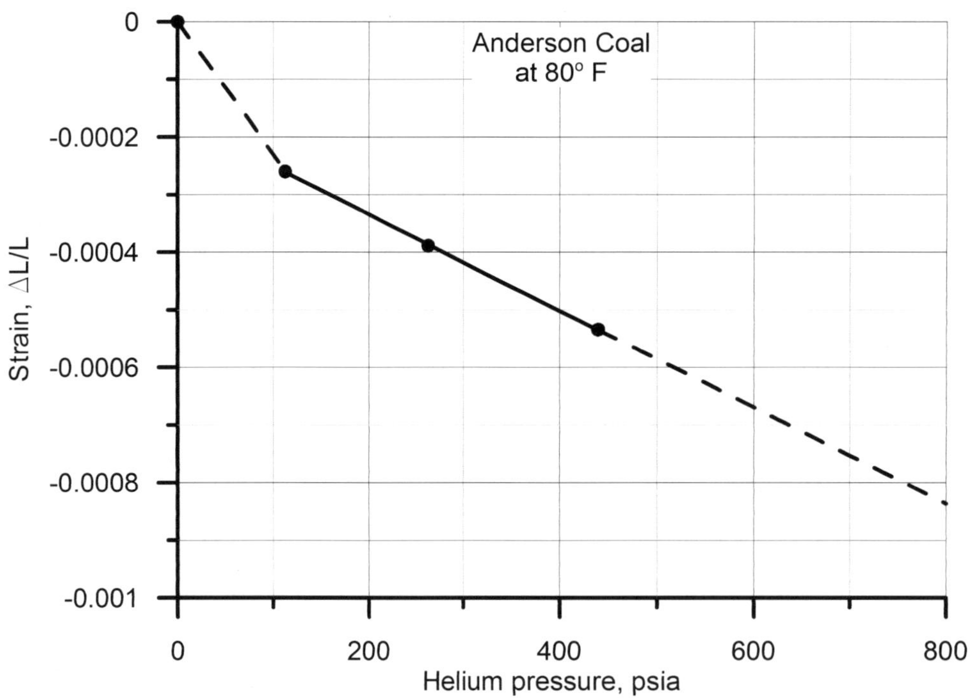

Figure 7. Longitudinal strain of Anderson (subbituminous) coal induced by various helium gas pressures at a constant temperature of 80 °F.

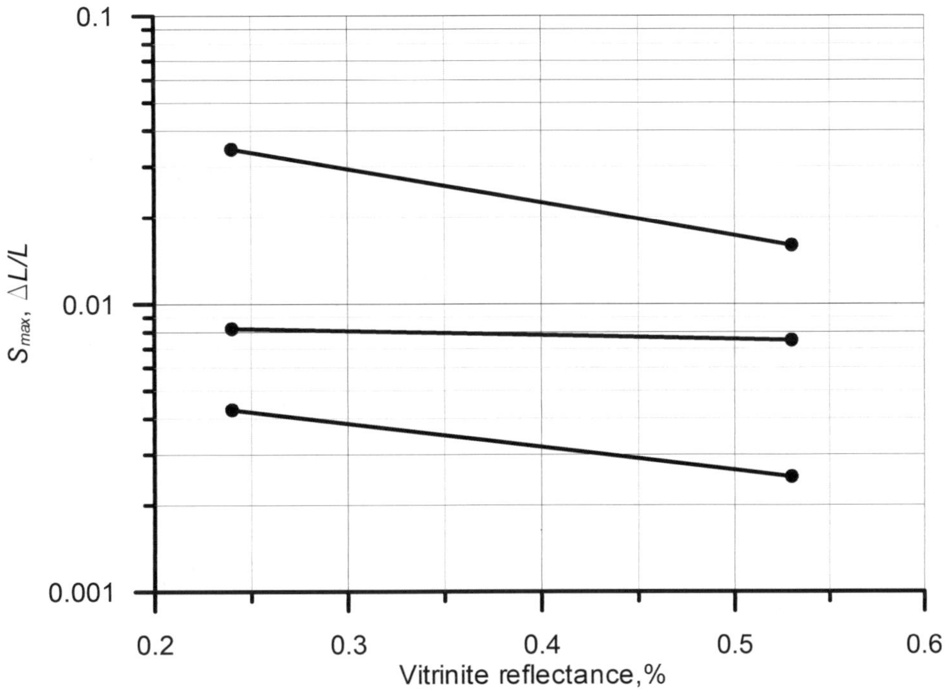

Figure 8. Graph of maximum strain and vitrinite reflectance showing trend toward lower strains as vitrinite reflectance (coal maturity) increases.

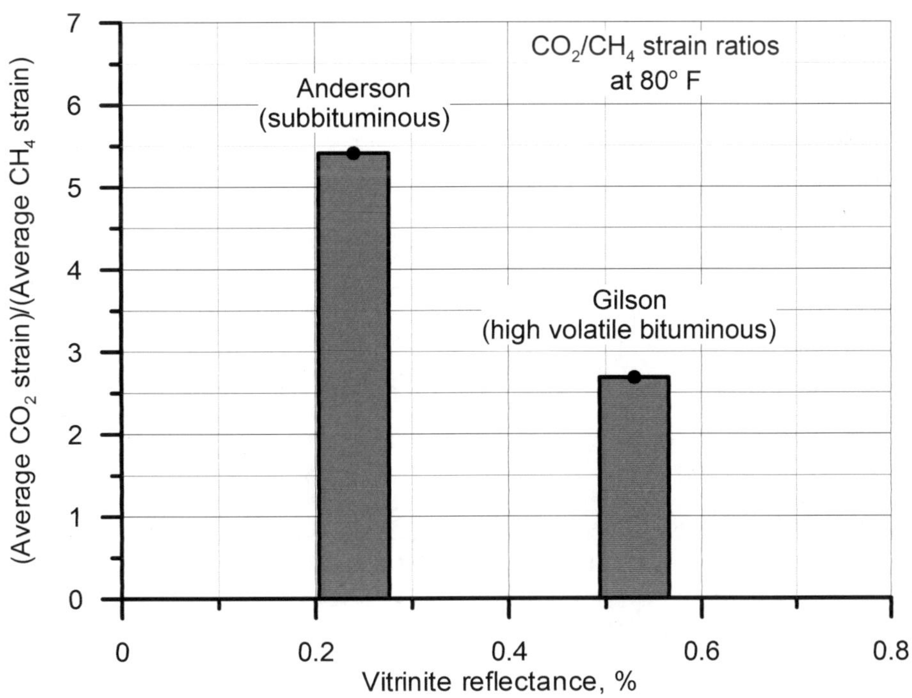

Figure 9. Average CO_2/CH_4 strain ratios for two coals of distinct ranks showing a decrease in the CO_2/CH_4 strain ratio with an increase in coal rank.

LOOKING FOR PERMEABILITY LOSS OR GAIN DURING COALBED METHANE PRODUCTION

Ian D. Palmer, Higgs Technologies, L.L.C., John R. Cameron
and Zissis A. Moschovidis, PCM Technical, Inc.

ABSTRACT

The Palmer-Mansoori (P-M) theory of permeability changes has been incorporated into a pseudo-steady-state equation to provide an approximate way to predict peak gas flowrate, and decline with depletion, when full reservoir modeling predictions are not yet warranted. The predictions are formulated in a risk-based model, which allows a range of values for each input parameter, and produces a range of results for gas flowrate. Screening model predictions can be made quickly and easily for different basins and different fields, to assess potential gas rates and profitability, at any stage of the exploration/exploitation process, as soon as ranges of the input data (even crude ones) become available. The model allows the prediction of P_{10} and P_{90} gas rates, for better understanding of the inherent risks. Absolute permeability in particular is difficult to determine, and the model predicts a spread in gas rates corresponding to the uncertainty in the permeability. We have assumed the full P-M theory applies, i.e. that both permeability loss and permeability gain are occurring. The gas rate predictions are applicable to horizontal wells, as well as vertical fracture-stimulated wells. The modeling can quickly and easily demonstrate the advantage of horizontal wells over vertical wells, for any given lateral length, coal thickness, or vertical permeability. The modeling here applies only to peak gas flowrate and subsequent decline, and we assume for saturated coals that peak flowrate occurs without much depletion at negligible water saturation. If there is a significant water saturation, all gas flowrates predicted would be overestimated by this residual rel perm effect, but the effect should not be large.

We present predictions of our risk-based model for the Cherokee and Piceance basins. These are mainly illustrative because of uncertainties in porosity, matrix shrinkage coefficients, and permeability. In hypothetical horizontal and vertical wells in the shallow Cherokee basin, permeability changes enhance gas production throughout depletion of the reservoir, as compared with production at constant permeability. In a 3 ft coal seam, the advantage of a 2000 ft horizontal well (with zero skin) over a vertical fractured well (with skin -3) is large: a factor of ~8 at initial reservoir pressure. If the vertical permeability is reduced by shale stringers, etc., it appears to be of no practical consequence to gas rates from horizontal wells in thin coal seams. In contrast to the Cherokee basin, in a hypothetical vertical well in the deep Piceance basin, permeability changes REDUCE gas rates substantially throughout depletion of the reservoir as compared with production at constant permeability. Permeability changes depend strongly on initial reservoir pressure. Using the same P-M parameters and porosities, we find in the shallow low-pressure Cherokee basin that permeability always increases after the peak gas production. But in the deep higher-pressure Piceance basin, permeability decreases at early depletions, then rebounds and increases at later depletions.

We also present a simple diagnostic approach to see if permeability is decreasing or increasing with depletion in the field. Applying this to the Cherokee basin example above, the approach confirms that permeability increases for all depletions, due to perm gain by matrix shrinkage dominating perm loss by stress-dependent perm. The overall increase is about the same for both horizontal and vertical wells. Using this diagnostic approach, we can adjust the P-M parameters in our risk-based model to allow more accurate predictions of future gas flowrates, and it becomes easier to history-match field data, forecast future flowrates, and estimate reserves. Once permeability losses/gains have been characterized in then field, it should also be easier to predict the upside of new or infill wells, or even a new play in the same basin. The permeability increase prize can be large: permeabilities in the San Juan basin can apparently increase by 10-100 times, and this has slowed gas rate declines, and given these wells a strong profit upside. Marginal fields may become decent commercial plays if perm increases occur with depletion.

INTRODUCTION

The Palmer-Mansoori (P-M) theory of permeability changes during CBM production includes a permeability loss term (due to an increase in effective stress), and a permeability gain term due to matrix shrinkage as gas desorbs from the coal [1]. This theory has been extended by Mavor et al. to several gas components, and also to injection of gases [2]. A summary of other papers in the area of permeability changes during depletion of CBM reservoirs has been provided in reference [3]. The Palmer-Mansoori equations are summarized in Appendix B.

As a brief summary of recent conclusions in this area of study, we note the following [3]:

- In general: if initial reservoir pressure is high enough, a theoretical permeability rebound should occur with depletion, but if it is not high enough perm may simply increase with depletion right from the start.
- Permeability increases have only been confirmed in the literature for the San Juan and Raton basins.
- Remarkable increases in absolute permeability have been observed with depletion (by 10-100 times) in the San Juan basin fairway, and have no precedent in the oil and gas industry. If perm increases like these can be forecast for prospective CBM plays, this clearly would increase their value.
- A majority of the data in the San Juan basin are consistent with an exponential increase of permeability with depletion. No perm decreases are evident.

In this paper, we discuss a simple method to calculate gas rates that result from permeability changes predicted by the full P-M theory. The P-M theory is used in a numerical pseudo-steady-state model to predict the effect of permeability changes on gas flowrates. With this model we can examine how gas rates are predicted to vary with drawdown and depletion. The results also show the benefits of horizontal wells in boosting gas flowrates, compared with vertical fractured wells. Finally, the model incorporates input data uncertainties (as ranges of input), and uses Monte Carlo methods to give probable ranges of gas production rates e.g., P_{10} and P_{90}. Model predictions can be made quickly and easily for different basins and different fields, to assess potential gas rates and profitability. This can be done at any stage of the exploration/exploitation process, as soon as ranges, even crude ranges, of the input data become available.

A useful diagnostic approach is first to find out, using field data, whether permeability decreases or increases with depletion. This information may not be obvious by simply trying to history-match production data. But once we know this, it becomes easier to match observed gas production using the model. Note that the permeability loss/gain information has implications for rates/reserves estimates, and possibly for choosing infill well sites.

PREDICTION OF GAS FLOWRATES

We have built a tool for rapid prediction of effects of permeability loss/gain on CBM production. This is based on the real gas pseudo-pressure formulation for gas production, but with permeability varying with pressure (from the P-M equation), as well as with viscosity and the gas compressibility (Z) factor. The basic equations for both vertical and horizontal wells are given in Appendix A. The radial gas flowrate under a pseudo-steady-state approximation can be expressed by equations (A-3) and (A-4), where m_k is a modified pseudo-pressure. Equations (A-3) and (A-4), plus (A-5) or (A-6), along with the P-M equation for permeability changes in Appendix B, are the basic equations we solve numerically to predict how gas flowrates vary with drawdown and depletion in CBM wells.

Note that k_o, in the expressions for Ck in Appendix A, is the original reservoir permeability, and is constant. Thus Ck should be constant if there are no changes in skin (S) during production.

Application of these equations assumes that the bulk of dewatering has occurred, the peak gas flowrate has been reached, and relative permeability effects are minimal [14,15]. This usually occurs without much depletion, unless the coals are substantially under-saturated [14]. ***Thus, our modeling here applies only to peak gas flowrate and subsequent decline, and we assume peak flowrate occurs without much depletion (ie, close to initial reservoir pressure).*** Note that all gas flowrates predicted below will likely be overestimated due to this residual relative permeability, but the effect should not be large. If a coal seam is under-saturated, the initial pressure (and flowrate) would correspond to the lower reservoir pressure at which the coal becomes saturated.

Note: the predictions of this paper are all given as functions of depletion, not time. To include time would require a material balance calculation, which we do not include here.

The predictions are risk-weighted in that input variables are specified by a range (ie., minimum, most-likely, and maximum). The output gas flowrate comes from a Monte Carlo solution based on hundreds of runs typically, and is also given as a range (e.g., as P_{10} and P_{90} predictions).

EXAMPLE PREDICTION FROM CHEROKEE BASIN

Parameters for P-M prediction of permeability changes

We will illustrate gas flowrate predictions for the Cherokee basin. These coals are quite a bit shallower than San Juan basin coals (see Table 1). For the gas flowrate predictions in Cherokee basin, we require a prediction of permeability changes using the P-M theory summarized in Appendix B.

The data required to predict permeability changes are listed in Table 1. 1/b is the pressure at a sorption strain of 0.5Co, due to matrix shrinkage, and this determines the shape of the sorption strain with pressure, which is generally assumed to be the same shape as the isotherm [12]. In fact the value of 1/b = 371 psi for Cherokee basin comes from the isotherm of coal samples in a well in Kansas just above the Oklahoma border [11]. Co is the sorption strain at infinite pressure (i.e., after it rises to a plateau during adsorption), and this should track the gas content. The gas content increases with rank, at least for coals between HVA and LV. Thus in the ***Cherokee basin,*** where coals are HVA rank and gas contents are 150-330 scf/ton [11,8], Co should be less than in the Colorado part of the San Juan basin (~500 scf/ton). Co should also be less since reservoir pressure will be lower in the Cherokee basin (400-800 psi using 0.4 psi/ft), because the coal seams are shallower (i.e., less gas adsorbed).

Despite this argument, we have decided to use for the Cherokee basin the same matrix shrinkage parameters Co and 1/b as in the San Juan basin [3], basically because the uncertainties in φo are so large (we are unaware of published values of porosity from the Cherokee basin). While recognizing that our P-M permeability increases in the Cherokee are likely to be over-optimistic, because Co is a bit high, when better data become available these predictions can be improved. Ranges of Young's modulus and Poisson's ratio have been discussed elsewhere [3], and the values in Table 1-3 are representative.

Cherokee Basin

To illustrate trends in gas flowrate during depletion in the Cherokee basin, hypothetical well and reservoir parameters shown in Table 2 have been used in our numerical risk-based model. Gas flowrates were calculated with depletion from an initial reservoir pressure of 800 psi, for both vertical and horizontal wells. This would correspond to a deeper well, based upon Po ~0.4 psi/ft [11]. We assume a constant bottomhole flowing pressure of 0 psi after unloading the well. Wells were assumed to be drilled on 160 acre spacing in a coal seam having a mean thickness of 3 ft +/- 1 ft [11]. The initial horizontal perm is taken to have a mean of 3 md in a range from 1.5 to 6 md, although no permeabilities have been published to our knowledge (i.e., actual permeabilities could be quite different).

The hypothetical ***vertical well*** in the Cherokee basin is assumed to be hydraulically fractured with an equivalent skin of -3 and wellbore radius of 0.354 ft (8.5" diameter). Figures 1 and 2 show gas flowrates without permeability changes (constant permeability), and Figures 3 and 4 show gas rates with permeability changes. Note: if the coal is a "dry" coal, the gas flowrate will decline from the start, relative permeability effects are small, and the prediction of Figure 1 at initial reservoir pressure corresponds to the well brought on line. However, if the coal is wet, a dewatering period often takes place and is accompanied by an incline in gas production to a peak, followed by a decline. ***At the gas peak and thereafter is where relative permeability effects should become small [14,15], and the prediction of Figure 1 at initial reservoir pressure will correspond to the time of peak gas production (i.e., depletion is generally small during dewatering [14]). This assumes a saturated coal.***

Figures 1 and 3 show the ranges of gas rates at 100 psi increments as the reservoir depletes for cases without and with permeability changes respectively. By comparing Figures 1 and 3, it is seen that ***the permeability changes do enhance gas production throughout depletion of the reservoir, i.e., there is no drop in gas production at early times due to permeability loss.*** That this can happen in low-pressure reservoirs was pointed out previously [5,6]. The most likely (P_{50}) peak gas rate without permeability changes is only 65% as great as the peak rate with permeability changes (i.e. 184 MCFD vs. 280 MCFD). Furthermore, the gas rate at final depletion is much smaller for the case of constant perm (i.e. 2.8 MCFD vs. 22 MCFD). With no permeability changes, the gas flowrate has declined by a factor of 65, as opposed to the case with permeability changes where gas flowrate has declined by only a factor of 13.

Note from Figures 1 and 3 that the ranges of predicted gas rates at a given depletion pressure can be substantial. These just reflect the impact of the ranges of input parameters assumed in Table 2. Figures 2 and 4 show actual probability distributions of the gas rates at ***initial*** reservoir pressure (800 psi) for the cases without and with permeability changes, respectively. Although Figures 1 and 3 give error bars at P_0 and P_{100} levels, Figures 2 and 4 enable P_{10} and P_{90} estimates to be read off easily. ***This is a major benefit of the risk-weighted model in decision-making: to reveal quickly and easily the uncertainties in the predictions of gas flowrate.***

As shown by Table 2, the hypothetical ***horizontal well*** was assumed to be 2,000 ft long with a wellbore radius of 0.354 ft (8.5" diameter). For the horizontal well, a standard vertical permeability range of k_v = 0.1, 0.3, 1.5 md is used. Again the horizontal permeability range is k_h = 1.5, 3, 6 md, the same as for the vertical well. The most likely ratio of k_v/k_h is thus 0.1. This ratio is typical for coals, but may be lower in the San Juan basin where shale stringers can be prevalent. However, the impact of vertical permeability on gas flowrate can be investigated by varying the mean k_h from 0.3 down to 0.000001 md (i.e., 0.001 μd). The other input reservoir data are as given in Table 2.

As an example for the horizontal well, Figure 5 shows gas flowrates with permeability changes. As for the vertical fractured well, at the gas peak and thereafter is where relative permeability effects should become small in the case of wet coals, and the prediction of Figure 5 at initial reservoir pressure will correspond to the time of peak gas production. By comparing Figures 3 and 5, it can be seen that the peak gas rate for a horizontal well (2.4 MMCFD) is 8 times greater than the peak for the vertical fractured well (0.3 MMCFD). ***In this 3 ft coal seam, the advantage of a 2000 ft horizontal well (with zero skin) over a vertical fractured well (with skin -3), is substantial.***

However, the advantage of a horizontal well should become less as the vertical permeability becomes less. Figure 6 shows the effect of k_v on gas flowrates (P_{50} values) with depletion for a horizontal well with permeability changes. Also shown is the vertical fractured well as a reference. Note that for k_v/k_h down to 0.01 (i.e. k_v = 0.03 md), gas rate is decreased by less than 15%. For k_v/k_h = 0.00033 (i.e., k_v as low as 0.001 md) gas rates are reduced by ~50%, but they are still 3 - 4 times greater than the fracture-stimulated vertical well shown for reference. For the extreme case of k_v= 0.000001 md (0.001 μd), gas rates are lower than that for a vertical well. In summary, a 2000 ft horizontal well with zero skin producing from a 3 ft coal seam appears to win out over a vertical fracture-stimulated well with skin of -3, even if k_v/k_h is as low as 0.0001. For practical purposes, ***if the vertical perm is reduced by shale stringers, etc, it appears of no practical consequence to gas rates from small coal seams.*** However, for thicker coal seams, eg. 20 ft, the impact of vertical perm is more pronounced. For instance, for k_v/K_h =

0.0001, the reduction in gas rate (compared with $k_v/k_h = 1$) is about 65% for a 3 foot coal seam but is about 90% for a 20 foot coal seam.

DIAGNOSTIC APPROACH

Our approach is to first find out, using field data, whether permeability decreases or increases with depletion. This information may not be obvious by simply trying to history-match production data [5]. With this knowledge, we can better attempt to match observed gas production using the model. We now illustrate the approach for a vertical fracture-stimulated well, but the approach applies to a horizontal well also. From (A-1) we can write:

$$q = C_p\,[P_b^2 - P_w^2] \qquad (1)$$

$$\text{where } C_p = \frac{0.703\, k_{av}\, h}{T\,[\ln(R_e/R_w) - 0.75 + S]} \qquad (2)$$

In this formulation, the effective permeability to gas k_{av} is the volumetric average of k(r). This depends on the radial variation of pressure, via the P-M equation, and therefore on Pb and Pw. Note: Pb is the average reservoir pressure, and is usually within 10% of the pressure at the drainage radius.

Thus if k_{av} is changing in the field during depletion, Cp should change to reflect this, assuming all other variables in (2) stay the same. If relative permeability effects are causing k_{av} (perm to gas) to change, Cp should reflect this too. Here we assume relative permeability effects are small (i.e., we focus on periods after dewatering, and after peak gas rate has been reached). ***In this situation, Cp measured from well performance in the field should approximately follow the k_{av} trend caused by P-M permeability changes.***

Thus Cp, obtained from field parameters in (1), can be charted against depletion, and reveal changes in permeability. At small depletions, Cp may fall initially, due to stress-dependent perm, but at larger depletions it can increase ever more strongly as a result of matrix shrinkage. Note that Cp depends on k_{av} and therefore primarily on Pb (via the P-M parameters), but it also depends on Pw. Cp usually reflects the permeability trend in the P-M equation, as it varies with depletion via Pb. The pressure Pw near the wellbore generally has only a small effect on k_{av} and Cp. However, ***if Pw drops suddenly (e.g., if a compressor is hooked up to a wellhead), Cp may depart from the permeability trend. Cp may also depart from the permeability trend if a workover is instigated, where the skin S in equation (2) is improved.***

To illustrate the approach for a vertical fractured well, we can diagnose the gas flowrate data of Figure 3, for the Cherokee well, where Po = 800+/- 50 psi, and Pw = 0 and fixed (see Table 2). Using corresponding values of q and Pb in equation (1), we calculate Cp as a function of depletion, and this is shown in Figure 7. **Cp actually increases for all depletions, and reflects the fact that permeability gain by matrix shrinkage dominates permeability loss by stress-dependent permeability.**

The approach also works for horizontal wells. Figure 8 shows Cp vs reservoir pressure corresponding to the gas flowrate results of Figure 5. ***The overall Cp increase (ie., the ratio of Cp at end to Cp at start) is about the same for both the horizontal and vertical wells.*** This ratio is ~8. However, recall that the permeability changes may be a bit over-estimated for the Cherokee basin, and so this ratio of 8 may be a bit high.

As a diagnostic, the tool provides a quick way to see if permeability is decreasing or increasing in the field. Permeability increases have been established for the San Juan basin [3], and reported for the Raton basin (but not published). It has also been argued for the Drunkards Wash field in the Uinta basin, but not confirmed [13]. It will be interesting to see whether permeability actually decreases or increases in other basins.

Field Cp values are not too hard to calculate from equation (1), where q is gas flowrate, but Pb and Pw must be known reasonably well. Pw can be directly measured by a downhole pressure gauge. If this is not available, it may be calculated from water levels in the casing right after a well is shutin, if production is up the tubing. Alternatively, it may be estimated from surface pressures, using frictional pressure calculations in the tubing/casing. Pb can be found from well shutins, after the pressure has built up towards a plateau.

Once permeability decreases or increases are established, it becomes easier to match gas rates vs depletion, because we have narrowed down one parameter: the perm change with depletion.

Note: after using the diagnostic tool to determine if permeability changes occur during depletion, we can try to match the field-derived Cp by the model. By varying the P-M parameters, and other well/reservoir parameters, we can adjust the model Cp results (eg, Figures 7 or 8) until we match the field data. The advantage of the error bars in understanding the uncertainties is retained.

DISCUSSION

Predictions for Piceance basin

Now we illustrate gas flowrate predictions for a deeper basin (>5000 ft), with higher reservoir pressure. For a hypothetical well in the ***Piceance basin***, we have the situation in Table 1. Gas content is expected to be high, as coal rank is high [9,19], and reservoir pressure will be high assuming a normally-pressured coal. Therefore in Table 1, we expect Co and 1/b to be similar to that in the San Juan basin, where gas content is ~500 scf/ton.

For the gas rate predictions, we have assumed the input variables in Table 3. We decided to keep the P-M parameter ranges (φo, E, v, b, Co) the same as for the Cherokee basin in Table 2, so that we could compare "apples with apples". Porosities from the Piceance basin are uncertain. The permeability range comes from estimates of field data [16], and the permeabilities are 10 times lower (0.11-0.5 md) than the Cherokee basin (1.5-6 md) because the coals are deeper. However, the hypothetical net coal thickness of ~26 ft for the Piceance is much greater than the Cherokee mean of 3 ft. Again we assume a constant bottomhole flowing pressure: this time 100 psi after unloading the well.

Results are shown for a vertical fracture-stimulated well in Figure 9, with no permeability changes, where peak and declining gas rates are plotted vs depletion. The gas rates are higher (~500 MCFD peak) than corresponding rates modeled for the Cherokee basin in Figure 1 (~200 MCFD peak). When permeability changes are included in the Piceance, we find Figure 10, where ***gas rates are substantially LESS than for the case of no permeability changes (~120 MCFD peak). This is the opposite of what happened in the Cherokee, where gas rates with permeability changes were GREATER than without permeability changes*** (compare Figures 1 and 3). For this hypothetical well in the Piceance basin, the permeability changes indicate that P_{50} gas rates could become less than 25 MCFD after only 400 psi depletion.

To again illustrate the diagnostic approach discussed earlier, we can plot Cp vs depletion, for the Piceancs case with permeability changes included. This is Figure 11, where it is clear that Cp decreases at early depletions, but increases at later depletions. This, in the Piceance basin, is indicative of permeability that rebounds after initially falling, and is in contrast to Figure 7 for the Cherokee basin, where Cp only increases with depletion. Again, this behavior has been discussed previously [5,6]: ***Cp (or average perm) tends to reveal rebound behavior in high-pressure reservoirs, but a monotonic increase with depletion in low-pressure reservoirs.***

Predictions for Powder River basin

For CBM reservoirs with very low gas contents, such as the Powder River basin (~50 scf/ton [17]), the perm increase due to matrix shrinkage will be small, because not much gas is adsorbed on the coal.

Although perm decrease due to stress-dependent perm could still act, this will also be small because the coals are shallow and initial reservoir pressure is low. Which one actually wins may depend on input P-M parameters which are not well known. Changes in permeability may be moot anyway, since permeabilities are so large: 100-1000 md [17].

BENEFITS

The diagnostic approach is first to find out, using field data, whether permeability decreases or increases with depletion. This information may not be obvious by simply trying to history-match production data (this point was made in ref. 5). Once permeability losses/gains have been characterized in the field, they can be matched by calibrating the P-M parameters. This makes it easier to forecast future flowrates, and estimate reserves. It should also be easier to predict the upside of new or infill wells, or even a new play in the same basin.

The prize can be large: permeabilities in the San Juan basin can increase by 10-100 times [3], and this has slowed gas rate declines, and given these wells a strong profit upside. Marginal fields may become decent commercial plays if permeability increases occur with depletion. Our results suggest that shallow plays, such as the Cherokee basin, may exhibit significant permeability increases, all through depletion, but that deeper basins such as the Piceance may not. It appears to depend on the initial reservoir pressure. Our results in the Cherokee and Piceance basins are to be regarded as illustrative, because they are hypothetical wells, and because of uncertainties in porosity, matrix shrinkage coefficients, and permeability. Furthermore, we may have over-estimated a little the matrix shrinkage coefficients in the Cherokee basin, as mentioned above.

With the pseudo-steady-state model, predictions can be made quickly and easily for different basins and different fields, to assess potential CBM gas rates and profitability. This can be done at any stage of the exploration/exploitation process, as soon as ranges of the input data (even crude ones) become available. The model allows the prediction of P_{10} and P_{90} gas rates, for better understanding of the inherent risks. Absolute permeability in particular is difficult to determine, and the model predicts a spread in gas rates corresponding to the uncertainty in the permeability.

Finally the gas rate predictions are applicable to horizontal wells, as well as vertical fracture-stimulated wells. The modeling can quickly and easily demonstrate the advantage of horizontal wells over vertical wells, for any given lateral length, coal thickness, vertical permeability, etc.

SUMMARY AND CONCLUSIONS

- The Palmer-Mansoori (P-M) theory of permeability changes has been incorporated into a pseudo-steady-state equation which provides an approximate way to predict peak gas flowrate, and decline with depletion, when full reservoir modeling predictions are not yet warranted. The predictions are formulated in a Monte Carlo risk-based model, which allows a range of values for each input parameter, and a range of results for gas flowrate. Screening model predictions can be made quickly and easily for different basins and different fields, to assess potential gas rates and profitability, at any stage of the exploration/exploitation process, as soon as ranges of the input data (even crude ones) become available. The model allows the prediction of P_{10} and P_{90} gas rates, for better understanding of the inherent risks. Absolute permeability in particular is difficult to determine, and the model predicts a spread in gas rates corresponding to the uncertainty in the permeability.
- We have assumed the full P-M theory applies, ie. that both permeability loss and perm gain are occurring.
- The modeling here applies only to peak gas flowrate and subsequent decline, and we assume peak flowrate occurs without much depletion in a coal seam that is saturated. If a coal seam is under-saturated, the initial pressure (and flowrate) would correspond approximately to the lower reservoir pressure at which the coal becomes saturated. Note that all gas flowrates predicted will

likely be overestimated by a residual relative permeability effect, but the effect should not be large.

- Our predictions for the Cherokee and Piceance basins are intended to be illustrative, because of uncertainty in (1) porosity and (2) matrix shrinkage coefficients (as well as permeability). Furthermore, we may have over-estimated a little the matrix shrinkage coefficients in the Cherokee basin.
- In a hypothetical well in the shallow Cherokee basin, the permeability changes appear to enhance gas production throughout depletion of the reservoir, ie. there is no drop in gas production at early times due to perm loss.
- In a 3 ft coal seam, the advantage of a 2000 ft horizontal well (with zero skin) over a vertical fractured well (with skin -3), is large: a factor of ~8 at initial reservoir pressure.
- If the vertical permeability is reduced by shale stringers, etc., it appears to be of no practical consequence to gas rates from thin coal seams.
- A simple diagnostic approach provides a quick way to see if permeability is decreasing or increasing with depletion in the field. The back-pressure coefficient Cp measured from well performance in the field should approximately follow the trend of average permeability caused by P-M permeability changes (assuming post-gas-peak relative permeability changes are small). However, if Pw drops suddenly (e.g., if a compressor is hooked up to a wellhead), Cp may depart from the permeability trend. Cp may also depart from the permeability trend if a workover is instigated, and the skin is suddenly improved.
- In the Cherokee basin example, Cp actually increases for all depletions, due to permeability gain by matrix shrinkage dominating permeability loss by stress-dependent permeability.
- The overall Cp increase (ie., the ratio of Cp at end to Cp at start) is about the same for both horizontal and vertical wells.
- For a hypothetical vertical well in the deep Piceance basin, gas rates with permeability changes are substantially LESS than for the case of no permeability changes. This is the opposite of what happened in the Cherokee basin, where gas rates with permeability changes were GREATER than those without permeability changes.
- Permeability changes depend strongly on initial reservoir pressure. In the shallow low-pressure Cherokee basin, permeability (and Cp) always increases after the peak gas production. But in the deep high-pressure Piceance basin, permeability (and Cp) decreases at early depletions, then rebounds and increases at later depletions.
- For CBM reservoirs with very low gas contents, such as the Powder River basin (~50 scf/ton), perm increase due to matrix shrinkage will be small, because not much gas is adsorbed on the coal. Perm decrease due to stress-dependent perm should be small also because the coals are shallow and initial reservoir pressure is low. Which one actually wins may depend on input P-M parameters for Powder River, which are not well known. It may be moot anyway, since initial permeabilities are so large.
- The diagnostic approach is first to find out, using field data, whether permeability decreases or increases with depletion. This information may not be obvious by simply trying to history-match production data. Once permeability losses/gains have been characterized in the field, they can be matched by changing (calibrating) the P-M parameters. This makes it easier to forecast future flowrates, and estimate reserves. It should also be easier to predict the upside of new or infill wells, or even a new play in the same basin.
- The perm increase prize can be large: permeabilities in the San Juan basin can apparently increase by 10-100 times, and this has slowed gas rate declines, and given these wells a strong profit upside. Marginal fields may become decent commercial plays if perm increases occur with depletion.
- Finally the gas rate predictions are applicable to horizontal wells, as well as vertical fracture-stimulated wells. The modeling can quickly and easily demonstrate the advantage of horizontal wells over vertical wells, for any given lateral length, coal thickness, vertical permeability, etc.

ACKNOWLEDGEMENTS

The platform of the risk-based model, StressMan, was designed while the author was employed with Amoco and BP. BP is acknowledged for providing the license for StressMan use.

SYMBOLS

E = Youngs modulus (psi)
v = Poissons ratio
M = constrained axial modulus (psi)
K = bulk modulus (psi)
f = fraction 0→1
β = grain compressibility (/psi)
Co, b = parameters of volumetric strain change due to desorption (depletion)
Co = strain at infinite pressure
1/b = pressure at strain of 0.5Co (psi)
φ = porosity
φo = initial porosity
P = reservoir pressure (psi)
Pb = average reservoir pressure (psi)
Po = initial reservoir pressure (psi)
Pw = bottomhole flowing pressure (psi)
q = gas flowrate (scfd)
k = permeability (md)
ko = initial permeability (md)
k(av) = permeability averaged over reservoir (md)
h = net coal thickness (ft)
μ = gas viscosity (cp)
Z = gas compressibility factor
T = reservoir temperature (R degrees)
Re = equivalent drainage radius (ft)
Rw = wellbore radius (ft)
S = skin factor
m(p) = real gas pseudo-pressure
$m_k(p)$ = modified real gas pseudo-pressure

REFERENCES CITED

1. Palmer, I.D. and Mansoori, J. "How permeability depends on stress and pore pressure in coalbeds: a new model", SPE Res. Eval. and Eng., p 539, Dec 1998.
2. Mavor, M. J. and Gunter, W. D. "Secondary porosity and permeability of coal vs. gas composition and pressure", SPE 90255, Ann. Tech. Conf., Houston, 26-29 Sept, 2004.
3. Palmer, I. D. "Permeability Changes in a CBM Reservoir during Production: an Update, and Implications for CO2 Injection", Paper 0403, International CBM Symp., Tuscaloosa, 3-7 May, 2004
4. Dake, L.P. in Fundamentals of Reservoir Engineering, ElSevier, New York, 1978.
5. Mavor, M., and Vaughn, J. "Increasing permeability in the San Juan basin Fruitland formation", Procs. 1997 Intl. CBM Symp., Tuscaloosa, Alabama, May 1997.
6. Shi, J. Q. and Durucan, S. "Changes in permeability of coalbeds during primary recovery – Part 2: model formulation and analysis", Paper 0342, Intl. CBM Symp., Tuscaloosa, AL, May 2003.
7. Ramurthy, M. et al: "Analysis of the Success of Cavity Completions in the Fairway Zone of the San Juan Basin", SPE 55603, Procs. SPE Rocky Mtn. Reg. Mtg., Gillette, Wyoming, May 1999
8. Cardott, B.J. "Introduction to coal geology of Oklahoma", in Fourth Annual Oklahoma Coalbed-Methane Workshop: Oklahoma Geological Survey, Open-File Report 9-2002, pp 34-55, 2002.
9. Scott, A.R. et al, "Geological and hydrological factors affecting coalbed methane producibility in the San Juan, Greater Green River, and Piceance basins", World Coal CBM Supplement, pp 6-10, January 1996.

10. Jones, A. H. et al. “A review of the physical and mechanical properties of coal with implications for coalbed methane well completion and production”, in Geology and Coalbed Methane Resources of the Northern San Juan Basin (ed. J.E.Fassett), Colorado and New Mexico Rocky Mtn. Assoc. Geologists, Denver, 1988
11. Lombardi, T. E. et al, “Western interior coal region gas resources evaluation: core hole results”, Paper 430, Intl. Coalbed Methane Symp., Tuscaloosa, May 2004.
12. Levine, J. “Model study of the influence of matrix shrinkage on absolute permeability of coal bed reservoirs.” In Gayer, R. and Harris, I. (eds): Coalbed Methane and Coal Geology, Geological Society Special Publication No. 109, pp 197-212, 1996.
13. Ramaswamy, G. “Production history provides CBM insights”, Oil & Gas Journal, p 49, 2 April, 2001.
14. Seidle, J. P. “A numerical study of coalbed dewatering”, SPE 24358, Rocky Mtn. Mtg., Casper, WY, May 1992.
15. Seidle, J.P. “Long-term gas deliverability of a dewatered coalbed”, SPE 21488, Gas Technology Symp. Houston, TX, January, 1991.
16. McKee, C. R. et al. “Stress-dependent permeability and porosity of coal”, Paper 8742, Coalbed Methane Symp., Tuscaloosa, AL, November 1987
17. Montgomery, S. “Powder River basin, Wyoming: an expanding coalbed methane (CBM) play”, AAPG Bulletin, V. 83, No 8, p 1207-1222, August 1999.
18. Joshi, S. in *Horizontal Well Technology*, published by PennWell of Tulsa, 1991.
19. Murray, D.K. “Natural gas from deep coalbeds: a promising resource of worldwide importance”, Paper 0335, Intl Coalbed Methane Symp., Tuscaloosa, May 2003.

APPENDIX A: Gas flowrate equations

We start with the pseudo-steady-state equation for radial gas flowrate into a vertical fracture-stimulated well (with skin factor S). For pressures <2,000 psi, this can be given by [4]:

$$q = \frac{0.703\ k\ h\ (P_b^2 - P_w^2)}{\mu ZT\ [\ln(R_e/R_w) - 0.75 + S]} \qquad \text{(A-1)}$$

However, when μ and Z vary with pressure, and their variation is significant as a reservoir depletes (ie., when pressure >2,000 psi), Darcy's law becomes a non-linear partial differential equation. This can be partially linearized by defining a real gas pseudo-pressure [4]:

$$m(p) = 2\int_{PR}^{P} [p\ /\ \mu(p)Z(p)]\ dp \qquad \text{(A-2)}$$

where PR is an arbitrary reference pressure. This leads to the radial diffusivity equation, which can be solved in the pseudo-steady-state regime to yield equation (A-1) above, but with $m(p)_{Pb} - m(p)_{Pw}$ replacing $P_b^2 - P_w^2$.

This concept can be extended to include other variables under the integral sign, eg., relative permeability to oil, to do the linearization for a gas-oil two phase flow situation (see chapter 8 of ref. 4). By analogy, in our situation with a permeability that varies with pressure (as in the P-M theory), then k(p)/ko can be included under the integral sign to define a modified pseudo-pressure $m_k(p)$:

$$m_k(p)_{Pb} - m_k(p)_{Pw} = 2\int_{Pw}^{Pb} [k(p)\ p\ /\ k_o\mu(p)Z(p)]\ dp \qquad \text{(A-3)}$$

By following the derivation in [4] for the standard m(p), we obtain the radial gas flowrate:

$$q = C_k\,[m_k(p)_{Pb} - m_k(p)_{Pw}] \qquad \text{(A-4)}$$

$$\text{where } C_k = \frac{0.703\,k_o\,h}{T\,[\ln(R_e/R_w) - 0.75 + S]} \qquad \text{(A-5)}$$

for a vertical well, and

$$C_k = \frac{0.703 k_o h}{T[\ln(R_e / Rw) - 0.75 + S_{mh} + S_h + S_{cah} - 1.386]} \qquad \text{(A-6)}$$

for a horizontal well. Here S_{mh} is the mechanical skin for a horizontal well, S_h is the skin due to the horizontal well which is a function of the horizontal well length and wellbore radius, and S_{cah} is the shape factor for the horizontal well which is a function of k_v/k_h, coal bed thickness, the well length, and the half dimension length of the horizontal well spacing [17].

Note that k_o is the original reservoir permeability in this formulation, and is constant, and therefore so should be Ck (if skin factors do not change).

This pseudo-steady-state formulation assumes that the bulk of dewatering has occurred, the peak gas flowrate has been reached, and relative permeability effects are minimal [14,15]. This usually occurs without much depletion, unless the coals are substantially under-saturated [14]. Thus, our modeling here applies only to peak gas flowrate and subsequent decline, and we assume peak flowrate occurs without much depletion. Note that all gas flowrates predicted below will likely be overestimated due to this residual relative permeability effect, but the effect should not be large. If a coal seam is under-saturated, the initial pressure (and flowrate) would correspond approximately to the lower reservoir pressure at which the coal becomes saturated.

Equations (A-3) and (A-4) plus (A-5) or (A-6), along with the P-M equation for permeability changes in Appendix B, are the basic equations we solve numerically to predict how gas flowrates vary with drawdown and depletion in CBM wells.

APPENDIX B: The Palmer-Mansoori equations

From Palmer and Mansoori [1], the porosity change is given by:

$$\frac{\varphi}{\varphi o} = 1 + \frac{Cm}{\varphi o}(P-Po) + \frac{Co}{\varphi o}(K/M-1)\left(\frac{bP}{1+bP} - \frac{bPo}{1+bPo}\right) \qquad \text{(B-1)}$$

$$k/ko = (\varphi/\varphi o)^3 \qquad \text{(B-2)}$$

$$Cm = 1/M - \beta\,(K/M + f - 1)$$

The term with Cm is the stress-dependent permeability term. M is the constrained axial modulus, and is a function of E and v. The term with Co is the matrix shrinkage term. The meaning of other symbols is given in the Symbols listing.

TABLES

Table 1: Hypothetical input data for P-M equation in three basins

Basin	Coal rank	Depth (ft)	Φo (%)	Co	1/b (psi)	E (psi)	v
San Juan [3]	MV, LV [7]	~3000	0.08,0.1,0.13	0.016	500	3.8-4.2 e5 [3]	0.35-0.42 [3]
Cherokee	HVA [8,11]	1000-2000 [11]	???	<0.016	371 [11]	???	???
Piceance	MV [9,19]	>5000	???	Same as SJ	Same as SJ	???	???

Table 2: Range of input data for hypothetical CBM well in Cherokee basin.

Input Summary Table:	Minimum	Most Likely	Maximum	Peak
READ ME !	0	5	10	0
Initial Reservoir Pressure (psi)	750	800	850	0
Gas Viscosity (cp)	0.0099	0.01	0.0101	0
Bottomhole Flowing Pressure (psi)	.	0	.	0
Compressibility Factor	.	0.95	.	0
Permeability (md)	1.5	3	6	2
Pseudo Skin Factor Vertical Well	-4	-3	-2	2
Temperature (deg F)	99	100	101	2
Wellbore Radius (ft)	.	0.354	.	0
Thickness (ft)	2	3	4	2
Perm loss coefficient	.		.	0
poissons ratio	0.35	0.39	0.42	5
Youngs modulus (psi)	380000	400000	420000	6
Initial porosity %	0.08	0.1	0.13	2
Shrinkage magnitude (Co/b)	.	8	.	0
shape parameter (b)	.	0.002	.	0
Type of Wellbore Pressure Strategy	.	1	.	0
Length of horizontal Well	1000	2000	2010	6
Vertical Perm (md)	0.1	0.3	1.5	6
Acre Spacing	.	160	.	0
Well Geometry (1=vertical, 2=horizontal)	.	1	.	0
Mechanical skin for horizontal well	.	0	.	0

Table 3: Range of input data for hypothetical CBM well in Piceance basin.

Input Summary Table:	**Minimum**	**Most Likely**	**Maximum**	**Peak**
Category: Gas Reservoir				
READ ME !	0	5	10	0
Initial Reservoir Pressure	2150	2200	2250	0
Gas Viscosity	0.0182	0.0192	0.0202	0
Bottomhole Flowing Pressure	.	100	.	0
Compressibility Factor	.	0.77	.	0
Permeability	0.11	0.21	0.5	2
Pseudo Skin Factor Vertical Well	-4	-3	-2	2
Temperature	148	150	152	2
Wellbore Radius	.	0.354	.	0
Thickness	13	26	35	2
Perm loss coefficient	.	0.003	.	0
poissons ratio	0.35	0.39	0.42	5
Youngs modulus	380000	400000	420000	6
Initial porosity	0.08	0.1	0.13	2
Shrinkage magnitude	.	8	.	0
shape parameter	.	0.002	.	0
Type of Wellbore Pressure Strategy	.	1	.	0
Length of horizontal Well	2700	3750	4000	6
Vertical Perm (md)	0.011	0.021	0.05	6
Acre Spacing	.	160	.	0
Well Geometry	.	1	.	0
Mechanical skin for horizontal well	.	0	.	0

FIGURES

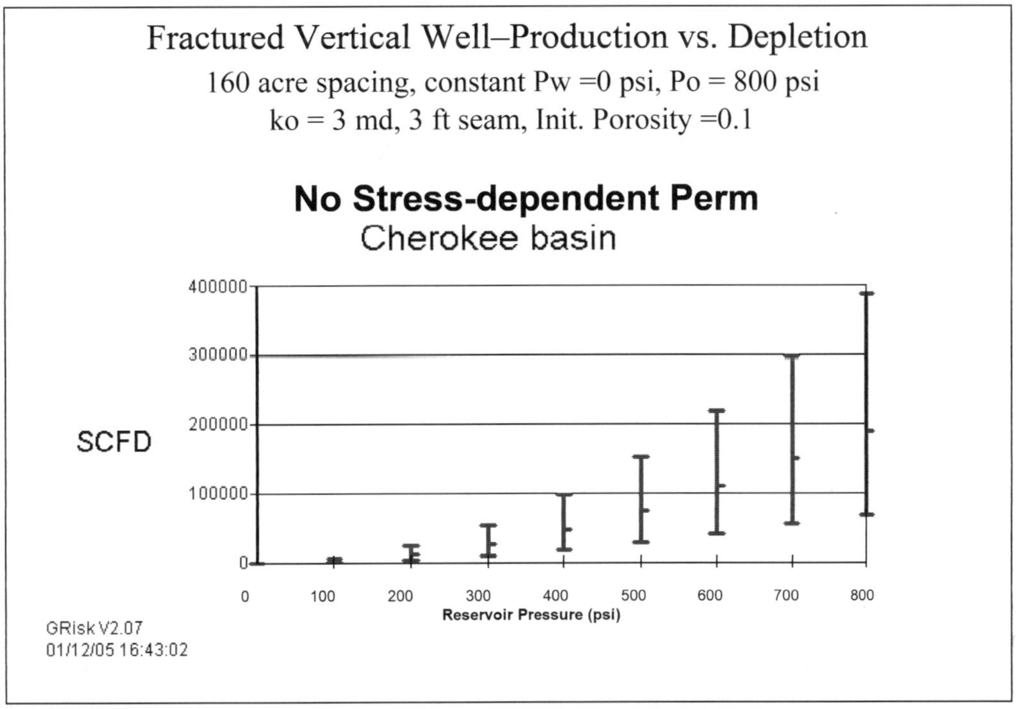

Figure 1: Cherokee gas flowrate prediction and decline without P-M permeability changes: vertical fractured well. Initial reservoir pressure is 800 psi, and this rate corresponds to the peak gas rate after dewatering.

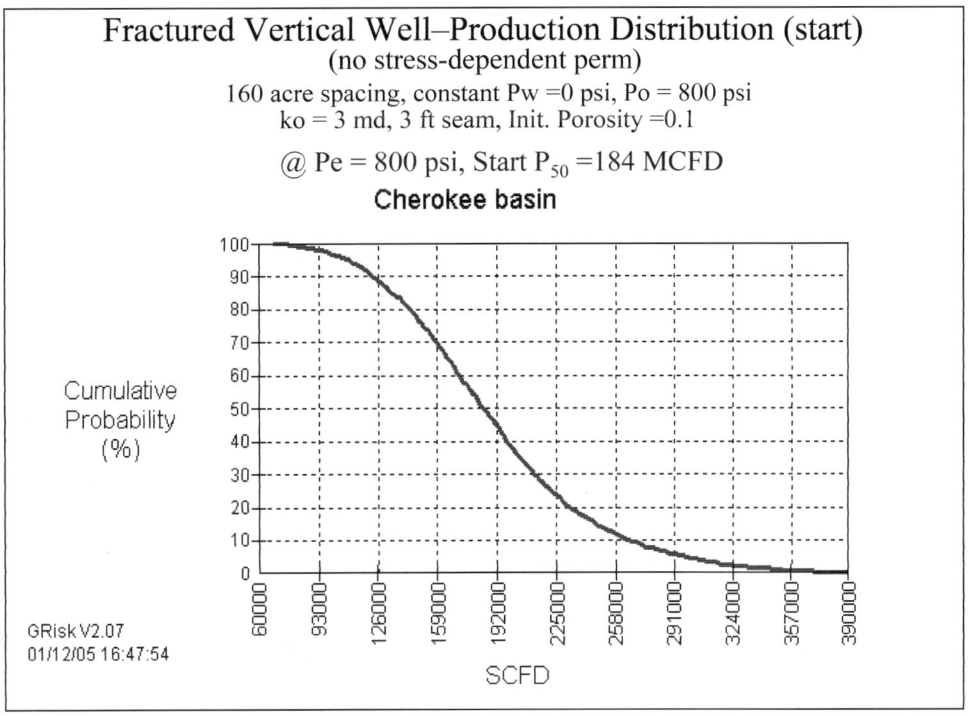

Figure 2: Cherokee probabilities of peak gas rates for a vertically-fractured well with no permeability changes: P10 and P90 can be read off the plot.

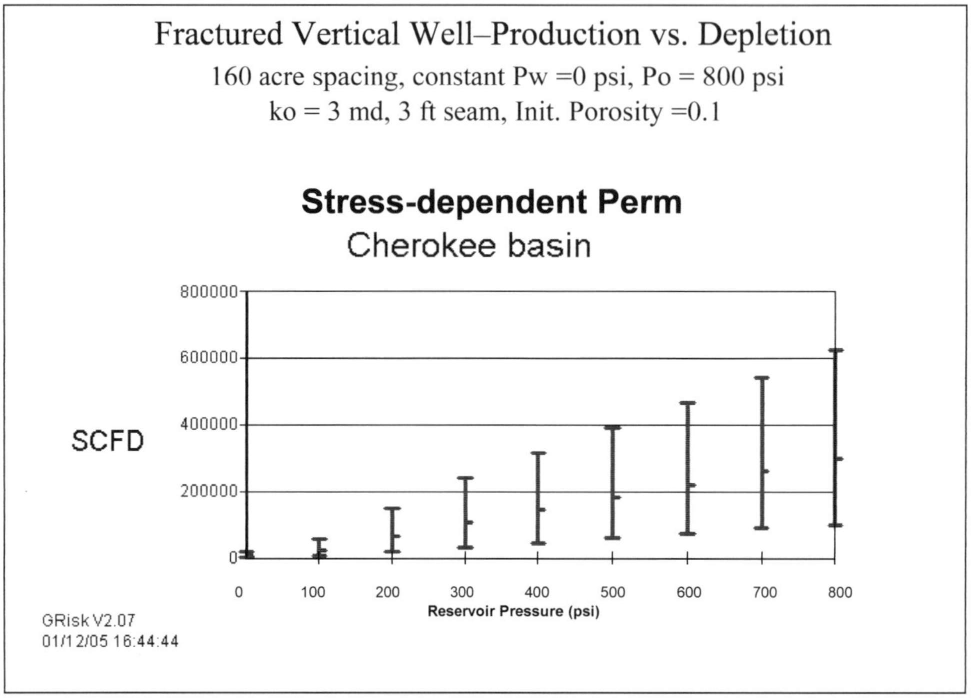

Figure 3: Cherokee gas flowrate prediction and decline with P-M permeability changes: vertical fractured well. Initial reservoir pressure is 800 psi, and the corresponding rate applies to the peak gas rate after dewatering.

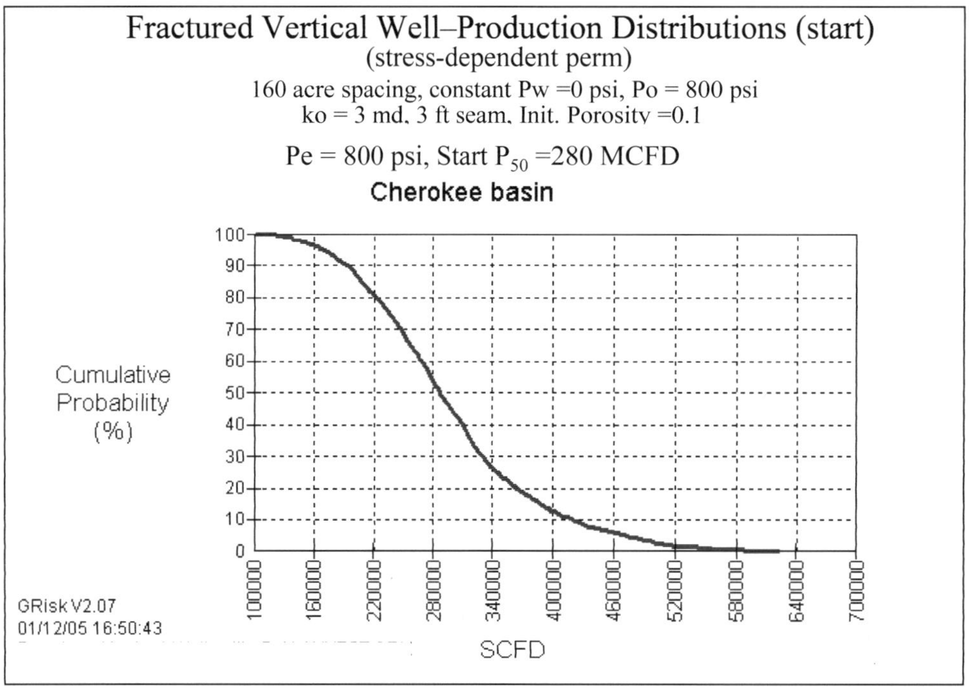

Figure 4: Cherokee probabilities of peak gas rates for a vertically-fractured well with permeability changes: P10 and P90 can be read off the plot.

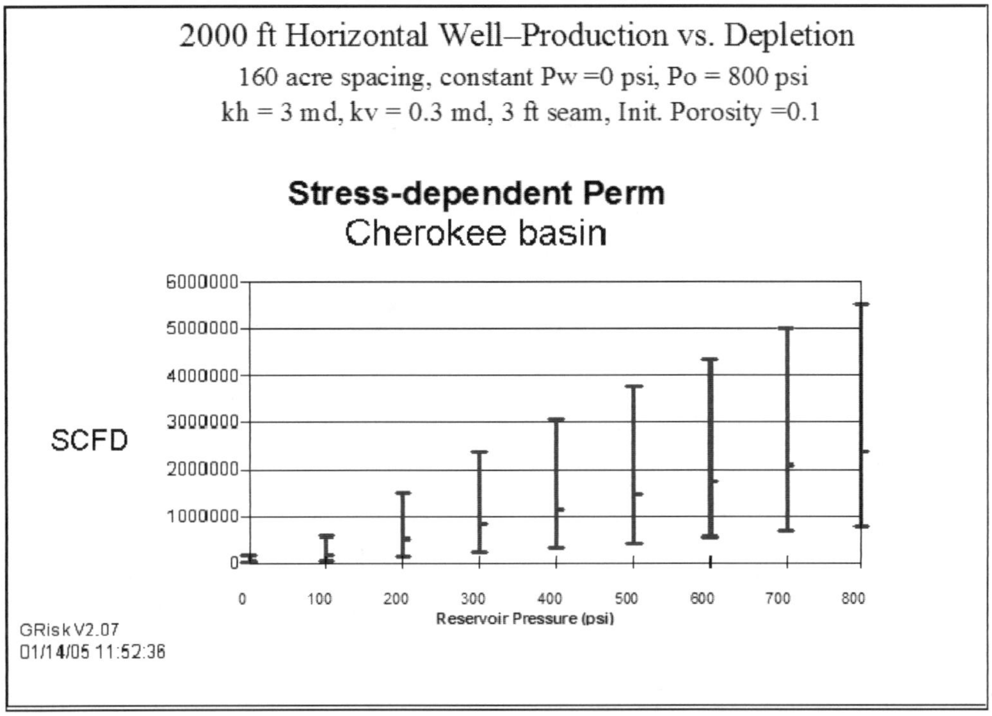

Figure 5: Cherokee gas flowrate prediction and decline with P-M permeability changes: horizontal well. Initial reservoir pressure is 800 psi, and the corresponding rate applies to the peak gas rate after dewatering.

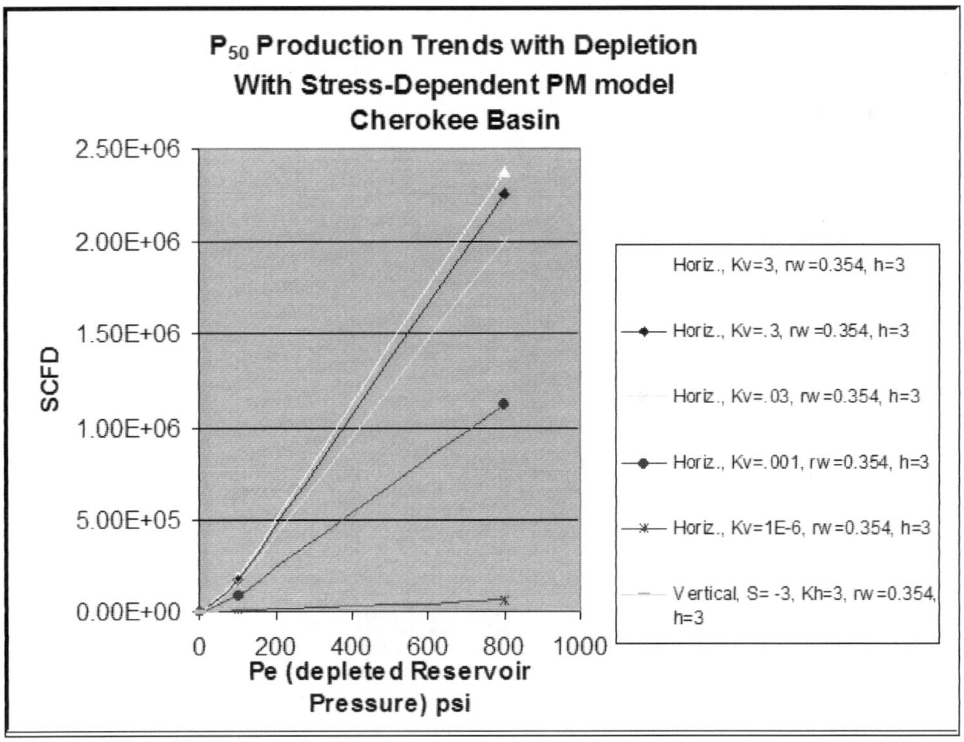

Figure 6: Cherokee gas flowrate declines with P-M permeability changes, for different k_v/k_h ratios: horizontal well with initial reservoir pressure 800 psi.

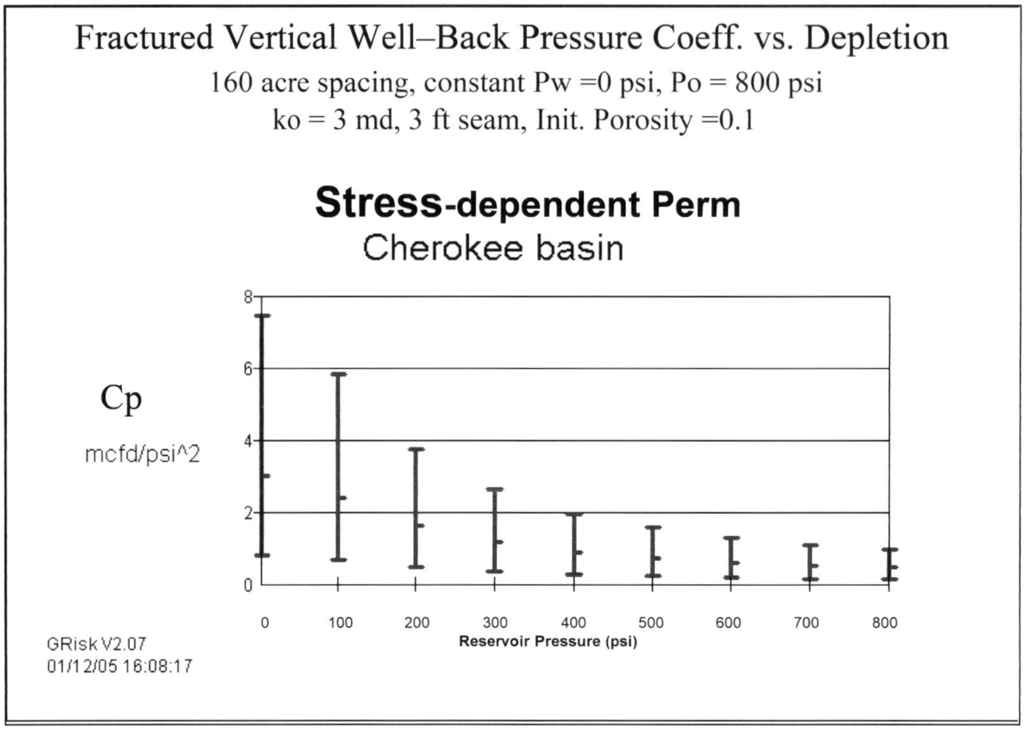

Figure 7: Cherokee basin plot of Cp vs reservoir pressure with permeability changes: vertical fractured well. Initial reservoir pressure = 800 psi.

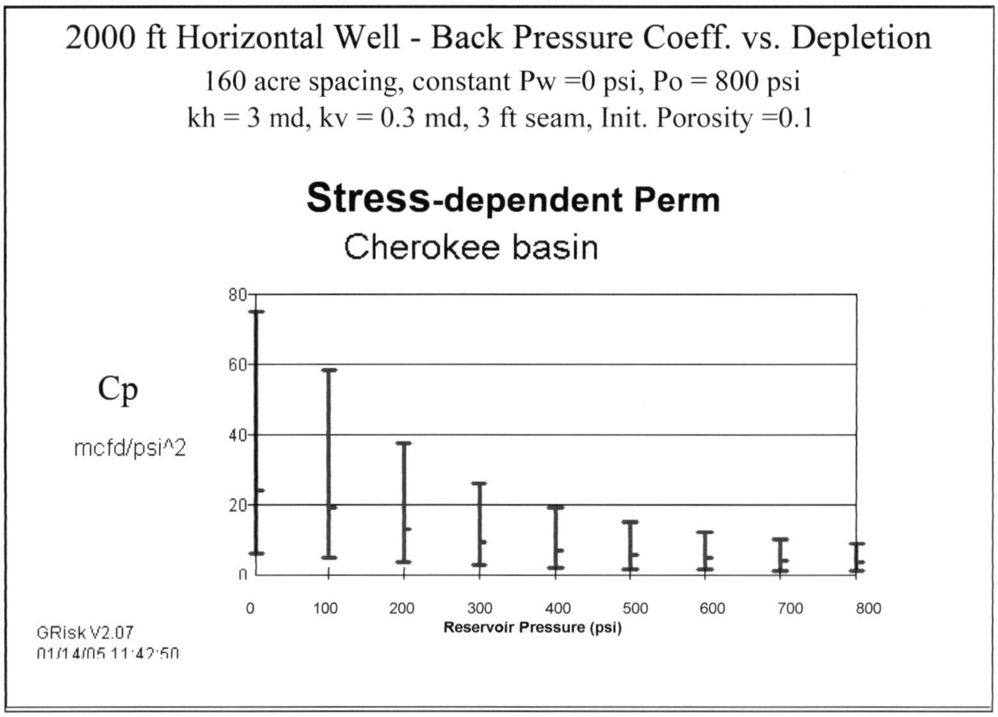

Figure 8: Cherokee basin plot of Cp vs reservoir pressure with permeability changes: horizontal well. Initial reservoir pressure = 800 psi.

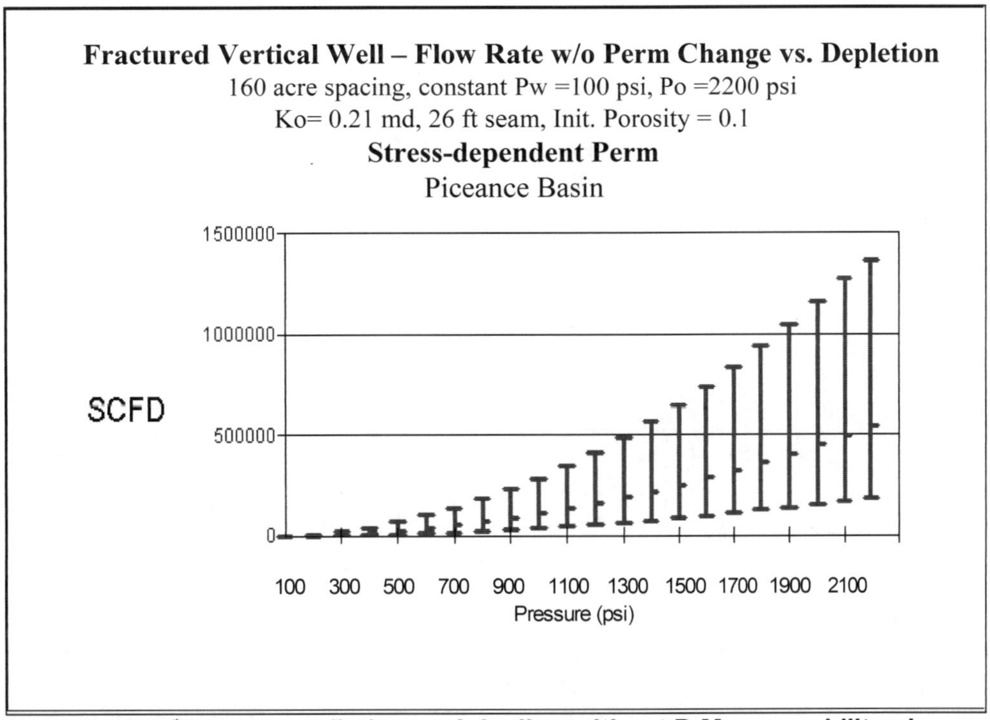

Figure 9: Piceance gas flowrate prediction and decline without P-M permeability changes: vertical fractured well. Initial reservoir pressure is 2200 psi, and the corresponding rate applies to the peak gas rate after dewatering.

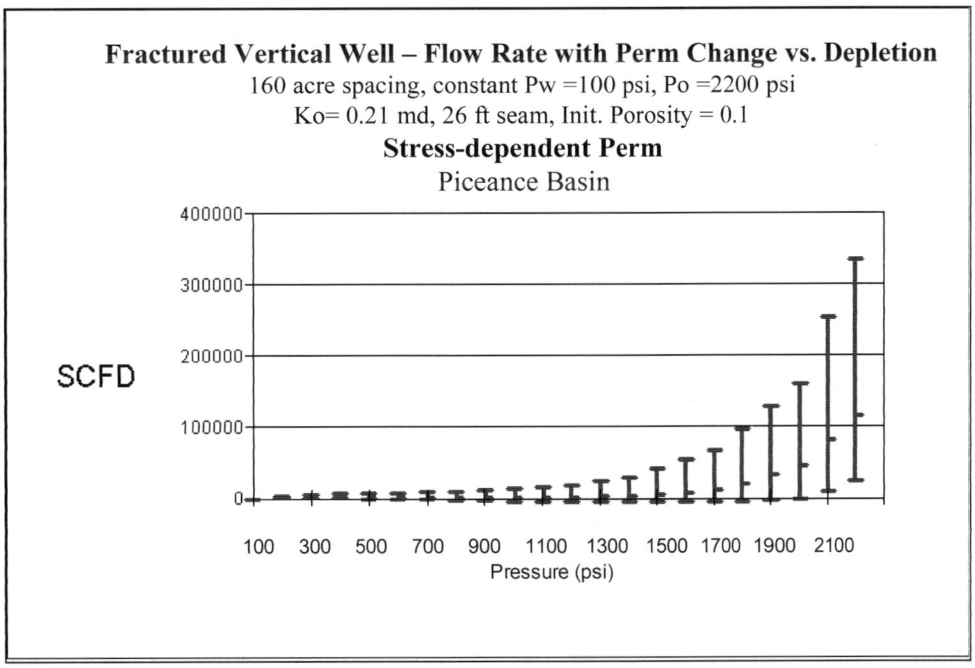

Figure 10: Piceance gas flowrate prediction and decline with P-M permeability changes: vertical fractured well. Initial reservoir pressure is 2200 psi, and the corresponding rate applies to the peak gas rate after dewatering.

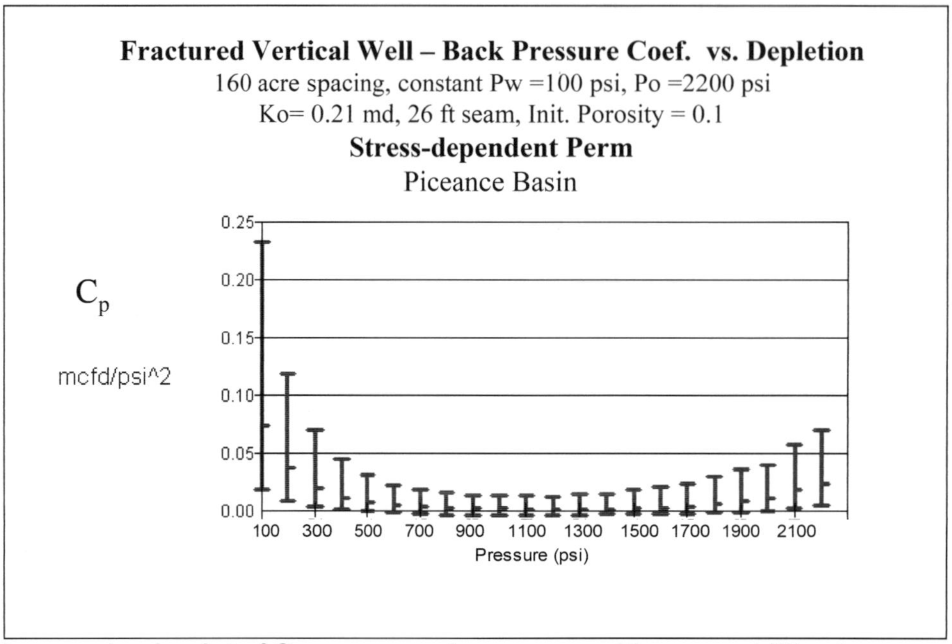

Figure 11: Piceance basin plot of Cp vs reservoir pressure with permeability changes: vertical fractured well. Initial reservoir pressure = 2200 psi.

0502

Modelling the Movement of Coalbed Gas for Different Coal Permeability

F. Sereshki[1], N. I. Aziz[1], I. Porter[1] and A. Godbole[2]

[1] School of Civil, Mining and Environmental Engineering, University of Wollongong, NSW 2522, Australia.

[2] School of Mechanical, Materials and Mechatronics Engineering, University of Wollongong, NSW 2522, Australia.

1. INTRODUCTION

The conventional reason for testing and measuring gas properties of coal was mine safety and the results were used to help in designing underground coal mine ventilation and gas drainage system. Drainage of methane and other gases is necessary for safe extraction of coal from underground operation. The effectiveness of the extraction is dependent on a clear knowledge and better understanding of the nature of gas accumulation and structural characteristics of the host coal and coal measure rocks. Also, recent studies have shown that the removal of gas from coal can contribute to coal shrinkage, which may change the coal permeability during operation.

2. GAS IN COAL

Figure 1 shows a magnified picture of a coal matrix. Coal is considered to be a dual-porosity rock. The matrix structure of coal is characterized by both micropores (< 2 nm) and macropores (> 50 nm) in size. The storage of methane in coal structure occurs in two different forms: (a) by sorption into pores and microfractures, and (b) as free gas. Almost 95% of the stored gas in coal is in the adsorbed state as monomolecular layers on the surfaces of fissures cracks and cleavages, and only a small percentage (<5%) is in free state as shown in Figure 2. Jolley, Morris and Hinsley (1968) estimated that the surface area of one kilogram of coal could be between 20,000 m^2 and 200,000 m^2; this corresponds to gas hold in the coal of 25 m^3/tonne (Creedy, 1991). The level and ease of gas desorption/retention from the coal seam is influenced by moisture, temperature, coal structure, coal porosity and permeability. The gas will flow out of these pores if there is a pressure gradient acting as a driving force.

Figure 1. Electron-micrograph of a Coal Matrix

Figure 2 shows how methane is stored in coal as free gas or adsorbed gas in the form of a monomolecular layer on the internal surface of coal (Harpalani and Schraufngel, 1990).

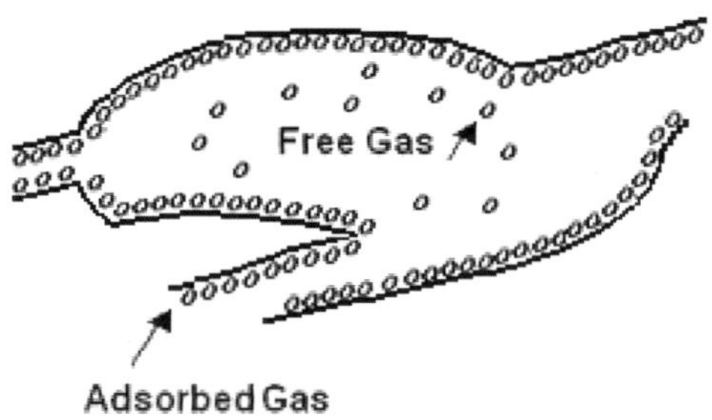

Figure 2. Gas Storage in Coal

The main mechanism of gas retention in coalbeds is physical adsorption. Adsorption depends on various factors such as moisture, temperature, mineral content, porosity, especially the micro-pore structure, because these pores can account for up to 85% of the total porosity (Mahajan, 1982). Gas will flow out of these pores if there is a pressure gradient acting as a driving force. When the pressure in the coal seam is reduced through mining, the gas begins to desorb and flow through the coal matrix. The gas transport in coal has two other distinct properties: diffusion through coal matrix and flow through the coal seam natural fracture system called "cleat". Desorbed gas is emitted to adjustment layers and mine workings through cleats. The movement of gas after desorbing from the coal surface is controlled by the concentration gradient of the gas molecules (diffusion). In diffusion, molecules move from a zone of higher concentration to one of lower concentration. This movement of methane continues until the molecule intersects the cleat or the fracture in the coal.

The most dangerous kind of gas emission in underground mine workings is the release of an unusually large volume of gas from the coal seam in a short period of time. Such incidents have caused considerable loss of life and equipment. This phenomenon is called "outburst".

3. OUTBURST

The presence of significant quantities of strata gas, if not effectively drained, may lead to the phenomenon of "outburst", a major safety problem in coal mining. An outburst has been defined as a sudden release of coal (or coal and rock), accompanied and followed by release of large quantities of gas, usually methane or carbon dioxide, into the working face or other mine workings (Harvey, 2001). This represents a major coal mining hazard, which may lead to a dust or gas explosion, with subsequent destruction of mine structure, and loss of equipment and possible injury or death to miners. The following circumstances may contribute to the occurrence of outbursts (Hargraves, 1983):

- Presence of sufficient quantity of gas in the coal seam (high gas content)
- Local fracturing and partial crushing of coal at faults, dyke intrusions in the seam, etc. (coal seam geology).

- Coal of low permeability to passage of gas
- Coal strength (coal characteristics)
- Stress regime

Some factors are easily controllable, while control of others has proven to be more difficult, requiring further research. For example, if there is no gas present, or if the gas pressure in the coal is reduced below a certain threshold value for a given physical coal strength, then a gas outburst cannot occur.

Gas drainage is a technique that has proved to be successful in reducing the outburst hazard by reducing the in-situ gas pressure. Influencing the effectiveness of gas drainage is the physical property of permeability. Permeability is a property of a porous medium and can be simply defined as the measure of the ability of the medium (rock) to convey fluids. Permeability of coal is influenced by the viscosity of the fluid contained in coal (gas type in this case), and changes to the coal matrix. Coal matrix changes can occur as a result of changes in the stress regime, which in turn may be influenced by release (drainage) of in-situ gas, i.e., all factors are inter-related.

4. PERMEABILITY

As mentioned before, the methane storage capacity of coal is a consequence of its high internal surface area. This surface area is the most important factor affecting the storage capacity of a coal bed. But when considering the drainage of methane from the coal bed, the permeability of the seam becomes the most important factor. Permeability is the measure of the ability of a porous rock to convey fluid. It determines the pressure gradient and the development of tensile stresses and it is one of the most important factors controlling the production of coal bed methane.

Permeability depends on various factors such as; the size and shape of the pores and their interconnections, type of the gas, and, most importantly, stress dependency. A number of researchers have shown the effects of stress on permeability (e.g. Gunter, 1965; Patching, 1965; Puri and Seidle, 1992). These researchers measured extremely wide ranges of permeability: for example, the measured permeability of the Bulli coal seam in Australia ranges from 0.001 mD (Lingard et al, 1984) to 0.15 mD (Somers, 1993) but their results show that in all tests the permeability decreases with increasing stress. According to Wallace (1990), the permeability of coal is usually in the range of 1 to 60 mD. In the case of drainage, Hughes and Logan (1990) stated that the minimum required permeability for coal is generally greater than 1 mD.

In this paper an attempt is made to investigate the variation in coal permeability with changing stress subject to different gas pressures and various kinds of gases. There are different methods for measuring the in-situ permeability of coal. For example, measurement from surface boreholes, which has been used for estimating the in-situ permeability of the coal seam before mining, or using in-seam sealed boreholes drilled from underground workings to the coal seam. Although permeability of coal as measured in the laboratory is not as accurate as the measuring the permeability of coal in-situ, laboratory experiments are less expensive, more controllable, and can produce rapid results. They are therefore favored by researchers.

4.1 Experimental Investigation

The following paragraphs outline a program of experimental work undertaken to evaluate the change in coal permeability under changing stress conditions, with different gases.

4.1.1 Coal Permeability Tests

Tests were conducted to determine the effect of stress as well as gas pressure on the permeability of coal samples. Coal samples were prepared from blocks of coal taken from different coalmines and cored to obtain 54 mm diameter cores. The cores were cut to 50 mm length and the end surface polished. A central hole of 5 mm diameter was drilled through the sample and the two ends were sealed with a special lock-tite. Coal permeability was investigated by carrying out a suite of laboratory tests using a special 'Permeameter', developed at the University of Wollongong. A general schematic diagram of this apparatus used for measuring the absolute permeability of coal under a range of in-situ stresses and various gases is shown in Figure 3. The apparatus was especially designed and built for this purpose, and includes the following features:

- Application of stress
- Measurement of strain of sample
- Measurement of flow rate
- Application of constant circumferential gas pressure
- Application of constant suction

Initially, the 54mm diameter, 50mm long core samples were placed between the loading plates of the Permeameter chamber. Axial force was applied to the coal sample via a universal torque device. Changes in the sample's axial and lateral dimensions due to gas sorption were monitored by two sets of strain gauges.

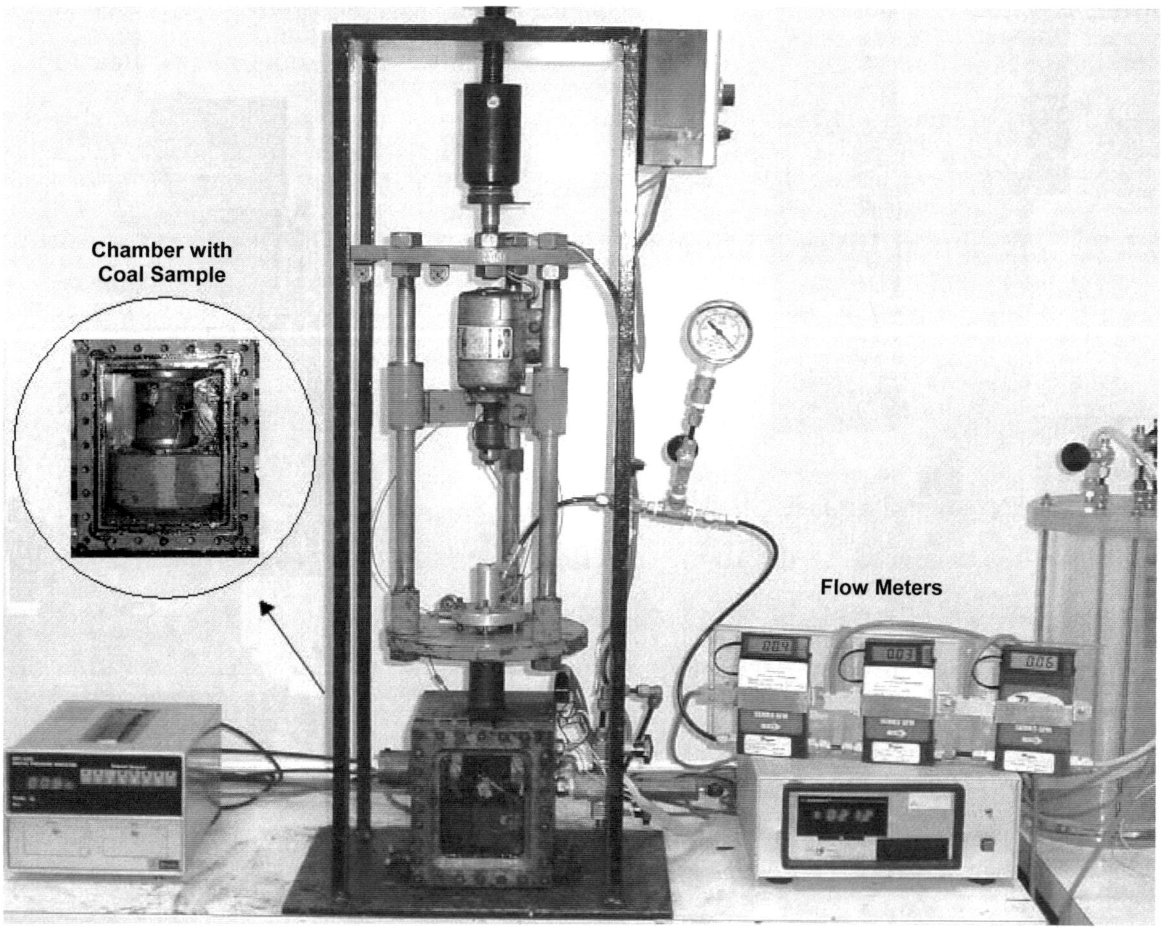

Figure 3. Equipment for Coal Permeability Tests

Permeability tests were conducted using various gases. The procedure of conducting the tests is as follows:

As shown in Figure 4, the sample is mounted in the pressure chamber, and the system was evacuated to remove air. The pressure in the chamber was then changed to a predetermined level. The level of pressurization was maintained constant. The gas was allowed to permeate the coal sample and flow out through the central hole. The released gas from the coal flows through the measuring system consisting of a vacuum pressure sensor and gas flow meters with different measurement ranges:

- High range: 0-15 L/min
- Medium range: 0-2 L/min
- Low range: 0-100 mL/min

The load cell, flow meters, pressure transducer and strain gauges were connected to a PC through a datataker (D500) for data collecting. About 30 minutes were allowed to elapse before steady-state conditions could be reached, and readings could be taken.

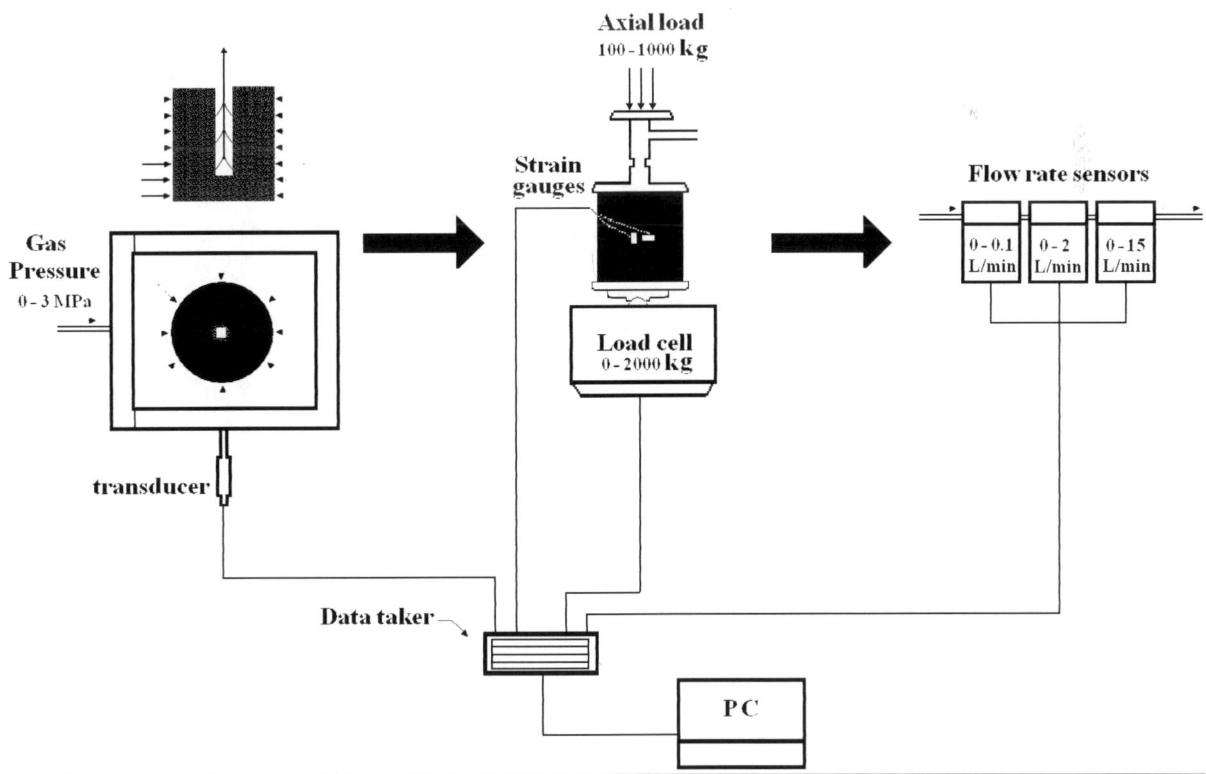

Figure 4. Schematic of the Equipment for Coal Sample Permeability Tests

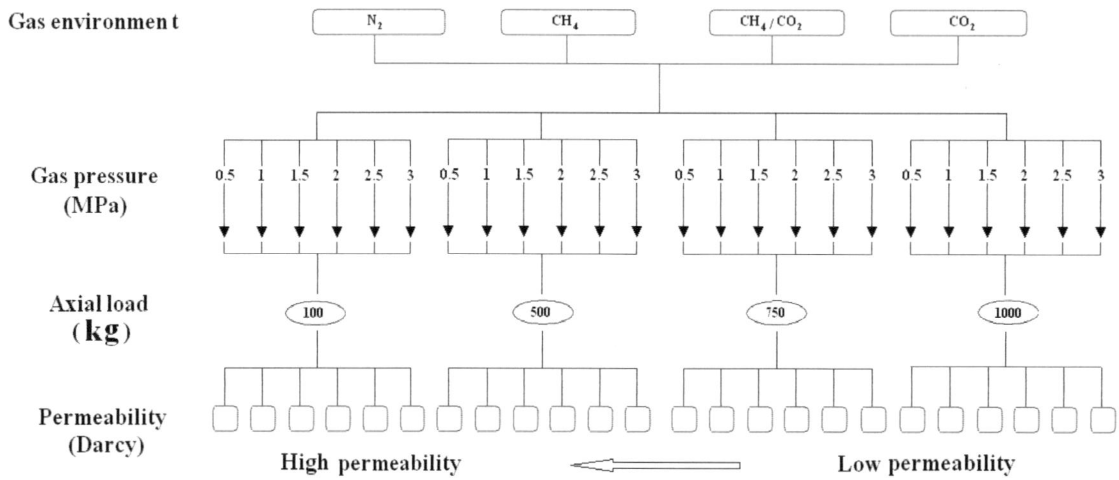

Figure 5 Sequence of changing Pressures and Loads in the Permeability Tests

The permeability of the sample was calculated using the following form of Darcy's equation (Lama, 1995):

$$K = \frac{\mu Q \ln\left(\frac{r_o}{r_i}\right)}{\pi \ell (P_o{}^2 - P_u{}^2)}$$

where:

K = Permeability (Darcy)(cm^2)
Q = Rate of flow of gas ($cc.sec^{-1}$)
r_o = External radius of sample (cm)
r_i = Internal radius of sample (cm)
ℓ = Height of the sample (cm)
P_o = Absolute pressure in the chamber (bar)
P_u = Absolute pressure in the outlet (bar)
μ = Viscosity of tested gas

4.1.2 Experimental Results

The permeability tests were started with using Nitrogen (N_2) as the flowing gas, followed by Methane (CH_4), Carbon Dioxide (CO_2) and finally a CO_2/CH_4 mixture (50%, 50%). Figures 6a-6d show the results of permeability measurements for Barton project coal samples with radial flow of the above-mentioned gases under in-situ stress conditions (Aziz et al., 2004). The permeability of coal in general was found to be highly stress-dependent and decreased with increasing stress for all tested gases. At lower mean gas pressures, the reduction in permeability was much greater than at higher pressures. It was observed that for the same in-situ stress, the permeability of coal to N_2 is more than for CH_4, CO_2 and the CO_2/CH_4 mixture. The lowest permeability measured was for CO_2.

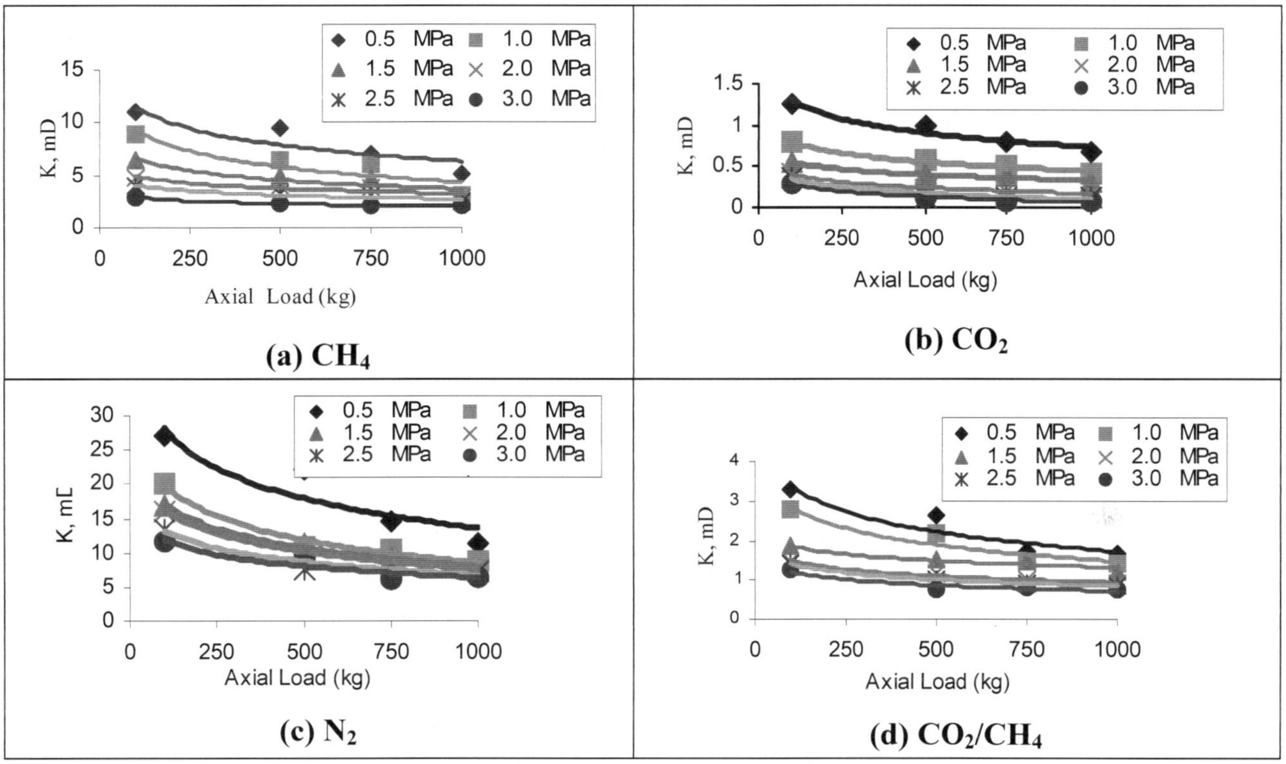

Figure 6 Effect of Stress on Permeability of Coal in different Gases

This is because of the higher adsorbtion capacity of coal to CO_2. Adsorbtion of CO_2 takes place at the internal surfaces of pores and cleats (cracks) in the coal matrix, resulting in lower flow rates. The adsorbtion capacity of coal for CH_4 is much lower than for CO_2, resulting in higher permeability to CH_4. For CO_2/CH_4 (50%, 50%) mixture the measured permeability was closer to that for CO_2 than for CH_4. According to Xue and Thomas (1995), varying the ratio of CO_2/CH_4 will result in different permeability.

5. GAS FLOW MODELLING

A preliminary Computational Fluid Dynamics (CFD) modeling exercise was carried out to "visualize" the gas flow in the porous coal sample.

5.1 Computational Domain and Mesh

Figure 7 shows the computational domain and the corresponding computational mesh used for the simulation. A thin slice of the coal sample was chosen as the computational domain, in order to take advantage of the axial symmetry of the sample geometry.

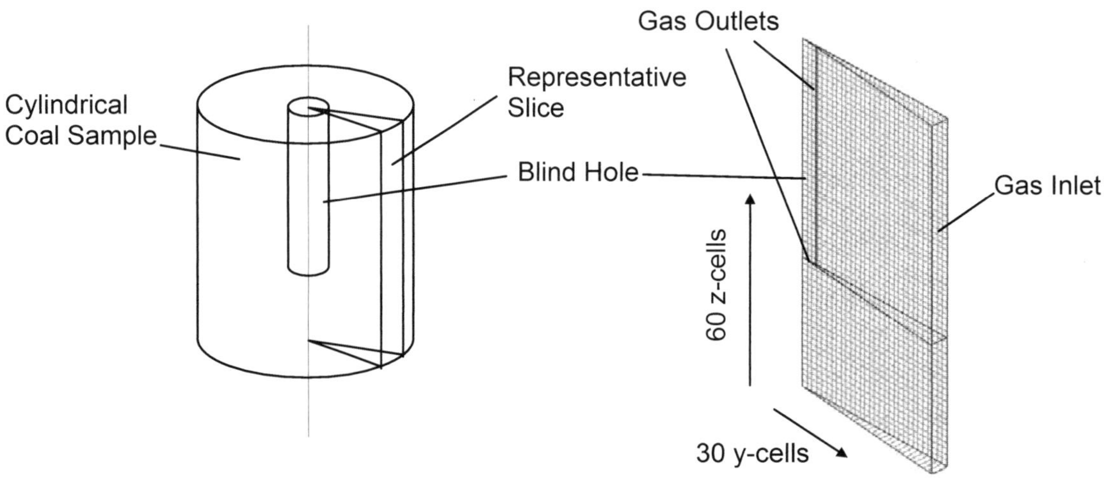

(a) Physical Domain (b) Computational Domain

Figure 7 Physical and Computational Domains for CFD Simulation

The computational domain is divided into a number of non-overlapping sub-domains called “cells”. Equations describing the conservation of mass, and the Darcy equation, which replaces the momentum equations in the fluid mechanics of porous media (Bejan, 1984), are solved iteratively until balances are achieved for each computational cell. Since the cells are contiguous, this implies balances for the entire computational domain. Results are presented in the form of velocity vector plots and pressure contour plots.

5. 1. 1 Typical Simulation Results

A typical experimental condition was chosen for the simulation. For this flow, a typical permeability value of 1 mD was used (Figure 6(b)). A stagnation pressure condition was applied at the inlet, and a zero gauge pressure at the outlets. For the flowing gas, CO_2, the following relevant properties at 300 K were used (Incropera and DeWitt, 1996):

Dynamic Viscosity μ = 149 (10^{-7}) N-s/m^2

Density ρ = 1.7730 kg/m^3

The inlet and outlet conditions were:

Inlet Stagnation (Total) Pressure ~ 5 (10^5) Pa (gauge)

Outlet Pressure = 0 Pa (gauge)

Figures 8(a) and (b) show the results in terms of the pressure contours through the coal sample, and the flow streamlines, respectively.

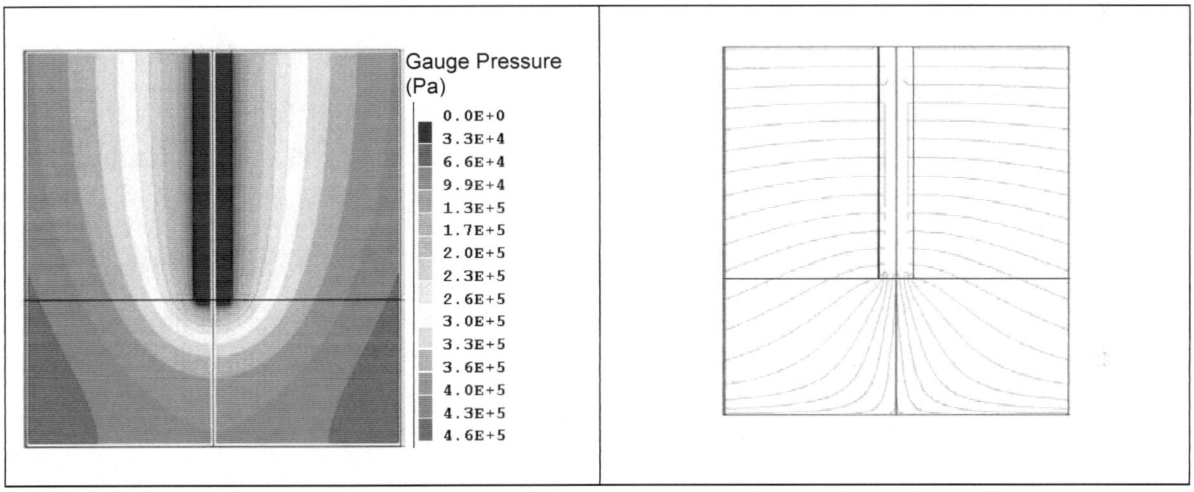

(a) Pressure Field (b) Flow Streamlines

Figure 8 Typical Simulation Results

The results suggest that for the above flow geometry (radially impressed flow with high stagnation pressure values, through a small axisymmetric sample with a small centrally located outlet), some of the gas exiting through the vertical sides of the central hole finds its way out from the top, while some may be trapped in the lower part of the hole. Also, it is very likely that the gas flow, under the above experimental conditions, reaches extremely high velocities as it flows through the tiny fissures and cracks in the coal sample.

6. Conclusions

The experimental study reported in this paper has demonstrated that stress tends to close the cleats and reduce permeability within the coal. Carbon dioxide gas appears to influence the coal permeability to the greatest extent, and Nitrogen gas the least. This is understandable in view of the fact that carbon dioxide gas has the greatest affinity to coal than the other gases used in the experiment.

The numerical simulations suggest that the gas can reach very high velocities locally, as it is forced through very small channels. It will be very interesting to carry out a CFD simulation of the flows through the micro-scale labyrinths that make up the coal matrix.

ACKNOWLEDGEMENTS

The financial support provided by the Ministry of Science, Research and Technology of the Iranian Government and Shahrood University of Technology is acknowledged. The assistance provided by the Mining Engineering Technical staff, Ian Laird, Alan grant, Bob Rowland, Greg Tillman and Nick Mackie is also appreciated. The authors wish to acknowledge the support provided by the management and staff of North Goonyella Mine, Queensland, Australia by providing coal samples for the experimental work in this project.

REFERENCES

Aziz, N. I., Porter, I. and Sereshki, F. (2004). The influence of gas environment on coal properties- Experimental study in outburst control., 5th Underground Coal Operators' Conference, Coal Mine Planning,Eds. Aziz, N. I. and Kininmonth, B., University of Wollongong, Australia, pp.195 - 201.

Bejan, A, (1984), Convection Heat Transfer, Wiley Interscience, p. 347-348

Creedy, D. P. (1991) An introduction to geological aspects of methane occurrence and control in British deep coal mines, *Quarterly Journal of Engineering Geology*. Vol. 24, pp. 209-220

Gray, I. (1987) Reservoir Engineering in coal seams: part I – The physical process of gas storage and movement in coal seams, *SPE Res. Eng.*, pp. 28-34

Gunter, J. (1965) Investigation of the relationship between coal and the gas contained in it. *Int. Conf. of dirs. of safety of mines*, Min. of power, Sheffield, Eng. SMRE translation No.5134.

Hargraves, A. J. (1983) Instantaneous outbursts of coal and gas, *a review in the proceeding, The Australian Institute of Mining and Metallurgy*, No.285, pp. 1-37

Harpalani, S. and Schraufngel, R. A. (1990) Shrinkage of coal matrix with release of gas and its impact on permeability of coal. Fuel, Vol.69, pp.551-556

Harvey, C. (2001) Analysis and management of outbursts, PhD Thesis, University of Wollongong, Australia

Hughes, B. D., Logan, T. L. (1990) How to design a Coalbed methane well. Petroleum Engineer International; **5** (62); 16-20

Incropera and DeWitt (1996), Fundamentals of Heat and Mass Transfer, Fourth Edition, John Wiley and Sons

Jolly, D. C., Morris, L.H. and Hinsley, F. B. (1968) An investigation into the relationship between the methane sorption capacity of coal and gas pressure. Mining engineer, Vol.127, No. 94, pp.539-548

Lama, R. D. (1995a). Effect of stress, gas pressure and vacuum on permeability of Bulli coal samples, *Int. Symp. cum Workshop on Management & Control of High Gas Emission & Outbursts in underground coal mines*, Ed. Lama, R. D., Wollongong, pp. 293-301

Mahajan O. P. (1982) Coal porosity. *Coal structure.* Meyers R.A. (ed.) New York, NY, USA, Academic Press, pp 51-86

Patching, T. H. (1965) Variations in permeability of coal. *Proceeding of Rock Mech. Symp.* , TN 292-59C, Toronto

Puri, R., and Seidle, J. P. (1992) Measurement of stress-dependent permeability in coal and its influence on coalbed methane production. *In situ*, Vol. 16(3), pp. 183-202

Wallace, J. A. (1990) Coalbed methane production- an operator's perspective. In: Coalbed methane in Alberta what's it all about? Nikols, D., Treasure, S., Stuhec, S., Goulet, D. (comps.) Edmonton, Alberta Canada, Alberta Research Council, ARC Information Series No. 108.

Xue, S. and Thomas, L. J. (1995) Laboratory investigation on the permeability of coal to mixture of methane and carbon dioxide. *Proceeding of the symposium on management and control of high gas emissions and outbursts in underground coal mines.* pp. 311-315

0521

Productivity of Fractured and Non-Fractured horizontal wells in Coalbed Methane Reservoirs

Hossein Jahediesfanjani[1], and Faruk Civan, the University of Oklahoma Mewbourne School of Petroleum and Geological Engineering, the University of Oklahoma, Norman, Oklahoma, 73019

Abstract

This paper investigates the effect of the fractured, non-fractured, and multi-lateral horizontal wells on the time required to attain non-economic production levels in the coalbed methane (CBM) reservoirs. A parametric study of the effects of characteristics of typical CBM reservoir and operation conditions on productivity of these completion techniques is carried out. This is accomplished by a series of 3-dimensional and 2-phase flow simulations. The efficiency and performance of these techniques are investigated for the effects of reservoir heterogeneity and anisotropy, change of the isotherm parameters, direction of butt and face cleats, the angle by which the horizontal wells and the induced fractures intersect cleats, the conductivity and length of each fracture along the horizontal well, and the number of the horizontal lateral and branches in the multi-lateral horizontal wells. A comparison of the various factors is made using a normalized cumulative gas production parameter defined as the ratio of the productivity index of the stimulated reservoir to that of a non-stimulated vertical well. The optimum number, length, and conductivity of the induced fractures, horizontal laterals and branches along the horizontal wellbore are determined by the present approach for various horizontal to vertical permeability ratios.

Introduction

Coal structure in CBM reservoirs is composed of bulk of very low porosity and permeability matrix surrounded by a series of natural fractures called cleats. The coalbed methane reservoirs are usually characterized as the dual-porosity and dual-permeability reservoirs. Cleats are distributed in both horizontal and vertical directions. The coal surface contains the adsorbed gas and the cleats are usually full of water at the initial reservoir condition. The gas production is only possible by dewatering of cleat system. The cleat pressure decline during dewatering causes the adsorbed gas to desorb from internal coal surface, diffuse throughout the matrix, enter the surrounding cleats, and flow to the wellbore and/or induced fractures. Therefore, creation of large pressure drawdown is essential to accelerate the reservoir dewatering and improve the CBM production rate.

Jahediesfanjani and Civan [1] studied the effect of the conventional completion and stimulation techniques on productivity of CBM reservoirs. They compared the performance of the open-hole cavity completion, fractured and non fractured vertical wells, frack and packs, and horizontal wells on the duration of the economic gas production rates and reservoir gas productivity indices. They concluded that horizontal wells are the best completion and stimulation techniques yielding the highest productivity index among the mentioned methods. This is because, horizontal wells cause higher pressure drawdown throughout the reservoir, cover a longer distance along the reservoir, and connect more butt and face cleats to the well bore in compared with the other completion techniques. However, horizontal wells can also be completed and stimulated with various techniques such as hydraulic, water, and hydrajet

[1] T-301 Sarkeys energy Center, 100 Boyds st, Norman, OK, 730119-1003; electronic Email: Hossein@ou.edu

fracturing. Depending upon the CBM reservoir special characteristics, economical parameters, butt and face cleats distribution throughout the reservoir, and reservoir anisotropy, the horizontal wells can be stimulated using any appropriate stimulation and completion technique.

This study reviews a parametric analysis and evaluation of the various completion methods applied in horizontal wells to improve reservoir productivity and determines the parameters which are most critical. A review of the potential methods for completion and stimulation of the horizontal wells especially in very low permeability gas reservoirs is also provided.

Single and multi-lateral horizontal wells

Maximizing the horizontal well productivity in CBM gas reservoirs required a fully understanding of the natural fractures and normal stress distributions throughout the reservoir. CBM reservoirs are naturally fractured and anisotropic formations. The reservoir horizontal permeability is usually higher than the vertical permeability because the face cleats are continuously distributed in the horizontal direction, where as the butt cleats are discrete fractures oriented in the vertical direction with lower permeability. Therefore, horizontal wells should be drilled orthogonally to the maximum permeability to increase the reservoir productivity [2]. To further maximize the horizontal well productivity, Sung and Ertekin [2] suggested increasing the number of horizontal wells until an optimum well spacing on reservoir productivity is attained. They recommended that drilling several horizontal wells orthogonal to the face cleats is a way to decrease well spacing and promote reservoir production rate. The horizontal wells should be completed in CBM reservoirs in a manner to intersect the face cleats in the most appropriate angles. The reservoir and well bore connectivity can be achieved by means of multi-lateral horizontal wells [3].Multi-lateral horizontal wells are usually composed of one or two primary horizontal or deviated laterals drilled from the main vertical borehole with some braches oriented from the horizontal lateral in the horizontal or vertical planes [4]. Depending upon the reservoir characteristics, especially the reservoir permeability distribution, the multi-lateral horizontal wells with different branches can be constructed to cover a wide range of the reservoir and connect the natural fractures to the multi-lateral network. The major advantages of multi-lateral horizontal wells are the ability to cover a large reservoir space, and the reduction in the number of platform slots and down-hole costs [3]. The disadvantages of this technique are the severe well bore failure and instability and difficulty in performing an accurate drilling plan [5].

Fractured horizontal wells

Depending on the designed future remedial work and the stimulation option and economic perspective, horizontal wells in CBM reservoirs are completed mostly in three categories. 1. Open-hole, 2. Slotted liner, and 3. Cased-hole completions [6]. Open-hole completion is the least expensive completion technology. It allows for the application of future stimulation techniques. Moreover, because there is not any casing or liner between the wellbore and reservoir, it allows for more production and avoids the perforation pressure drop [7]. Liners are mostly used to isolate the most critical sections of the wellbore. Casing the wellbore with perforations makes it possible to stimulate and analyze different segments of the wellbore separately [8].

Water-frac and hydrajet fracturing in open-hole horizontal wells

High velocity water is used in many industries for different purposes. In petroleum industry, the water jet has been applied in drilling, perforating, cutting, and slotting operations. Using high injection rate water to induce fractures along with horizontal wells that need no proppant placement have been recently practiced in low and very low permeability gas reservoirs [8]. This fracturing technique does not cause any formation damage in the near well bore area because of proppant placement and fracturing fluid invasion. Therefore, it can be very useful in stimulating horizontal wells in CBM reservoirs. Hydrajet fracturing is a combination of hydraulic and jet fracturing so that fracturing fluid is pumped through a jet instead of perforation holes. To accomplish hydrajet fracturing, first the fracture-extension direction and plate are determined and a jet is placed near the pointed spot, and the annulus is pressurized until bottom-hole pressure at the special point is the same as the fracture extension pressure. Then, the tubing pressure is increased until the jet differential pressure is about 4000 to 5000 psi. Fracture is therefore induced and extended along the fracture-extension plate [9]. One of the advantages of this method is the technique can be applied in open-hole completed horizontal wells. Therefore, the fractures can be selectively placed in even or uneven intervals along the horizontal lateral [10].

Hydraulic fracturing

One of the best methods available to increase reservoir productivity in CBM reservoirs is a method connecting the butt and face cleats to the horizontal wellbore. The reservoir production rate will increase if the horizontal wellbore is drilled parallel to the minimum permeability direction in a manner to induce fractures so that they intersect with the high permeability naturally existing fractures at an appropriate angle. Therefore, hydraulic fracturing is frequently resorted to promote gas production in tight gas reservoirs [11]. Depending on the reservoir characteristics, the shape and direction of the induced fractures vary from case to case. The following is a brief review of the most common form of fractures that can be induced along with the horizontal lateral.

Transverse fractures

It is assumed that the principles stress components σ_v, σ_H and σ_h are the vertical, maximum horizontal, and minimum horizontal normal stresses applied throughout a CBM reservoir. In rather deep CBM basins the normal stresses indicate a trend of [12].

$$\sigma_v > \sigma_H > \sigma_h \tag{1}$$

A horizontal well drilled orthogonally to the face cleats practically penetrates in the direction of the minimum horizontal stress. The induced fractures along the horizontal lateral propagate orthogonally to the minimum normal stress, and therefore, orthogonally to the wellbore and butt cleats. A fracture propagating orthogonally to the horizontal wellbore is called a transverse fracture. However, the size and direction of the perforated interval will affect the propagation direction. The perforated interval size should be larger than the quadruple of the wellbore size in order to induce a successful transverse fracture [13]. The induced transverse fracture may twist or reshape as it propagates far from the wellbore when the wellbore is not drilled exactly parallel to σ_h, or the stress distribution around the wellbore changes due to the reservoir anisotropy or wellbore penetration in the reservoir, leading to a misestimating of the direction of σ_h. Consequently, the effectiveness of the induced fracture reduces and the production rate does not increase to anticipated levels [14]. Moreover, the production rate after stimulation depends on the fracture dimension, shape, length, and conductivity.

Longitudinal fractures

The wellbore propagates orthogonally to the minimum normal stress when a horizontal well in CBM reservoir is drilled parallel to the face cleats. Therefore, if the perforated interval is more than four times the wellbore diameter, the induced fractures along the horizontal lateral will propagate parallel to the wellbore [15].These fractures can be easily formed in open-hole horizontal wells. Creating several small parallel fractures along the entire horizontal lateral is economical [16]. The productivity of the induced fracture depends on their volume coverage in the reservoir, fracture size, and conductivity. The best method to compare the performance of the transverse and longitudinal fractures is to assume the same volume of each fracture induced in different directions and under various reservoir characteristics. Such comparison may yield different results at different reservoir conditions [17].

CBM Reservoir Simulation Procedure

Figure 1 shows a schematic of a naturally fractured dual-porosity and dual-permeability reservoir considered for the present study. This model assumes matrix blocks surrounded by a net work of butt and face cleats. Therefore, each grid block is introduced with two representative pressures called the matrix and fracture pressures. The fracture pressure is composed from the gas and water pressures, whereas the matrix pressure is only the gas pressure inside the matrix. An isotherm, relating the ability of the coal to adsorb gas at various pressures, is used to describe the gas adsorption/desorption processes from the internal coal matrix surface. The most common isotherm used in CBM gas reservoirs is the Langmuir isotherm defined as:

$$C_m = \frac{V_L P}{P_L + P}\left(0.031\rho_B\right) \tag{2}$$

Three mechanisms have been identified for diffusion of an adsorbing gas in the macrospores: 1. Molecular diffusion (molecule-molecule collision dominates), 2. Knudsen diffusion (molecule-wall collision dominate), and 3. Surface diffusion (transport through physically adsorbed layers). Basically gas diffusion in CBM reservoirs is a combination of these three diffusion phenomena expressed with Fick's law as:

$$q_{gm} = \frac{8\pi D V_m}{S_f^{\,2}}\left[C_m - C(P)\right] \tag{3}$$

The fluid flow inside the cleat system can be modeled using the naturally fractured flow models. For two phase flow, three forces should be considered: 1. Viscous forces, 2. Gravity forces, and 3. Capillary forces. Fracture gas and water pressures are calculated at each time step using Darcy's law and material balance equations as following:

The mass balance equation for gas flow inside the fracture system is given as:

$$\Delta\left[Tr_{gf}\left(\Delta P_{gf} - g_{gf}\Delta D_f\right)\right] + q_{gmf} - q_g = \frac{V_b}{\Delta t}\Delta_t\left(\frac{\phi S_g}{B_g}\right)_f \tag{4}$$

Similarly for water flow:

$$\Delta\left[Tr_{wf}\left(\Delta P_{wf} - g_{wf}\Delta D_f\right)\right] - q_w = \frac{V_b}{\Delta t}\Delta_t\left(\frac{\phi S_w}{B_w}\right)_f \tag{5}$$

Where the gas and water transmissibilities inside the fracture grid-blocks are given respectively as:

$$Tr_{\alpha f} = 0.001127\left(\frac{k k_{r\alpha}\phi}{\mu_\alpha B_\alpha}\right)_f\left(\frac{\Delta z \Delta y}{\Delta x}\right) \tag{6}$$

After estimating the gas and water production rates and the fracture pressure distribution at each time step, the matrix gas pressure is estimated applying a material balance equation and the Langmuir isotherm for the grid-block at each time step.

In addition, the following auxiliary equations are used for saturation and capillary pressure, respectively:

$$S_{gf} = 1 - S_{wf} \tag{7}$$

$$P_{cf} = P_{gf} - P_{wf} \tag{8}$$

The wellbore constraints required for q_g and q_w in equations 4 and 5 are calculated using the appropriate equations and procedures described elsewhere by Jahediesfanjani and Civan (1).

To simulate the gas and water production rates in a CBM gas reservoir and solve the above equations under various scenarios, a series of two-phase and three-dimensional flow simulators have been carried out using MATLAB® [18] as described in the following sections.

1. Multi-lateral horizontal wells

It is assumed that the horizontal lateral section of the multi-lateral horizontal wells operates at a constant pressure that is, P_{bh}. However, the gas and water flow in the horizontal or vertical branches separated from the main lateral is assumed to be one dimensional radial pipeline flow under the different boundary conditions described in **Figure 2**. The two-phase flow of water and gas pressure drop in an inclined pipeline is calculated using the Beggs and Brill correlation [19]:

$$\frac{dP}{dL}=\frac{\frac{g}{g_c}\sin\theta\left[\rho_w H_w+\rho_g\left(1-H_w\right)\right]+\frac{fG_m v_m}{2g_c d}}{1-\frac{\left[\rho_w H_w+\rho_g\left(1-H_w\right)\right]v_m v_{sg}}{g_c P}} \tag{9}$$

The Beggs and Brill [19] correlations are used to calculate the water hold-up (H_w), and friction factor (f) for an inclined pipe-line. Also, note that the mixture mass flux rate ($G_m = G_g+G_w$), mixture velocity ($v_m = (q_{wp}+q_{gp})/A_P$), superficial gas velocity ($v_{sg} = q_{gp}/A_P$) are function of the pipe-line length as shown in **Figure 3**.

2. Hydraulic fractures along with horizontal well

The reservoir flow is called a two-dimensional when a fracture propagates throughout the reservoir and reaches the reservoir boundaries from all dimensions regardless of the type or number of the induced fracture. However, if it does not reach the reservoir boundaries the flow in the reservoir is called a three-dimensional flow. **Figures 4** and **5** show various boundary conditions and flow equations necessary to simulate a two-phase flow in a fractured reservoir.

Performance of Different Completion Techniques

To analyze the simulation data and compare the performance of various horizontal well completion and stimulation techniques in CBM reservoirs, a normalized cumulative gas production and a normalized permeability are defined as:

$$Qn(t)=\frac{Q_m(t)}{Q_v(t)} \tag{10}$$

$$K_n=\sqrt{\frac{K_{Face\ Cleat}}{K_{Butt\ Cleat}}} \tag{11}$$

where the subscript *m* refers to various completion and stimulation techniques.

The plots of normalized cumulative production versus the production time are generated for various reservoir and stimulation conditions. Therefore, if the cumulative gas production after a specific time for a vertical well located in a CBM reservoir is known, the possible cumulative gas production from the same reservoir can be predicted for the other stimulation techniques. To generate such plots, a reservoir in the San Juan basin with a coal depth of 2000 ft is chosen. This reservoir has a coal thickness of 30 ft with initial pressure of 1300 psia. The cleat and matrix porosities are 3.95% and 8.0%, respectively. The coal cleat permeability is 6.5 mD with the initial gas saturation of 2.0%. Several case studies presented in the following describe the various stimulation results.

Case 1. Single lateral horizontal wells

The productivity of horizontal wells under various conditions are considered. Several scenarios are presented here to demonstrate the reservoir productivity. The cumulative gas production for various cases is expressed by the normalized cumulative production defined by Equation 10.

Scenario 1. Horizontal wellbore length. Depending upon the reservoir stresses distribution, internal characteristics, and economical parameters, the length of the horizontal well may be different than the reservoir length. This ratio is expressed as:

$$R_a = \frac{Horizontal\ Well\ Length}{Reservoir\ Length(L_X)} \quad (12)$$

Figure 6 shows the gas normalized cumulative production for different R_a at any time after the production starts. It is observed that even a small horizontal well can increase production rate as high as 10 times of the vertical wells. However, a longer well can increase it up to about 70 times. The normalized production rate increases as time of the production increases showing that horizontal wellbores delay attaining non-economical production rate in compared with the vertical ones. The higher the pressure draw down the higher is the production rate, the longer is the production time, and therefore the more economical is the process for any completion method.

Scenario 2. Reservoir anisotropy. CBM reservoirs are anisotropic reservoirs. Therefore, the face to butt cleat ratio are considered to evaluate the effect of the reservoir anisotropy on the normalized gas production rate. **Figure 7** shows that normalized gas production decreases as the normalized permeability described by Equation 12 increases, meaning that the face cleats permeability is higher than the butt cleats permeability. When the normalized permeability is greater than unity, the reservoir horizontal permeability is higher than the vertical permeability, and therefore the fluid transmissibility in the horizontal direction is faster than the vertical direction. Practically the connectivity of the horizontal well bore with the upper and lower layers is less than the connectivity of the wellbore and the same layer that the horizontal lateral is distributed. This is usually the case in CBM reservoirs. Therefore, it is expected from **Figure 6** that a horizontal wellbore will increase the reservoir productivity by 10-20 times than a vertical well.

Scenario 3. Change in the isotherm parameters. Due to the reservoir heterogeneity, and coal rank, and type, and therefore capacity to adsorb and hold gas may be different throughout the reservoir. As described by Equation 2, the Langmuir isotherm contains two parameters that are functions of the coal properties and components. **Figures 8** and **9** compare the horizontal wellbore normalized cumulative production rate for various values of these two parameters. As P_L decreases the ability of the coal to retain gas according to Equation 2 increases and therefore the production rate will increase. In the contrast, as V_L increases production rate will increase.

Scenario 4. Number of horizontal wells. Drilling several horizontal wells decreases well spacing and increases the cumulative gas production rate. Finding the optimum number of the horizontal wells that can be drilled in a limited volume of the reservoir is a factor of mainly the economical parameters and reservoir characteristics. **Figure 10** shows that drilling more horizontal wells increase the reservoir productivity until the optimum number is obtained. Rather than the productivity increase, the economical conditions are also important in deciding whether to drill more horizontal wells or choose other completion techniques. This will be discussed later in more detail.

Scenario 5. Angle of the horizontal well and face cleats intersection. The angle by which the horizontal wells intersect each other and face cleats in the horizontal plane affects the reservoir performance. **Figure 11** shows that drilling 2 horizontal wells orthogonal to each other and face cleats do not significantly increase the reservoir productivity. Further, drilling two separated horizontal wells parallel to each other increases the reservoir productivity by several fold. However, drilling two separate wellbores, requires two different platforms which may not be very economical. The other choice is to drill more horizontal wells intersecting with each other by a 45 degree angle. Drilling more horizontal wells from one single spot reduces the initial drilling costs and increases reservoir performance significantly. However, reservoir anisotropy affects the comparison of the results. **Figure 11** shows that decreasing the value of normalized reservoir permeability increases reservoir performance for two diagonally drilled horizontal wells like the single horizontal wells.

Case 2. Multilateral horizontal wells

It was previously mentioned that multilateral horizontal wells with several horizontal and vertical branches drilled from the main horizontal lateral is one of the best methods to increase the reservoir productivity. **Figure 2** shows a series of typical multilateral horizontal wellbores with different branches. **Figure 12** compares the reservoir productivity for various multilateral wellbores with multiple branches. The reservoir performance increases as the number and/or length of the branches increases. However, the maximum increase in the reservoir performance occurs when the number of the multilaterals increases from 1 to 2. **Figure 13** shows that the reservoir productivity does not increase for the CBM reservoir with normalized permeability being more than unity as for an isotropic reservoir. An economical evaluation is necessary to determine whether multi-lateral horizontal should be employed.

Case 3. Hydraulic fracturing horizontal wells.

Hydraulic fracturing in horizontal wells can be very useful if the fractures are successfully induced in a proper direction so that they connect the reservoir vertical layers effectively to the horizontal wellbore. Both the longitudinal and transverse fractures are considered in this study to evaluate their effect on the reservoir productivity.

3.1. Longitudinal fractures. The longitudinal fractures can propagate parallel to the wellbore at different angles to the wellbore centerline. Two extreme cases as zero (parallel) and ninety (orthogonal) degree fractures are considered in this study.

Scenario 1. Parallel longitudinal fractures. Parallel longitudinal fractures propagate in the horizontal plane containing the wellbore and therefore only connect the face cleats together. **Figure 14** shows the effect of the fracture size on the reservoir productivity. As the length of the fracture increases, the productivity of the reservoir will increase as well. **Figure 15** shows the effect of the fracture conductivity on the reservoir productivity. As fracture conductivity ($\lambda_f=b_f\,k_f$) increases reservoir productivity increases as well. However, an economical analysis may further reveal the best fracture size or conductivity which increases the reservoir productivity more than the other methods.

Scenario 2. Orthogonal longitudinal fractures. These fractures propagate in the vertical plane containing the horizontal wellbore and therefore connects only some vertical planes located above and below the horizontal wellbore to the wellbore. Similar to the parallel longitudinal fractures, the orthogonal longitudinal fractures also increase the reservoir productivity depending upon the fracture size and conductivity. **Figure 16** shows the effect of the fracture conductivity on the reservoir productivity. The difference here is that the normalized gas cumulative production rate decreases as the production time increases. This expresses that even though orthogonal longitudinal fractures are capable of increasing reservoir productivity, the non-economical time is shorter to achieve with these fractures than with the parallel ones. However, the longitudinal fractures can propagate in different sizes and different angles that can significantly change the results of the comparison.

3.2. Transverse fractures. Transverse fractures usually propagate vertical to the horizontal wellbore in both vertical and horizontal planes. Therefore, practically these fractures connect the butt and face cleats together and to the wellbore. The number, size, and conductivity of the induced fractures are the critical factors to be considered in studying these fractures.

Scenario 1. Number of the transverse fractures. **Figure 17** shows that increasing the number of the fractures increases the reservoir productivity. However, there is always an optimum number of fractures beyond which increasing the number of the fractures do not significantly increase the reservoir productivity.

Scenario 2. Size of the fractures. These fractures propagate in both *z*- and *y*-directions. Because, usually the height of the reservoir is much shorter than its width, fracturing operation should be managed so that the fracture is expanding in the y-direction more than the *z*-direction. **Figure 18** compares various possible fracture size effects on the reservoir productivity for a system of 7-fractures intersecting a horizontal well in a reservoir.

Scenario 3. Fracture conductivity effect. Inducing wide and more open fractures is the most demanding issue in fracturing operations. **Figure 19** shows that increasing the fracture conductivity increases the reservoir performance.

Economical Analysis

An economical analysis is necessary to complete the comparison analysis between the various horizontal well stimulation and completion techniques, and also to be able to make the right decision in selecting the most appropriate stimulation method. A computer code is linked to the previous series of CBM reservoir simulators. The net present value *(NPV)* for any stimulation and completion technique is estimated based on the following equations [20], Cumulative Gas Production at the *n*'th year $[C.G.P]_n$, Initial drilling and treatment Cost *(I.C)* and discount rate *i*.

$$NPV = \sum_{n=1}^{N} \frac{R_n}{(1-i)^n} - I.C \tag{12}$$

$$R_n = [C.G.P]_n [Cost] \tag{13}$$

$$(NPV)_n = \frac{(NPV)_m}{(NPV)_v} \tag{14}$$

Table 1 summarizes the economic data used to estimate NPT for various stimulation and completion techniques. The following example will illustrate the determination of the best possible technique that should be used to stimulate and complete horizontal wells in a CBM reservoir based on the production and economic analysis.

Comparison between Various Stimulation and Completion Techniques

Consider that the cumulative gas production for a vertical well at the middle of the described reservoir after 10 years is 30 MMscf. Using the series of Figures 6-19 the cumulative gas production for various stimulation and completion methods after 10 years of production can be predicted as following. Economic analysis data from Table 1 is used in the computer code to compute the net present value for the following completion techniques.

1. 3 horizontal wells separately drilled at equal intervals.
2. 2 Multi-lateral horizontal wells with 12 branches on each of them.
3. 4 horizontal wells intersecting each other at the center of the reservoir.
4. 7 transverse fractures propagated at equal space along with the horizontal well.
5. Series of longitudinal fractures totally occupied ½ of the horizontal well.

Figure 20 compares the net present value ratio estimated for each of these completion methods. It is concluded from this figure that completion technique number 2 (multi-lateral horizontal wells with several branches) is the best method resulting in the best and highest cumulative gas production and *NPV*. Completion number 3 (several horizontal wells intersecting each other at one point) is the second best completion method. However, the use of transverse fractures is not economical.

Summary and Conclusions

Increasing reservoir dewatering and gas production rate are critical in producing from a CBM gas reservoir. This can be accomplished by the means of various stimulation and completion techniques. Various reservoir parameters and their effects on the reservoir productivity rate were considered in this study. These parameters are the reservoir anisotropy, methane hydraulic diffusivity, isotherm parameters, and several other factors affecting the performance of any completion and stimulation techniques in CBM reservoirs.

Increasing the number of the horizontal wells, transverse and longitudinal fractures along the horizontal wellbore, drilling multi-lateral wellbores throughout the reservoir with several branches, and varying the angle of the drilled horizontal wells affect the reservoir productivity significantly. Series of plots were generated as the ratio of the cumulative gas production of the various completion and stimulation methods to the cumulative gas production for an undamaged vertical well versus the production time. Knowing the cumulative gas production from a vertical well in the middle of a portion of a CBM reservoir, these charts enable estimation of the reservoir productivity increase of other stimulation techniques.

The economic analysis shows that two multi-lateral horizontal wells with several branches provide the best method to increase the reservoir performance. This method covers a larger volume of influence inside the reservoir and creates a tremendous pressure draw down in the reservoir. Moreover, the method has more flexibility to adjust the angle of the horizontal laterals and branches. The comparison between longitudinal and transverse fractures shows that, if induced in the same volume, longitudinal fractures outperform the transverse fractures by 1.5-2.0 fold increase in the production rate. Drilling several horizontal wells intersecting with each other at various angles increases the reservoir productivity more than the parallel horizontal wells.

Nomenclature

A_P	=	Pipe-line, branches induced from the horizontal lateral area	ft^2
b	=	Width of fracture	ft
B	=	Fluid formation volume factor	RB/STB
C	=	Gas concentration	Scf/ft^3
$C.G.P$	=	Cumulative Gas Production	MMScf
D	=	Diffusion Coefficient	ft^2/Day
D_{qg}	=	Non-Darcy flow coefficient	Dimensionless
f	=	Friction factor	Dimensionless
G	=	Mass flux rate	$Lb/ft^2.day$
G_m	=	Mixture mass flux rate	$Lb/ft^2.day$

g	=	Acceleration due to gravity	ft/sec^2
g_c	=	Gravitational constant	ft/sec^2
g_f	=	Fluid gradient	Psi/ft
h	=	Reservoir height	ft
H_w	=	Water holdup	fraction
i	=	Discount rate	%
$I.C$	=	Initial Drilling and completion Cost	$
J	=	Productivity Index	MScf/psi
k	=	Reservoir permeability	md
k_r	=	Relative permeability	Dimensionless
k_e	=	Effective formation permeability	md
k_H	=	Horizontal permeability	md
k_{max}	=	Matrix permeability	md
k_v	=	Vertical permeability	md
l_x	=	Reservoir length in x-direction	ft
l_y	=	Reservoir length in y-direction	ft
l_z	=	Reservoir length in z-direction	ft
NPV	=	Net Present Value	$
N	=	Total Number of production years	Years
n	=	Year of production	year
P_b	=	Base pressure	psia
P_{bh}	=	Flowing bottomhole pressure	psia
P_C	=	Gas/water system capillary pressure	psia
P_L	=	Langmuir pressure constant	psia
q_g	=	Gas production rate	Scf/day
qgp	=	Pipe-line gas flow rate	ft^3/day
q_w	=	Water production rate	STB/Day
q_{wp}	=	Pipe-line water flow rate	ft^3/day
r_e	=	Drainage radius	ft
r_i	=	Outer radius of first node	ft
r_w	=	Well bore radius	ft
R_a	=	Ratio of horizontal wellbore length to the reservoir length	dimensionless
R_n	=	Incremental gas revenue	$
s	=	Skin factor	dimensionless
S	=	Saturation	fraction
S_f	=	Fracture spacing	ft
t	=	Time	Days
T	=	Reservoir temperature	^{o}F
T_r	=	Fluid transmissibility	STB/day-psi
V	=	Volume	ft^3
V_L	=	Dry, ash-free Langmuir volume	Scf/ton
V_r	=	Ratio of fracture height to the reservoir height	dimensionless
z	=	Real gas compressibility factor	dimensionles
σ	=	Normal stress	psia
λ	=	Fracture Conductivity	mD.Inch
Δ	=	Finite-difference operator	
ρ	=	Density	Lb/ft^3
ρ_B	=	Bulk density	gr/cm^3
Δx	=	Grid block dimension in x-direction	ft
Δy	=	Grid block dimensionless in y-direction	ft
Δz	=	Grid block dimensionless in z-direction	ft
ϕ	=	Porosity	fraction

μ	=	Fluid viscosity	Cp
υ_m	=	Mixture velocity	ft/day
υ_{Sg}	=	Superficial gas velocity	ft/day

Subscripts

bh	=	Bottom-hole
f	=	Fracture
g	=	Gas
h	=	Minimum horizontal stress
H	=	Maximum horizontal stress
i	=	Initial
ma	=	Matrix
w	=	Water
Vw	=	Vertical Well
x	=	x- Cartesian direction
y	=	y- Cartesian direction
z	=	z- Cartesian direction

Acknowledgement

The first author gratefully acknowledges the fellowship provided by the ConocoPhillips Company at Bartlesville, Oklahoma.

References

[1] Jahediesfanjani, H., and Civan, F.: "Damage Tolerance of Well-Completion and Stimulation Techniques in Coalbed Methane Reservoirs", Journal of Energy Resources Technology, special issue on formation damage, June 2005, Vol. 127, 1-9.

[2] Sung. W., and Ertekin, T.: "An Analysis of Field Development Strategies for Methane Production from Coal Seams," paper SPE 16858 presented at the SPE Annual Technical Conference and Exhibition, Dallas, Texas, September, 27-30, 1987.

[3] Konopczynski, M.R., Hughes, John, and Best, J.E., "A Novel Approach to Initiating Multi-Lateral Horizontal Wells", paper SPE 29385, proceedings from SPE/IADC Drilling Conference, held in Amsterdam Netherlands, 28 February-2 March, 1995.

[4] Chambers, M.R.: "Making Multi-lateral Wells Cost Effective" paper SPE 51244, proceedings from the Annual International Conference and Exhibition, held in Lagos, Nigeria, 5-7 August, 1998.

[5] Brister, A.: "Analyzing a Multi-Lateral Well Failure in the East Wilmington Field of California", paper SPE 38268, proceedings from the SPE Western Regional Meeting, held in Long Beach, California, 25-27 June, 1997.

[6] Osisanya, S.M., and Schaffitzel, R.F.: "A Review of Horizontal Drilling and Completion Techniques for Recovery of Coalbed Methane," paper SPE 37131 presented in the SPE International Conference on Horizontal Well Technology held in Calgary, Alberta, Canada, November, 18-20, 1996, P 13.

[7] El-Sayed, A.A.H.: "Maximum Allowable Production Rates from Open-hole Horizontal Wells", proceedings from the SPE Miffle East Oil Show held in Bahrain, 16-19 November, 1991.

[8] East, L.E., William Grieser, jr., McDaniel, B.W., Johnson, B., Jackson, R., Fisher, K.: "Successful Application of Hydrajet Fracturing on Horizontal Wells Completed in a Thick Shale Reservoir", paper SPE 91435, proceedings from the SPE Eastern Regional Meeting, held in Charleston, West Virginia, 15-17 September, 2004.

[9] Love, T.G., McCarty, R.A., Surjaatmadja, J.B., Chambers, R.W., Grundmann, S.R.: "Selectively Placing Many Fractures in Open-hole Horizontal Wells Improves Production", paper SPE 74331, proceedings from the SPE Production & Facilities, Vol. 16, Number 4, November Pages219-224.

[10] Surjaatmadja, J.B., Grundmann, B., McDaniel, B., Deeg, W.F.J., Brumley, J.L., and Swor, L.C.: "Hydrojet Fracturing: An Effective Method for Placing Many Fractures in Open-hole Horizontal Wells", proceedings from the 1998 SPE International Conference and Exhibition in China held in China, 2-6 November 1998.
[11] Guo, g., and Evans, R.D.: "Inflow Performance and Production Forecasting of Horizontal Wells With Multiple Hydraulic Fracturing in Low-Permeability Gas Reservoirs" proceedings from the SPE gas technology Symposium held in Calgary, Alberta, Canada, 28-30 June 1993.
[12] Sparks, D. P., McLendon, T. H., Saulsberry, J. L., and Lambert, S. W.: "The Effects of Stress on Coalbed Reservoir Performance, Black Warrior Basin, U.S.A", paper SPE 30734, proceedings from the SPE Annual Technical Conference and Exhibition, held in Dallas, Texas, 22-25 October, 1995.
[13] El Rabaa, W.: "Experimental Study of Hydraulic Fracture Geometry Initiated From Horizontal Wells", paper SPE 19720, proceedings from the SPE Annual Technical Conference and Exhibition, held in San Antonio, Texas, 8-11 October, 1989.
[14] Abass, H.H., Hedayati, S., Meadows, D.L.: "Nonplanar Fracture Propagating from a Horizontal Wellbore: Experimental Study", paper SPE 24823, SPE Production & Facilities, August 1996.
[15] Soliman, M.Y., Hunt, J.L., and Azari, Mehdi.: "Fracturing Horizontal Wells in Gas reservoirs", paper SPE 59096 proceedings from the SPE gas technology Symposium held in Calgary, Canada, 28 April-1-May 1999.
[16] Soliman, M.Y., and Boonen, P.: "Review of the Fractured Horizontal Wells Technology", paper SPE 36289, proceedings from the SPE Asia and Pacific Oil and Gas Conference and Exhibition held in Kuala Lumpur, Malasia, April 14-16 1997.
[17] Lietard, O., and Hegeman, P.: "Optimum Development of a Thin Box-Shaped Reservoir with Multiply Fractured Horizontal Wells", paper SPE 50420, proceedings from the SPE International Conference on Horizontal Well Technology held in Calgary, Alberta, Canada, 1-4 November 1998.
[18] Matlab® Version 7.0, The Math Works, Inc., Natick, MA, 2005.
[19] Beggs, H.D., and Brill, J.P.: "A Study of Two-Phase Flow in Inclined Pipes", Journal of Petroleum Technology, May 1973, 607-617.
[20] Crosby, D.G., Yang, Z., and Rahman, S.S.: "Transversely Fractured Horizontal Wells: A Technician Appraisal of Gas Production in Australia", paper 50093 proceeding from the 1998 SPE Asia Pacific Oil and Gas Conference and Exhibition held in Perth, Australia, 12-14 October 1998.

Table 1. **Economic Evaluation Data**

Type of charges	Value
Fixed horizontal drilling cost	$100,000
Horizontal wellbore drilling and completion/each horizontal well drilled	$500,000
Fixed fracturing costs	$100,000
Each transverse fracture treatment costs	$100,000
Each longitudinal fracture treatment costs	$150,000
Each branch drilling cost	$100,000
Gas price	$4.5/Mscf
Discount Rate	10%

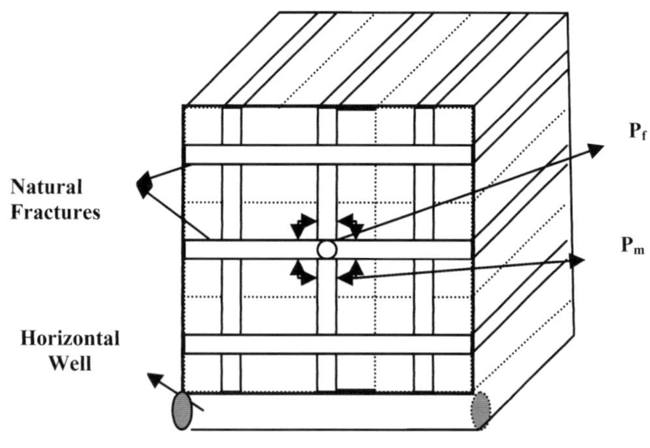

Figure 1. **Naturally fractured reservoir with a horizontal well used to simulate a CBM gas reservoir.**

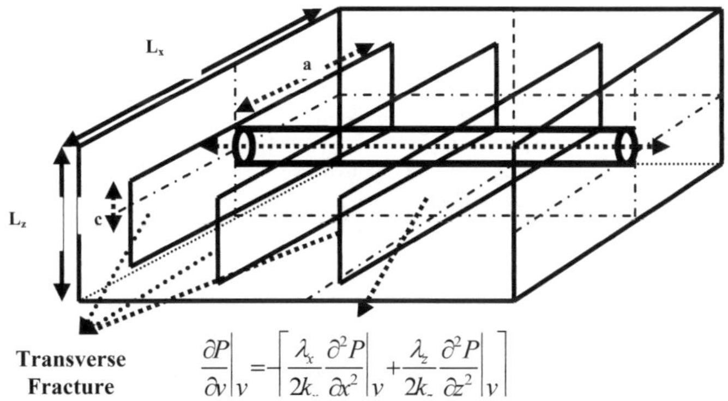

Figure 4. Transverse fractures and flow conditions.

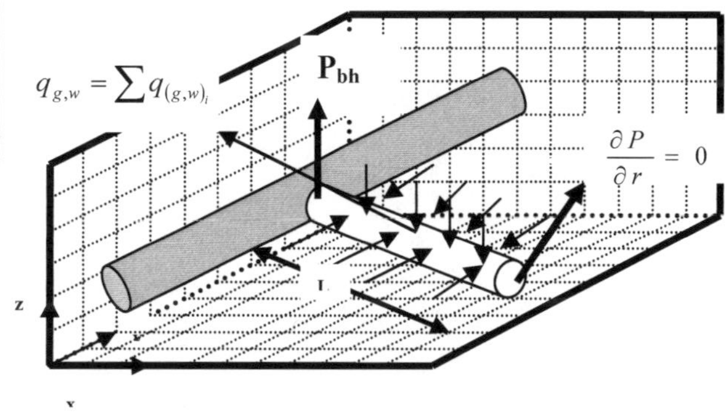

Figure 3. Flow of gas and water inside a branch induced from the horizontal lateral

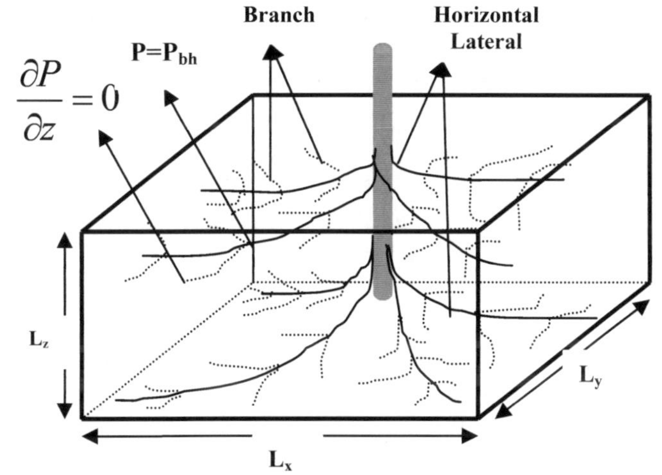

Figure 2. Multi-lateral horizontal wells with different branches

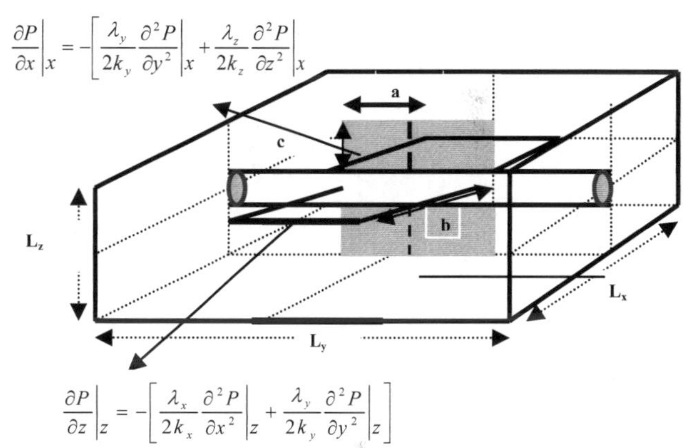

Figure 5. Different longitudinal fractures along with the horizontal well.

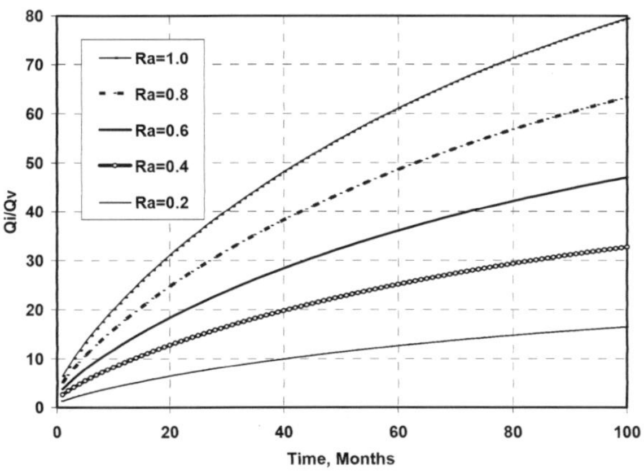

Figure 6. Effect of R_a on cumulative gas production versus time

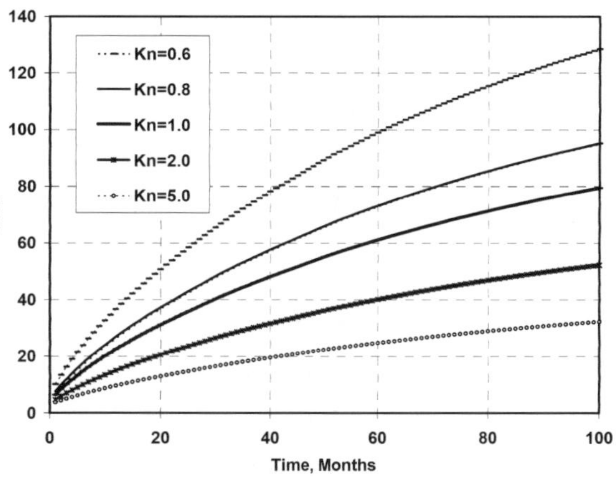

Figure 7. Effect of Normalized permeability on normalized gas production

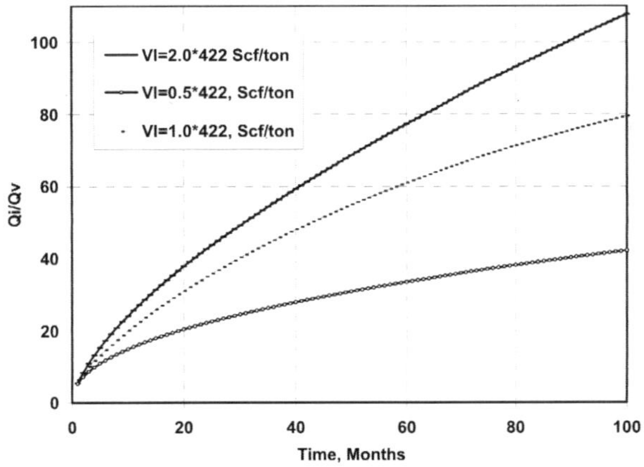

Figure 8. Effect of P_L on cumulative gas production from a horizontal wellbore

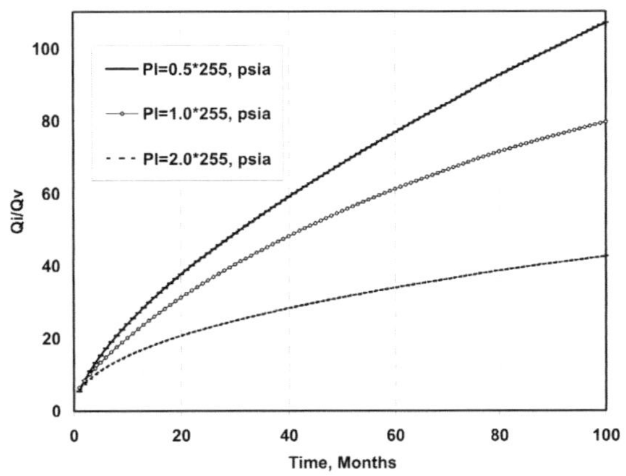

Figure 9. Effect of V_L on cumulative gas production versus time from a horizontal wellbore

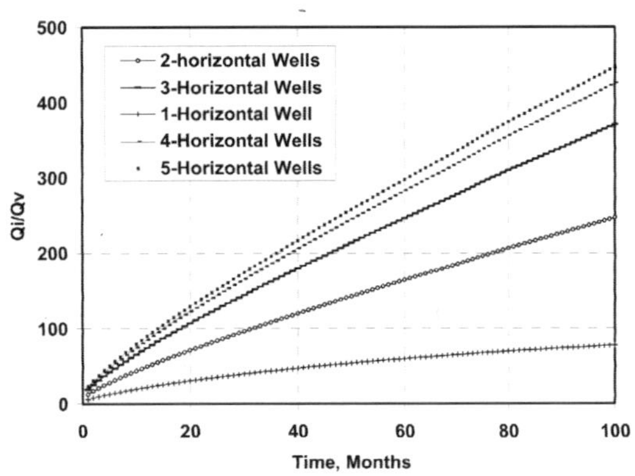

Figure 10. Effect of number of horizontal wells on normalized gas production versus time

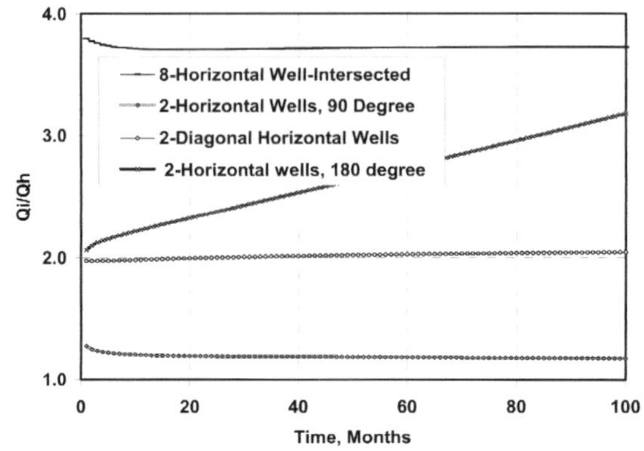

Figure 11. Effect of angle by which horizontal wells intersect horizontal face cleats on cumulative gas production versus time

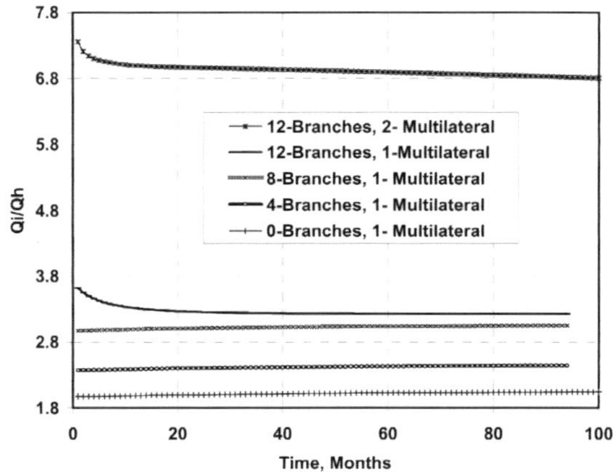

Figure 12. Effect of number of branches and laterals in multi-lateral horizontal wellbores on cumulative gas production versus time.

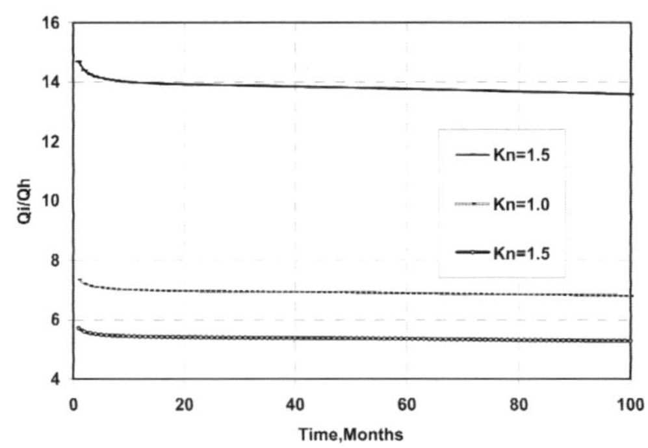

Figure 13. Effect of normalized permeability on normalized cumulative gas production versus time in multi-lateral horizontal wellbore

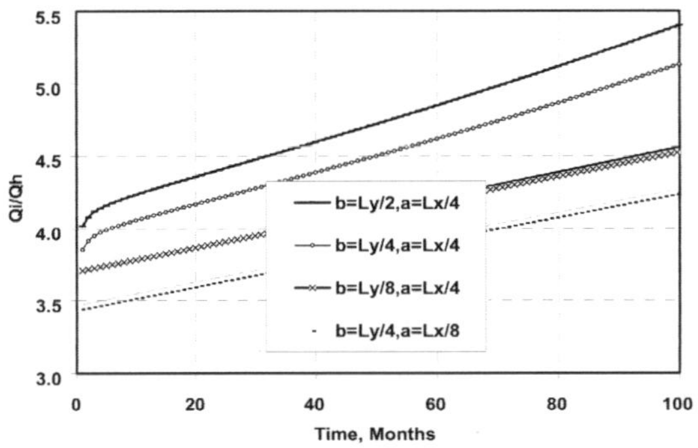

Figure 14. Effect of length of a parallel longitudinal fracture on normalized cumulative gas production versus time

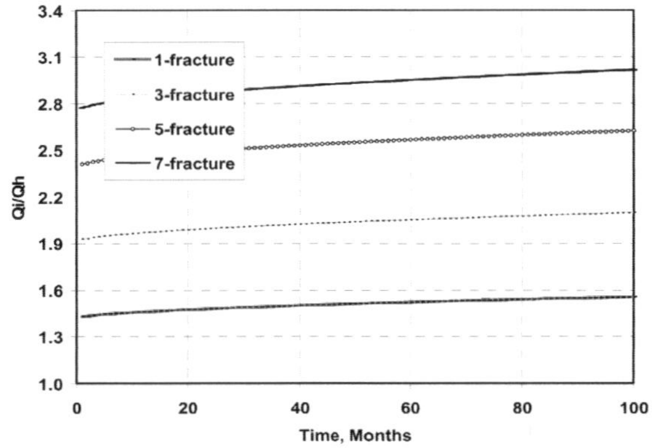

Figure 17. Effect of number of transverse fractures on normalized cumulative gas production versus time.

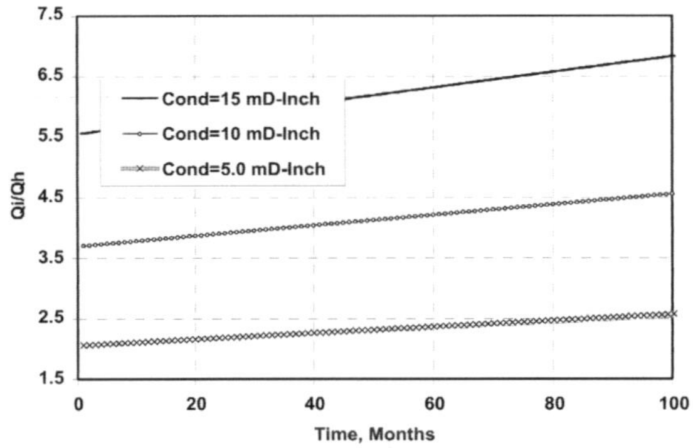

Figure 15. Effect of fracture conductivity on cumulative gas production versus time for parallel longitudinal fracture.

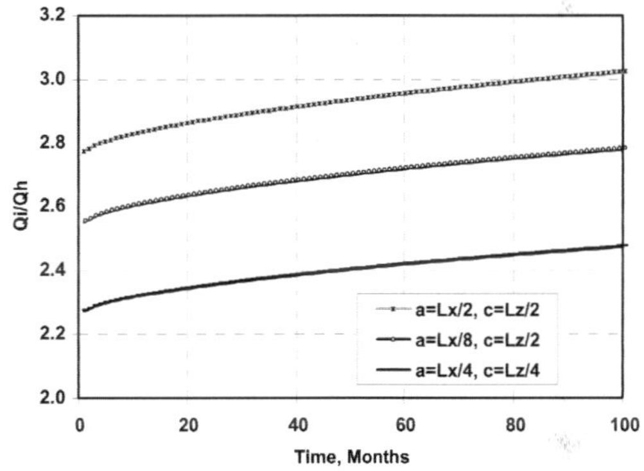

Figure 18. Effect of transverse fracture length on normalized cumulative gas production versus time

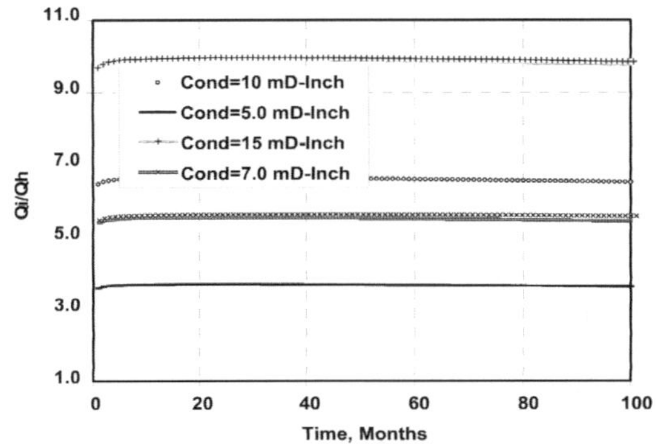

Figure 16. Effect of fracture conductivity on normalized gas production versus time for an orthogonal longitudinal fracture.

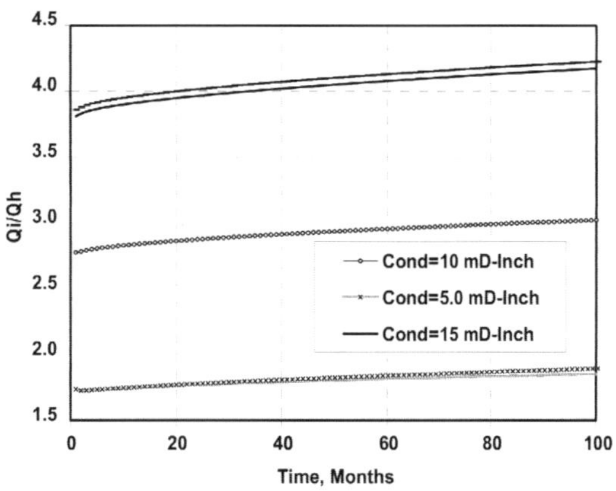

Figure 19. Effect of transverse fracture conductivity on normalized gas production versus time.

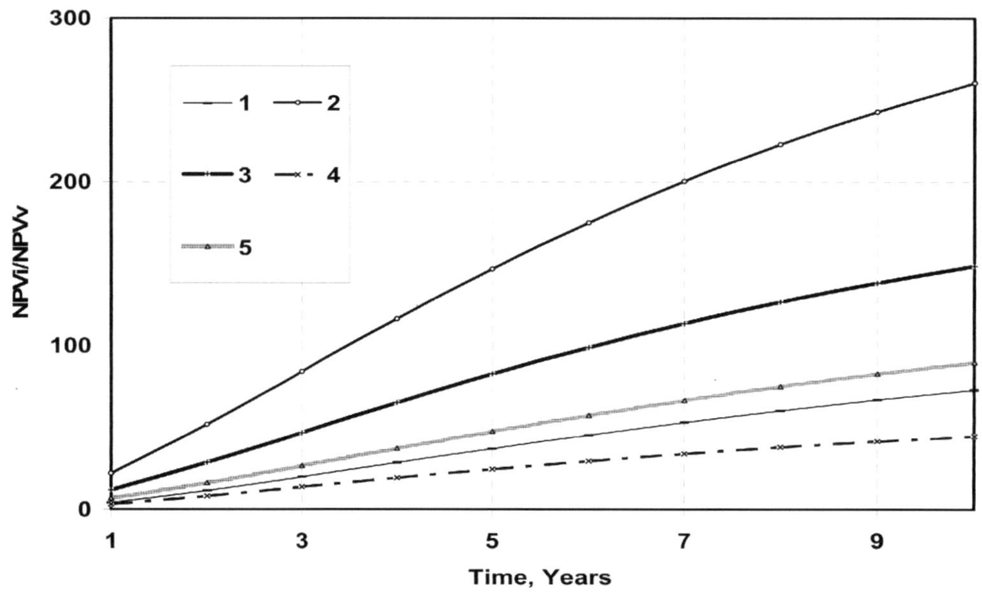

Figure 20. Comparison of normalized NPV for various cases described in the text versus time. 1. Three-parallel horizontal wells, 2. Two-multi-lateral horizontal wells, 3. Four-intersecting horizontal wells, 4. Seven-transverse fractures, 5. Series of longitudinal fractures.

0513

Coal Mine Methane Reservoir Testing and Production Methodologies

Peet M. Sööt, PhD; Dale R. Jesse; Michael E. Smith; Aldeane M. Sööt
Northwest Fuel Development, Inc.

ABSTRACT

Numerous Coal Mine Methane (CMM) development projects stumble because a significant fraction of installed equipment is idled by the lack of CMM supply. Some installations are even dismantled when the lack of turndown capabilities keep the equipment from being operated at all. These problems are a direct result of inadequate reservoir testing by the developers before equipment is installed. Application of basic scientific and engineering principles allows one to measure the size of the reservoir and make estimates of the recharge rate. Application of the concepts described in this paper will help CMM developers avoid these costly mistakes.

Northwest Fuel Development, Inc. (NW Fuel) has tested and operated CMM wells for over 20 years. This experience has provided numerous unusual results reservoirs. Forensic engineering skills have helped translate observed flow characteristics into models for various types of wells. The following categories have been established for such CMM wells: Check Valve, Trap, Flow Regulator and Island Reservoir Wells.

COAL MINE METHANE (CMM) RESERVOIR TESTING

There are two primary variables one needs to define for a CMM reservoir:

1) the size of the reservoir cavity, and

2) the rate of recharge into the cavity.

The reservoir size is important in order to define the flexibility the operator will have in planning short term production from the well. If the reservoir is large enough, then the operator can produce more CMM gas for short periods of time by pulling it into a vacuum. Subsequent lowering of the production rate would allow the reservoir to recharge back to its "equilibrium" condition. This flexibility may be desirable for various purposes. Providing fuel for a peaking power generating unit would be one example of such an application.

The reservoir recharge rate is the more important variable for determining the long term production rate that a reservoir can sustain. If the operator tries to produce more CMM from a CMM reservoir than it receives from the recharge then one, or both, of the following situations will occur (neither of them good):

1) the reservoir will be pulled into a deep vacuum, and/or

2) air will be pulled into the reservoir, contaminating the indigenous CMM.

Reservoir Size Estimation

There are simple methods by which one can estimate coal mine methane (CMM) reservoir size and make an estimate of the rate at which the reservoir is being recharged. The difficulties arise when an operator does not account for some simple issues related to such low pressure - or even vacuum - operations.

Exhibit 1 shows a schematic of a simplified CMM wellhead and reservoir configuration. In order to define the mine reservoir size (V_m) one need only perform some simple reservoir production tests. One simply needs to produce CMM from the reservoir for some period of time and record the amount of gas produced during that period.

The subsequent data analysis is based on the ideal gas law, which is applicable since the reservoir is at such a low pressure. The ideal gas law states that

$$P_m V_m = n_m R T_m \qquad (1)$$

where,

P_m = mine reservoir pressure
V_m = mine reservoir volume
n_m= quantity of gas in the mine reservoir volume
R = ideal gas law constant
T_m = mine reservoir temperature

One can apply this equation before the production test begins, as follows:

$$P_{m0} V_{m0} = n_{m0} R T_{m0} \qquad (2)$$

where,

the subscript "0" denotes the respective variable value at time = 0.

After some period of time (t_1) the ideal gas law indicates that

$$P_{m1} V_{m1} = n_{m1} R T_{m1}. \qquad (3)$$

Assuming the production rate (Q_{P1}) during that increment of time is constant, then the cumulative total of gas (CMM) produced from the reservoir during the first time increment (t_1) is

$$CMM_1 = Q_{P1} * t_1. \qquad (4)$$

If the production rate is not constant, one need only integrate the production chart from time zero to t_1 in order to define CMM_1.

The following material balance can also be applied to the reservoir during the production test:

$$\text{CMM In} - \text{CMM Out} = \text{CMM Accumulation}. \qquad (5)$$

Gas coming into the reservoir is defined by

$$\text{CMM In} = Q_R * t_1. \qquad (6)$$

Gas leaving the reservoir is the sum of the leakage form the reservoir plus the gas produced during the test, as follows:

$$\text{CMM Out} = (Q_L * t_1) + CMM_1. \qquad (7)$$

The CMM that accumulates in the reservoir is simply the difference between the amount of gas in the reservoir after the test minus the amount of gas in the reservoir before the test and can be written as

$$\text{CMM Accumulation} = n_1 - n_0. \quad (8)$$

Combining Equations 4 through 8,

$$[Q_R * t_1] - [(Q_L * t_1) + (Q_{P1} * t_1)] = n_1 - n_0. \quad (9)$$

Some assumptions can simplify this equation and make it easily solvable.

Since the reservoir was probably at equilibrium when the flow tests was started, then

$$Q_R = Q_L \quad (10)$$

If one produces the gas at a high enough rate, then

$$Q_{P1} >> Q_R. \quad (11)$$

Using Equations 10 and 11, then Equation 9 simplifies to the following:

$$-(Q_{P1} * t_1) = n_1 - n_0. \quad (12)$$

Substituting from Equations 2 and 3 into Equation 12,

$$CMM_1 = Q_{P1} * t_1 = \frac{P_{m0} V_{m0}}{R T_{m0}} - \frac{P_{m1} V_{m1}}{R T_{m1}} \quad (13)$$

One can assume that the reservoir boundaries are fixed* so that

$$V_{m0} = V_{m1} = V_m. \quad (14)$$

The reservoir temperature also stays constant, which in the case of reservoirs in the US Appalachian Basin is about 510 °R (50 °F).

$$T_{m0} = T_{m1} = 510\ ^\circ R = T_m. \quad (15)$$

Equation 13 can then be solved for the reservoir volume (V_m):

$$V_m = \frac{[(R * T_m) * (Q_{P1} * t_1)]}{(P_{m0} - P_{m1})} \quad (16)$$

All of the variables on the right side of the equation are either known values or are measured during the flow test(s). Hence, the reservoir volume can be calculated.

It is possible to obtain a better estimate of the reservoir size if production data are disaggregated into several time periods along with the corresponding pressure data. That provides several separate sets of data that can be used in Equation 16.

A major cautionary note must be made about applying Equation 16. One cannot necessarily use the blower inlet pressure for the P_m values in Equation 16. The reservoir pressure (P_m) is only equal to the

* In the "Island Reservoir Well" case study below one can see that this assumption is not necessarily accurate, but it should suffice for most cases.

blower inlet pressure (P_B) when there is no flow from the reservoir. Any time that there is flow, even flow by natural convection, these pressures are not the same.

If the blower inlet pressure readings are the only ones available, then one can calculate the reservoir pressure from the blower inlet pressure. This calculation must allow for friction losses in the well production casing along with the differences in elevations between the surface and the mine reservoir. Even if this calculation is made, it must be repeated for any changes in the flow rate or changes in the pressures.

A much easier method is to take mine reservoir pressure readings directly. If the reservoir is shut in, then $P_M = P_B$. This is the case before the test begins and at the end of the test. One can even take direct measurements of the reservoir pressure during the flow tests. NW Fuel has developed simple and relatively inexpensive methods for taking such readings.

Reservoir Recharge Rate Estimation

The second major variable that can be defined from production tests is the reservoir recharge rate. Actually, measurements for this calculation are taken after the flow tests described above are completed. If one assigns the following notations for the reservoir variables after the completion of the flow test:

$$P_{m2}\ ,\ V_{m2}\ ,\ n_{m2}\ ,\ T_{m2}$$

and the following notations after some period of time:

$$P_{m3}\ ,\ V_{m3}\ ,\ n_{m3}\ ,\ T_{m3}$$

then one can apply another material balance to the reservoir:

$$\text{CMM In} - \text{CMM Out} = \text{CMM Accumulation} \qquad (5)$$

which in this case can be defined by

$$[Q_R * (t_3 - t_2)] - [Q_L * (t_3 - t_2)] = n_3 - n_2. \qquad (17)$$

During the recovery phase, the reservoir is typically drawn down into a significant vacuum so the leakage rate is likely diminished to the point that

$$Q_R >> Q_L \qquad (18)$$

In which case, Equation 17 can be simplified to

$$Q_R * (t_3 - t_2) = n_3 - n_2. \qquad (19)$$

Applying the ideal gas law (Equation 1) to Equation 19 results in

$$Q_R * (t_3 - t_2) = \frac{P_{m3}\, V_{m3}}{R\, T_{m3}} - \frac{P_{m2}\, V_{m2}}{R\, T_{m2}}. \qquad (20)$$

Knowing that R is a constant and assuming that the reservoir volume is constant (see footnote * above) and the reservoir temperature is constant, Equation 20 can be simplified to

$$Q_R = \frac{(P_{m3} - P_{m2}) * V_m}{(t_3 - t_2) * R * T_m}. \qquad (21)$$

The value for V_m comes from Equation 16 and all of the other variables on the right side of the equation are either known values or are measured during the reservoir recovery test. Hence, the reservoir recharge rate can be calculated using Equation 21. During the recovery test there is no flow in the well casing so the difference in pressure readings for the reservoir (P_{m3} - P_{m2}) are the same as the wellhead pressure differences (P_{B3} - P_{B2}).

As with the reservoir size calculations, it is possible to obtain a better estimate of the recharge rate if data are disaggregated into several time periods along with the corresponding pressure data. That provides several separate sets of data that can be used in Equation 21. It is even better if other separate flow tests are performed and recovery tests are run after each flow test. Then results from the various tests can be compared. If there is consistency in the results, then one can conclude that the recharge rate estimates are quite reliable.

COAL MINE METHANE (CMM) WELL CASE STUDIES

NW Fuel has encountered a broad range of CMM vents and wells. These include vents that have been left behind after mining while other vents or wells are drilled solely for the purpose of producing the CMM. Some types of wells provide an interesting challenge in trying to understand the mechanism by which certain conditions exist at the well. The descriptions below show some of these types of wells, and the well model that NW Fuel staff members have hypothesized to explain the behavior of the respective well. One can apply a modern term to this exercise of developing a model for each type of well that exhibits unusual characteristics: "forensic engineering". Forensic engineering applied to CMM projects requires some sleuthing but it also requires a knowledge of the scientific and engineering principles associated with CMM production.

The models presented in this paper are but one explanation for the observed phenomena for each type of well. The readers are encouraged to come up with their own interpretations. If feasible, the authors encourage you to contact them to share your observations if they differ from the presented models.

Case A – CMM Check Valve Well

This type of CMM well allows air movement into the well, but does not yield any CMM flow out of the well. The well acts like a check valve that one would find in piping systems, or a diode in electrical systems. Exhibit 2 shows the conditions that one can observe at the wellhead for the Check Valve Well, as follows:

1. Under Natural Convection
    a. the wellhead pressure is less than atmospheric pressure,
    b. air flows into the well bore if the wellhead valve is open, and
    c. water injected into the well flows unobstructed.
2. Under Forced Convection there is no flow out of the well.

The model that explains this behavior is presented in Exhibits 3 & 4. The well has most likely been drilled into a flooded part of the coal mine (Exhibit 3). The bottom of the well is not far from the water surface within the mine. This allows air to be pulled into the mine, but CMM cannot escape. One can establish that the water height (h) near the bottom of the well is less than the water head equivalent of the pressure difference between the atmosphere and the mine ($h < P_{atm} - P_{mine}$). If $h > P_{atm} - P_{mine}$ then the well would not be much of an oddity. There would not be any gas flow in either direction. The only thing that would be unusual is that water injected into the well would easily disappear. Exhibit 4 shows the situation when a blower is put on a Check Valve Well. The water at the bottom of the well would be pulled up the casing to a height that is equal to the blower suction head. It would still not allow any CMM gas to be produced.

Some developers have encountered such a situation and argued against this model because the driller pulled a dry drill bit to the surface. They did not test for water at the bottom of the well by any other means. The possible explanation for that result is that the drill bit was hot from drilling and any water that it touched was evaporated as the bit was pulled to the surface.

Case B – CMM Trap Well

The Trap Well is capable of producing CMM up to some production level, but if one tries to produce too much gas, the well closes off like a trap and the gas production drops to zero. NW Fuel staff encountered such a well in Ohio. Under natural convection conditions the well was able to produce a good CMM gas flow of about 100 MCFD (see Exhibit 5). When a blower was put onto the wellhead additional CMM gas was produced – even double the natural flow case. However, when the blower was run faster, and the wellhead pressure was lowered further, the wellhead would rapidly go into a deep vacuum and gas production stopped altogether. The high wellhead vacuum condition had closed the trap.

This condition was easily reversed. As soon as the blower was stopped, the wellhead pressure came back to atmospheric levels. Within a few minutes gas again started to flow out of the well of its own accord.

The model for this type of CMM well is presented in Exhibits 6 & 7. It is likely that a roof fall has occurred in the mine at the bottom of the CMM well (Exhibit 6). One can hypothesize that there is water flowing into the mine through the rubble zone. The rubble zone is permeable enough to allow the countercurrent flow of water and gas under low to moderate gas flow conditions. The well can even produce significant amounts of gas when a blower is placed on the wellhead.

The problem occurs when the vacuum imparted on the well bore pulls too much water into the casing and the water totally seals off any gas production (Exhibit 7). The harder the blower pulls on the wellhead, the higher the water level rises within the casing and the better the water seal becomes. As soon as the blower is stopped, the vacuum is released, and water falls out of the well casing. Then the natural convection case is re-established.

The idea for this type of well behavior comes from conventional natural gas wells that have a water drive. If one overproduces such a well, the production can be killed by water coning into the wellbore. In conventional gas wells, this can permanently kill the well. Fortunately, NW Fuel has been able to reverse the problem in the CMM well by simply eliminating the wellhead vacuum as noted above.

Case C - CMM Flow Regulator Well

This type of well behaves like a flow regulator in a piping system. CMM can be produced from such a well at all times, but it has a maximum flow rate. No matter how deep a vacuum one imparts on the wellhead, the gas production rate can not be made to exceed a maximum that the well will produce. NW Fuel is operating such a well in West Virginia. The observed conditions for this well are presented in Exhibit 8. About 600 MCFD of total gas is emitted from this well when it is opened to the atmosphere. When a blower is used to pull gas from this well, it will produce about 1 MMCFD of gas. This is the same flow rate whether one imparts 5" Hg vacuum at the wellhead or imparts 10" Hg vacuum at the wellhead.

The model that was initially developed for this behavior is described in Exhibit 9. The hypothesis was that an obstruction in the pipe had created a condition where the gas flow would achieve sonic velocity when the blower was turned on. Since such a condition causes a physical limit to the gas flow, it was felt that this was a reasonable explanation for the flow characteristics witnessed at the surface. Calculations for the measured gas flows indicated that a 1.5" ID round opening would have sonic velocity in the throat of the opening. Any number of things may have happened to create an obstruction in the well casing that would leave an opening that was 2 in^2 or smaller For example,

a. There might have been a partial collapse of the casing caused by subsidence after the mining.
b. The miners may have left a small fitting at the bottom of the well. (This well had been drilled before the mine was sealed.)

It was subsequently discovered that this was not the correct model for this well. There was no permanent obstruction within the casing nor at the bottom of the casing. A video camera survey showed that there were breaks in the casing (Exhibit 10) where ground water was flowing into the well. The water did not interfere with the gas production as long as it was at low to moderate levels. When the gas flow was raised to high levels two things happened:

a. The gas velocity exceeded 100 ft/sec.
b. The added vacuum probably increased the water flow into the well from the casing breaks.

The combination of these two conditions may have established a fluidized water bed within the casing (Exhibit 11). Gas could flow through the bed but the bed limited the maximum gas throughput. An equilibrium production condition was established once the water bed was formed. This is a new hypothesis that could not be visually verified. The video camera had to be removed during the subsequent blower flow tests and was not available for the high vacuum tests.

Case D - CMM Island Reservoir Well

The CMM Island Reservoir Well is similar to wells in Cases B and C in the first two stages where natural convection produces CMM gas and this can be increased by pulling a modest vacuum. It is in Stage 3 where this type of well is different. While in Stage 3 the production from the well is inversely proportional to the vacuum. The gas flow slowly decreases as the imparted vacuum is increased.

This type of well exhibits the following characteristics (see Exhibit 12):

Stage 1: After the well is drilled, it emits CMM at some level (Q_1) while $P_{B1} > P_{atm}$.

Stage 2: After a blower is put on the well one is able to extract more gas (Q_2). At that point the blower is pulling a moderate vacuum on the wellhead and $P_{B2} < P_{atm}$. This leads to more CMM production, $Q_2 > Q_1$.

Stage 3: When attempting to produce more gas the blower speed would be increased. In that case the vacuum on the wellhead rises so the wellhead pressure is much less than atmospheric pressure, $P_{B3} << P_{atm}$. However, the higher vacuum actually leads to a lower production than the previous condition, $Q_3 < Q_2$.

The model developed for these conditions has a reservoir that is totally surrounded by water and unmined coal, as presented in Exhibit 13. In Stage 1 the water is at its equilibrium level and recharge into the reservoir provides the basis for producing CMM at a rate of Q_1. By increasing the vacuum in Stage 2 the recharge rate is increased modestly and CMM can be produced at a higher rate, Q_2. The water level is slightly higher than in Stage 1. One can even quantify the change in water level by the simple conversion that 1 pound per square inch (psi) of additional would raise the water level by over 2 vertical feet. The continuing rise in water level is the explanation why the production rate decreases when moving to Stage 3. At that point the rise in water level has covered up too much of the reservoir perimeter surface. The water has apparently blocked enough of the recharge sources so that there is less CMM that can flow into the reservoir.

This model is readily understood when considering that the rise of 2 vertical feet for the water (for each additional psi of vacuum imparted on the reservoir) corresponds to a much greater length along the floor of the mine entries. Given a modest slope of a few percent in the floor of the mine entries, the 2 foot rise in water level can lead to tens of additional feet of mine floor being under water. A well producing from such a formation is appropriately called an Island Reservoir Well since it is completely surrounded by water and unmined coal.

As with the other models, there may be other explanations for these conditions but this is at least one possible resolution of the observed events.

Exhibit 1.

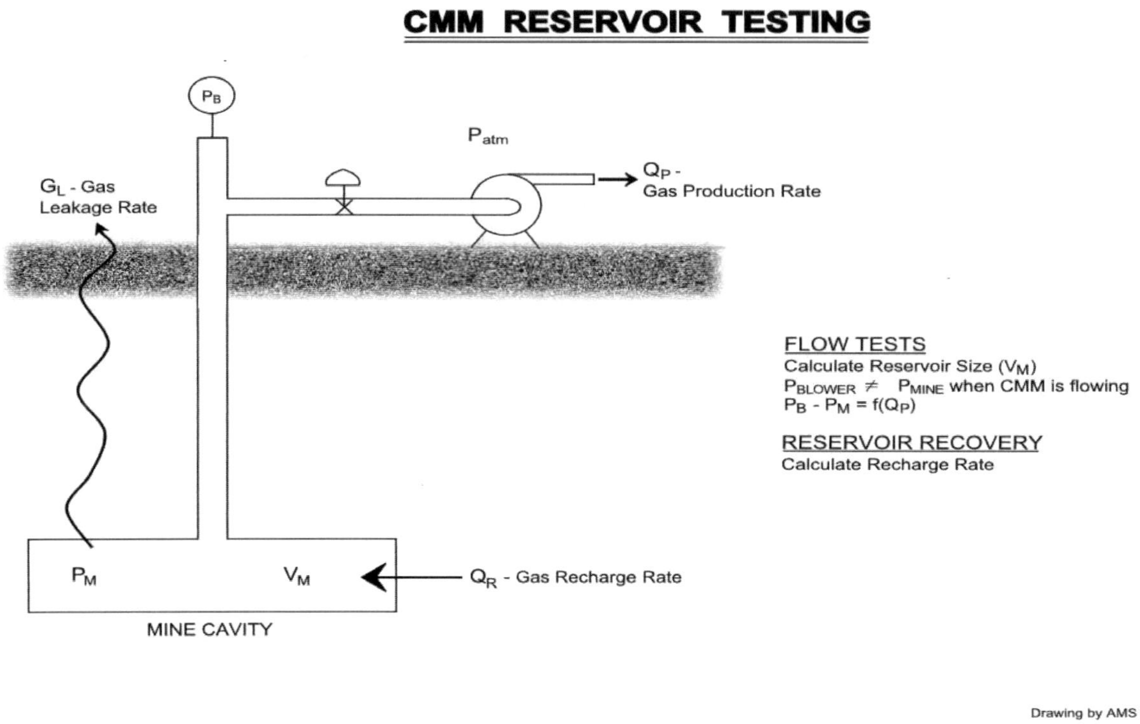

Exhibit 2.

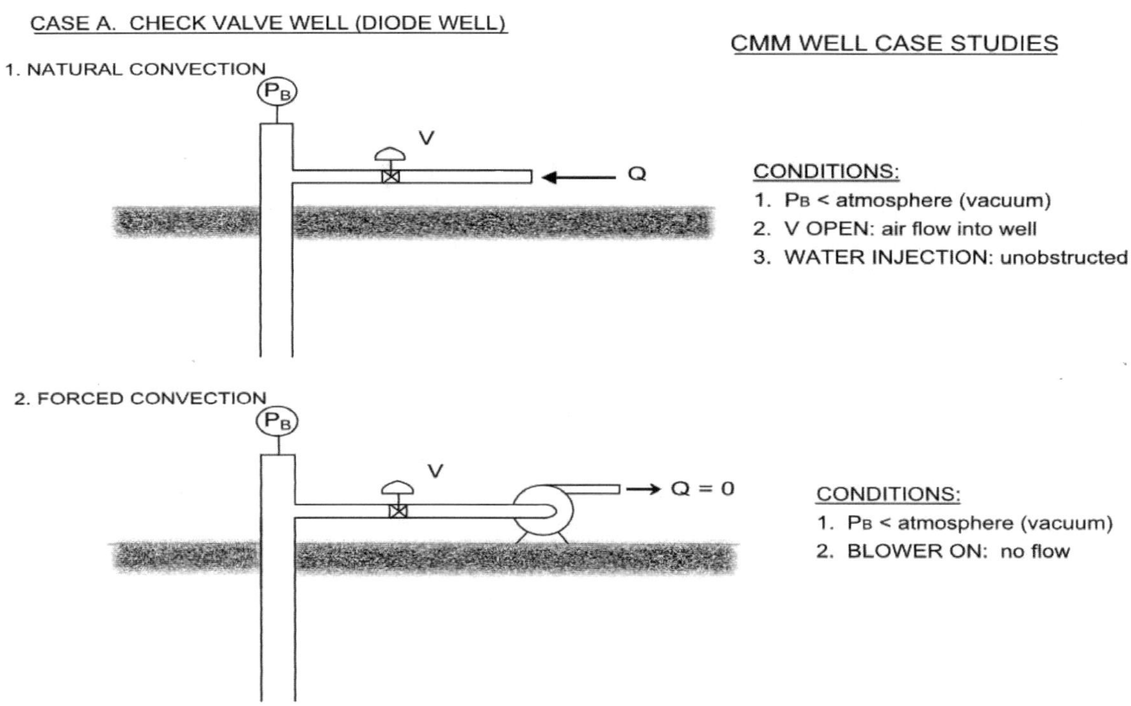

Exhibit 3.

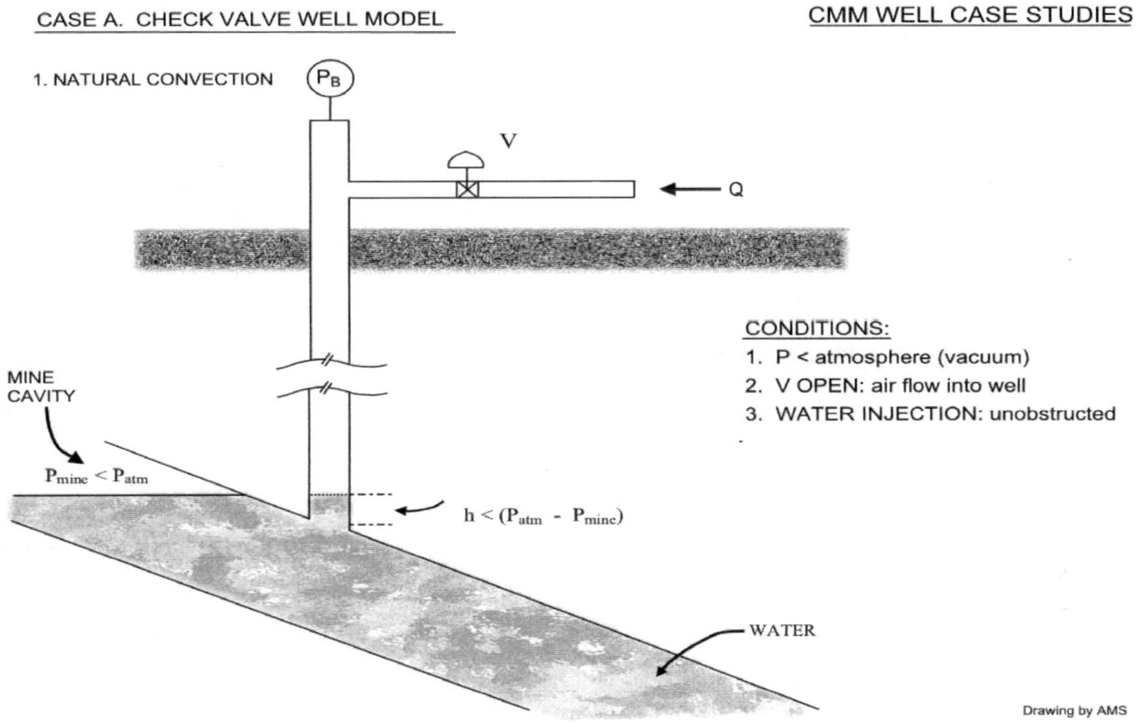

Exhibit 4.

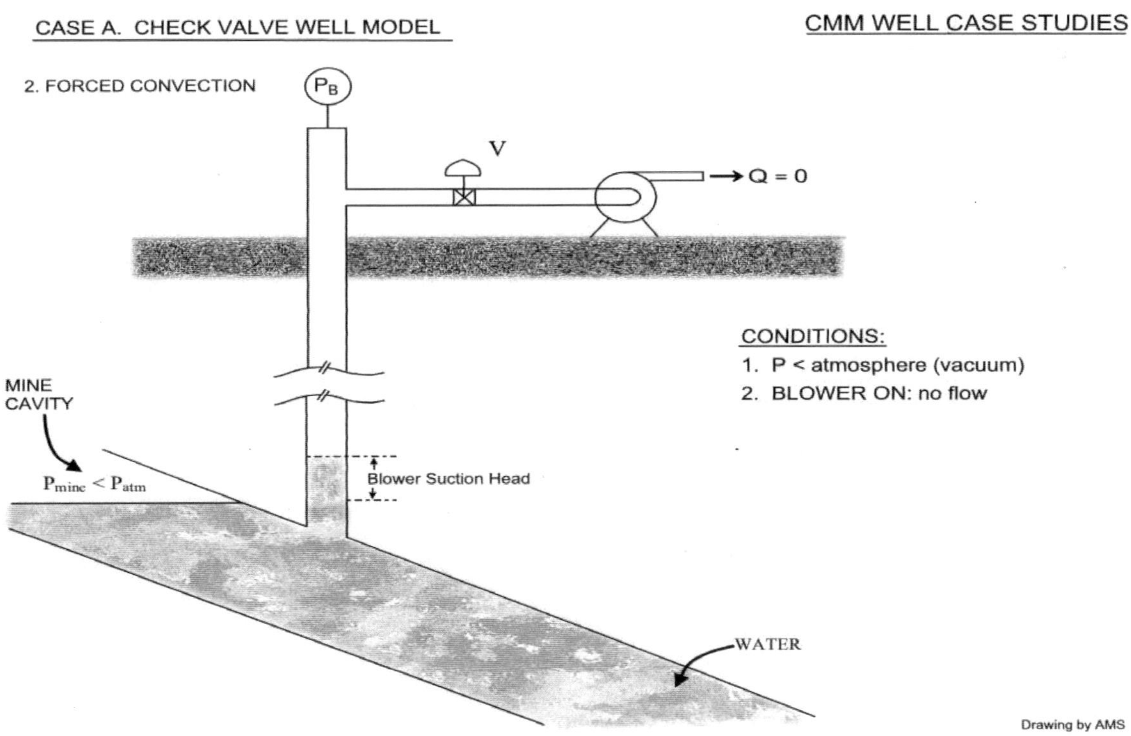

Exhibit 5.

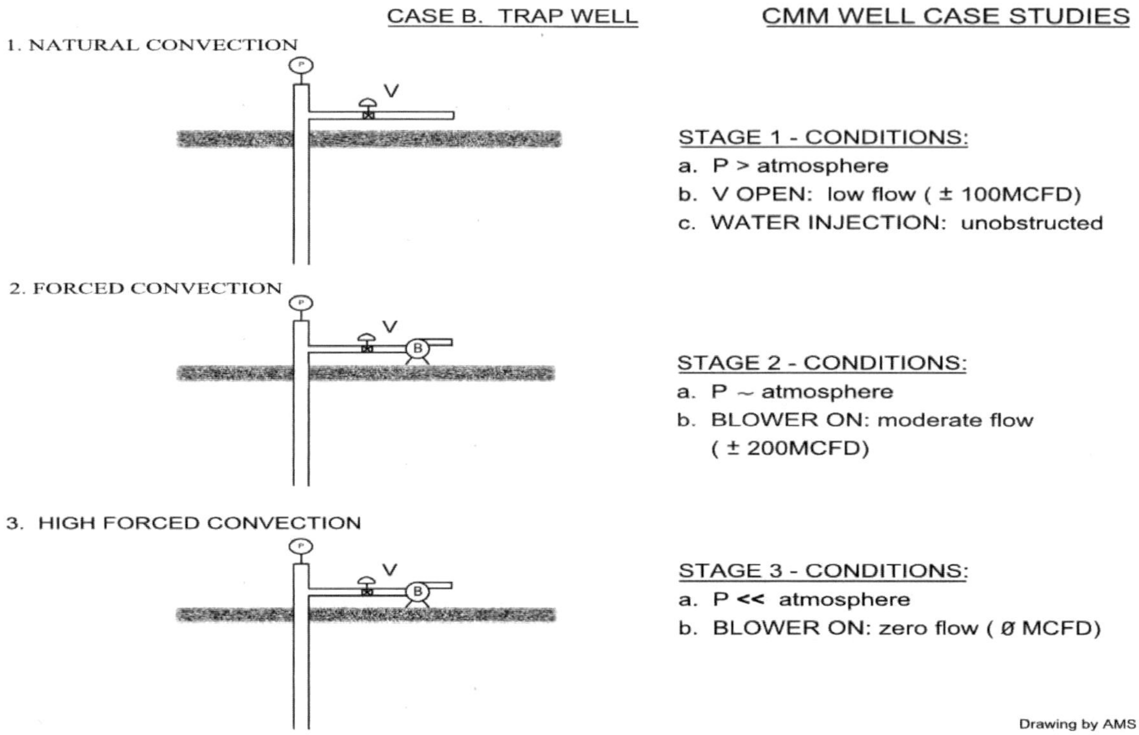

Exhibit 6.

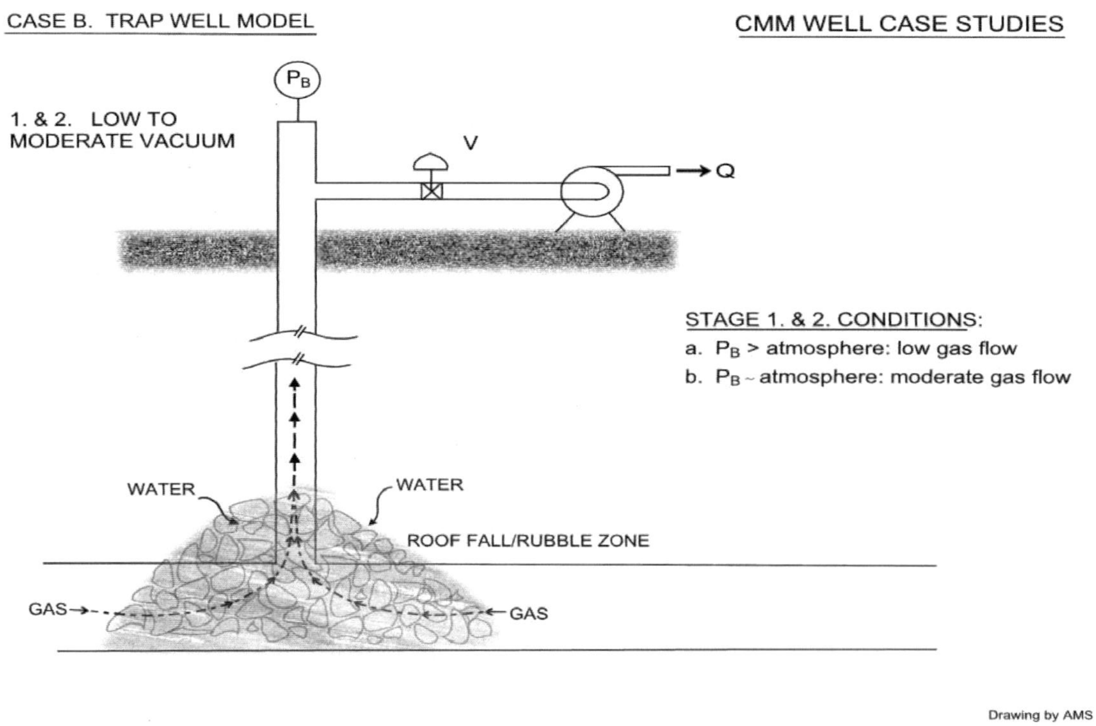

Exhibit 7.

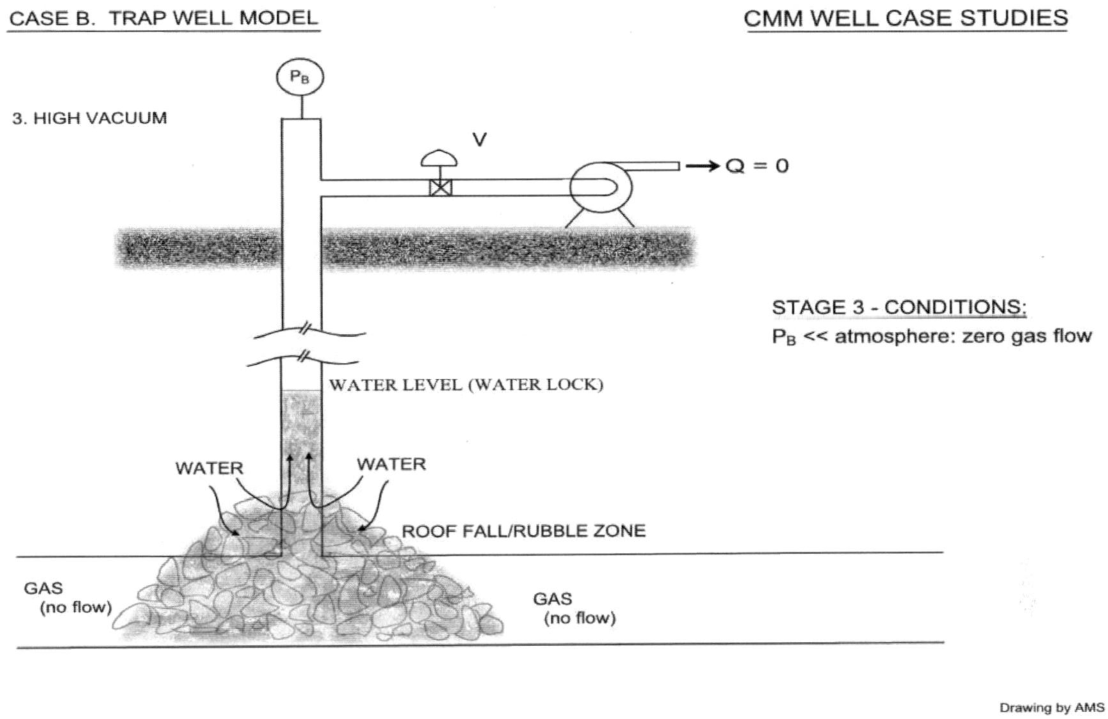

Exhibit 8.

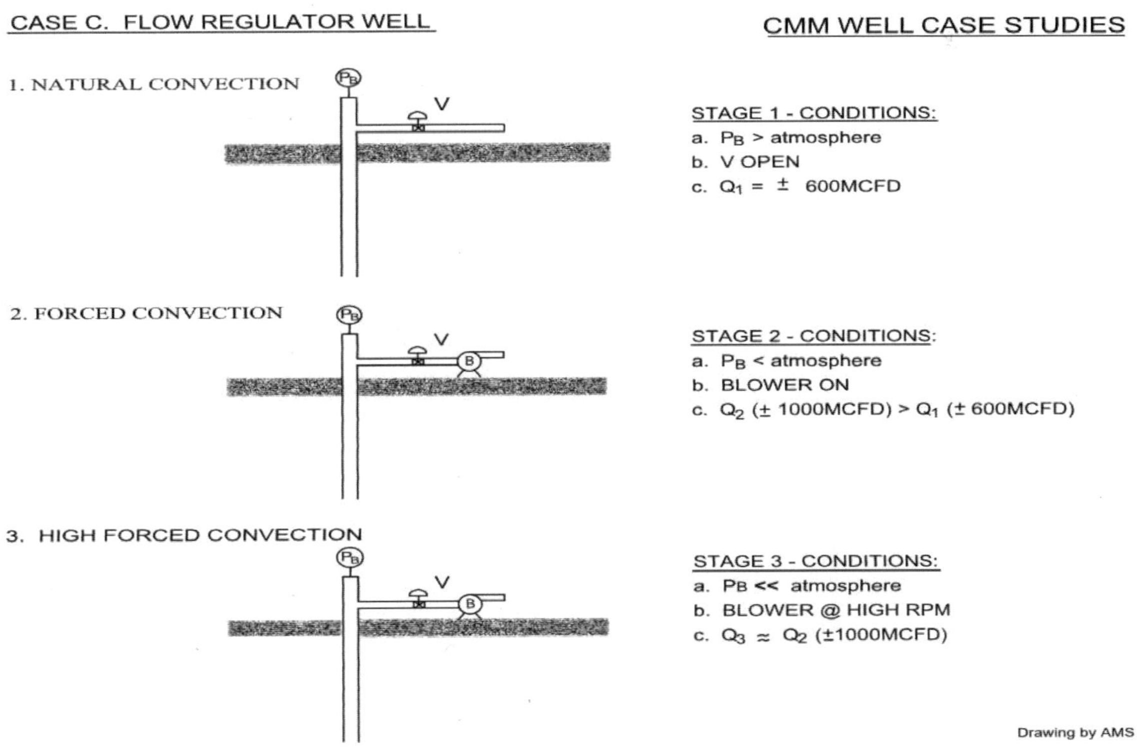

Exhibit 9.

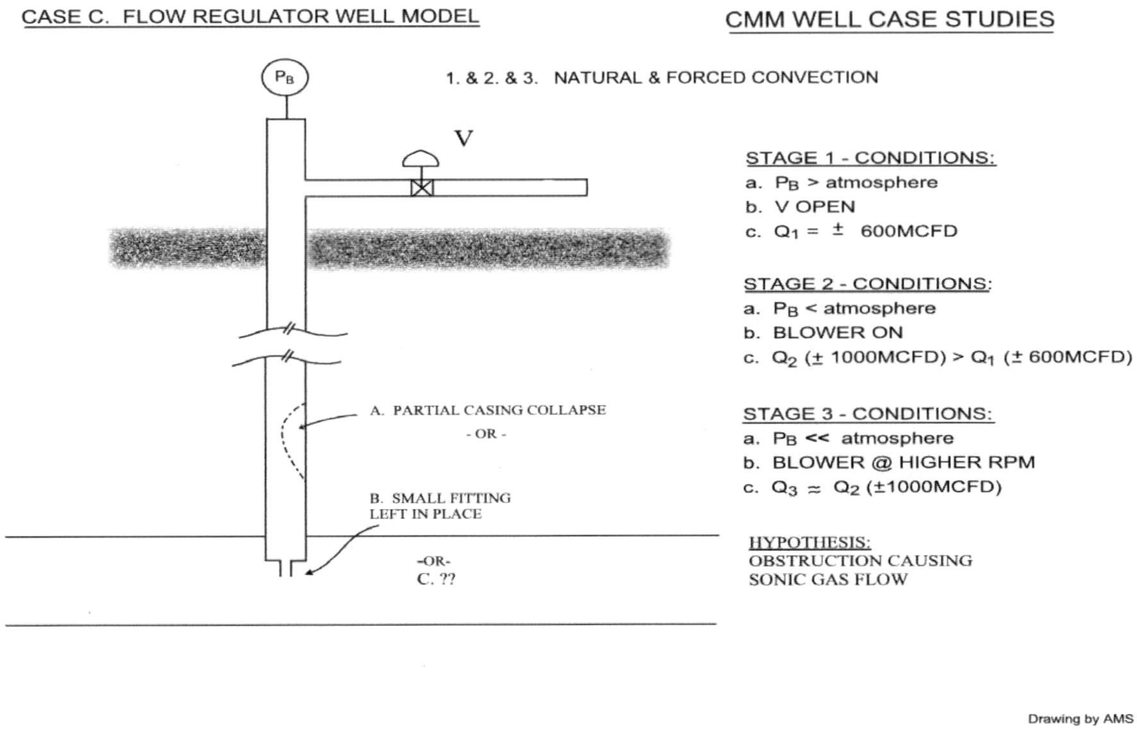

Exhibit 10.

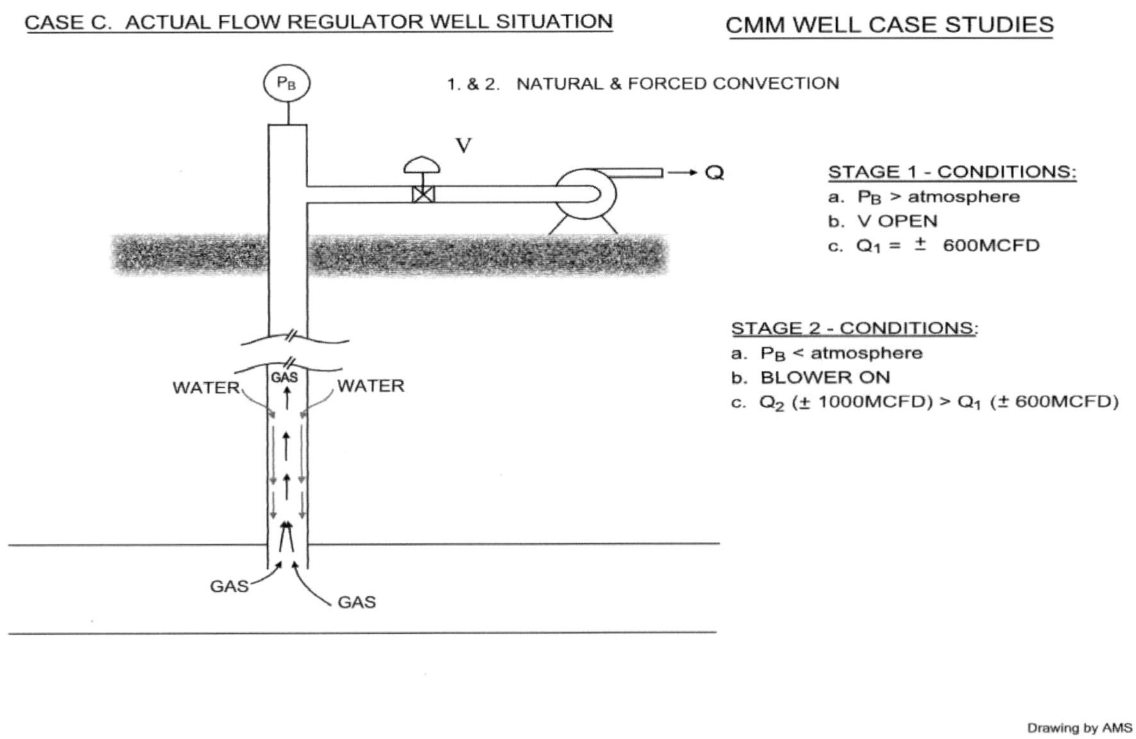

Exhibit 11.

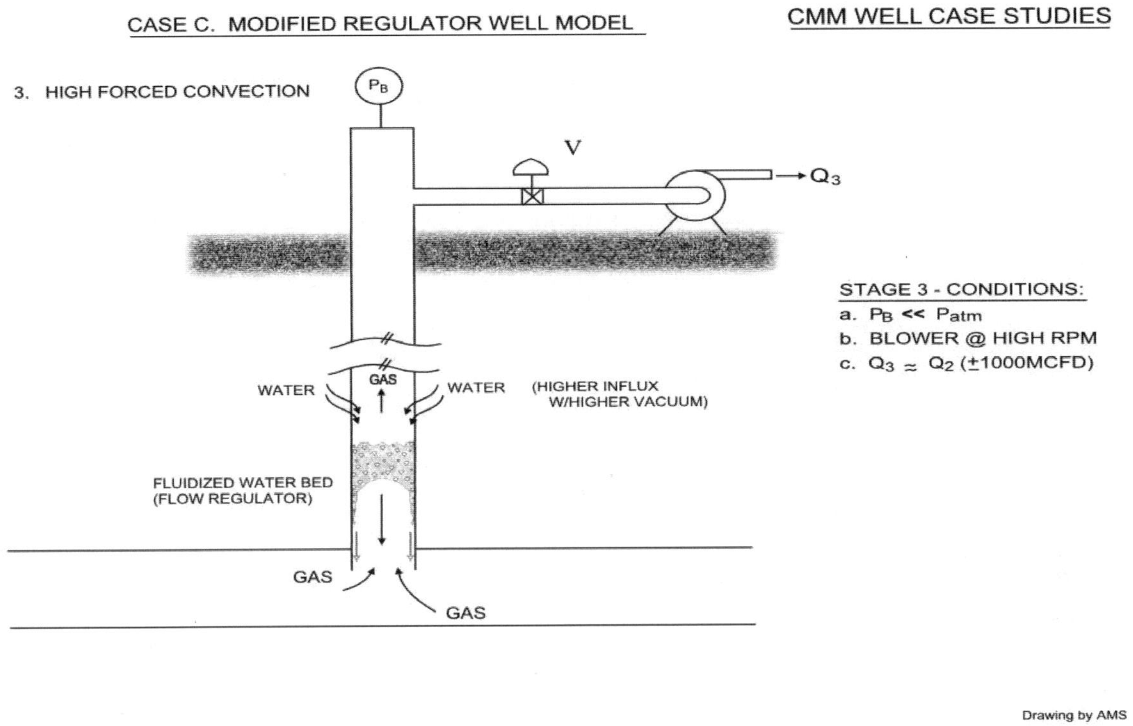

Exhibit 12.

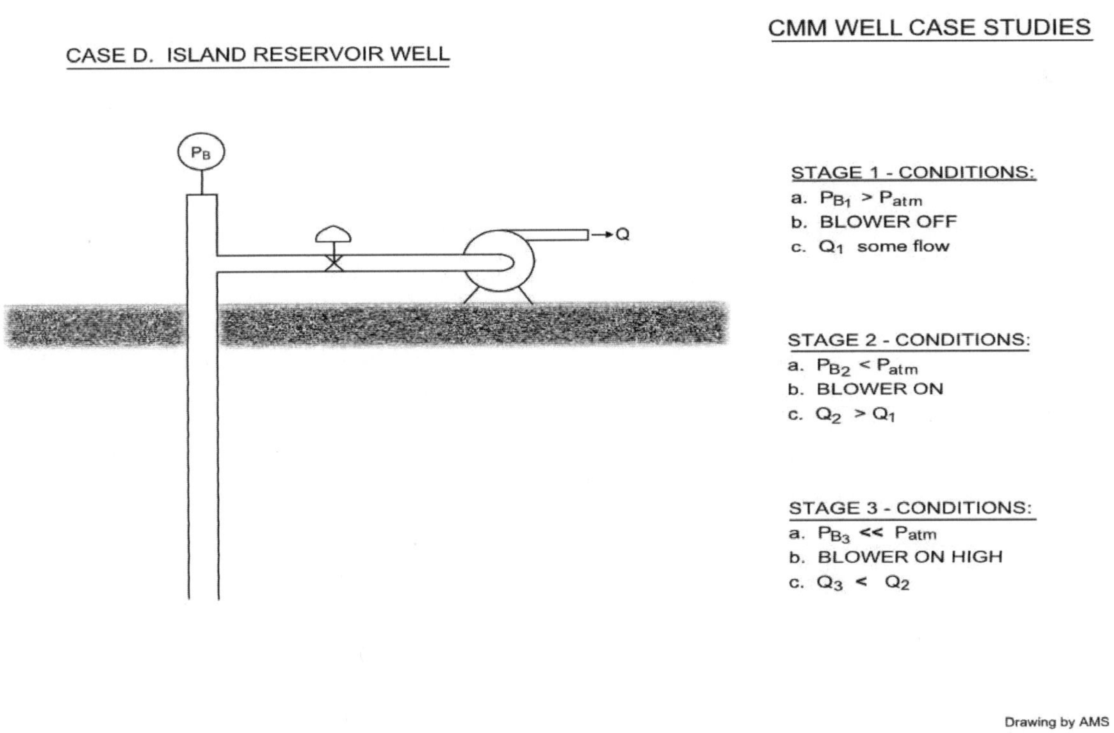

Exhibit 13.

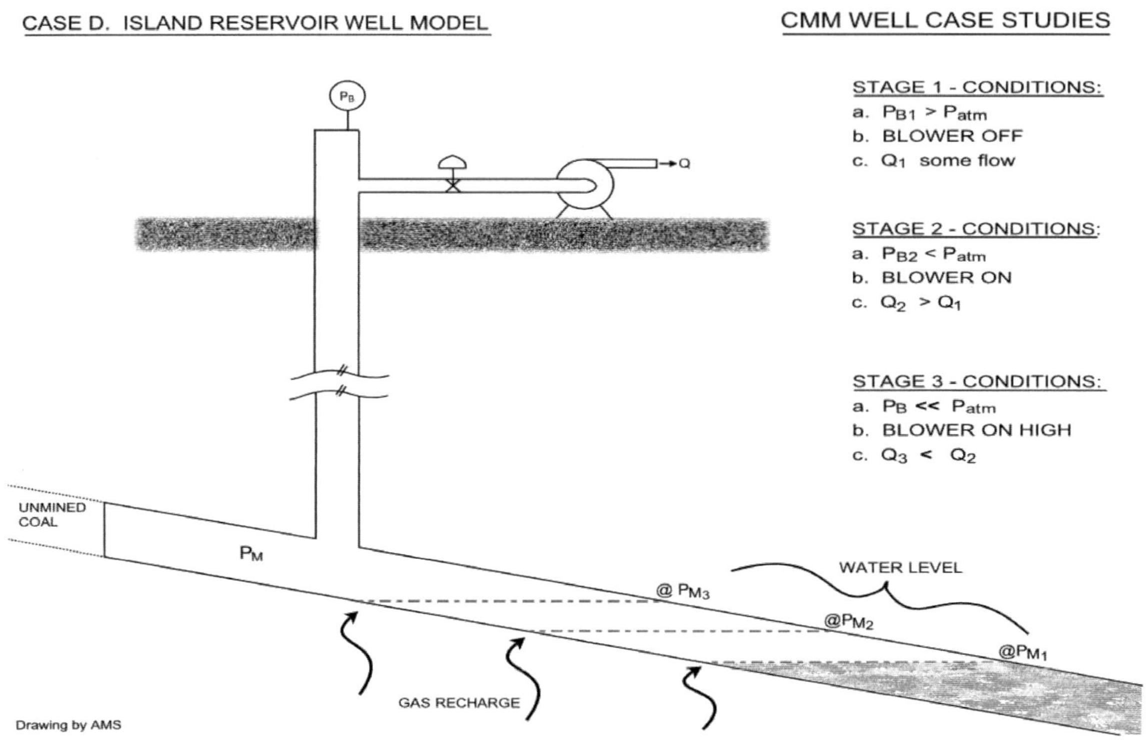
CASE D. ISLAND RESERVOIR WELL MODEL
CMM WELL CASE STUDIES
P_B
Q
STAGE 1 - CONDITIONS:
a. P_B1 > P_atm
b. BLOWER OFF
c. Q_1 some flow
STAGE 2 - CONDITIONS:
a. P_B2 < P_atm
b. BLOWER ON
c. Q_2 > Q_1
STAGE 3 - CONDITIONS:
a. P_B << P_atm
b. BLOWER ON HIGH
c. Q_3 < Q_2
UNMINED COAL
P_M
WATER LEVEL
@ P_M3
@P_M2
@P_M1
GAS RECHARGE
Drawing by AMS

0514

Gas Flow Characterization of Illinois Coal: Assessment for Recovery of Coalbed Methane and Carbon Sequestration Potential

A. Zutshi and S. Harpalani
Southern Illinois University
Carbondale IL 62901

ABSTRACT

This paper presents the results of an experimental investigation undertaken to generate some of the basic data on flow characterization of selected Illinois Basin coals. The experimental study included measurement of permeability of coal cores subjected to *in situ* conditions of stress, temperature and pressure. The results obtained to date suggest that the permeability of most Illinois coals should be considered medium to high compared to other CBM producing basins in the US. Furthermore, the permeability of coal reservoirs should be expected to vary significantly with continued production of CBM as a result of changes in the effective stress and pore pressure. The variation would be even more significant if CO_2 is injected to enhance the production of CBM and/or for carbon sequestration since the permeability of coal to CO_2 is lower to begin with. Hence, with continued gas recovery, CBM production would change significantly and this should be taken into consideration when planning such operations. Finally, the capacity of coal to retain methane and CO_2 was measured for the coals tested. The results show that, for the pressure range permissible in the region, CO_2 sequestration potential of the coals is three to five times the amount of incremental methane recovered.

INTRODUCTION

Economical viability of a coalbed reservoir depends primarily on the gas content of coal and permeability of the reservoir. Coal permeability, being dynamic in nature, requires that a good knowledge about its variation over the life of a reservoir be acquired prior to making any long-term gas production projections. This is even more critical in situations where a second gas is injected to enhance the production of gas, further affecting the dynamics of the coal permeability parameter.

Limited information available for the Illinois Basin suggests that the gas content of the coal varies significantly across the basin. Although there are several "sweet spots" suitable for commercial production of CBM, there are also regions within the basin with low gas content. However, with tremendous success with recovery of methane in the Powder River Basin, where the gas content is low to very low, there is increased optimism that CBM can also be recovered commercially in the Illinois Basin. As for the second factor, the coal is well cleated and the permeability is expected to be good.

Since the gas content of Illinois coal is low compared to other US basins with significant CBM activity, enhanced coalbed methane (ECBM) techniques can have a significant economic impact on commercial CBM production, and these must be considered when planning CBM, or CO_2 sequestration, operations. With enormous coal reserves in the basin, ECBM, apart from providing excellent potential to increase the recovery of CBM, also presents a vast potential to sequester large quantities of CO_2. However, prior to any full-scale commercial CBM production, it is critical to carry out elaborate testing of coal in order to obtain accurate assessment of flow characteristics of the coal.

This paper presents the results of the first part of an experimental study carried out to measure the permeability of coal in order to establish and predict its variation over the life of a reservoir, with and without CO_2 injection. As a supporting activity, methane/CO_2 sorption isotherms were also established for the coal type. As a final step, an analysis of the results was carried out to determine the impact of variable permeability on long-term methane production/CO_2 sequestration and its viability in the basin.

BACKGROUND

Permeability is perhaps the most critical parameter controlling the flow of gas in a coalbed reservoir and determining the gas productivity. Accurate determination of the *in situ* permeability is not easy, and when estimated, it simply provides a static snapshot. The permeability of coal, however, is a dynamic property depending on several parameters like effective stress, gas pressure, water saturation, and matrix shrinkage/swelling due to desorption/adsorption. Thus, an understanding of the physical process of gas flow in coal, primarily the permeability and different parameters affecting it, is critical for economic methane recovery.

Several studies conducted in the past have shown that permeability is highly stress-dependent. Somerton *et al* [1] conducted permeability experiments on bituminous coals by flowing methane gas axially through cylindrical test specimens under a specific sequence of stress conditions. Based on the experimental results, Somerton summarized that the permeabilities of the coal specimens were strongly stress-dependent, higher permeability specimens showing a decrease of over a full order of magnitude, while lower permeability specimens exhibiting a decrease of over two orders of magnitude within the stress range of tests. Durucan *et al* [2] conducted a series of loading-unloading permeability experiments under different stress conditions and found that the permeability of coal in general was highly stress-dependent, decreasing with increasing stress level. His results concluded that the rate of reduction in permeability under stress is characteristic of each type of coal. The effects of stress on gas permeability of mid-western US coals were studied by Harpalani and McPherson [3], where samples were subjected to cycles of loading and unloading (hydrostatic and triaxial) during which permeability was measured. Repeated cycling showed strong hysteresis and an exponential decrease in the permeability of coal.

Ian Gray *et al* [4] showed that permeability of cleat structure within the coal may vary in two basic ways. The first is by phase relative permeability effects, whereby the degree of saturation affects the gas and water relative permeabilities of the reservoir. The second, and probably more important, effect is by a change in the effective stress within the seam. Since permeability is a function of the effective stress, it may increase or decrease with stress changes associated with drainage and influence the coal seam reservoir performance.

Considering pore compressibility as one of the primary parameter, Mckee *et al* [5] developed a theoretical relationship between coal permeability and effective stress under hydrostatic stress conditions. The measurement of brine permeability as a function of pore pressure of coal samples subjected to reservoir stresses, temperature and pressure was studied by Puri and Seidle [6]. The results showed that, at reservoir conditions, permeability decreases by a factor of five during depletion, that is, the permeability decreases logarithmically with reservoir depletion. Considering equations for permeability and porosity for the matchstick geometry, as discussed by Reiss [7], Seidle *et al* [8] developed a theoretical permeability stress relationship, again showing a decrease in permeability with increase in stress.

Studies have also demonstrated that changes in coal matrix and cleat system due to desorption can have a profound effect on the reservoir permeability, thus affecting the production performance [9, 10, 11]. Gray *et al* [4] first proposed the impact of shrinkage of coal matrix associated with desorption of gas, and its impact on permeability, which was later demonstrated by Harpalani and Schraufnagel [10, 11]. Harpalani and Chen [12, 13] measured the changes in the volume of coal matrix associated with gas desorption, and calculated the corresponding changes in coal porosity. Palmer and Mansoori [14] developed a shrinkage model that describes permeability as a function of matrix shrinkage and showed that, apart from matrix shrinkage coefficient and the elastic moduli, cleat porosity also plays an important role.

It is evident from the brief discussion presented above that the permeability varies with changes in stress and volume of coal matrix, affecting the gas flow characteristics of coal. This becomes even more critical when a second gas is injected in a coal reservoir, either to enhance the production of CBM, or to sequester CO_2. The primary objective of the study was to determine the technical feasibility of producing coalbed methane from Illinois coal on a commercial level by evaluating the permeability of coal, its variation over the life of production wells, and its impact on long-term gas production potential in Illinois. Although changes in the physical structure of coal associated with injection of CO_2, the mechanism of binary adsorption of methane and carbon dioxide on coal, and the ability of carbon dioxide to displace methane from coal were also studied, this paper is limited to the first part of the study involving permeability testing.

EXPERIMENTAL STUDY

Cores of coal were obtained from the ongoing exploratory drilling program of the Illinois State Geological Survey (ISGS) in the Herrin, Springfield and Davis seams in southern Illinois. Cores were preserved in their native state to prevent any damage due to weathering by storing them in an environmental chamber, with no source of light, and under controlled conditions of temperature and humidity.

Permeability Experiments

Sample Preparation: Just prior to testing, appropriate length of cores were cut and their top and bottom ends polished to enable proper placement in the triaxial cell.

Experimental Setup and Procedure: In order to simulate the *in situ* conditions, controlling and monitoring of the external stress conditions and gas pressure are very important. The experimental setup for permeability measurement was, therefore, designed to have independent control of stress conditions, gas pressure (upstream and downstream), and capability of gas flowrate measurement. The setup consisted of a triaxial cell, a loading system, and a means to monitor and measure flowrate. A schematic of the experimental setup is shown in Figure 1. The setup enabled applying both confining and axial stresses independently to simulate the *in situ* conditions. The axial stress was applied by placing the cell in a load frame. The triaxial cell was connected to a hydraulic pump to provide the confining stress. The perforated steel disks and porous metal plates were placed at both ends of the sample to distribute (and collect) the gas and prevent small particles from entering the tubing. The temperature of the triaxial cell was kept constant using a heating tape and temperature controller. The gas containers at the inlet and outlet were placed in a water bath, set at the same temperature as that of the triaxial cell. The gases entering and coming out of the triaxial cell were, therefore, at the same temperature. To keep the downstream pressure constant, a relief valve was used.

Experimental Conditions: Cores used for the permeability experiments were taken from a depth of 880 feet, with the gas pressure of coalbed reservoir estimated to be 600 psi. The estimated reservoir temperature at the depth was estimated to be 72°F. Hence, all tests were carried out at these conditions of temperature, maximum gas pressure, and levels of external stresses. The tests were carried out on cylindrical cores of coal, 3 inches in diameter and ~3 inches long, depending on the length of the cores received.

Prior to testing, the core, along with the perforated steel disks and porous metal plate, was sealed using PVC shrinkage tubing in order to avoid seepage of oil into the core during application of stress. The core setup was then placed in the cell, which in turn, was placed in the load frame. The sample was then stressed triaxially, using the loading frame for axial stress and hydraulic pump for confining stress. Before injecting gas into the sample, the triaxial cell was properly insulated to maintain constant temperature throughout the experiment. The core sample was then saturated with gas, at the desired pressure, by injecting gas through the inlet, keeping the outlet closed. After attaining pressure equilibrium, a pressure gradient of 40-60 psi was applied across the sample using a relief valve. The pressure, temperature, and flowrate at both ends (upstream and downstream) were monitored continuously. When all parameters became, and remained, constant (equilibrium condition), flowrate measurements for the particular conditions were made. Using the measured flowrate, permeability of the sample for the gas was calculated. The testing procedure was the same for methane and CO_2.

Methane and CO_2 Sorption

Adsorption isotherms for the coal type were established using the standard gas expansion technique, where a known quantity of gas is allowed to expand into a sample container and equilibrate. By monitoring the pressure before and after the expansion, the amount of "free" gas in the sample container is calculated. The difference between the two quantities, that entering the container and that appearing as free gas, gives the volume of gas adsorbed at the equilibrium pressure. By a step-wise increase in gas pressure, a complete isotherm is established. For this part of the experimental work, methane/CO_2 isotherms were established.

RESULTS AND DISCUSSION

Permeability Experiments

In order to test proper working of the setup and gain some experience with the operation of the setup prior to using valuable cores, a sample from the Herrin seam was used for one permeability test. Minor modifications were made to the setup as a result of this test run in order to stabilize the stress and pressure conditions for the duration of the experiment.

Permeability Calculation: Darcy's equation, modified for compressible fluids, was used to calculate the permeability of coal. The equation is given as:

$$k = \mu \frac{Q_0}{A} \frac{LP_0}{\Delta P P_m} \quad (1)$$

where, μ is the viscosity of the gas, Q_0 is the volumetric flow rate at the outlet, A is the cross sectional area of the sample, L is the length of the sample, P_0 is the gas pressure at the outlet, ΔP is the pressure difference between upstream and downstream, and P_m is the mean gas pressure $[(P_0 + P_i)/2]$.

Permeability Results: The first permeability test was initiated using a well preserved core taken from the Springfield seam. The core was taken from a depth of 882 ft. The test was conducted in two phases. In the first phase, the sample was subjected to a constant axial stress of 930 psi and confining stress of 600 psi respectively, which are both representative of the stresses *in situ*. The temperature throughout the test was maintained at 72 ± 1^0F. Figures 2 and 3 show the level of stability attained in the stress and temperature conditions during the test.

The gas flow measurements were carried out for varying effective stress levels by changing the mean gas pressure, maintaining the applied axial and confining stresses constant. Using the flow measurements, permeability of the sample to methane was determined using equation (1). For a change in the horizontal effective stress from 180 psi to 480 psi, the change in permeability of the sample to methane was calculated to be between 75 and 6 md. The permeability variation for methane is shown as the upper plot in Figure 4. It is evident from the graph that permeability decreases with increase in effective stress. This is expected since increased effective stress compacts the cleats in the coal, resulting in cleat closure and reduced permeability.

In the second phase of the experiment, the sample was subjected to a constant effective stress throughout the experiment. The axial and confining effective stresses were maintained at approximately 860 and 510 psi respectively. The gas flow measurements were made by varying the mean gas pressure, and permeability of the sample to methane was determined. The permeability results for methane are shown as the upper plot in Figure 5. It is evident from the results that the permeability decreases linearly with decrease in mean gas pressure. This finding was somehow unexpected since a reduction in pressure results in shrinkage of coal matrix, and thus in increased permeability. From these results, it appears that there is a coupled effect of external applied stress and volumetric strain induced by desorption of methane. At high stresses, like those encountered in the San Juan Basin, the permeability does increase with reduction in pore pressure.

As a continuation of the above experiment under same experimental conditions of temperature, pressure and stress, the core was tested using CO_2. The test was, once again, conducted in two phases, maintaining the applied axial and confining stress constant in the first phase, and the effective stress constant in the second phase. The gas flow measurements were made by varying the mean gas pressure. The change in permeability with changes in mean gas pressure for CO_2 is shown as the lower plot in Figure 5. From these permeability trends, it is evident that, at zero mean gas pressure, the permeability becomes zero – as expected.

For the second phase of experiment using CO_2, the sample was subjected to a constant axial and confining stress of 940 and 610 psi respectively. Using the flow measurements, permeability of the sample to CO_2 was determined. For a change in horizontal effective stress from 330 to 515 psi, the permeability decreased from 3.2 md to 1 md. The variation in permeability with changes in horizontal effective stress for CO_2 is shown as the lower plot in Figure 4.

It is evident from the graph that the permeability decreases with increase in effective stress. The trend of permeability variation vs. effective stress also correlates well with the findings of some of the earlier studies illustrating the exponential relationship of the form [3]:

$$k = Ae^{B\sigma_h} \tag{2}$$

where, σ_h is the horizontal effective stress and A and B are constants. The values of A and B depend on the particular coal type. For the core used to obtain the results shown, at zero effective stress, the permeability of coal sample to methane is 300 md while that to CO_2 is ~30 md – difference of an order of magnitude. This may provide an explanation of the field observations made at Burlington Resources' Allison site, where a reduction of *in situ* coal permeability of two orders of magnitude, is suspected. Of course, the CO_2 injection site was at a depth of ~1800 feet and the injection pressures used were very high [15].

Methane and CO_2 Sorption

This part of the investigation was divided into two parts: single gas sorption analysis using methane and CO_2, and CO_2 flooding. Only single gas sorption results are relevant for this paper. For Illinois coals, isotherms were completed for methane and CO_2. The Gibbs adsorption was determined experimentally. The absolute sorption, which is more representative at higher pressures, was calculated from the Gibbs adsorption using the equation below:

$$V_{abs} = \frac{V_{Gibbs}}{1 - \frac{p_{gas}}{p_{ads}}} \tag{3}$$

where, V_{abs} and V_{Gibbs} are the absolute and Gibbs sorption respectively, and p is the density of the gas phase. The value for the sorbed density for methane and CO_2 were taken from Arri et al [16] to be 0.421 g/ml and 1.18 g/ml respectively. The experimental sorption data were analyzed using the Langmuir isotherm model [17], given as:

$$V = \frac{PV_L}{P + P_L} \tag{4}$$

where, P is the equilibrium gas pressure, V is the volume of gas adsorbed, V_L, is the Langmuir monolayer volume, and P_L is the Langmuir pressure constant. The absolute sorption isotherms are shown in Figure 6. It is apparent that all isotherms are of Type I, according to Brunauer's classification [18]. It is also apparent that the relative absolute sorption (V_{CO2}:V_{CH4}) varies from 2.2:1 to 3:1. Finally, the Herrin seam not only has the largest capacity to sorb methane/CO_2, but also has the highest relative sorption ratio.

CONCLUSIONS AND RECOMMENDATIONS

The specific conclusions based on the work completed to date are as follows:

1. The permeability measurement results suggest high to medium permeability values making several Illinois coals suitable for CBM operations. The measured variation in permeability further suggests that, with continued drainage, CBM production would change significantly.

2. The permeability trend suggests that change in effective stress due to changes in gas pressure is the main contributing factor in bringing about the permeability reduction. Matrix shrinkage due to desorption of methane, or swelling due to injection of CO_2, does not seem to have a significant influence on the permeability variation. This is probably due to the relatively low levels of stress.

3. There is a linear decrease in permeability with pressure decline. This is perhaps due to decreased "push" for the gas to continue moving. This favors application of an ECBM technique, since the pressure decline would be prevented.

4. All Illinois coals have a preferential sorptive affinity for CO_2 over methane; the relative sorption ratio varies from 2:1 to 3:1.

Based on the above specific findings, it is concluded that an enhanced coalbed methane recovery technique should be considered in Illinois throughout the production period, as opposed to following the primary depletion method for initial production. The increased pressure resulting from the injected CO_2 would have two benefits: Firstly, it would reduce the increase in effective stress, and dampen the negative impact of increased stress on permeability. Secondly, it would promote the flow of gas due to increased driving force pushing the methane out. Of course, it would result it sequestration of large quantities of CO_2.

From a sequestration point of view, more CO_2 can be sequestered if relative sorption capacity of coal for CO_2 is higher. This is also true from a carbon credits point of view. However, from an economic point of view, lower the relative sorption capacity, smaller is the amount of CO_2 that must be injected to obtain incremental gas recovery. For Illinois coals, the relative sorption at low pressures is high and this favors CO_2 sequestration and incremental methane production in the state.

ACKNOWLEDGMENTS

This work has been conducted by the authors with support by grants made by the Illinois Department of Commerce and Economic Opportunity through the Office of Coal Development and the Illinois Clean Coal Institute. The authors wish to thank these organizations for their support.

REFERENCES

1. Somerton, W.H., 1975: "Effect of Stress on Permeability of Coal", International Journal of Rock Mechanics, Mining Science and Geological Abstracts, 12, p. 129–145.
2. Durucan, S and J.S. Edwards, 1986: "The Effect of Stress and Fracturing on Permeability of Coal", Mining Science and Technology, 3, p. 205-216.
3. Harpalani, S. and M.J. McPherson, 1985: "Effect of stress on Permeability of Coal", Proceedings of the 26th US Symposium on Rock Mechanics, South Dakota School of Mines and Technology.
4. Gray, I., 1987: "Reservoir Engineering in Coal Seams: Part I – The Physical Process of Gas Storage and Movement in Coal Seams", SPE Reservoir Engineering, V. 2, No. 1, p. 28–34.
5. McKee, C.R., Bumb, A.C., and R.A. Koening, 1987: "Stress Dependent Permeability and Porosity of Coal", Proceedings of the 1987 Coalbed Methane Symposium, Tuscaloosa, AL, p. 183-193
6. Puri, R. and Seidle, J. (1991) Measurement of stress dependent permeability in coals and its influence on coalbed methane production. Proceedings of the 1991 Coalbed Methane Symposium, Tuscaloosa, Alabama, pp. 415–24.
7. Reiss, L. H., 1980: The Reservoir Engineering Aspects of Fractured Formations, Editions Technip, France
8. Seidle, J.P., 1992: "Application of Matchstick Geometry to Stress Dependent Permeability in Coals", paper SPE 24361, Proceedings of the SPE Rocky Mountain Regional Meeting, Casper, Wyoming, p. 433–45,
9. Seidle, J.P. and L. G. Huitt, 1995: "Experimental Measurement of Coal Matrix Shrinkage Due to Gas Desorption and Implications for Cleat Permeability Increases", paper SPE 30010, Proceedings, International Meeting on Petroleum Engineering, Beijing, China, p. 575-582.
10. Harpalani, S. and R. A. Schraufnagel, 1990: "Influence of Matrix Shrinkage and Compressibility on Gas Production from Coalbed Methane Reservoirs", paper SPE 20729, Proceedings, 65th Annual Technical Conference and Exhibition of the Society of Petroleum Engineers, LA, p. 171-179.
11. Harpalani, S., R. A. Schraufnagel, 1990: "Shrinkage of Coal Matrix with Release of Gas and its Impact on Permeability of Coal', Fuel, V. 69, No. 5, p. 551–556.
12. Harpalani, S. and G. Chen, 1995: "Estimation of Changes in Fracture Porosity of Coal with Gas Emission", Fuel, V. 74, No. 10, p. 1491–1498.
13. Harpalani, S. and G. Chen, 1997: "Influence of Gas Production Induced Volumetric Strain on Permeability of Coal", Intl. Journal of Geotechnical and Geological Engineering, V. 15, No. 4, p. 303-325.
14. Palmer, I. and J. Mansoori, 1996: "How Permeability Depends on Stress and Pore Pressure in Coalbeds: A New Model," paper SPE 36737, Proceedings, 71st Annual Technical Conference, Denver, CO.
15. Reeves, S.R., 2003: "The Allison Unit CO_2 – ECBM Pilot: A Reservoir Modeling Study", Topical Report, USDOE Award Number- DE-FC26-0NT40924.
16. Arri, L.E., Yee, D., Morgan, W.D., and M.W. Jeansonne, 1992: "Modeling Coalbed Methane Production with Binary Gas Sorption." paper SPE 24363, Proceedings of the SPE Rocky Mountain Regional Meeting, Casper, WY.
17. Langmuir, I. 1918: "The Adsorption of Gases on Plane Surfaces of Glass, Mica and Platinum." Journal of American Chemical Society, 40(9), p. 1361-1403.
18. Brunauer, S., Deming, L.S., Deming, W. E., and E. Teller, 1940: "On a Theory of the Van der Waals Adsorption of Gases." Journal of American Chemical Society, 62(7), p. 1723-1732.

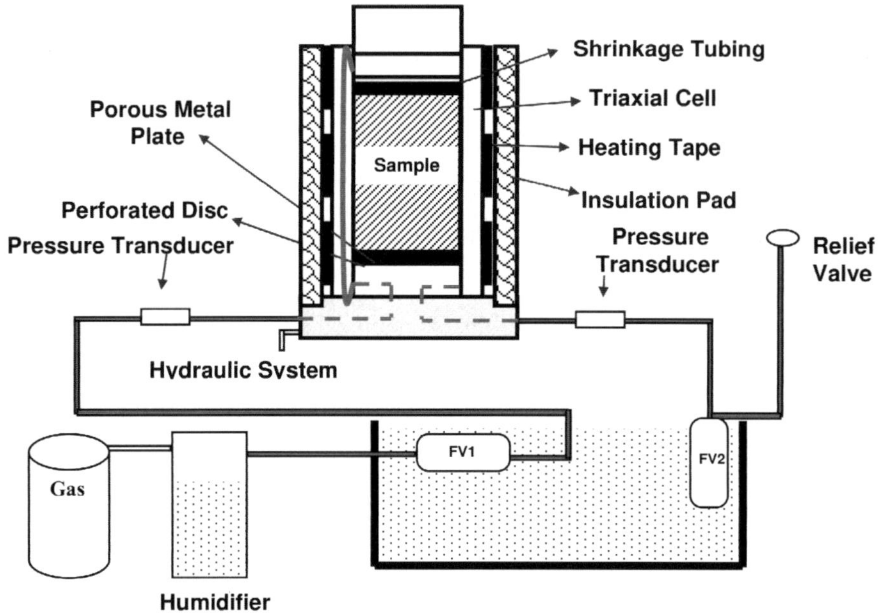

Figure 1: Schematic of the permeability setup.

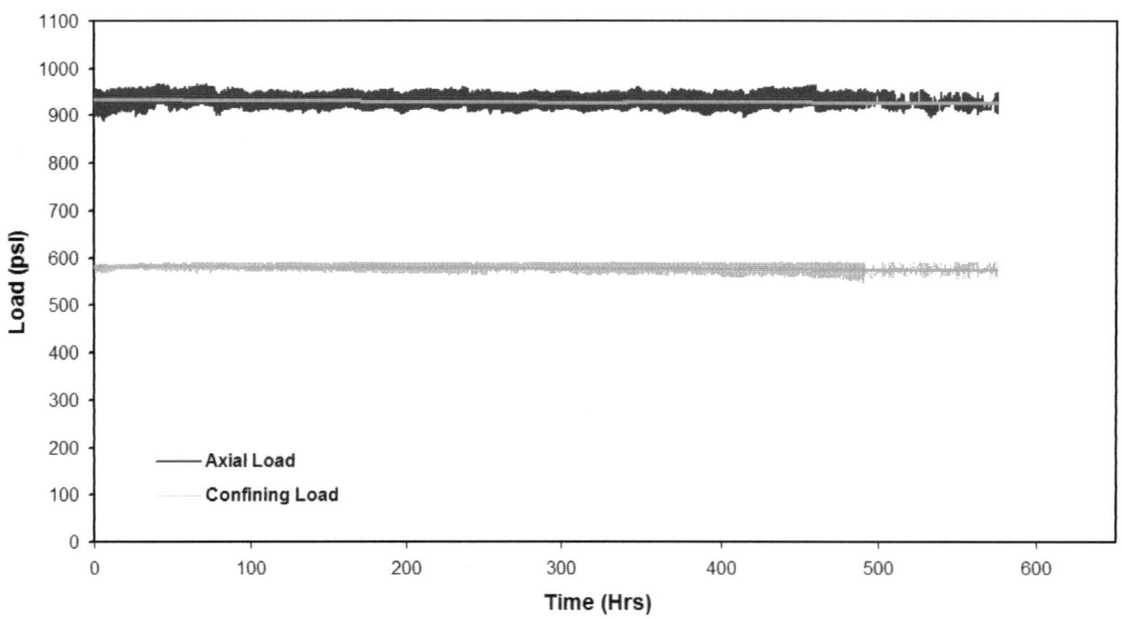

Figure 2: Stability of axial and confining stress over time.

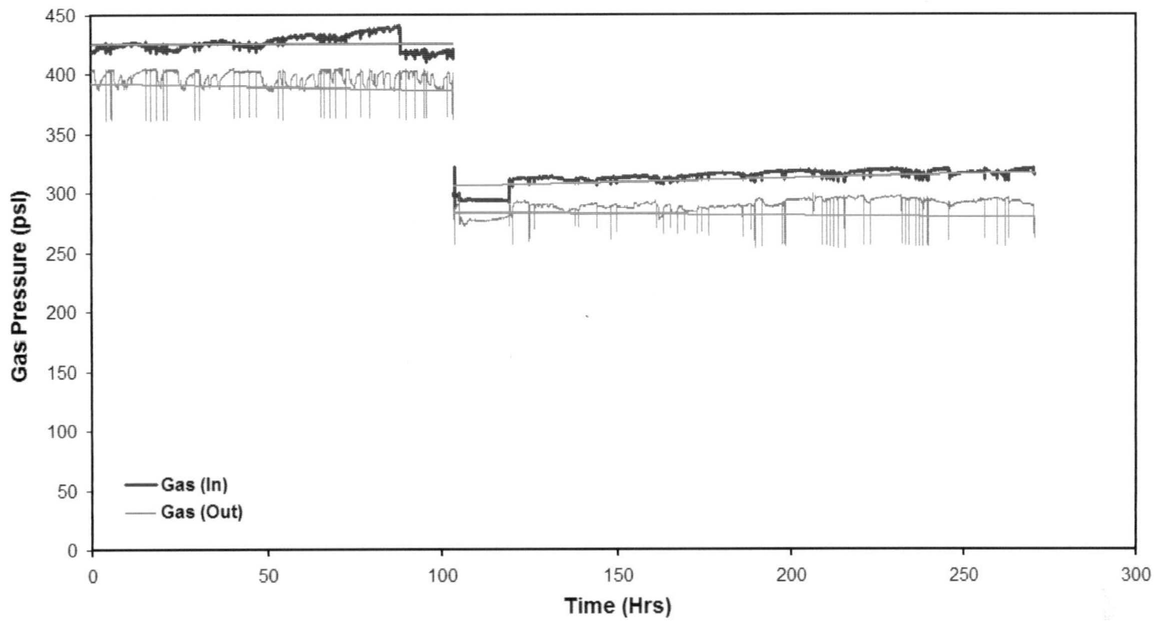

Figure 3: Stability of gas pressure over time.

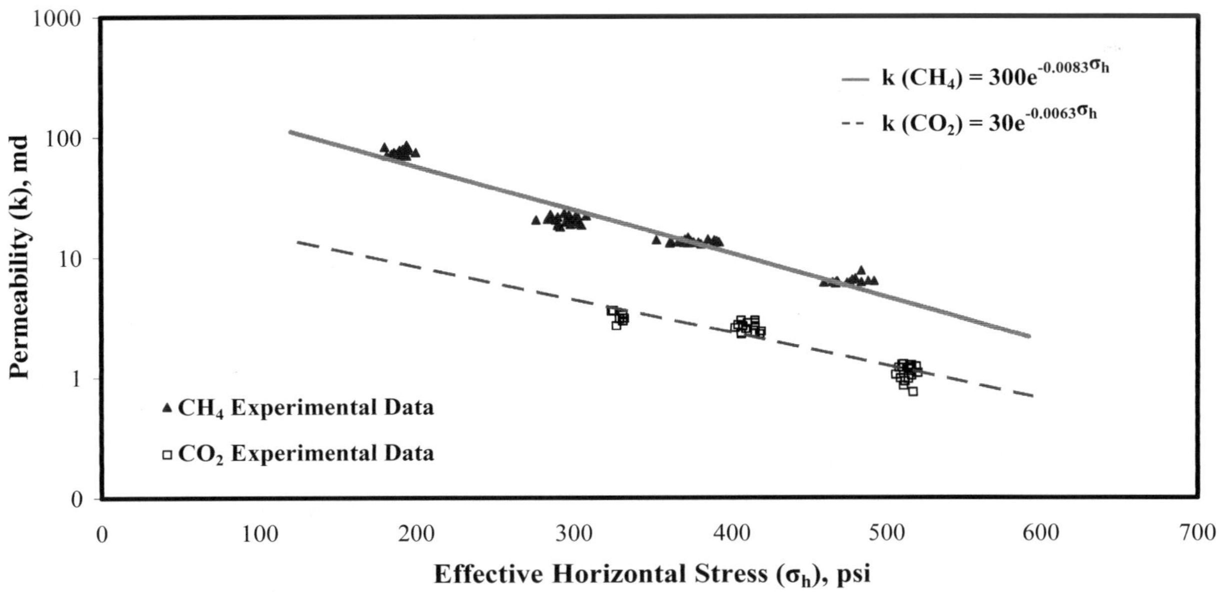

Figure 4: Variation in permeability with changes in horizontal effective stress.

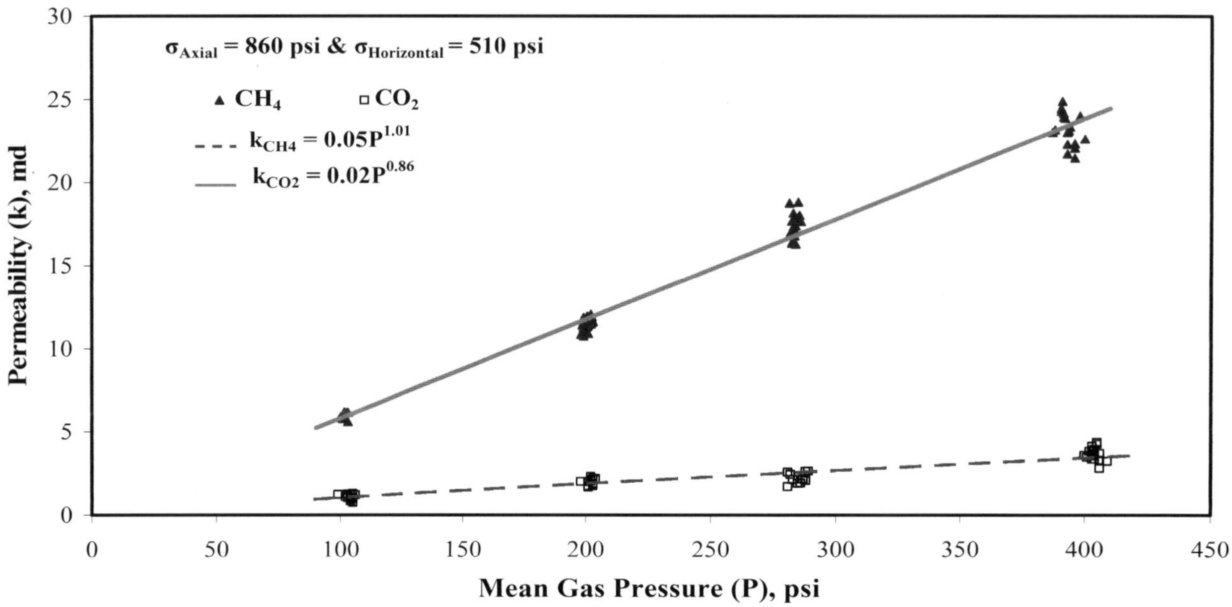

Figure 5: Variation in permeability with changes in mean gas pressure – constant effective stress.

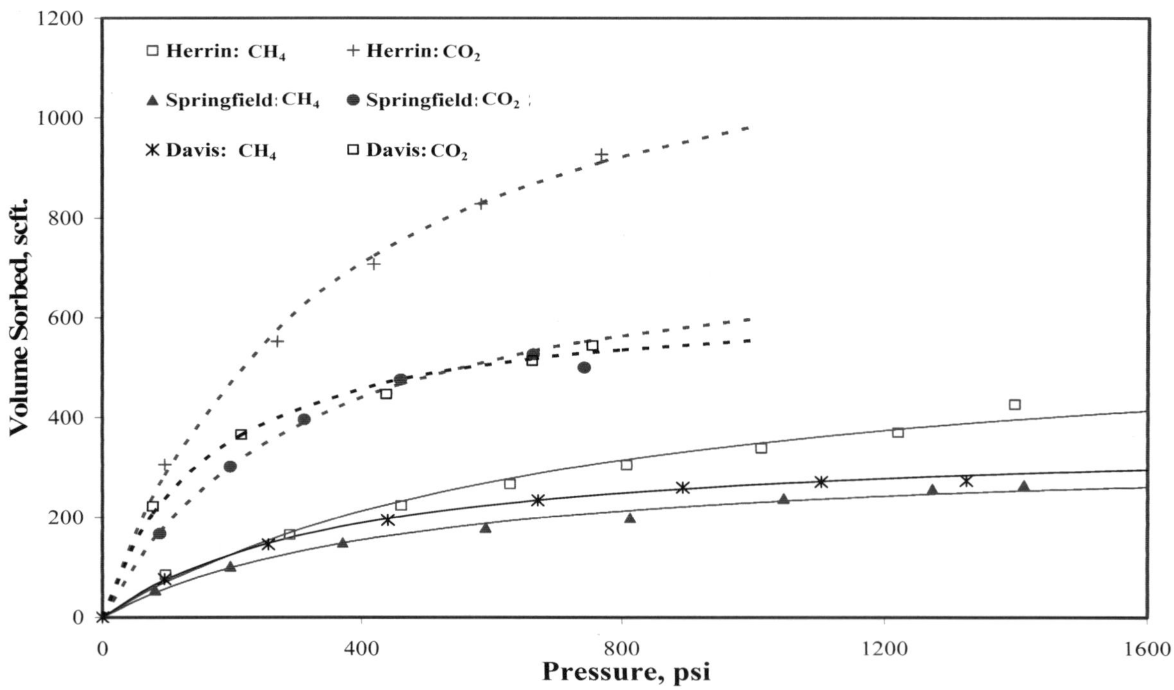

Figure 6: Sorption isotherms for Illinois coals.

0501

Effect of Coal Properties on Gas Drainage

A.Hutton[2], D.Bruggemann[1], N. I. Aziz[1], F.Sereshki[1]

[1]School of Civil, Mining and Environmental Engineering; [2]School of Earth and Environmental Sciences; University of Wollongong, Wollongong 2522, New South Wales, Australia

ABSTRACT

The current practice of draining gas from coal measure rocks is through drilling, both in coal seams as well as in the surrounding stratification. The effectiveness of the drainage operation changes from location to location because of the changing geological conditions. A number of factors influence the success of gas drainage. Intrusions, faults and folding can influence the effectiveness of drainage operations. The recent ineffective drainage operations in some parts of the Bulli seam have contributed to slowing downing down of the mining operation particularly during heading development for longwall operations. Some mines have resorted to such methods as remote mining,, face grunching and on one occasion, large diameter augering, particularly in areas of heavy dyke intrusion.

A recent study was undertaken by the mining and geological group of the University on Wollongong, ain, aimed to relate the changes in the ground geological conditions to the permeability and the coal volume changes and the type of gasses present being released from the tested sites. A development site at a section along its path, which is heavily faulted and geologically disturbed, was studied with respect to drainage characteristics. Coal samples were collected from different parts of the district. Parameters studied included coal geology, ground stress regime, coal permeability and coal shrinkage.

Gases used for the study included N2, CO2, CH4 and 50% mixture of CH4/CO2. Coal samples were tested in gas environments up to 3MPa pressures. The findings from the study demonstrated clearly that there were clear relationships between the extents of local ground stresses and the level of coal permeability. The type of gas and pressure confinement affected coal volume change. It is considered that the findings from the programme of study could be extended to defining poor drainage areas in coal seams for effective future mining operations.

Keywords: Bulli seam; gas content; sorption; macerals; permeability; coal shrinkage; carbonate minerals, clay minerals;

INTRODUCTION

Over the past 100 years, observations of 'coal mine outbursts' that have occurred around the world have led to an understanding of the concepts and nature of their occurrence and the factors which play a dominant role in the triggering and sustaining processes, and eventual cessation, of outburst incidences. Because of the quite variable geological conditions that occur across coalfields, and in collieries on a smaller scale, and the changing technologies used for the mining of coal over the history of outburst incidences, there are differing views on the importance of the various parameters associated with the occurrence of outbursts.

Coalmine outburst control in the Southern Coalfield has largely been obtained through experimental results as well as mine observations. In this paper we present a case study of a mine site in New South Wales where coal shrinkage, permeability and coal petrography techniques have produced data that permit a better understanding of the geotechnical gas regime. The results are important to the further understanding of the inter-relationship between gas sorption, the coal matrix and permeability in what is considered to be 'normal' and 'tight' coal (locally referred to as disturbed coal), the latter being coal that degases slowly compared to normal coal. For this study, coal samples were obtained from the Bulli seam southwest of Wollongong.

Figure 1. Location plan of Southern Coalfield of New South Wales, Australia

GEOLOGICAL SETTING OF THE SOUTHERN COALFIELD

The Sydney Basin is a large sedimentary basin located on the east coast of Australia. The Southern Coalfield contains the economically important Illawarra Coal Measures, which is Permian to Triassic in age (Sherwin and Holmes, 1986). The Southern Coalfield is the southern portion of the Sydney Basin, covering an area south of Sydney and bounded approximately by the towns of Campbelltown and Mittagong to the west, and Wollongong and Helensburgh to the east (Figure 1).

There are a number of coal seams in the Illawarra Coal Measures. These are in descending order the Bulli, Balgownie, Cape Horn, Wongawilli, American Creek, Tongarra and Woonona seams. Of these it is only the Bulli and Wongawilli seams which are economically viable through underground mining. In particular for the Bulli seam, the seam varies from 1.6m to greater than 2.5m generally. The depth of workings can range from 380m to 450m below the surface. The Bulli seam section usually occurs as a single unit of coal, although shale and other dirt bands can occur from time to time in various portions near the floor and roof of this seam.

The Bulli seam is conformably overlain by shale, mudstone, laminite, siltstone or sandstone, depending on the palaeochannel geometry and other palaeofluvial deposit systems. The shale, mudstone and laminite are generally widely distributed, whereas the sandstone is more confined to former stream channels. The floor rocks are mostly shale and mudstone, with little reactivity to water, due to their apparent lack of reactive clay minerals.

BULLI SEAM AT THE MINE SITE

Locally the Bulli seam is mined to a thickness of up to 2.4 m. Geological structures are typically strike–slip, normal, and thrust and sometimes reverse faults. Manifestation of these common structures lead to disturbed ground surrounding the main features, resulting in slickensided, sheared, and juxtaposed coal and roof/floor strata for some distance away from the faults. In addition, an igneous dyke is present in the working section of this mine, and thus further disturbance of the seam surrounding this intrusion is experienced. These factors are particularly important in the site description of the local gas regimes and the coal properties with respect to gas drainability .Other areas of the mine do not contain such structural complexities from a geological standpoint, and have corresponding differences in gas drainage efficiency.

Run-of-mine coal from the Bulli seam at this location is a high volatile bituminous coal with inherent moisture of <2%, <28% volatile matter and <10% ash; fixed carbon is 87 to 88%. The coal is generally characterised by 50% vitrinite, 45% inertinite and has a mean maximum reflectance ($R_o$max) of 1.04%.

GAS REGIME OF THE SOUTHERN COALFIELDS

Outbursts in Australian coalmines have been a common occurrence. Lama and Bodziony (1996) reported there had been over 700 recorded outbursts in Australia in the 100 years up to 1996. Coalmine outbursts in the Illawarra Coal Measures have typically occurred along strike slip faults, sometimes in conjunction with igneous dykes (Harvey 1995).

Many researchers have used a variety of methodologies to elucidate the origin of in-seam gases and to understand the distribution and accumulation of the gases. Faiz (1993), Faiz and Hutton (1995) and Faiz and Hutton (1997) researched the geological controls on the distribution of CH_4 and CO_2 in the Illawarra Coal Measures and found that the volume of CO_2 increases towards structural highs with the highest concentrations in anticlines and near faults with isolated pockets near some faults. CH_4 dominates in structural lows.

Relevant to this paper, Crosdale and Beamish (1993) found that methane sorption rate and capacity is related to coal type and coal rank with vitrinite-rich coals desorbing more slowly than inertinite-rich coals but have gas sorption capacities up to 30% greater than inertinite-rich coals of the same rank. Crosdale and Beamish (1995) also found that crushed coal (-0.212 mm) from South Bulli (also in the Southern Coalfield) show little correlation between desorption rate and petrography. Lump samples (-5.60 to +2.00 mm) also showed little relationship between effective diffusivity and maceral content. However, Central Queensland lump sample coals with increased inertinite content are associated with more rapid desorption.

The gas environment of the case study mine is dominated by carbon dioxide, where the relative quantities of methane to carbon dioxide are generally 10% to 90%. Historically, for all coal mined to date at this mine, this relative quantity has averaged 65%CO_2, and 35%CH_4 (Wynne 1992).

MECHANISM OF COAL MINE OUTBURSTS

Many definitions of outbursts have been given but the most concise to date, and perhaps most relevant to the study site, is 'a sudden and violent release of coal and gas from a coal face being worked' (Hargraves, 1983). The mechanisms of outbursts have been the subject of much research over the past four decades and hence several theories for the mechanisms of outburst have been developed.

From a mechanistic viewpoint, outbursts can be considered to be the result of mechanical failure of coal due to the combined effects of field stress and pore gas pressure, each of which contributes, to varying extents, depending on the in-situ mechanical conditions. In a simplest form, if the effective stress exceeds the tensile strength of the coal, the coal fails under tension and the accompanying shear causes the coal to spall from the host coalface. The detached coal is further broken under tension to very smaller debris and may be completely shattered by gas pressure, leading to very fine coal particles. This has been observed by Bruggemann (2002 a; 2002b) at the mine study site.

There appears to be several fundamental parameters appear to be crucial to outburst phenomena Hanes (2003) stated that three conditions must be met:

- a large volume of gas which also serves as a transporting medium for the pulverised coal;
- fractured of pulverised coal produced by a geological disturbance, high stress or under the effects of gas; and
- a trigger mechanism (such as mining or blasting) that alters the equilibrium.

It has been suggested by a number of researchers that coal maceral type can also be important in the generation of gas in coal. Coals that are inertinite-rich (especially fusinite and semifusinite) may exhibit higher gas emission rates, whereas vitrinite-rich coals show slower emission rates (Crosdale *et al.*, 1998). Other studies (Gurba and Weber, 2001) suggested that desmocollinite (matrix vitrinite) might play a key role in thermogenic methane generation, whereas telocollinite, which contains micro-cleats, may contribute to increased seam permeability.

In-seam gas is in part a product of the chemical reactions associated with the coalification process (Smith and Sloss, 1992). The gas mainly resides in the coal as adsorbed layers on the internal coal surfaces or alternatively as free gas in large pores and fractures (Gray, 1987). In coalmine outburst cases, the gas released will depend on the state of the coal and the gas. If the coal is brecciated or mylonised, such as around fault zones, gross interconnecting pore spaces can be expected to exist along with diminished mechanical strength of the coal.

Where the coal seams are disturbed, such as where mylonisation caused by tectonism is found, the coal matrix is damaged or completely obliterated and these coals are prone to outbursts (Lama and Bodziony, 1996). Cao *et al.* (2000) stated that two important geological observations have been related outburst occurrence:

a. nearly all outbursts occur in deformed coal layers within tectonically deformed zones; and
b. outburst-prone coals have higher gas contents than coals not prone to outbursts.

In 'normal' or nondeformed coal, the opposite is considered to be true, and the coal matrix is largely intact. This has implications for the effectiveness of gas drainage, permeability characteristics of the coal and effectiveness of coal mine outburst control methods.

SITE INVESTIGATION

One of the current methods for reducing the risk of outbursts in high gas or areas that do not drain effectively with drilling and in-seam mine drainage is grunching. This is a relatively expensive method and the rate of coal production in drive advancing is low. Grunching is a method of roadway development advance, similar to metaliferous mining practices, in which explosives are used to advance a heading. After grunching, the coal structure is disturbed to a point where the contained gas moves freely from high in-seam fluid pressures to the drainage lines. Grunching is commonly used in 'restricted' or tight areas (12 m per day in these) or 'normal' (25 m per day advance in these) areas, particularly where gas contents are greater than the allowable gas threshold limits allowable by mining law.

To provide better insight into poor draining areas, a program of mine site studies has been undertaken to attempt to relate changes in geological conditions to gas storage and/or drainage characteristics of the Bulli seam and to better understand and possibly predict outbursts. As part of this study, samples from a tight, poor draining area (900 Panel; Fig 2) and a better draining area (800 Panel) were collected and subjected to shrinkage tests.

Both of these geologically different underground areas that had been monitored and mapped by the senior author. The properties of the coal coals from the two areas, at hand specimen scale, are significantly different:

800 Panel – undisturbed or 'normal' coal in terms of cleat spacing (regular spacing) and orientation (orthogonal to bedding), ordered bright and dull layers with no visible deformation or fracturing.

900 Panel – 'disturbed' or deformed coal with cleat orientation not vertical, cleat spacing irregular, occasional small scale dislocation of both bright and dull layers; calcite vein mineralization commonly found towards the top of seam, usually oblique to bedding plane but tending towards parallel to bedding planes in the lower parts of each vein.

From a practical perspective, gas drainage has been exceedingly difficult in the 900-panel area compared to the 800-panel area. Management has resorted to the 'grunching' method of heading development

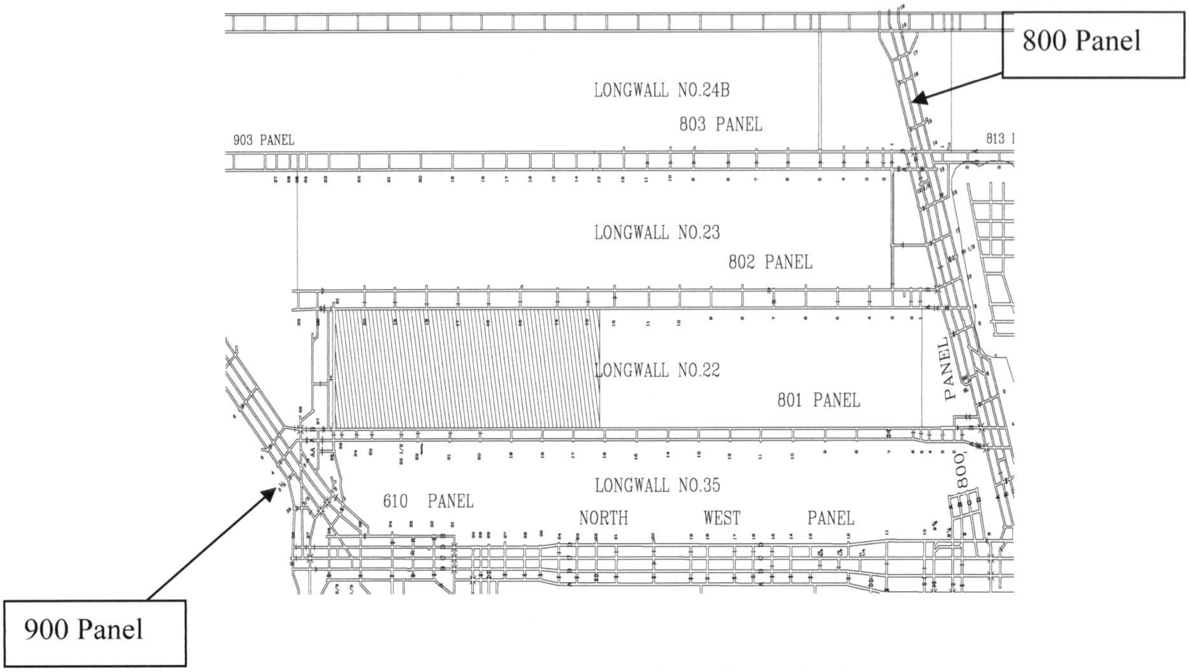

Figure 2. Working plan of mine site showing the location of sample sites.

COAL SHRINGAGE TESTS

Equipment and Methodology

The experiment was set up to measure the volumetric strain within the coal matrix under changing gas pressures, as described by Lama and Bartosiewics (1982), and later by Aziz and Ming-Li (1999). A pressure transducer as shown in was mounted on each bomb. The coal samples were sealed in the gas bombs and pressurized to a saturation level of 3Mpa. The sample containers were immersed in a water bath, but were isolated from the water bath by a copper sleeve to keep them dry. This thermostatically controlled water bath allowed the coal samples to be kept at the desired temperature of $25C^0$.

The experimental apparatus is shown in figure 3. Four strain gauges were adhered to each sample surface, two for axial and two for radial strains, placed 180^0 apart, as recommended by the International Society for Rock Mechanics (ISRM). The strain gauge wires were then fitted onto the sample, with the other end coming out of the pressure cell.

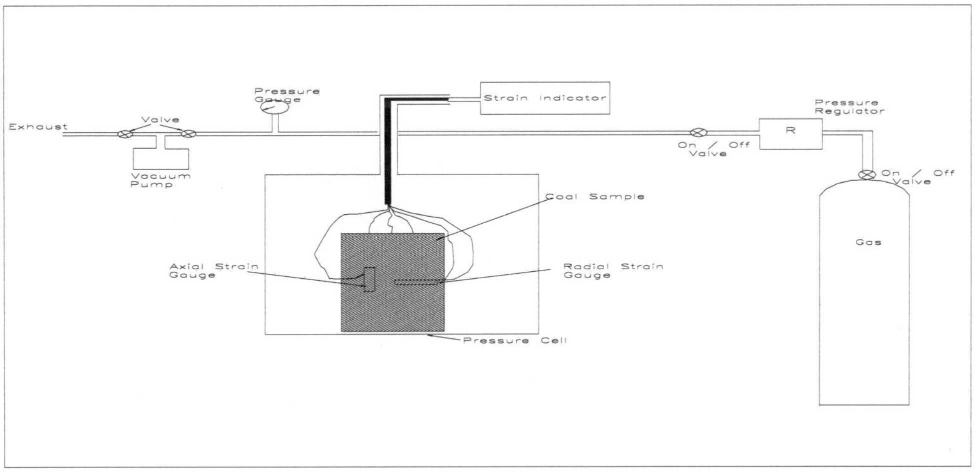

Figure 3. Schematic diagram of apparatus for testing volumetric change in coal

At the beginning of the test, the strain gauge readings were set to zero, and the sample bomb was evacuated, via a pump connected to the manifold where suction was applied to a vent line, which expelled any gas or air from the system. The bomb lid was attached by six bolts with the lid attaining a perfect seal with the use of an 'O' ring. The changes in strain were recorded continuously via a data logger connected to a PC. Pressure meters were used to indicate the bomb inlet gas pressure.

After gas / air evacuation, each sample was left in this vacuum for 24 hours. Each bomb was then charged with the appropriate gas type until a maximum pressure of 3MPa was attained. Once the coal was saturated, the gas was then discharged at incremental steps of 0.5MPa, where the pressure was held for a period of 2 hours at each successive step down of pressure. The changes in the volume of coal were then recorded via the data logger and the PC. On the last increment, the pressure was reduced to atmospheric and the sample was allowed to desorb until it reached equilibrium with respect to strain. After finishing one set of tests for a gas type, the bomb was evacuated and the same procedure was repeated for the other gasses.

The grain compressibility, C_m is defined as the rate of change of coal matrix volume to the change in gas pressure and it is called the shrinkage coefficient for desorbing gas, and is given by (Harpalani and Chen, 1997):

Vm = Matrix Volume

dV_m = Change in Volume

$$C_M = \frac{1}{V_M}\frac{dV_M}{dP}$$

.

dP = Change in applied pressure

C_m = Shrinkage coefficient

Results

From Figures 4 and 5, the incremental change in volumetric strain can be seen, which corresponds to the progressive reduction of gas pressure for the coal samples. The greatest changes in volume occurred for Carbon Dioxide, and interestingly for sample 800051 which originates from 800 panel. This means that for this sample, and its corresponding local geological environment of the mine, the coal will have the greatest capacity for inseam gas drainage.

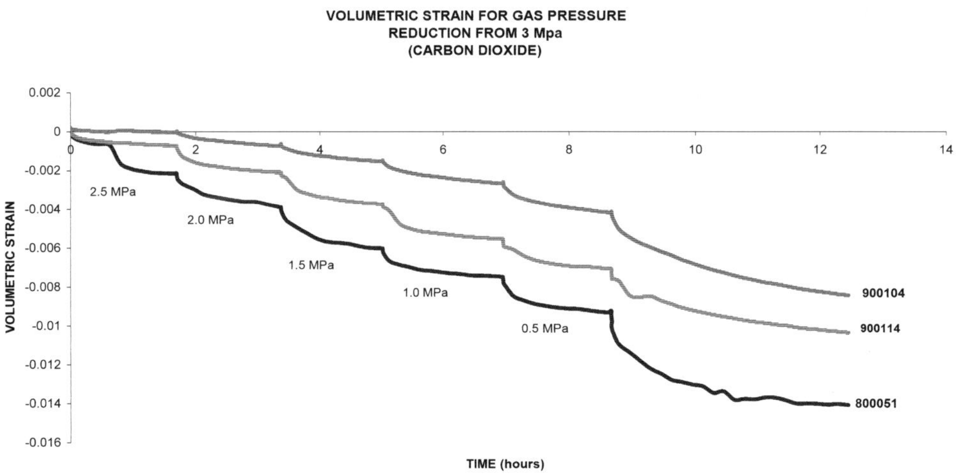

Figure 4

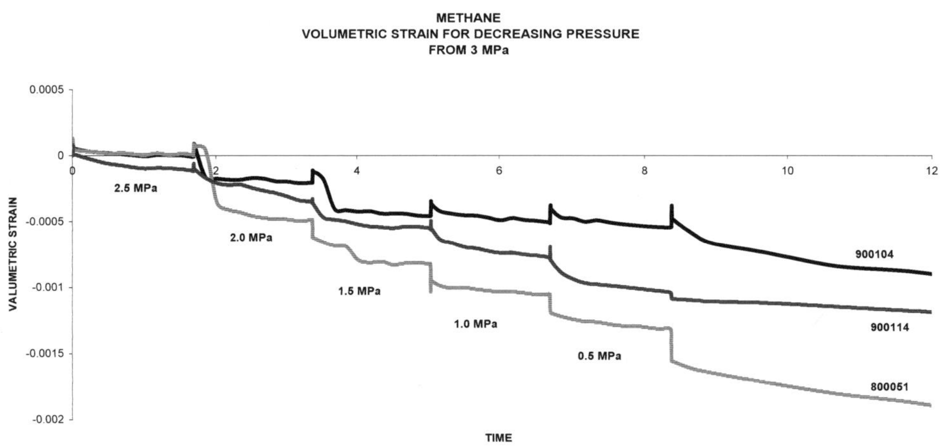

Figure 5

PERMEABILITY TESTS

Equipment and Methodology

The laboratory test equipment used in this study incorporates a high-pressure chamber, details of which have been reported by Lama (1995) and Aziz and Ming-Li (1999). The high-pressure test unit is composed of a triaxial pressure cell, which enables the effect of stress, as well as gas pressure, to be measured. For the tests, blocks of coal were taken from the coalface, two from 900 Panel and four from 800 Panel. A 50 mm diameter core, with the long axis parallel to bedding, was extracted from each sample. A 50 mm length was cut from the central section of each core and the ends polished. A 6 mm hole was drilled in the centre of each sample and the ends of the residual core sealed with a *Lock-tite* adhesive. The core was then placed between loading plates of the chamber.

The chamber consists of a rectangular prism of cast iron with removable front and back plates, with dimensions of 110 mm x 110 mm x 140 mm. The viewing window is 20 mm thick glass set in the cast iron frame. Twenty four bolts secure the frame to the chamber. The chamber is made leak proof by inserting rubber packers or washers between the frame and the box as well as fitting 'O' rings around the loading shaft located at the top of the chamber.

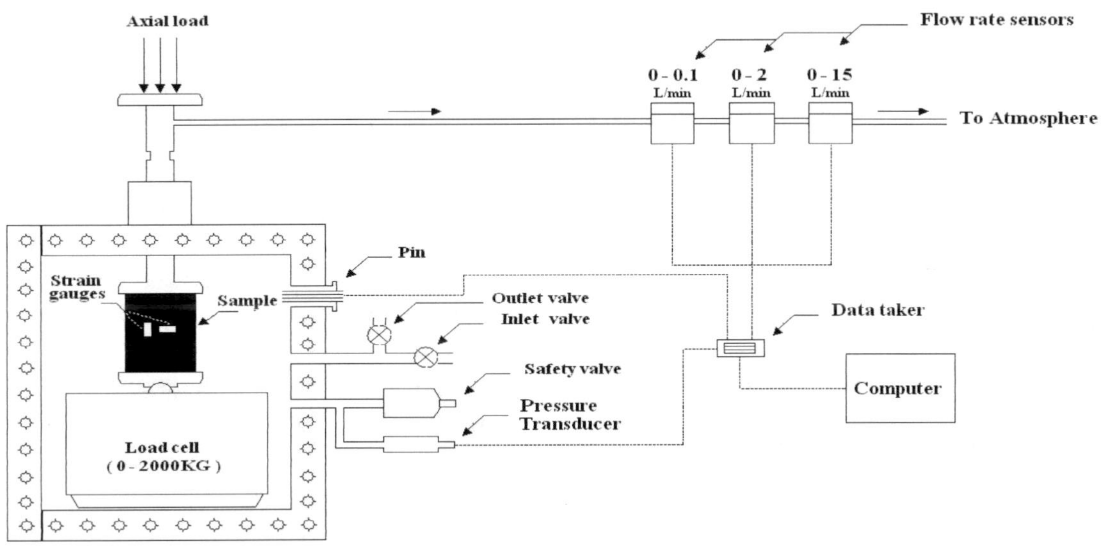

Figure 6. Schematic diagram of permeability test equipment (after Sereshki *et al.*, 2004)

The side of the chamber and the top-loading plate are connected to the universal thrust connector via a central axis, which transfers the applied axial load to the sample. The bottom loading plate has a hemispherical, concave seat that rests on a ball bearing at the top of a 40 kN capacity load cell. The bottom-loading plate assemblage transfers the load from the sample to the load cell and also pivots and aligns the coal sample in the event that the top and bottom surfaces of the coal sample are not parallel. Lips on the loading plates prevent any lateral movement of the sample during testing (Aziz et al. 2004).

Axial strain is applied to the core via a universal torque. Changes in the sample axial and lateral load dimensions due to gas sorption are monitored by two sets of strain gauges. Other parameters monitored or measured include:

a. Application of stress
b. Measurement of strain on the sample
c. Measurement of gas flow rate
d. Application of constant circumferential gas pressure
e. Application of constant suction.

Each core was subjected to a range of axial loads ranging from 100 to 2000 kg and charged in a sealed pressure chamber at a pressure (for these tests 3 Mpa) and maintained at constant pressure for a period of one week to allow the coal to be sufficiently saturated. Three gasses were used - methane, carbon dioxide, and a 50/50 mixture of both gases. Data from the load cells, strain gauges and flow meters is passed directly to a data logger connected to a computer. The strain is recorded over this period. Once the sample is sufficiently saturated, the release valve is opened and the released gas passed through a set of three flow meters to cover a wide range of flow rate to avoid the need to remove a given meter and insert a second if the flow rate goes off scale.

Meter 1	Low flow range	- 0 – 100 mL/minute
Meter 2	Medium flow range	- 0 – 2 L/minute
Meter 3	High flow rate	- 0 – 15 L/minute

The permeability of each sample is calculated using the Darcy flow equation:

$$K = \frac{\mu Q \ln(r_0 / r_i)}{\pi l (P_0^2 - P_u^2)} \quad (1)$$

where: K = permeability (mD)
Q = rate of flow of gas ($ccsec^{-1}$)
r_0 = external radius of sample (cm)
r_I = internal radius of sample (cm)
l = height of sample (cm)
P_0 = absolute pressure in chamber (bars)
P_u = absolute pressure in outlet (bars)
μ = viscosity of CH_4

Results

The results have shown a marked difference in permeability (in mD) between the 800 and 900 panel coals. The difference in permeabilities between 800 panel and the 900 panel coal for each of carbon dioxide and methane is quite significant. 800 panel had approximately three times greater permeability when compared to the 900 panel coals (Figs 7 and 8).

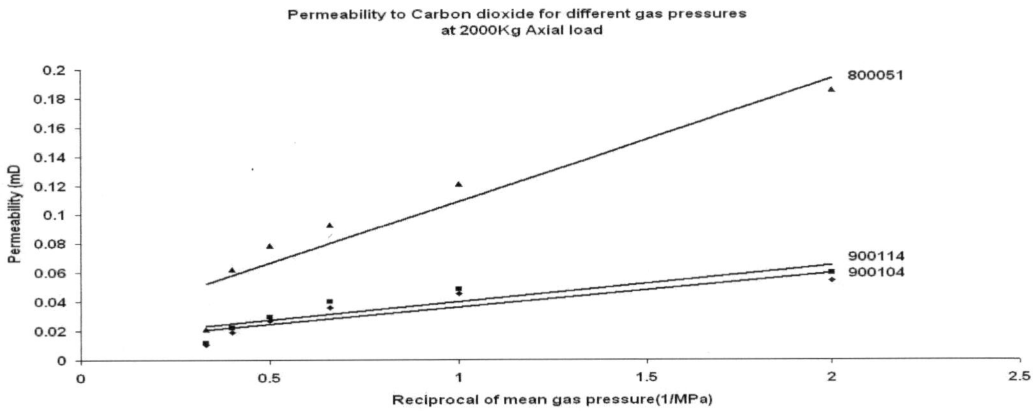

Figure 7. Permeability to Carbon Dioxide for different gas pressures

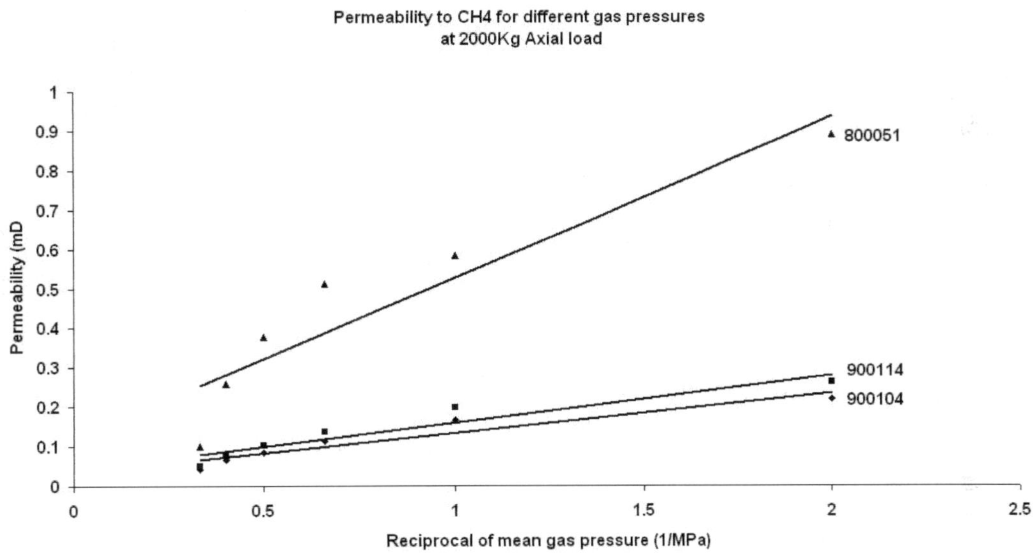

Figure 8. Permeability to Methane for different gas pressures

COAL PETROGRAPHY

After the experimental testing each sample was crushed to –2 mm and a representative subsample mounted in polyester resin to form a block. Each block was cut in half perpendicular to any density separation and the cut face from one of the halves polished. The maceral composition of each sample was determined as outlined in the Australian Standard AS2856.2-1998: Coal Petrography – Maceral Analysis (Table 1) Maceral analyses for the samples are given in Table 1.

Table 1. Maceral analyses of coal samples used in tests

Location	Vitrinite	Inertinite	Liptinite	Mineral	Carbonate	Cavity	Fracture	Total
800051	70.4	21.8	1.8	2.0	0	3.8	0	99.8
900104	68.4	19.0	1.0	2.8	2.8	5.4	1	100.4
900114	67.8	19.8	1.8	6.6	1.0	2.4	0	99.4

RESULTS

Petrographically the samples have similar organic components. However the mineral content of the samples is quite different, with carbonate in the form of calcite being the main component.

CONCLUSIONS

As can be seen from the forgoing results, there are clear differences in the type of coals present at mine study site, which lead to differences in coal shrinkage, permeability and coal petrography. These differences also relate to what influence stress, and thus strain has had on the coals, that is, if coal is highly stressed, it will relax more quickly than coals which are not so highly stressed, if we assume that the material behaves elastically.

In the 900 panel area, the coal is considered to be a 'difficult to drain' area, which is highly stress due to geological structures and visual indicators from underground mapping, therefore upon inseam drainage, the effective stress will change by a lesser magnitude as does the pore pressure, but not to the extent that free flowing drainage could be achieved. Unlike in the 800 panel area, which has a shrinkage coefficient of nearly twice that of the 900 panel area, this causes good drainage and greater change is effective stress from pore pressure is envisaged, particularly in the CO_2 environment.

Permeability and petrographic data, also confirm that the coals from the 900 and 800 panel area are markedly different. This has had consequential results for the efficiency of the mining cycle in these two areas of the mine. Further work is required to further investigate the relationship between mineral matter and the gas regime.

REFERENCES

Aziz, N.I. and Ming-Li, W., 1999. The effect of sorbed gas on the strength of coal – an experimental study. *Geotechnical and Geological Engineering,* 17, 387-402.

Aziz, N.I, Porter, I. and Sereshki, F., 2004. The influence of gas environment on coal properties - Experimental study in outburst control. *5th Underground Coal Operators' Conference, Coal Mine Planning,* 4-6 February, University of Wollongong, Australia, 195-201.

Bruggemann, D., 2002a. Case study – Central Colliery, QLD. *3rd Australasian Coal Operators Conference Tribute to Ripu Lama.* The Australian Institute of Mining and Metallurgy in conjunction with the University of Wollongong, Wollongong, February 2002, 89-97.

Bruggemann, D., 2002b. Geology of the Central Colliery Outburst. *ACARP Outburst and Drainage Workshop,* 10th May 2002, Mackay, QLD, Australia, 43-48.

Cao, Y., Mitchell, G.D., Davis, A. and Wang, D., 2000. Deformation metamorphism of bituminous and anthracite coals from China. *International Journal of Coal Geology,* 43, 227-242.

Crosdale, P and Beamish, B., 1993. Maceral effects on methane sorption by coal; in Beeston, J.B., New developments in coal geology. A symposium, Brisbane, November 1993, Coal Geology Group, 95-98.

Crosdale, P.J and Beamish, B.B., 1995. Methane diffusivity at South Bulli (NSW) and Central (Qld) Collieries in relation to maceral composition; *in* Lama, R. (ed.) *Proceedings International Symposium-Cum-Workshop On Management And Control Of High Gas Emissions And Outbursts In Underground Mines.* 20-24 March, 1995, Wollongong NSW, Australia, 363-368.

Crosdale, P.J., Beamish, B.B. and Valix, M., 1998. Coalbed methane sorption related to coal composition. *International Journal of Coal Geology,* 35, 147-158.

Faiz, M.M., 1993. Geological controls on the distribution of coal seam gases in the southern Sydney Basin. PhD Thesis, Wollongong 326pp (unpublished).

Faiz, M.M. and Hutton, A.C. 1995. Geological controls on the distribution of Ch_4 and CO_2 in coal seams of the Southern Coalfield, NSW, Australia; *in* Lama, R. (ed.) *Proceedings International Symposium-Cum-Workshop On Management And Control Of High Gas Emissions And Outbursts In Underground Mines.* 20-24 March, 1995, Wollongong NSW, Australia, 375-383.

Faiz, M.M. and Hutton, A.C., 1997. Coal seam gas in the southern Sydney Basin, New South Wales. *APEA Journal,* 37(1), 415-428.

Gray, I., 1987. Reservoir engineering in coal seams: Part 1 – The physical process of gas storage and movement in coal seams. *SPE Reservoir Engineering*, 28-34.

Gurba, L., 2002. Microscopic studies of low permeability coal. ACARP Outbursts and Gas Drainage Workshop, Mackay, 10th May, 2002.

Gurba, L.W. and Weber, C.R., 2001. The relevance of coal petrology to coalbed methane evaluation, using Gloucester Basin, Australia as a model. *Proceedings of 2001 International Coalbed Methane Symposium,* The University of Alabama, Tuscaloosa, Alabama, May 14-18, 371-382.

Harpalani, S. and Chen, G., 1997. Influence of gas production induced volumetric strain on permeability of coal. *Geotechnical and Geological Engineering,* 15, 303-325.

Hargraves, A.J., 1983. Instantaneous outbursts of coal and gas, a review. *Proceedings Australasian Institute Mining and Metallurgy,* 285, March 1983, 41-42.

Harvey, C., 1995. The development of an outburst database for the Bulli coal seam; in Lama, R.D. (ed), *International Symposium – Cum Workshop on Management & Control of High Gas Emissions & Outbursts in Underground Coal Mines,* Wollongong, NSW, 20-24 March 1995, 27-35.

Lama, R.D., 1995. Effect of stress, gas pressure and vacuum on permeability of Bulli coal samples; in Lama, R.D. (ed), *International Symposium - Cum Workshop on Management & Control of High Gas Emissions & Outbursts in Underground Coal Mines,.* Wollongong, NSW, 20-24 March 1995, 293-301

Lama, R.D., and Bartosiewicz, H., (1982) Determination of gas of coal seams. *Proceedings of the symposium on seam gas drainage with particular reference to the working seam. The Australian Institute of Mining and Metallurgy, Illawarra Branch, Wollongong, May, pp. 36 – 52*

Lama, R.D. and Bodziony. J., 1996. Outburst of gas, coal and rock in underground coal mines. R.D. Lama and Associates, Wollongong, NSW, Australia, 499 pp.

Sherwin, L. and Holmes, G.G., 1986. Geology of the Wollongong and Port Hacking 1:100000 Sheets. *Geological Survey of New South Wales, Department of Mineral Resources,* 16-32.

Smith, I.M. and Sloss, L.L., 1992. Methane emissions from coal. *IEA Coal Research, IEAPER/04, 26.*

Wynne, P., 2002. Case Study – Management of Outburst Risk at Tahmoor Colliery. *3rd Australasian Coal Operators Conference, Tribute to Dr Ripu Lama,* The *Australian Institute of Mining and Metallurgy in conjunction with the University of Wollongong,* Wollongong, February 2002, 98-104.

0511

STUDIES FOR COALBED METHANE EVALUATION IN NORTHERN SPAIN

Fernando Pendás; Jorge Loredo; Juan José Fernández; Pablo Cienfuegos.
E.T.S. Ingenieros de Minas. University of Oviedo. Independencia, 13. 33004 Oviedo. Spain

ABSTRACT

The Asturian Central Coal Basin has been a very important coalfield in production since the 16th century up to nowadays, when a program for mine closure is being in action from some years ago. The stratigraphic column of coal basin has been traditionally divided in unproductive and productive carboniferous. The productive carboniferous contains two great coastal ensembles of 3000 m thickness each one, over a extension bigger than 1500 km^2. The stratigraphic formations of the productive carboniferous are named "Paquetes": Caleras, Generalas, San Antonio, María Luisa, Sotón, Entreregueras, Sorriego and Modesta- Oscura and their age is mainly comprised between Namurian and Westphalian D. They are sediments of plain, deltaic front and off-shore with vertical pilling of para-sequences of marine levels, coastal levels with planar, festooned and hummocky stratification and continental levels with coal beds, which correspond to successive elevations of the sea level. The coal seams thickness ranges from 0.8 to 1.5 meters. The maturation grade of the coals varies from north zone of the basin where the volatile content ranges from 35%-40% to 20%-30%, to the south zone of the basin where the volatile content ranges from 20% to less of 10%. At the beginning of the 90s two exploration drills 2000 m deep have been made on the Asturian Central Coal Basin and although this first test did not result into commercial explotation of methane, it does not mean that the zone lack of interest. New technological development and a best knowledge of the basin allow to have justified expectation to find coalbed methane reserves. Currently, new studies are being undertaken in the most favourable areas on shallow depth of the basin in order to evaluate in detail the real possibilities for coalbed methane (CBM) and coalmine methane (CMM) from the closed mines. Between these studies it must be included coal quality, genetical and sequential stratigraphy, and deep hydrogeology, in order to define the 3D geological and hidrogeological model of the basin. These studies will be complemented with new data obtained from an exploration drill of 400 m deep located in a selected area of the basin. Permeability measurements in situ on this drill will allow obtained a first assessment of the permeability of the Asturian Central Coal Basin for coalbed methane development. Measurements of gas content in unexploited coal seams are made in order to estimate the CBM and CMM could revive the economical interest of the Asturian Central Coal Basin.

INTRODUCTION

The Asturias Central Carboniferous Coal Basin, in northern Spain, is an important area for coal production from the 18th century up to now, when a coal mining closure plan is being applied from the beginning of the 1990´s. The Asturian Central Carboniferous Coal Basin constitutes the biggest productive carboniferous area of the Iberian Peninsula, with an extension of 1400 km^2. It is enclosed in the Iberian Massif [1], inside the unit named Cantabrian Zone. This Carboniferous Basin extends between the Aramo Mountain Range to the west, and the Peña Mea to the east, forming an ensemble of mountains of medium altitude, shaped by the fluvial network over the carboniferous rocks, which are mainly of detrital nature. The Asturian Central Coal Basin, is a fore-deep basin with a lot of simililarities with the Appalachian basin in the U.S.A.

In this Coal Basin, in the beginning of the 20th century natural gas has been found in various reconnaissance drills accomplished to locate hidden coal deposits. These findings together with the surface gas manifestations in some areas of the Coal Bassin (an example of these manifestations is the locally named "Mechero de Saús"), have driven a exploration campaign in the north of Asturias with not very successful results. This exploration campaign has been carried out at the beginning of the 1990s, due to the interest offered by the CBM recuperation projects in the world. The exploration was the result of a collaborative research between Unión Texas España and HUNOSA national enterprise with the objectives to known the possibilities for economical recuperation of coalbed methane (CBM) in the Asturian Central Carboniferous Coal Basin.

In this project a total of 63 coal beds with thickness above 0.25 m. have been perforated with two research drills. From the whole of the coal beds, 53 of them belong to the "Paquete Sotón" with an accumulated coal thickness of 50.9 m. The other 10 coal beds belong to the "Paquete María Luisa", with an accumulated coal thickness of 6.7 m. Both Paquetes or rock formations are the most important in the carboniferous basin from a point of view of coal beds. Currently the British Company Heritage Petroleum PLC has research permission for hydrocarbons in all the area of the Asturian Central Carboniferous Coal Basin.

GEOLOGY AND HYDROGEOLOGY

The stratigraphic sequence is divided in two big groups named Lena Group and Sama Group. The first one is also known as unproductive group and it is characterized by a great abundance of calcareous materials and for the scarce presence of coal beds. In contrast to the before mentioned group, the second one contains scarce calcareous levels and a great number of coal beds which have been exploited since the 18th century. Traditionally, the different groups of coal beds and enclosing rocks have been subdivided in an ensemble of litostratigraphic associations which have been named "mining packages" [2]. For the Units of Caudal-Nalón and La Justa-Aramil the different "mining packages" are from footwall to hanging wall: Levinco, Llanón, Tendeyón, Caleras, Generalas, San Antonio, María Luisa, Sotón, Entrerregueras, Sorriego and Modesta-Oscura (Figure 1).

GEOLOGICAL UNITS	MINING "PAQUETES" (from top to base)	"PAQUETE" THICKNESS (m)	COAL SEAMS	ACUMULATE COAL THICKNESS (m)
CAUDAL - NALON	MODESTA-OSCURA	575	7	7,50
	SORRIEGO	340	5	8,00
	ENTRERREGUERAS	340	3 - 5	4,50 □ 6,00
	SOTON	400	8 - 12	10,50 □ 14,00
	MARIA LUISA	300	6 - 9	6,50 □ 8,50
	SAN ANTONIO	315	2	1,70
	GENERALAS	300	2 - 4	2,50 - 4,50
	CALERAS	310	2	2,50
RIOSA - OLLONIEGO	ESPERANZA	350	3 - 6	3,50 □ 6,50
	PUDINGAS	700	3 - 5	5,00 □ 7,00
	CANALES	800	8 - 12	12,00 □ 15,00

Figure 1. Coal seams number and thickness, and total thickness of coal for the different "paquetes" of the Asturian Central Coal Basin (HUNOSA, unpublished).

The Lena Group, with 3500 metres of estimated thickness is formed by alternance of sandstones, limolites and lutites with intercalations of coal beds, some of them sporadically exploited. The lowest part of the Group is constituted by a slaty level which laterally replaces the Mountain Limestone, named "Fresnedo Package". Immediately after are situated the Packages Levinco, Llanón and Tendeyon. This Group contains abundant calcareous levels which are rich in faune, furthermore of algae which have a scarce interest from a chronostratigraphic point of view. These calcareous levels decrease in thickness according as they approach to the hanging wall. This fact is indicative of a transit to sedimentary conditions increasingly more superficial.

The Sama Group or productive ensemble has about 2000 metres thickness and it is formed by alternances of sandstones, limolites and lutites, together with a great quantity of coal bed economically exploitable. The content in limestones is very scarce and its presence in the estratigraphic sequence is in the form of levels or bed of reduced thickness. In the middle part of the stratigraphic sequence stands up the presence of two great conglomeratic ensembles. The packages which form the Group are, from the footwall to the handwall: Caleras, Generalas, San Antonio, Mª Luisa, Sotón, Entrerregueras, Sorriego and Modesta-Oscura. From all of them the María Luisa y Sotón packages are whose presenting the biggest density of coal beds, with average bed thickness between 1 and 1.5 metres. In some zones, as in the north subzone, the accumulated coal bed thickness exceeds 60 metres. The deposit environment corresponds with a shallow coastal line. Currently, it is under study the sequential stratigraphy of the productive coal packages.

The structural arrangement of the carboniferous materials corresponds with a great basin formed during the Hercynian orogeny. From a tectonic point of view three units separated by important structural accidents can be differentiated (Figure 2). From west to east they are: Riosa-Olloniego Unit, La Justa Aramil Unit and Caudal-Nalón Unit.

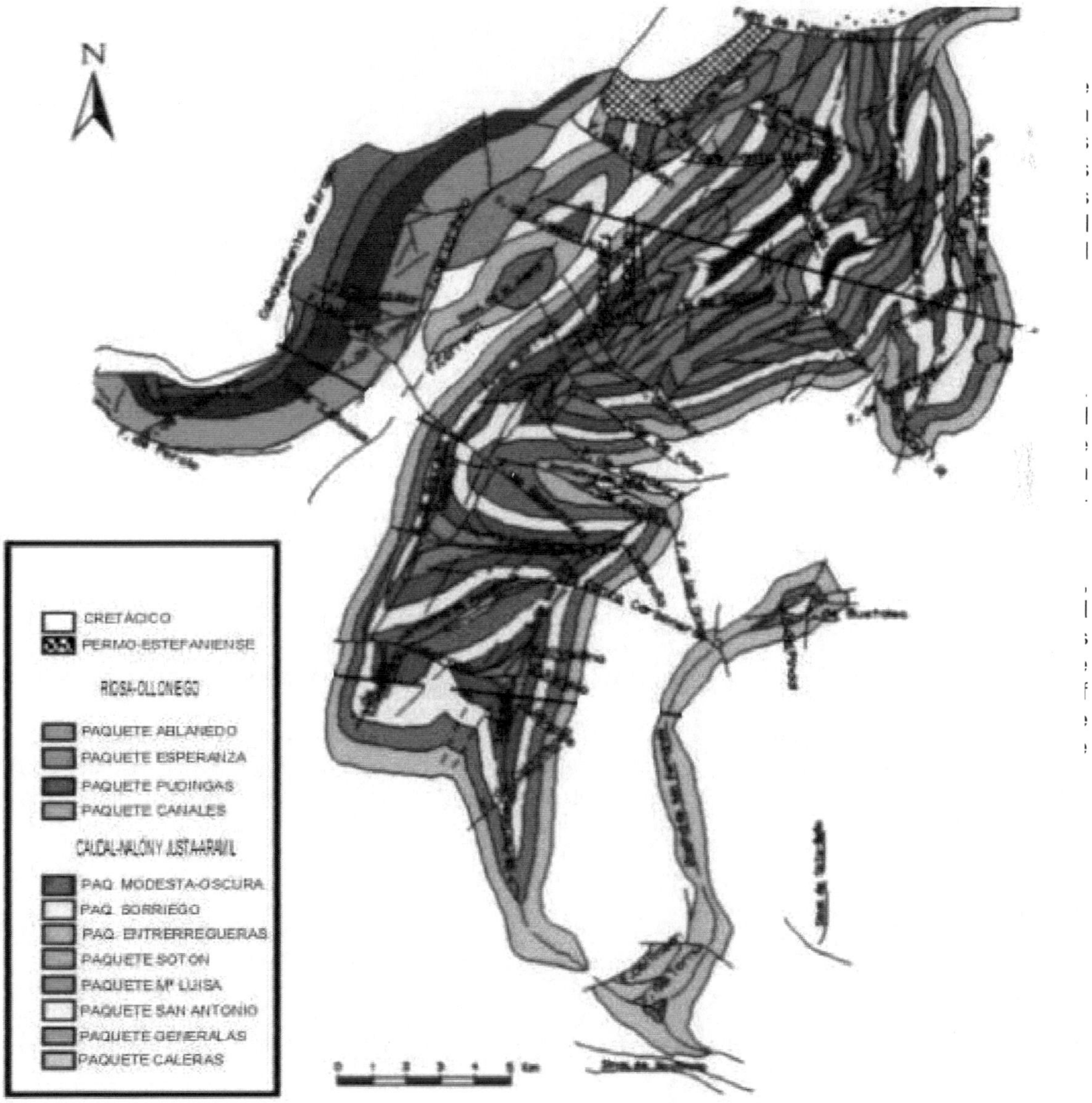

Figure 2. Geological map of the productive "paquetes" or Sama Group (HUNOSA, unpublished)

The volatile content of coals is variable, from percentages in the order of 10% into the southern zone to percentages in the order of 35% into the northern zone. It is observed a clear decrease in content of volatile materials according as pass from south to north of the basin. Coal from the northern part of the basin is more evolved, living place to coals of the type hard coal. Then, in the northern part of the basin there are coals of the bituminous and sub-bituminous type with high content in volatile (Figure 3).

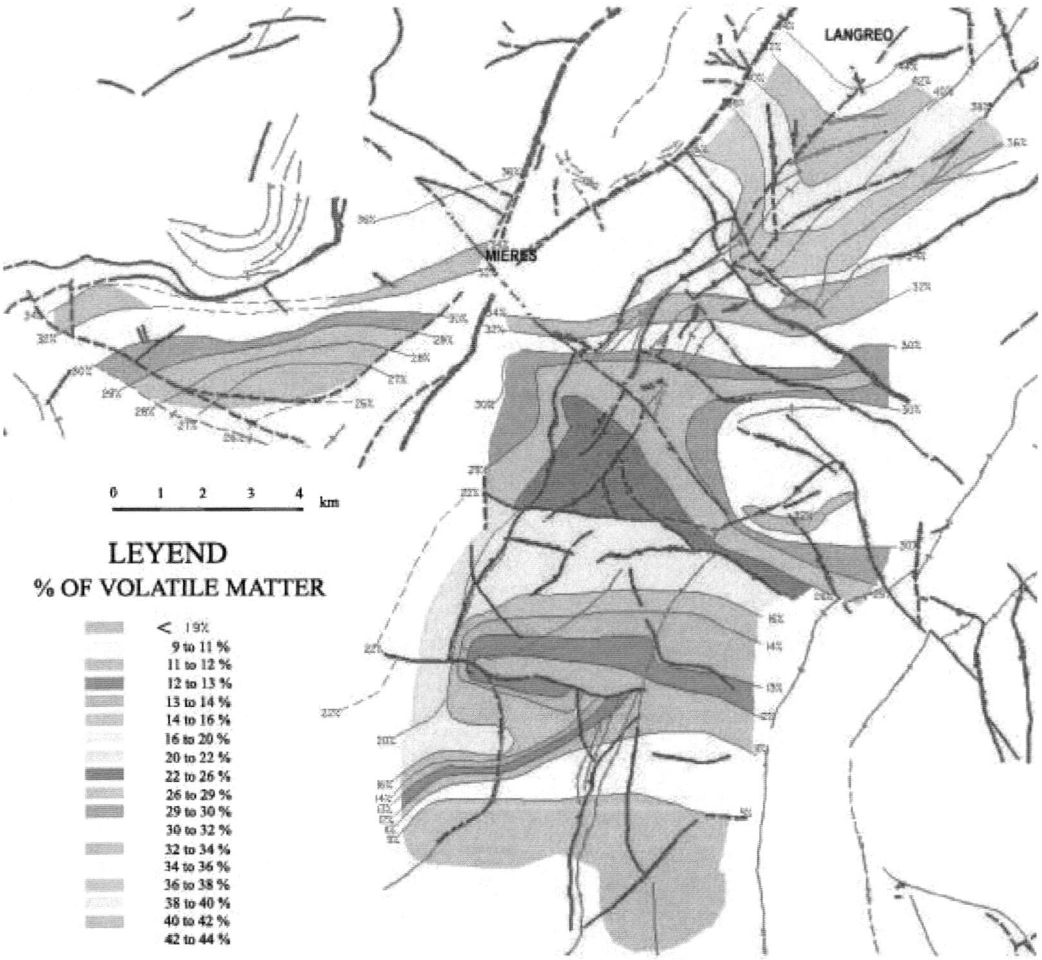

Figure 3. Volatile map for the productive "paquetes" or Sama Group (modificated from HUNOSA)

The reflectance of the vitrinite shows a progressive increase from north to south, with values in the order of 0,78% Rr in Pumarabule Mine, in the south of the basin, with average values of 2,54% as in the case of San Antonio Mine. The ash content is very variable inside each zone, and in general in the whole of the basin.

CBM RESOURCES ESTIMATION

For resources estimation the subdivision in zones made in 1984 by ENADIMSA in collaboration with the IGME (Geological and Mining Institute of Spain) has been applied. At this time and with the purposes to evaluate the coal resources of the Asturian Central Coal Basin, they established five subzones inside the coal basin (the distribution of subzones is represented in figure 4):

- Subzone Nalón or Subzone North. It is divided in three areas named respectively Sama, El Entrego and San Mamés.
- Subzone Turón or Subzone Centre. It is divided in three areas named respectively Polio, Barredo-Turón and Urbiés.
- Subzone Aller or Subzone South. It is divided in six areas named respectively Caborana, Moreda, Pontones East, Pontones West, Melendreros and Cobertoria.
- Subzone Riosa-Olloniego or Subzone West. It is divided in three areas named respectively Monsacro, Nicolasa and Olloniego.
- Área of La Justa-Aramil.

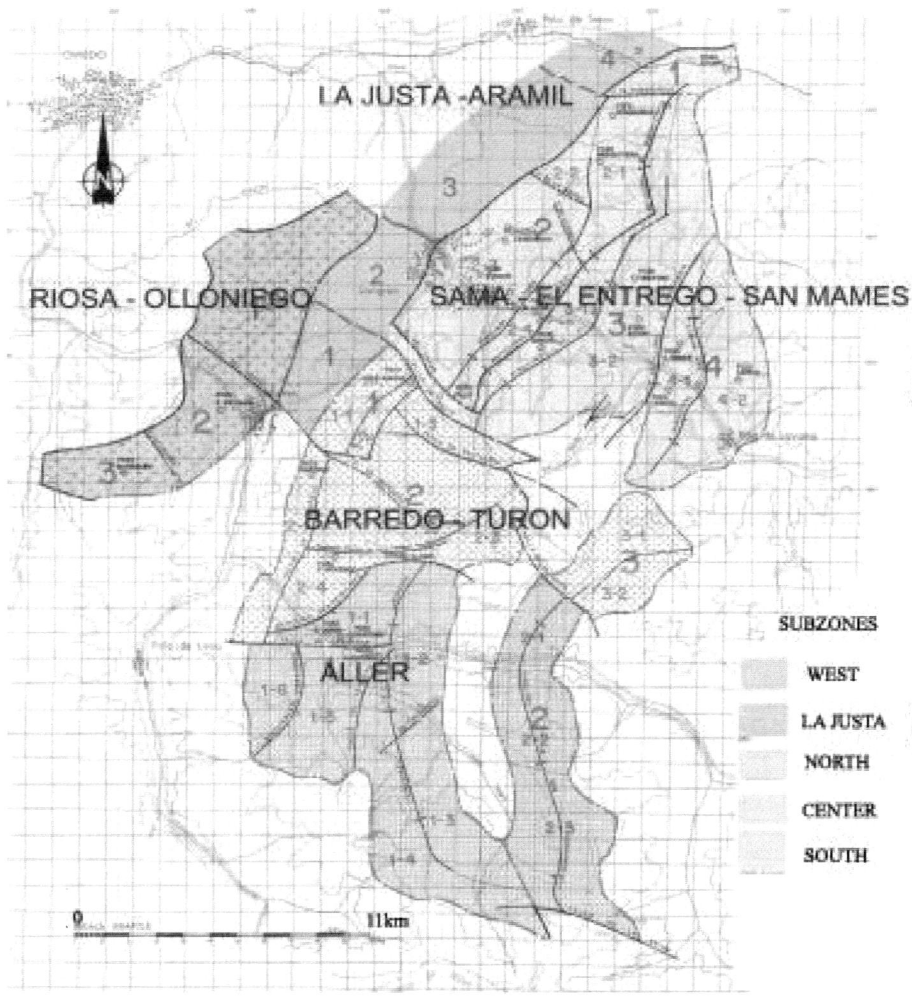

Figure 4. Subdivision in Subzones of the Asturian Central Coal Basin

Some factors to consider for the estimation of the Coalbed methane resources in the Basin, have been:

- Areas of the Coal Basin with big coal resources. Number and importance of the unexploited coal beds have been considered as a point of interest
- Coal characteristics.
- Favourable geological formations and structures for methane retention in coal beds.
- Historical data about gas concentrations in coal mines.

The possibility of access to these coal beds either from surface as is the case of "La Justa – Aramil" or either from inside the mine as is the case of "Riosa – Olloniego" has been considered too as another point of interest.

The CBM resources estimation for each one of the before mentioned zones has been determined by use of the expression:

$$\text{CBM (m}^3) = \rho \cdot S \cdot e_m \cdot G$$

Where:
ρ = Average coal density. A value of 1.6 t/m^3 has been considered for all coal beds.
e_m= Coal bed thickness (m)
S = Average coal bed surface (m^2).
G = Average methane content in coal bed (m^3/t).

The resources have been estimated until a maximum depth of 2000 metres from surface [3], according the data on coal reserves estimated for the actualization document of the Inventory of Coal National Resources. In the particular case of La Justa-Aramil Unit, the equispaciated parallel sections method where has been applied. The coal bed thickness has been obtained from the stratigraphic columns in the different coal mines. For average methane content in coal bed, the average data obtained in the mining degasification essays made in active coal mines managed by HUNOSA National Company, have been used. The resources estimation has been made for equispaciated intervals of 200 metres, as well by coal packages as by depth. The evaluated CBM resources are as follow.

NORTH SUBZONE: SAMA-EL ENTREGO-SAN MAMES.

Main attributes characterizing the potentiality of coal beds for methane production in this Subzone, are as follow:

- The average coal bed thickness is in the order of 57 metres
- The average volatile content in coals from this Sector is 30% [4], with bituminous coals of high volatile content.
- The average gas content ranges from 3,88 m^3/t to 9,35 m^3/t for coals without ash, with average values of 6,60 m^3 / t.
- The zone with highest volume of CBM resources is Sama, and immediately behind it El Entrego and San Mamés.

ÁREA	CBM RESOURCES (Mm^3)
Sama	7579,44
El Entrego	5849,64
San Mamés	863,65
Total	**14292,73**

Table 1. CBM resources by areas in "Sama – El Entrego – San Mamés"

CENTER SUBZONE: TURÓN

The different attributes characterizing the potentiality of coal beds for methane production in the Turón Sector, are as follow:

- The average coal thickness in the Turón Sector, calculated from mining data is 31.67 metres.

- The volatile content present in coals is comprised between 22% and 32%, with average values of 27 % . These data show an area with a medium grade of evolution, with bituminous to sub.bituminous coals.
- The average gas content in the Sector is in the order of 5.6 m^3/t to 10.81 m^3/t for coals without ash, being the average values of 7.25 m^3 / t.

SOUTH SUBZONE: ALLER

The different attributes characterizing the potentiality of coal beds for methane production for the Aller Sector, are as follow:

- The average coal thickness in the Aller Sector, calculated from mining data, are in the order of 21,70 metres.
- The volatile content present in coals is comprised between 7% and 15 %, showing that it corresponds to an over mature area with very evolutioned hard coals.
- The average gas content in the Sector is comprised between 7 m^3/t and 9 m^3/t for coals without ash.

WEST SUBZONE: RIOSA-OLLONIEGO.

The different attributes characterizing the potentiality of coal beds for methane production for the west Subzone, are as follow:

- The average coal bed thickness, calculated from the database of HUNOSA National Company is in the order of 20 metres.
- The volatile content of coals is between 27% and 33 %, showing an area with a high evolution degree with bituminous coals with high volatile content.
- The average gas content is between 5 m^3/ t and 10 m^3/ t for coal without ash.

LA JUSTA-ARAMIL AREA

The attributes characterizing the potentiality of coal beds for CBM production in the area are :

- The coal thickness is: Río Miñera (4.96 m), La Justa (17.13 m), Barros/Tablado (19.50 m) and Aramil (6.03 m).
- Coals are bituminous with high volatile content according to ASTM coal classification.
- The abandoned mining Works in La Justa are flooded and drain to the Nalón River by the general transversal located in proximity to Lada village. The subarea Barros-Tablado drains to the Nalón River on its left border [5].
- The volatile content is comprised between 32% and 37%, which are higher than 30% considered the limit for oil generation. In this Unit a detailed study has been made. For this purpose the Unit has been divided in some subunits, named from south to north: Rio Miñera, La Justa, Barros-Tablado and Aramil. From all of them, the most important from the CBM point of view are La Justa (Figure 5) and Barros-Tablado (tables 2 to 5). In these tables C means CBM resources, expressed in Mm^3, in coal without ash, and Cb means CBM resources, expressed in Mm^3, in coal with ash.

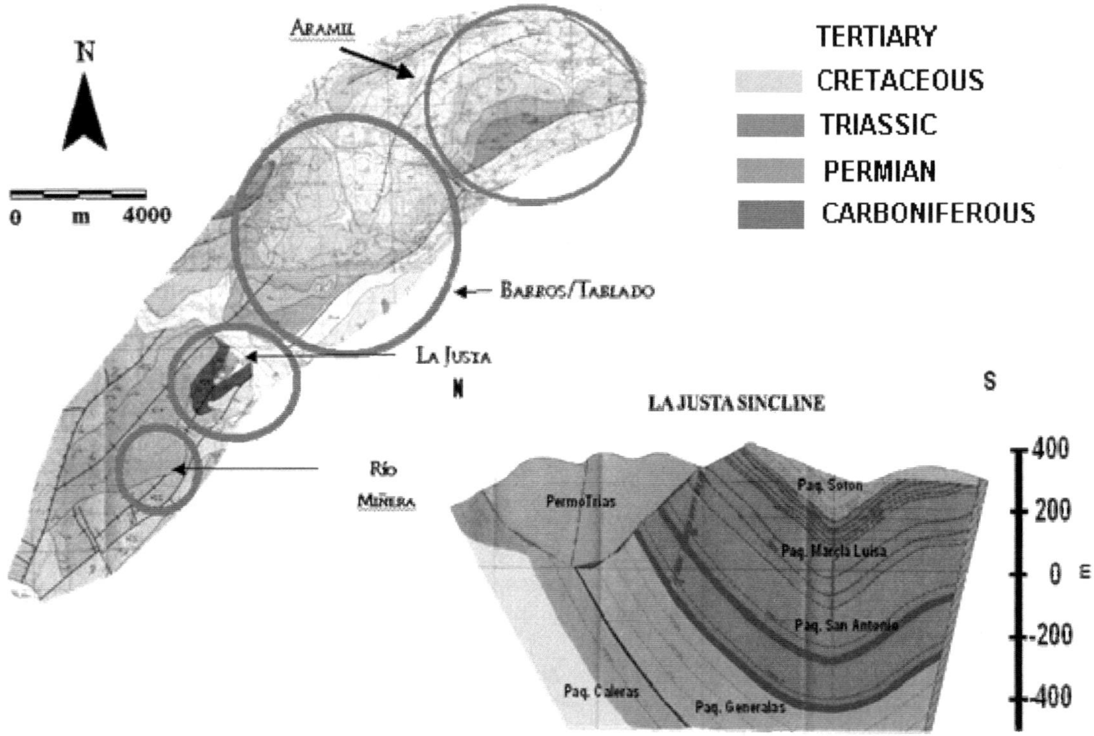

Figure 5. Geological map of La Justa-Aramil Unit and cross section of La Justa Sincline, one of the most important structures of the La Justa-Aramil Area, with the situation of the more thickness coal seams.

Package Name	Coalbed Name	Average thickness (m)	Accumulated thickness (m)	Theoretical coal tonnage (Mt)	C (Mm^3)	Cb (Mm^3)
Sotón			11.80	1,39	11,40	10,32
Mª Luisa				26,70	218,98	198,38
San Antonio	Rosario 2ª	0.57	1.11	5,43	44,57	40,34
	Rosario 1ª	0,54				
Generalas	Pedregala	1,44	3.37	22,19	181,97	164,87
	Llamerona	1,12				
	Vaca	0,81				
Caleras	Anchona	0.99	0.99	7,64	63,68	56,76
TOTAL			**17,3**	**63,35**	**520,6**	**470,67**

Table 2. CBM resources in La Justa – Aramil Zone.

Depth	Total Coal (Mton)	C (Mm^3)	Cb (Mm^3)
Mountain-0	24,03	197,04	178,54
0-(-200)	12,89	105,69	95,77
(-200)-(-400)	8,32	68,22	61,81
(-400)-(-600)	11,62	95,28	86,33
(-600)-(-800)	3,12	25,58	23,18
(-800)-(-1000)	1,76	14,43	13,07
TOTAL Mountain- (-1000m)	61,74	506,26	458,72
(-1000)-(-1500)	1,61	13,20	11,96
(-1500)-(-2000)	0	0	0
TOTAL (-1000m/ -2000m)	1,61	13,20	11,96
TOTAL	**63,35**	**519,46**	**470,68**

Table 3. CBM resources by depths in La Justa Zone.

<table>
<tr><th>Package Name</th><th>Coalbed Name</th><th>Average thickness (m)</th><th>Accumulated thickness (m)</th><th>Theoretical coal tonnage (Mt)</th><th>C (Mm³)</th><th>Cb (Mm³)</th></tr>
<tr><td rowspan="2">Entrerregueras</td><td>Guillermo</td><td>1,41</td><td rowspan="2">2.50</td><td>9,63</td><td>78,98</td><td>71,55</td></tr>
<tr><td>San Pedro</td><td>1,09</td><td></td><td></td><td></td></tr>
<tr><td rowspan="7">Sotón</td><td>1</td><td>1,08</td><td rowspan="7">14.96</td><td rowspan="7">54,02</td><td rowspan="7">443,00</td><td rowspan="7">401,36</td></tr>
<tr><td>2</td><td>1,7</td></tr>
<tr><td>3</td><td>0,76</td></tr>
<tr><td>4</td><td>1,93</td></tr>
<tr><td>5</td><td>2,63</td></tr>
<tr><td>6</td><td>1,56</td></tr>
<tr><td>7</td><td>3,2</td></tr>
<tr><td>María Luisa</td><td>8</td><td>2,1</td><td>2.1</td><td>7,79</td><td>63,89</td><td>57,87</td></tr>
<tr><td>TOTAL</td><td></td><td></td><td>19,56</td><td>71,44</td><td>585,87</td><td>530,78</td></tr>
</table>

Table 4. CBM resources by packages in Barros-Tablado Zone.

Depth	Total Coal (Mton)	C (Mm^3)	Cb (Mm^3)
Montaña-0	6,76	55,43	50,22
0-(-200)	8,30	68,06	61,66
(-200)-(-400)	21,67	177,69	161,00
(-400)-(-600)	17,04	139,72	126,60
(-600)-(-800)	13,43	110,12	99,78
(-800)-(-1000)	3,96	32,47	29,42
TOTAL Montaña- (-1000m)	71,16	583,51	528,71
(-1000)-(-1500)	0	0	0
(-1500)-(-2000)	0	0	0
TOTAL (-1000m/ -2000m)	0	0	0
TOTAL	**71,16**	**583,51**	**528,71**

Table 5. CBM Resources by depths in Barros-Tablado Zone.

In this zone a research drill named CBM#1 has been made in 2004 (Figure 6). The drilling reached a depth of 350 m, and the perforation has been made by rotopercusion method in the first 110 metres with a diameter of 180 mm and casing of 150 mm. The next 240 metros have been drilled with continuous core by wire-line system with diameter of 116 mm. Maximum values obtained for gas concentration have been in the order of 10 m^3/t, at 290 metres depth.

Figure 6. CBM#1 investigation borehole

Considering as CBM productive only the Mª Luisa and Sotón packages, a hypothetical rectangular production drills network could be developed for the synclinal of La Justa-Aramil and Barros Tablado, where drill should could be located each 400 metres. Then, in an area in the order of 5.80 km^2 could be placed 36 vertical drills. On the basis of a efficiency of a 50%, from the 1000 Mm^3 evaluated for this depth could be exploited in the order of 500 Mm^3.

CONCLUSIONS

Alter the preliminary study accomplished about the potential exploitation of CBM in the Asturias Carboniferous Central Coal Basin it can be inferred that this basin constitutes an interesting area in order to carry out an in deep study in some of the sectors (Figures 7 and 8). According to first estimations based on the studies accomplished, the minimum resources of coalbed methane in the whole of the Asturian Central Coal Basin are in the order of 25000 Mm^3 and the gas content of the coal seams range from 6 m^3 to 14 m^3 by ton.

From all sectors, the most interesting is the Barros Unit, given the scarce mining activity developed in the area, the presence of a great density of coal beds close to surface, the favourable rank of coals and the simple geological structure. The new technologies in guided drilling, which were not posible to apply on the first CBM exploration campaigns developed in the Basin on the 1990s, supposes on the moment a new way for the development of the CBM industry in the zone. The existence of a thermal power on the proximities increases the interest on the research in order to combine the CBM exploitation with CO2 sequestration in coal beds [6], [7].

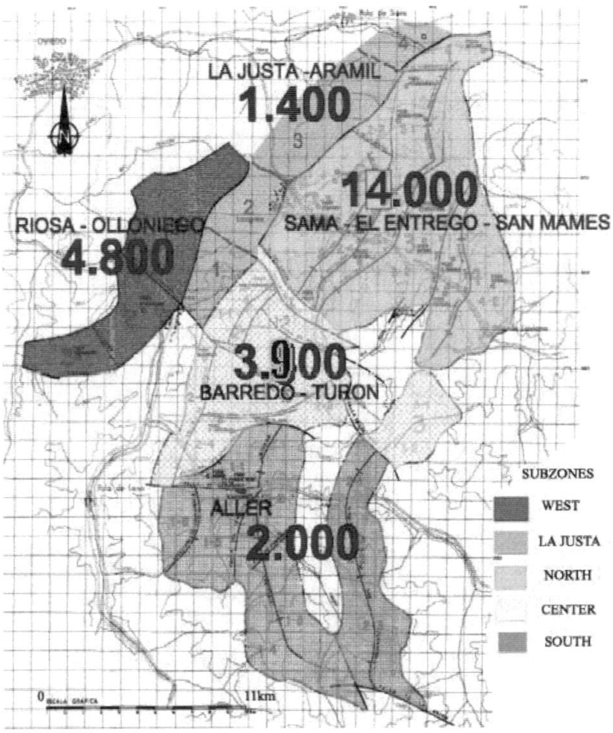

Figure 7. CBM resources (Mm^3) of the different Subzones of the Central Asturian Coal Basin

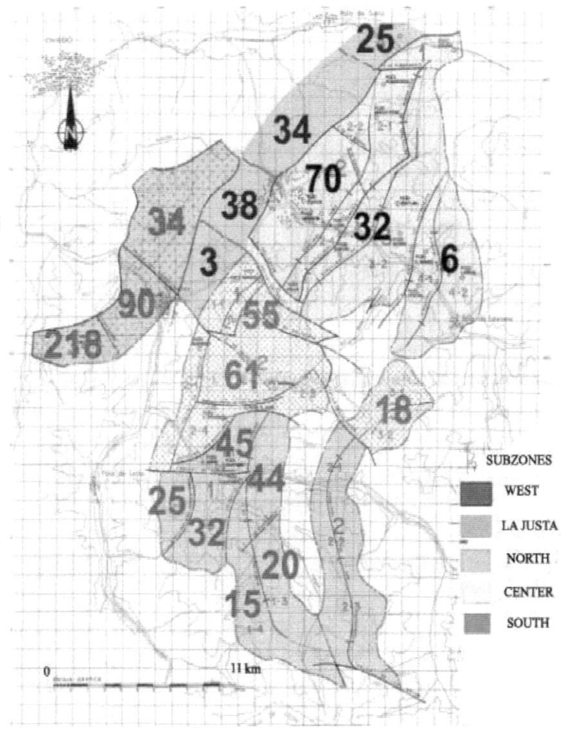

Figure 8. CBM resources (Mm^3/km^2) of the different Subzones of the Central Asturian Coal Basin

REFERENCES

1 Lotze, F., 1945: "Zur Gliederung der Variszíden der Iberischen Meseta". Geotekt. Forsch. 6, 78-92

2 García-Loygorri, A., Ortuño, G., Caride de Liñán, C., Gervilla, M., Greber, Ch. & Feys, R. 1971: "The Carboniferous of Northwest Spain". Trabajos de Geología, 3, pp 101-150. Oviedo Spain

3 Fernández, J. J.,2004: "Estimación de recursos de CBM en la Cuenca Carbonífera Central de Asturias y zonas adyacentes". Memoria de Investigación, Uni. Oviedo 258pp. Inedit

4 ENADIMSA, 1970: Coal Database Unpublished.

5 Pendás, F., Loredo, J., Fernández, J. J.; Cienfuegos, P., 2004: "Posibilidades de CBM en la Unidad La Justa-Aramil de la Cuenca Carbonífera Central de Asturias". Reunión Científico-Técnica sobre aprovechamiento de CBM en la Cuenca Carbonífera Central de Asturias, 16pp. Proceedings Volume.

6 Cienfuegos Suárez, Pablo, 2004: "Posibilidades de almacenamiento de CO_2 en capas de carbón en la Cuenca Carbonífera Central de Asturias". Memoria de Investigación. Uni. Oviedo, 43pp. Inedit.

7 Loredo, J., Pendás, F., Zapatero, M. A., Mansilla, H., Cienfuegos, P., Fernández, J. J., Martínez, R., Suárez, I., Arenillas, A., 2004: "El subsuelo: una opción para el secuestro del CO_2 emitido a la atmósfera por las industrias". Congreso Nacional de Medioambiente Madrid. Proceedings Volume.

0519

COALBED METHANE EXPLORATION IN THRUST BELTS: EXPERIENCE FROM THE SOUTHERN APPALACHIANS, USA

Jack C. Pashin
Geological Survey of Alabama, P.O. Box 869999, Tuscaloosa, AL 35486

ABSTRACT

Reservoir performance in the southern Appalachian thrust belt reflects a complex interplay among stratigraphic, structural, and hydrologic variables. Productive coal beds are distributed through more than 1,000 m of section, and in each well, between 5 and 20 coal beds ranging in thickness from 0.3 to 4 m are completed. Thrust-related folds in the southern Appalachian coalbed methane fields have steep forelimbs, gently dipping backlimbs, and contain multiple hinge zones. Fresh-water recharge is most effective in steep fold limbs, and fresh-water plumes extend from the outcrop to reservoir depth in thick, continuous coal beds. Late-stage bacterial methanogenesis within the fresh-water plumes facilitates high gas saturation in coal. Linear productivity sweet spots occur along fold hinges. Hinges in proximity to fresh-water recharge areas tend to be water-productive, whereas those that are distal to recharge contain sweet spots with exceptional gas production. Within broad, gently dipping fold limbs, linear gas- and water-production sweet spots appear to mark the positions of conjugate shear zones.

INTRODUCTION

Geologic structure is an important control on the performance of coalbed methane reservoirs because it is a major determinant of the distribution of permeability and basin hydrodynamics [e.g., 1-3]. To date, most coalbed methane development has been in large coal basins with gently deformed strata, such as the San Juan, Powder River, and Black Warrior basins [e.g., 4-6], but as these basins approach maturity and natural gas prices remain high, small basins with diverse structural styles are becoming increasingly attractive exploration targets. As exploration moves into regions with significant tectonic deformation, such as orogenic thrust belts, it is important to understand how geologic structures influence reservoir performance.

The southern Appalachian thrust belt of Alabama (fig. 1) has been the site of coalbed methane production since the infancy of the modern coalbed methane industry [7, 8], and the coal basins of the Alabama thrust belt remain the focus of intensive exploration and development [e.g., 3, 9-11]. Accordingly, experience with coalbed methane development in the southern Appalachians provides insight that facilitates the formulation of exploration and production strategies for deformed regions, and this paper summarizes the major stratigraphic, structural, and hydrogeologic information that are important to consider as exploration moves toward small, structurally diverse basins.

Economic coal-bearing strata in the southern Appalachian thrust belt are concentrated in the Lower Pennsylvanian Pottsville Formation of the eastern Black Warrior basin, the Cahaba synclinorium, and the Coosa synclinorium (figs. 1, 2). Coalbed methane is generally produced from the upper part of the Pottsville Formation in strata shallower than 1 km. The Black Warrior basin and Cahaba synclinorium have been the sites of intensive exploration and production activity [3, 9, 10], and coalbed methane reserves in these areas are thought to exceed 115 billion standard cubic meters [12]. Production has yet to be established in the Coosa synclinorium, and exploration efforts have focused on the northeastern part of the synclinorium [11, 13].

STRATIGRAPHY

The Pottsville Formation of the southern Appalachian thrust belt is of Lower Pennsylvanian (Westphalian A; Langsettian) age and is dominated by interbedded shale, sandstone, and coal. Coal

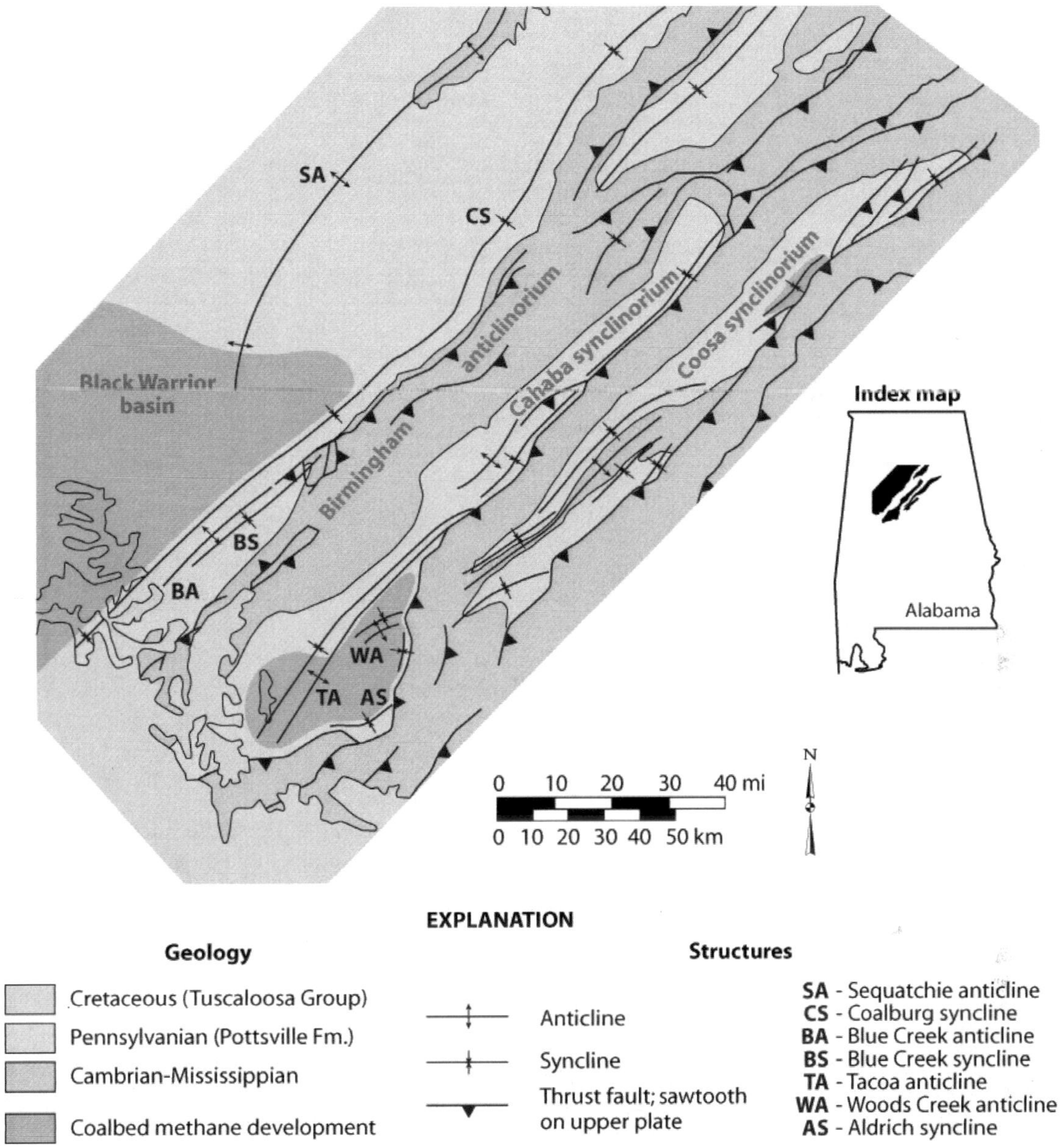

Figure 1.—Geologic map of the southern Appalachian thrust belt in Alabama showing major structures and areas of coalbed methane development. Geologic map modified from [37].

beds are distributed through up to 2.5 km of section (fig. 2), and individual coal beds are typically thinner than 3 m. Coalbed methane is produced from beds as thin as 0.3 m. In general, 5 to 20 coal beds are completed in Alabama coalbed methane wells, and most production is from the upper part of the Pottsville Formation in strata shallower than 1 km.

Numerous coal beds distributed through a thick stratigraphic section is a common characteristic of the Pottsville Formation throughout the study area, but the stratigraphic and depositional style of the Pottsville Formation in each area is different. The Black Warrior section (fig. 2) contains numerous stacked fluvial-deltaic depositional cycles that constitute 3rd- through 5th-order parasequences [14, 15], and coalbed methane is produced from multiple seams mainly in the Black Creek through Utley coal zones [16]. Accumulation of the thickest, most widespread coal beds that form the backbone of the Alabama coalbed methane and mining industries is thought to be associated with 5th-order base-level rises driven by glacial eustasy [14, 15].

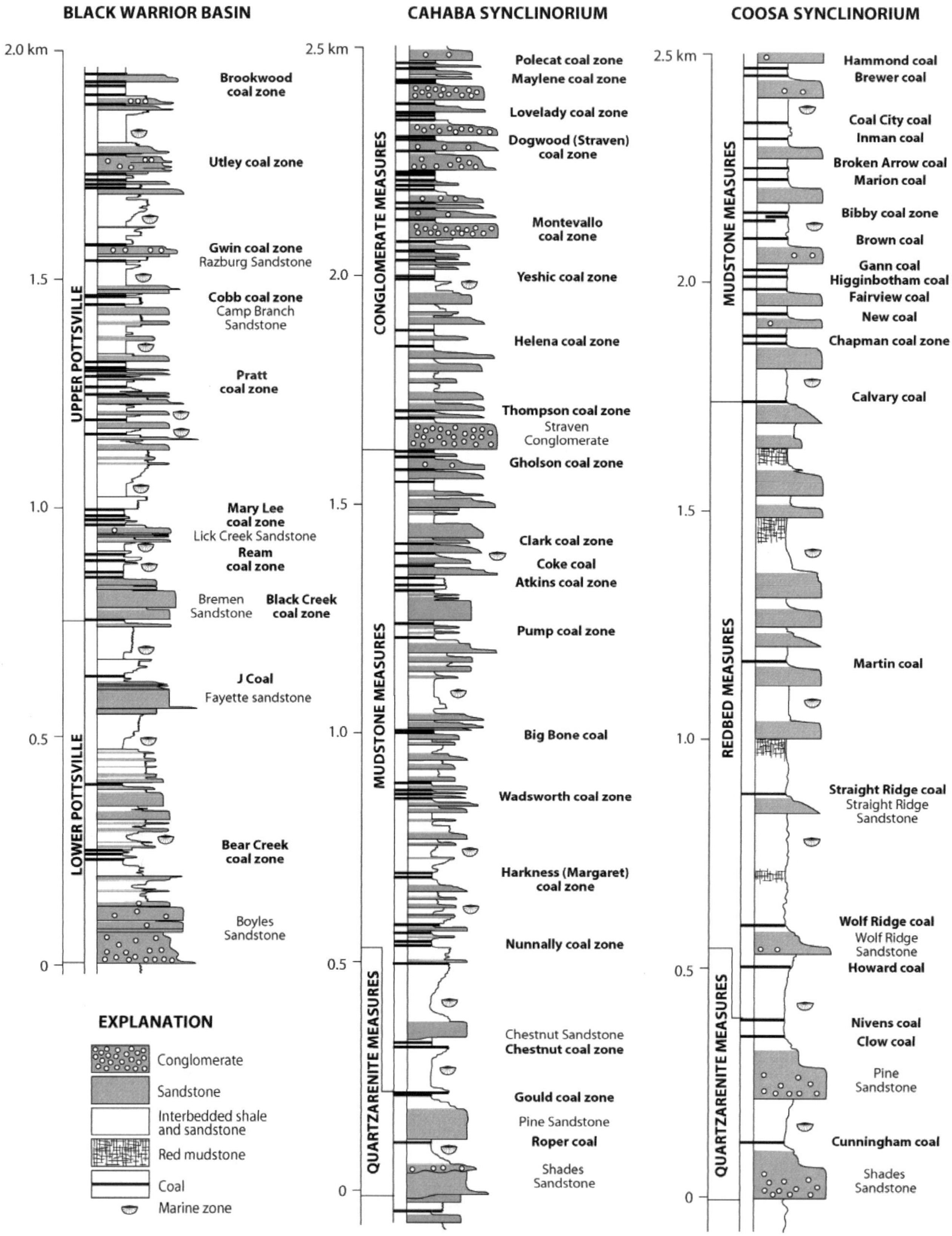

Figure 2.—Stratigraphic sections of the Pottsville Formation in the Black Warrior basin, Cahaba synclinorium, and Coosa synclinorium.

The Pottsville section in the Cahaba synclinorium is subdivided into the quartzarenite measures, mudstone measures, and conglomerate measures (fig. 2), and coalbed methane is produced mainly from the mudstone and conglomerate measures [16]. The mudstone measures resemble the fluvial-deltaic parasequences in the Black Warrior basin, but the frequency of marine strata decreases upward in section. The conglomerate measures contain only one significant marine zone and are interpreted to contain mainly bedload-dominated and anastomosing fluvial deposits. Although changing base level in response to glacial eustasy may have affected deposition of the mudstone measures, it is unclear whether facies patterns in the conglomerate measures were sensitive primarily to eustatic events, tectonic events, or a combination thereof.

In the Coosa synclinorium, Pottsville strata are subdivided into the quartzarenite measures, the redbed measures, and the mudstone measures [17] (fig. 2). Few coal beds have been identified in the quartzarenite and redbed measures, so the coalbed methane potential of this part of the section may be limited. Strata in the quartzarenite and redbed measures of the Coosa synclinorium resemble strata in the Black Warrior basin and the lower part of the Cahaba section, although the redbed measures contains a red mudstone facies that is unique in the Lower Pennsylvanian of the eastern U.S. and has been interpreted as lateritic paleosols [17]. The mudstone measures in the Coosa synclinorium contain marine through terrestrial facies similar to those in the mudstone measures of the Cahaba synclinorium, and the abundant, closely spaced coal beds in this part of the section constitute the most attractive coalbed methane prospects in the Coosa synclinorium [11, 13].

STRUCTURAL GEOLOGY

The southern Appalachian thrust belt (fig. 1) comprises thin-skinned compressional structures that formed during the Carboniferous-Permian Alleghanian orogeny and were transported northwestward above a master detachment in Cambrian shale [e.g., 18, 19]. The Cambrian shale is a ductile layer that in places is deformed into thick, blind duplexes that form the cores of large anticlines, such as the Birmingham anticlinorium [20]. Above the ductile shale, a thick Cambrian-Devonian carbonate section forms a stiff layer that is deformed into a series of broad anticlines and synclines. The Carboniferous section, which includes the Pottsville Formation, contains a thick succession of synorogenic siliciclastic rocks that forms an upper weak layer. Consequently, multiple upper-level or secondary detachments are developed in the synorogenic stratigraphy [3, 19].

The Pottsville Formation is exposed at the surface throughout the eastern Black Warrior basin and is exposed in broad, northeast-elongate synclines in the Cahaba and Coosa synclinoria (fig. 1). The Pottsville Formation contains numerous folds and thrust faults throughout the study area, and the folds tend to be broad, open structures that verge northwest and southeast and have forelimbs that dip more steeply than the backlimbs. Within this general framework, the geometry of the folds is highly variable.

In the Black Warrior basin, the Sequatchie anticline forms the frontal structure of the Appalachian thrust belt (figs. 1, 3). The anticline in this area is an extremely subtle, open structure with about 120 m (400 ft) of structural relief and an interlimb angle of 178° [3, 21]. The anticline plunges southwest and terminates at a swarm of normal faults that forms a transtensional tear fault system. The anticline shares a common limb with the Coalburg syncline, which closely follows the southeastern margin of the Black Warrior basin. Strata dip most steeply in a crestal uplift near the axial trace of the Sequatchie anticline, and dip decreases noticeably along a hinge zone northwest of the Oak Grove Mine. Interestingly, a northeast-striking synclinal flexure is developed immediately northwest of the Oak Grove Mine at the southwestern end of the hinge zone.

Strata along the southeastern edge of the Black Warrior basin have been folded to form a steeply dipping basin margin in the forelimb of the Birmingham anticlinorium (figs. 1, 4). Thus, the Coalburg syncline is a footwall syncline that defines the southeastern margin of the Black Warrior basin. Structural style along strike of the Birmingham anticlinorium varies considerably [19, 20] (fig. 1). Pre-Pottsville strata contain an array of forward thrusts and backthrusts, and Pottsville strata exhibit a series of corresponding changes in structural geometry. For example, steeply dipping Pottsville strata in the southeast limb of the Coalburg syncline typically form the edge of the Black Warrior basin, but locally these strata have been folded into an anticline-syncline pair (Blue Creek anticline and syncline) (figs. 1, 4). Strata within the Black Warrior basin tend to be nearly flat lying, whereas the steep dip in the southeast limb of the Coalburg syncline brings reservoir coal beds to the surface (fig. 4), which has a significant impact on the hydrogeology and production characteristics of the coalbed methane reservoirs [4, 22, 23].

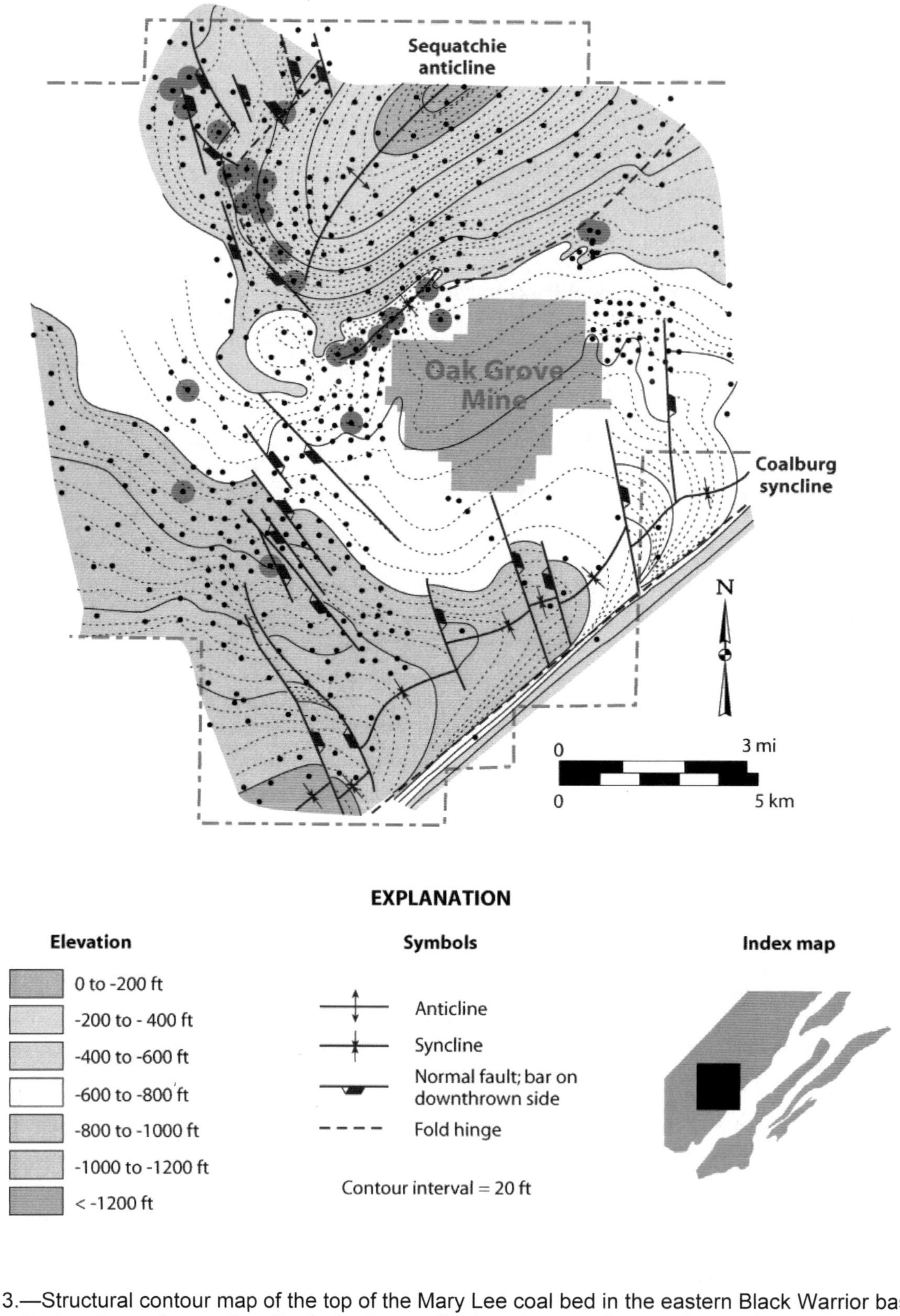

Figure 3.—Structural contour map of the top of the Mary Lee coal bed in the eastern Black Warrior basin showing the locations of exceptionally productive coalbed methane wells near the southwest terminus of the Sequatchie anticline (modified from [4]).

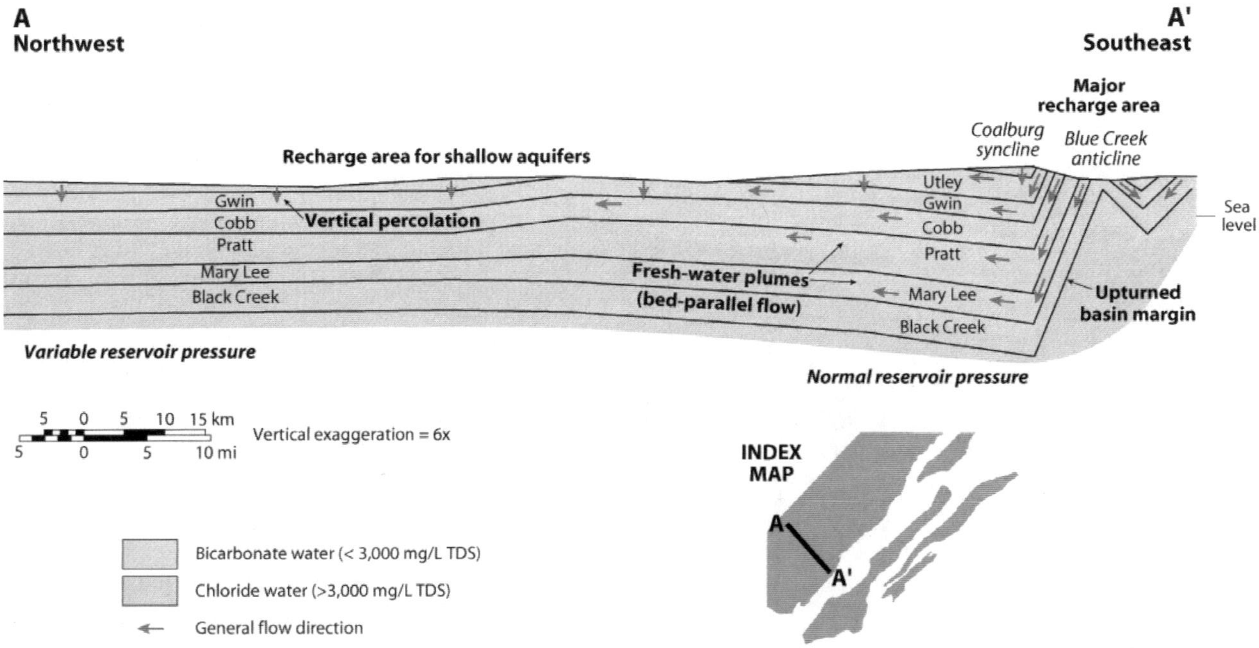

Figure 4.—Generalized structural cross section showing relationship of formation-water chemistry to the frontal structures of the Appalachian thrust belt (modified from [22]).

The Cahaba synclinorium is a northeast-elongate structure that is bounded on the northwest by the Birmingham anticlinorium and on the southeast by the Helena thrust fault (figs. 1, 5). Pottsville strata generally dip southeast at less than 20°and contain a series of open folds, including the Tacoa and Woods Creek anticlines (fig. 5). Folds and thrust faults in the Pottsville section are generally developed above upper-level or secondary detachments within the synorogenic stratigraphy [16]. In the area of coalbed methane development, the Tacoa anticline has a broad forelimb and backlimb, as well as a crestal uplift that verges southeast opposite the regional structural trend (fig. 5, cross-section B-B'). The overall structure appears to be detached below the Pottsville Formation, and the crestal uplift appears to be detached within the Pottsville. The Woods Creek anticline, by contrast, contains a thrust fault that is exposed in the crestal region and is developed above a detachment within the Pottsville Formation that may be a splay of the Helena thrust fault (fig. 5; cross-section C-C').

As in the Black Warrior basin, steeply dipping strata are commonly developed in the southeast limb of a footwall syncline that is associated with the Helena thrust fault (figs. 1, 5). However, footwall synclines in the Cahaba synclinorium tend to be discontinuous, and some wells have penetrated a thick Pottsville section below the Helena fault [24]. In general, the thickest steeply dipping sections are adjacent to embayments of the Helena thrust, and a prominent example is in the southwestern part of the synclinorium where a nearly complete Pottsville section has been overturned in the southeast limb of the Aldrich syncline.

Overall, the structural style of the Coosa synclinorium resembles that of the Cahaba synclinorium, although the Coosa section tends to dip more steeply and be more intensely folded and faulted [25, 26]. A series of structural cross sections have recently been published by Bearce [26] and Osborne [13]. Coalbed methane development in the northeastern part of the synclinorium is in a footwall syncline that has been overridden by the Eden fault, which juxtaposes intensely deformed Mississippian shale on top of the Pottsville Formation. Pottsville strata dip gently adjacent to the Eden fault [25], although steeply dipping strata have been preserved adjacent to other thrust faults in synclines that have yet to be explored.

Much of the literature on coalbed methane reservoirs focuses on orthogonal cleat system, but coal is a weak rock type that is sensitive to shear stress associated with flexural slip during folding and can thus contain a diverse array of shear fractures in addition to cleats [27, 28]. In strata that are nearly horizontal, such as in the backlimb of the Sequatchie anticline, a northeast-trending face-cleat orientation is

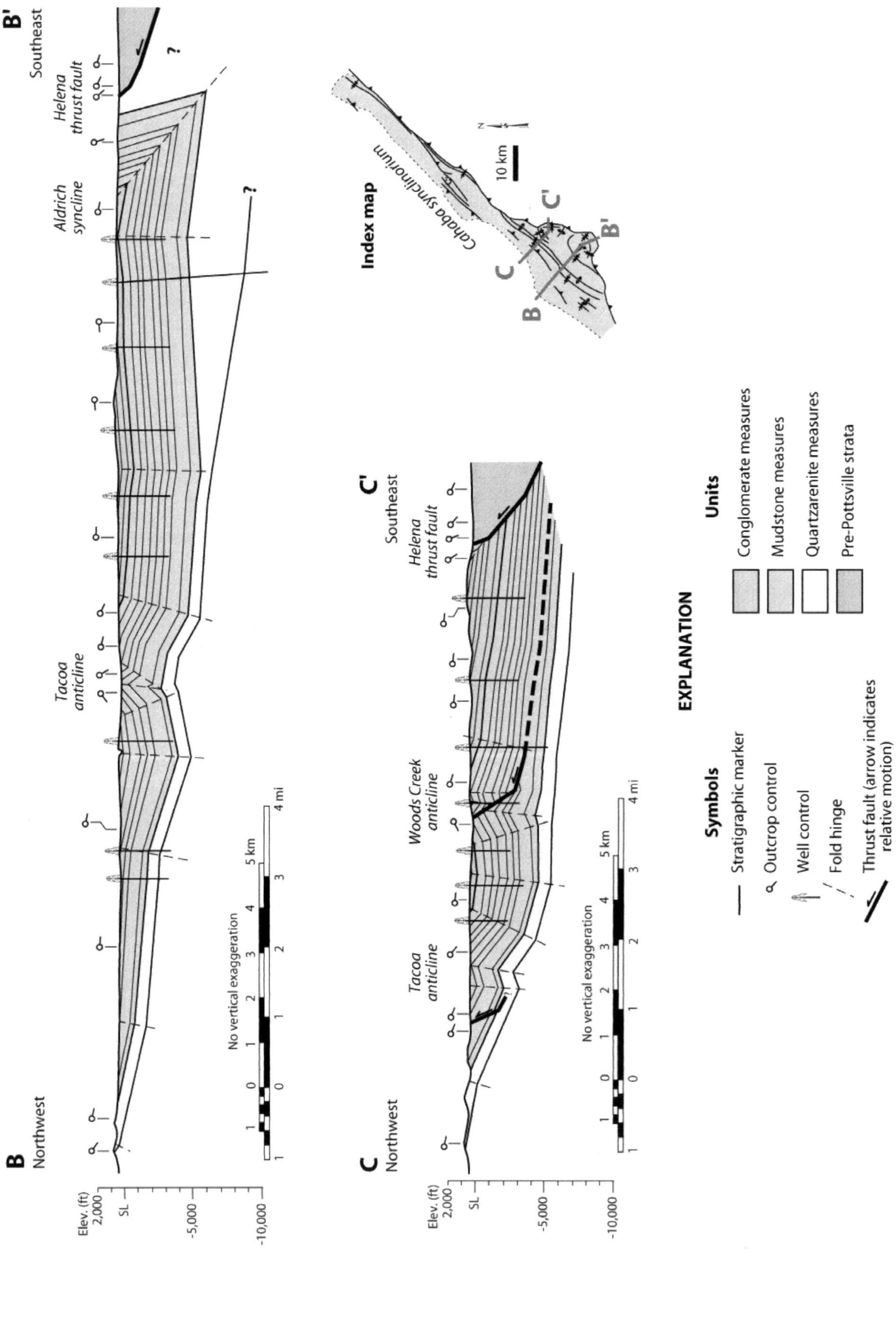

Figure 5.—Structural cross sections of the Pottsville Formation in the Coosa synclinorium (modified from [16]).

consistent with regional cleat patterns that extend far beyond the thrust belt (fig. 6). In fold limbs dipping steeper than 10°, such as in the northwest flank of the Cahaba synclinorium, face cleats tend to strike northwest, or perpendicular to fold axes, indicating that stresses associated with folding determined cleat orientation. Shear fractures are curvilinear and tend to be oriented along the strike of bedding. The shear fractures can be extensional or compressional structures, and geometries range from small-displacement normal faults, thrust faults, and conjugate fractures to mélange-like shear zones. Interestingly, one outcrop of a coal bed can contain mainly compressional fractures, whereas a nearby outcrop of the same coal can contain mainly extensional fractures. This suggests significant strain partitioning within fold limbs during deformation, and Pashin and others [28] suggested that irregularities in the contacts between coal and more competent beds can function as restraining and releasing fault bends during flexural slip (fig. 7).

HYDROGEOLOGY

Strata associated with coalbed methane production in the Alabama Pottsville have effectively no matrix permeability to water, so closely spaced cleats and shear fractures make coal the most permeable rock type, and hence, the best aquifer and reservoir rock in the coalbed methane fields [16, 28]. Well testing in the backlimb of the Sequatchie anticline indicates that permeability in coal can have a varied expression in thrust belts [29, 30]. The permeability of coal in this area is highly stress-sensitive and decreases exponentially from as much as 1 darcy at the surface to about 1 millidarcy at depths exceeding 700 m [29]. In shallow, well-cleated coal beds, pressure-buildup testing indicates that permeability is strongly anisotropic and that flow is favored in the face-cleat direction [30]. In deep, low-permeability coal beds, however, permeability is weakly anisotropic, and flow is favored in the systematic joint direction, indicating that flow is difficult to isolate in tight coal deeper than 700 m. The Blue Creek coal of the Mary Lee coal zone (fig. 1) is sheared within the backlimb of the anticline, and well test results indicate that permeability is anisotropic in this bed.

These results indicate that significant permeability exists in sheared coal in thrust belts. The maximum horizontal compressive stress in the Black Warrior basin is directed east-northeast [31], which favors flow in the northeast-striking cleats and shear fractures in the southern Appalachian thrust belt. Importantly, cleats and extensional shear fractures were probably conductive of fluids during deformation, but compressional fractures may have been closed and functioned mainly as slip planes during folding. Thus, compressional shear fractures may not be sources of permeability in thrust belts where the stress field has not been reoriented following orogenic deformation.

At a larger scale, fold limbs can play an important role in fluid flow, the development of the pressure regime, and late-stage gas generation [2, 23, 32]. In the Black Warrior basin, reservoir coal beds are exposed at the surface in the southeast limb of the Coalburg syncline where they can accept meteoric recharge (figs. 4, 8). Sodium bicarbonate formation waters with total dissolved solids (TDS) content less than 3,000 mg/L occur at reservoir depth adjacent to the southeast basin margin, whereas sodium chloride waters with higher TDS content occur at reservoir depth in the interior of the basin, indicating that recharge is dominated by the southeast limb of the Coalburg syncline. Mapping TDS content of the Mary Lee coal zone defines a series of fresh-water plumes that extend northwestward from the basin margin (fig. 8). Importantly, normal faults strike at a high angle to the southeast basin margin and thus do not block recharge. Recharge supports a normal hydrostatic pressure regime as far as 20 km northwest of the basin margin, whereas reservoirs tend to be underpressured farther northwest [23]. Fresh to weakly saline water at reservoir depth is thought to host bacterial consortia that facilitate late-stage bacterial methanogenesis, and a combination of thermogenic and late-stage biogenic gases helps explain the geochemistry and high gas content of the Pottsville Formation in the eastern Black Warrior basin [32].

Whereas the hydrologic system in the Black Warrior basin is dominated by steeply dipping strata at the southeast basin margin, the structure of the Cahaba synclinorium indicates that the hydrologic system in the interior parts of thrust belts can be more complex and include multiple recharge areas (fig. 5). For example, steeply dipping strata in footwall synclines, such as the southeast limb of the Aldrich syncline, may facilitate recharge similar to that observed in the Black Warrior basin (fig. 5; cross-section B-B'). But in other areas, footwall synclines are not present or are sheltered by the Helena thrust fault (fig. 5; cross-section C-C'), indicating that meteoric recharge along the southeast margin of the Cahaba synclinorium is discontinuous. In addition, dipping strata on the northwest margin of the Cahaba synclinorium may also facilitate meteoric recharge, and additional sources of recharge probably exist on fold limbs within the synclinorium. Although reservoir pressure and water chemistry data are few in the Cahaba synclinorium,

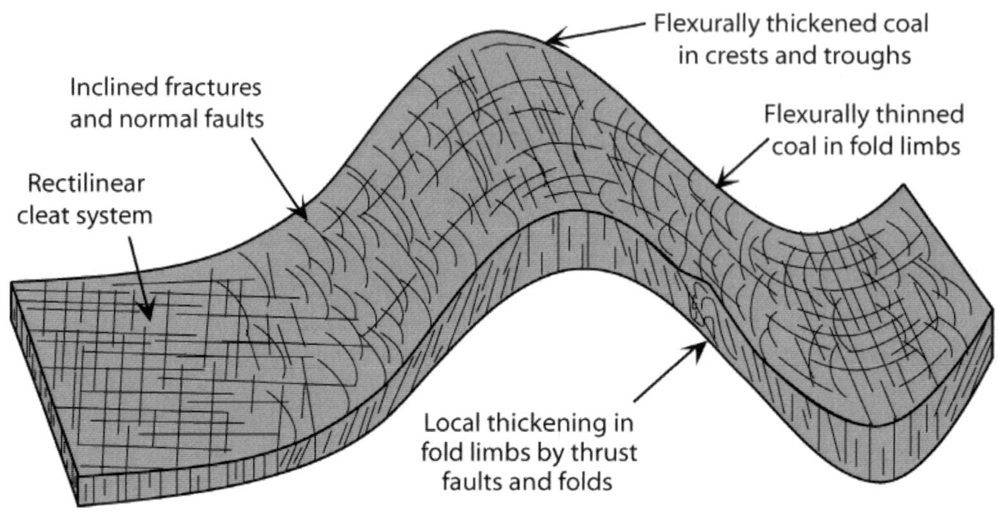

Figure 6.—Generalized model of fracturing in folded coal beds (modified from [16]).

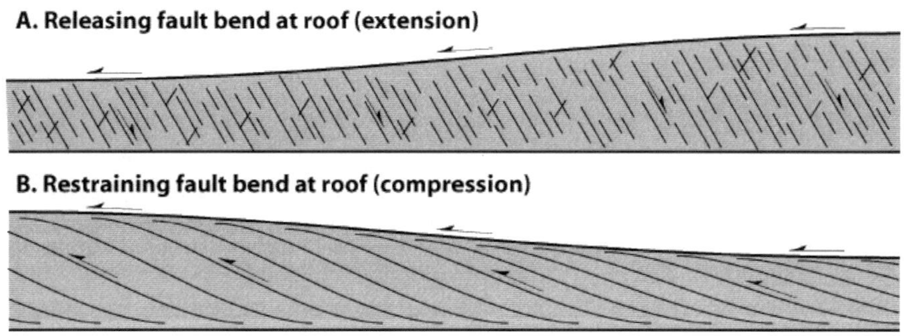

Figure 7.—Relationship of extensional and compressional shear structures in coal to irregularities of roof geometry (modified from [28]).

multiple sources of recharge may introduce fresh water into deep coalbed methane reservoirs, and considering the small size of the synclinorium relative to the Black Warrior basin, recharge can be predicted to support a normal hydrostatic pressure regime over large areas.

PRODUCTION TRENDS

A series of distinctive production trends can be related to structure in the southern Appalachian thrust belt. Exceptionally productive wells with peak production exceeding 300 Mcfd form linear trends associated with the Sequatchie anticline (fig. 3). Five exceptionally productive wells are distributed along the small syncline immediately northwest of the Oak Grove Mine, and this trend was originally identified by Pashin and others [4]. Another linear trend of exceptionally productive wells extends northwest from the southern terminus of the Sequatchie anticline into the en echelon tear fault system. Most of the wells in these two trends are aligned along fold hinges. Increased well productivity in fold hinges suggests that flexure has enhanced fracturing, thus increasing permeability and making the reservoir easier to depressurize. Increased water production has been identified locally along the axis of the Coalburg syncline [3, 33]. Importantly, the Coalburg syncline is proximal to the main area of meteoric recharge, whereas the Sequatchie anticline is distal to the main recharge area. One possibility is that fracturing in

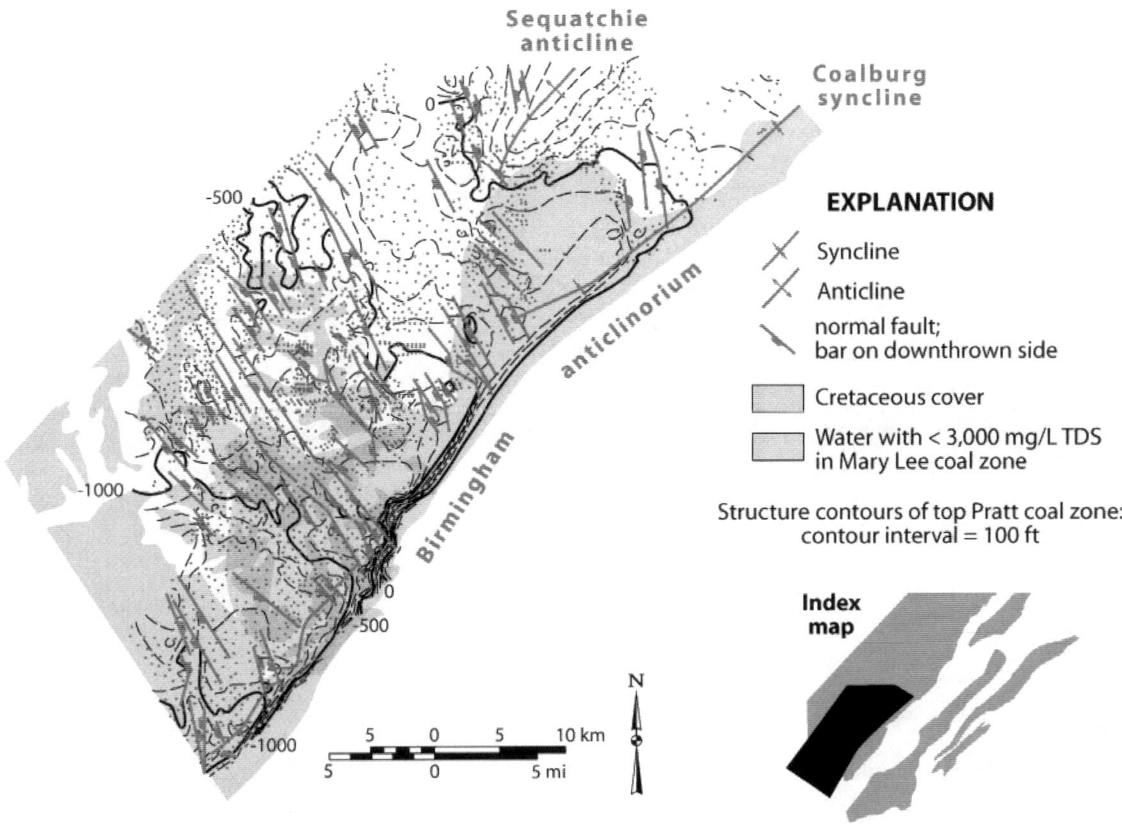

Figure 8.—Map showing position of fresh-water plumes relative to the Appalachian frontal structures in Alabama (modified from [23]).

fold hinges near recharge areas result in increased water production, whereas hinges that are removed from active recharge areas show better promise for the development of gas-productive sweet spots.

In the Cahaba synclinorium, Pashin and Groshong [3] observed that production rates were locally higher on folds like the Woods Creek anticline, suggesting locally increased permeability in folds. However, wells with the highest production rates are located in the gently dipping strata between the Tacoa anticline and the Helena thrust. Productivity appears to be influenced by tectonic stress [34], and an explanation for increased productivity in weakly deformed strata is that residual stresses are reduced in weakly deformed strata between the major folds.

In the Cedar Cove area of the Black Warrior basin, strong linear production anomalies are developed adjacent to the Appalachian frontal structures [35, 36] (fig. 9). A structural contour map of the top of the Gwin coal zone shows a strong fold hinge that trends northeast along the basin margin and is broken by a series of north- to northwest-striking normal faults (fig. 9A). Mapping peak gas and water production reveals a series of linear production trends that are oblique to some normal faults and are parallel to others (figs. 9B-9E). Cates et al. [36] interpreted the linear trends as the products of conjugate shear zones that developed in response to compressional stresses associated with Appalachian folding (fig. 9F). Parallelism of some of the linear trends to normal faults may also indicate that the orientation of the normal faults was also influenced by conjugate shear stress in the Cedar Cove area. However, the precise structural expression of the conjugate shear zones remains unclear.

Ten coalbed methane wells were drilled in the Coosa synclinorium, but to date, no commercial production has been established. Gas content estimates based on desorption of well cuttings indicate that the coal contains between 6.7 and 20.9 cc/g (214-670 scf/t) of gas [11, 13]. Although core analysis is more reliable than cuttings analysis, these numbers indicate that commercial production potential exists.

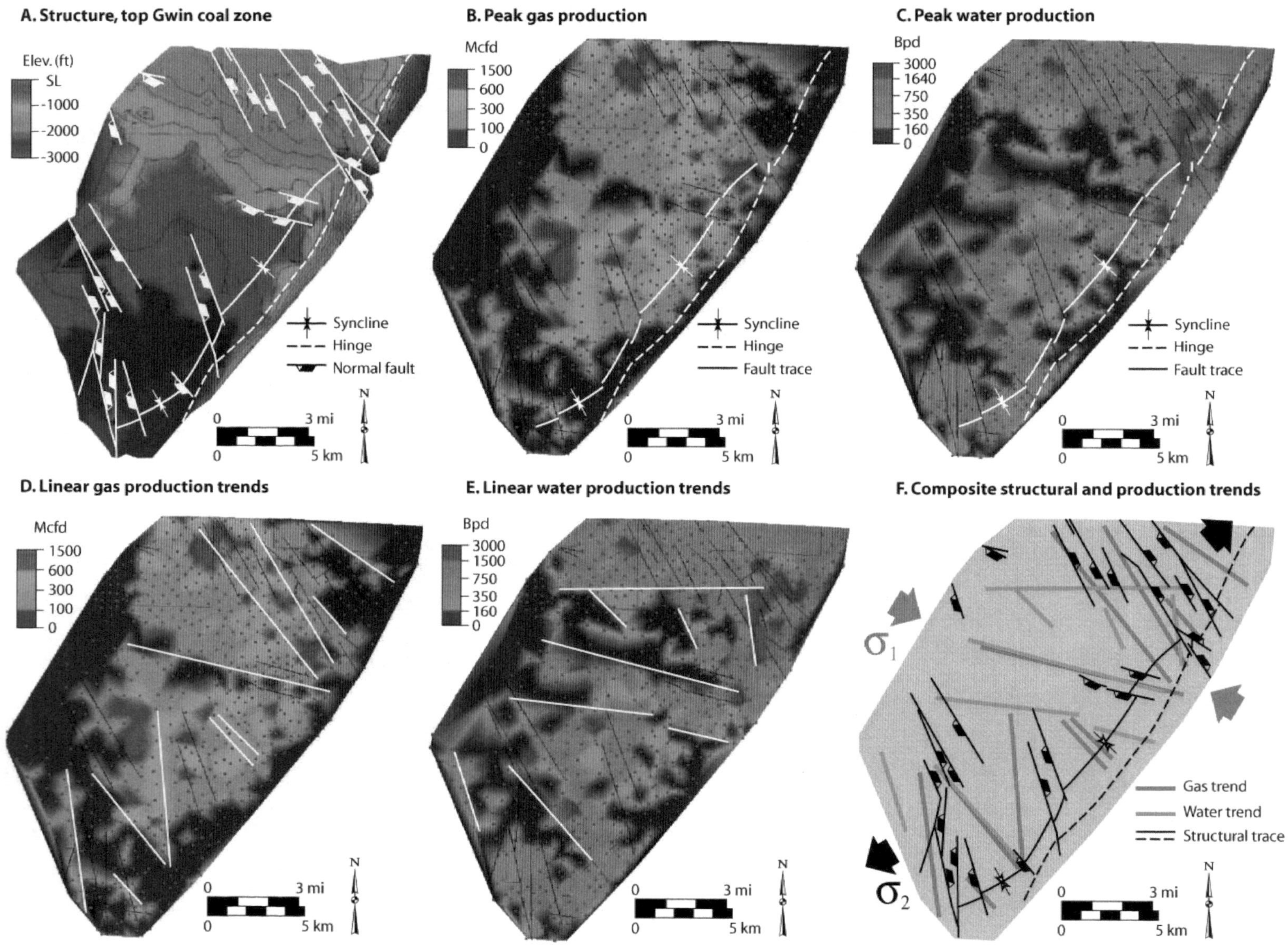

Figure 9.—Maps showing structure and linear gas and water production trends in the Cedar Cove area of the Black Warrior basin (modified from [36]). Linear trends of gas and water production are interpreted to mark the positions of conjugate shear zones adjacent to the southern Appalachian thrust belt.

With the sustained high price of natural gas, reevaluation of the coalbed methane potential of the Coosa synclinorium would be timely.

SUMMARY AND CONCLUSIONS

Numerous coalbed methane wells have been drilled in the southern Appalachian thrust belt of Alabama, and geologic and fluid production data from this area provides important insight into coalbed methane exploration strategies for thrust belts. In the southern Appalachian coalbed methane fields, reservoir performance reflects a complex interplay among stratigraphic, structural, and hydrologic variables. Productive coal beds are of Lower Pennsylvanian age, are distributed through more than 1,000 m of section, and are of high volatile A bituminous to low volatile bituminous rank. Completed coal beds range in thickness from 0.3 to 4 m.

Thrust-related folds in the southern Appalachian coalbed methane fields typically have steep forelimbs, gently dipping backlimbs, and contain multiple hinge zones. Meteoric recharge is most effective in steep fold limbs, and fresh-water plumes extend from the outcrop to reservoir depth in thick, continuous coal beds. Late-stage bacterial methanogenesis within the fresh-water plumes facilitates high gas saturation in coal, and the fresh-water plumes help maintain a normal hydrostatic pressure regime in parts of basins proximal to recharge areas.

Economic coalbed methane production is proven in the Black Warrior basin and the Cahaba synclinorium. Production has yet to be established in the Coosa synclinorium, but limited gas content data indicate that the economic potential of coal in the Coosa synclinorium should be reevaluated. Linear productivity sweet spots for gas and water are associated with geologic structures in the southern Appalachian thrust belt. Fold hinges that are proximal to fresh-water recharge areas tend to be zones of exceptional water production, whereas those that are distal to recharge areas host sweet spots with exceptional gas production. In weakly deformed strata between major fold hinges, linear gas- and water-production sweet spots appear to mark the positions of vertical conjugate shear zones. However, the precise geologic expression of the shear zones remains unclear.

REFERENCES CITED

[1] Ayers, W. B., Jr., and Kaiser, W. A. (eds.), 1994, Coalbed methane in the Upper Cretaceous Fruitland Formation, San Juan Basin, Colorado and New Mexico: Texas Bureau of Economic Geology Report of Investigations 218, 216 p.

[2] Scott, A. R., Kaiser, W. R., and Ayers, W. B., Jr., 1994, Thermogenic and secondary biogenic gases, San Juan basin, Colorado and New Mexico—Implications for coalbed gas producibility: American Association of Petroleum Geologists Bulletin, v. 78, p. 1186-1209.

[3] Pashin, J. C., and Groshong, R. H., Jr., 1998, Structural control of coalbed methane production in Alabama: International Journal of Coal Geology, v. 38, p. 89-113.

[4] Pashin, J. C., Ward, W. E., II, Winston, R. B., Chandler, R. V., Bolin, D. E., Richter, K. E., Osborne, W. E., and Sarnecki, J. C., 1991, Regional analysis of the Black Creek-Cobb coalbed-methane target interval, Black Warrior basin, Alabama: Alabama Geological Survey Bulletin 145, 127 p.

[5] Flores, R. M., 1993, Coal-bed and related depositional environments in methane gas-producing sequences: AAPG Studies in Geology, v. 38, p. 13-37.

[6] Ayers, W. B., Jr., 2002, Coalbed gas systems, resources, and production and a review of contrasting cases from the San Juan and Powder River Basins: AAPG Bulletin, v. 86, p. 1853-1890.

[7] Elder, C. H., and Deul, Maurice, 1974, Degasification of the Mary Lee coalbed near Oak Grove, Jefferson County, Alabama, by vertical borehole in advance of mining: U.S. Bureau of Mines Report of Investigations 7968, 21 p.

[8] Murrie, G. W., Diamond, W. P., and Lambert, S. W., 1976, Geology of the Mary Lee group of coalbeds, Black Warrior coal basin, Alabama: U.S. Bureau of Mines Report of Investigations 8189, 49 p.

[9] Telle, W. R., and Thompson, D. A., 1987, Preliminary characterization of the coalbed methane potential of the Cahaba coal field, central Alabama: Tuscaloosa, Alabama, University of Alabama, 1987 Coalbed Methane Symposium Proceedings, p. 141-151.

[10] Telle, W. R., and Thompson, D. A., 1988, The Cahaba coal field: a potential coalbed methane development area in central Alabama: Tuscaloosa, Alabama, University of Alabama, School of Mines and Energy Development Research Report 88-1, 174 p.
[11] Osborne, T. E., Moore, D. K., Kidd, J. T., and Pescatore, F. T., Jr., 1991, Coalbed methane potential of the northern Coosa basin in Alabama: Journal of Coal Quality, v. 10, p. 95-103.
[12] Hatch, J. R., Pawlewicz, M. J., Charpentier, R. E., Cook, T. A., Crovelli, R. A., Klett, T. R., Pollastro, R. M., and Schenk, C. J., 2003, Assessment of undiscovered oil and gas resources of the Black Warrior Basin province, 2002; U.S. Geological Survey Fact Sheet FS-038-03, 2 p.
[13] Osborne, T. E., Stratigraphy and structure of sections 1 and 12, T. 16 S., R. 3 E., of the Coal City basin within the Coosa synclinorium of Alabama, *in* Bearce, D. N., Pashin, J. C., and Osborne, W. E., eds., Geology of the Coosa Coalfield: Alabama Geological Society 34th Annual Field Trip Guidebook, p. 39-49.
[14] Pashin, J. C., 2004, Cyclothems of the Black Warrior basin in Alabama: eustatic snapshots of foreland basin tectonism: AAPG Studies in Geology 51, p. 99-217.
[15] Pashin, J. C., and Raymond, D. E., 2004, Glacial-eustatic control of coalbed methane reservoir distribution (Pottsville Formation; Lower Pennsylvanian) in the Black Warrior basin of Alabama: Tuscaloosa, Alabama, University of Alabama College of Continuing Studies, 2004 International Coalbed Methane Symposium Proceedings, Paper 0413, 15 p.
[16] Pashin, J. C., and Hinkle, Frank, 1997, Coalbed methane in Alabama: Alabama Geological Survey Circular 192, 71 p.
[16] Pashin, J. C., Carroll, R. E., Barnett, R. L., and Beg, M. A., 1995, Geology and coal resources of the Cahaba coal field: Alabama Geological Survey Bulletin 163, 49 p.
[17] Pashin, J. C., 1997, Stratigraphy and depositional environments of the Pottsville Formation (Lower Pennsylvanian) in the Coosa coalfield, *in* Bearce, D. N., Pashin, J. C., and Osborne, W. E., eds., Geology of the Coosa Coalfield: Alabama Geological Society 34th Annual Field Trip Guidebook, p. 19-28.
[18] Rodgers, John, 1950, Mechanics of Appalachian folding as illustrated by the Sequatchie anticline, Tennessee and Alabama: American Association of Petroleum Geologists Bulletin, v. 34, p. 672-681.
[19] Thomas, W. A., 1985, Northern Alabama sections, *in* Woodward, N. B., ed., Valley and Ridge thrust belt: balanced structural sections, Pennsylvania to Alabama: University of Tennessee Department of Geological Sciences Studies in Geology 12, p. 54-60.
[20] Thomas, W. A., 2001, Mushwad: Ductile duplex in the Appalachian thrust belt of Alabama: AAPG Bulletin, v. 85, p. 1847-1869.
[21] Pashin, J. C., 1994, Coal-body geometry and synsedimentary detachment folding in Oak Grove coalbed-methane field, Black Warrior basin, Alabama: AAPG Bulletin, v. 78, p. 960-980.
[22] Pashin, J. C., Carroll, R. E., Groshong, R. H., Jr., Raymond, D. E., McIntyre, M. R., and Payton, J. W., 2004, Geologic screening criteria for sequestration of CO_2 in coal: Quantifying potential of the Black Warrior coalbed methane fairway, Alabama: Final Technical Report, U.S. Department of Energy, contract DE-FC26-00NT40927, 254 p.
[23] Pashin, J. C., and McIntyre, M. R., 2003, Temperature-pressure conditions in coalbed methane reservoirs of the Black Warrior basin, Alabama, U.S.A: implications for carbon sequestration and enhanced coalbed methane recovery: International Journal of Coal Geology, v. 54, p. 167-183.
[24] Raymond, D. E., 1991, New subsurface information on Paleozoic stratigraphy of the Alabama fold and thrust belt and the Black Warrior basin: Alabama Geological Survey Bulletin 143, 185 p.
[25] Rothrock, H. E., 1949, Geology and coal resources of the northeast part of the Coosa coal field, St. Clair County, Alabama: Alabama Geological Survey, Bulletin 61, 163 p.
[26] Bearce, D. N., 1997, Geology of the northeastern part of the Coosa coalfield in the Coosa synclinorium, St. Clair County, Alabama, *in* Bearce, D. N., Pashin, J. C., and Osborne, W. E., eds., Geology of the Coosa Coalfield: Alabama Geological Society 34th Annual Field Trip Guidebook, p. 1-17.
[27] Hathaway, T. M., and Gayer, R. A., 1996, Thrust-related permeability in the South Wales coalfield: Geological Society of London Special Publication 109, p. 121-132.
[28] Pashin, J. C., Carroll, R. E., Hatch, J. R., and Goldhaber, M. B., 1999, Mechanical and thermal control of cleating and shearing in coal: examples from the Alabama coalbed methane fields, USA, *in* Mastalerz, M., Glikson, M., and Golding, S., eds., Coalbed Methane: Scientific, Environmental and Economic Evaluation: Dordrecht, Netherlands, Kluwer Academic Publishers, p. 305-327.

[29] McKee, C. R., Bumb, A. C., and Koenig, R. A., 1988, Stress-dependent permeability and porosity of coal and other geologic formations: Society of Petroleum Engineers Formation Evaluation, March 1988, p. 81-91.

[30] Koenig, R. A., 1989, Hydrologic characterization of coal seams for optimal dewatering and methane drainage: Quarterly Review of Methane from Coal Seams Technology, v. 7, p. 30-31.

[31] Zoback, M. L., and Zoback, M. D., 1989, Tectonic stress field of the continental U.S, in Pakiser, L., and Mooney, W., eds., Geophysical framework of the continental United States: Geological Society of America Memoir 172, p. 523-539.

[32] Pitman, J. K., Pashin, J. C., Hatch, J. R., and Goldhaber, M. B., 2003, Origin of minerals in joint and cleat systems of the Pottsville Formation, Black Warrior basin, Alabama: implications for coalbed methane generation and production: American Association of Petroleum Geologists Bulletin, v. 87, p. 713-731.

[33] Pashin, J. C., Groshong, R. H., Jr., and Wang, Saiwei, 1995, Thin-skinned structures influence gas production in Alabama coalbed methane fields: Tuscaloosa, Alabama, University of Alabama, InterGas '95 Proceedings, p. 39-52.

[34] Sparks, D. P., Lambert, S. W., and McLendon, T. H., 1993, Coalbed gas well flow performance controls, Cedar Cove area, Warrior basin, U.S.A.: Tuscaloosa, Alabama, University of Alabama, 1993 International Coalbed Methane Symposium Proceedings, p. 529-548.

[35] McIntyre, M. R., Groshong, R. H., Jr., Pashin, J. C., and Yin, H., 2003, Structure of Cedar Cove and Peterson coalbed methane fields and correlation to gas and water production: Tuscaloosa, Alabama, University of Alabama College of Continuing Studies, 2003 International Coalbed Methane Symposium Proceedings, Paper 0312, 14 p.

[36] Cates, L. M., McIntyre, M. R., Hawkins, W. B., and Groshong, R. H., Jr., 2004, Structure and oil and gas production in the Black Warrior basin: Tuscaloosa, Alabama, University of Alabama College of Continuing Studies, 2004 International Coalbed Methane Symposium Proceedings, Paper 0440, 34 p.

[37] Osborne, W. E., Szabo, M. W., Copeland, C. W., Jr., and Neathery, T. L., 1989, Geologic map of Alabama: Alabama Geological Survey Special Map 221, scale 1:500,000.

0526

COAL-BED GAS POTENTIAL IN PENNSYLVANIAN COAL BEDS OF INDIANA

Solano-Acosta W., Mastalerz, M., Rupp, J.A.,
Indiana Geological Survey;
Strapoc, D., and Schimmelmann, A.,
Indiana University Bloomington.

ABSTRACT

The content of coal-bed methane (CBM) gas within the Pennsylvanian coal beds of the Illinois Basin varies widely depending on a series of factors including depth and thickness of the coal bed and lithologies of adjoining strata. Along the eastern margin of the Illinois Basin in Indiana, such a variable distribution is evident both laterally and vertically. In Daviess, Gibson, and Sullivan counties, gas saturation can be generally related to the depth of coal beds, with higher gas volumes occurring at greater depths, although variations from this trend are common. In other areas, for example, western Posey County, gas volumes show erratic changes with depth, partly related to the lithologic composition of the strata overlying the coals, and specifically the proximity of porous and permeable sandstone-dominated sequences. The carbon and hydrogen stable isotopic compositions of coal-bed gases, $\delta^{13}C$ (-54 to -60 ‰) and δD (-180 to -210 ‰), from the Springfield and the Seelyville Coal Members from Gibson, Knox, and Sullivan counties indicate that gas is of mixed thermogenic and biogenic origin, and that the contribution of biogenic gas ranges from 60 to more than 90 percent of the total CBM. The combination of variable origins and different degrees of gas saturation necessitates multiple exploration strategies for CBM in the Illinois Basin.

INTRODUCTION

Original coal reserves in Indiana have been recently estimated as 59.53 billion short tons out of which, 56.68 billion short tons remain unmined (Mastalerz, Drobniak et al. 2004). 17.54 billion short tons of coal remain available for either surface or underground mining; and 39.13 billion short tons are unminable because of either technological limitations or due to land-use restrictions. Given this large volume of potentially unminable coal resources, the most continuous, thick, and widespread coal seams in the state of Indiana are currently under investigation for their potential as coal-bed methane sources.

A net CBM resource for the Illinois Basin has been estimated at 20 Tcf (trillion cubic feet of gas) by Archer and Kirr (1984) and at 21 Tcf by the Gas Research Institute (1999). The gas potential of the Seelyville Coal Member, Linton Formation (Pennsylvanian) in Indiana was estimated to be 1.1 Tcf (Drobniak, Mastalerz et al. 2004). Not being mined because of high ash yield and high sulfur content

(Mastalerz and Harper 1998), the Seelyville Coal is one of the major CBM targets in Indiana because of its depth and thickness. Gas has been commercially produced from the Seelyville Coal in Sullivan County, Indiana, for several years.

Other coal beds have not yet been evaluated within the state for their gas potentials. The Danville and Springfield coals have been the primary targets for mining in Indiana, because of their thickness and quality, but neither of them have yet been evaluated and economically exploited for coal-bed methane. Other thinner coal beds, such as the Houchin Creek, Survant, Colchester, or those in the Brazil Formation, complementing the thicker coal beds, might also contain significant CBM potential.

In order to understand coalbed gas distribution in Indiana, we are evaluating existing data from the perspective of coalbed methane exploration. This paper combines measured canister desorption CBM data from drilling operations, archived data sets, and laboratory analyses as tools to quantify the occurrence of CBM in this portion of the Illinois Basin.

PENNSYLVANIAN COAL STRATIGRAPHY

At the end of the Mississippian Period, major erosion and uplift rendered an unconformable surface upon which Pennsylvanian sediments were laid down onto an inclined surface that received sediments from the Canadian Shield and the Appalachian mountains (Archer and Kirr 1984). The predominantly siliciclastic sequence that characterizes the Pennsylvanian System shows evidence of various sea level fluctuations that, in addition to coal intervals, ended with the deposition of alternating sands and shales, with minor interpositions of limestone and black shales indicating open marine conditions. Subaerial exposure is also evident in areas that exhibit good paleosol development.

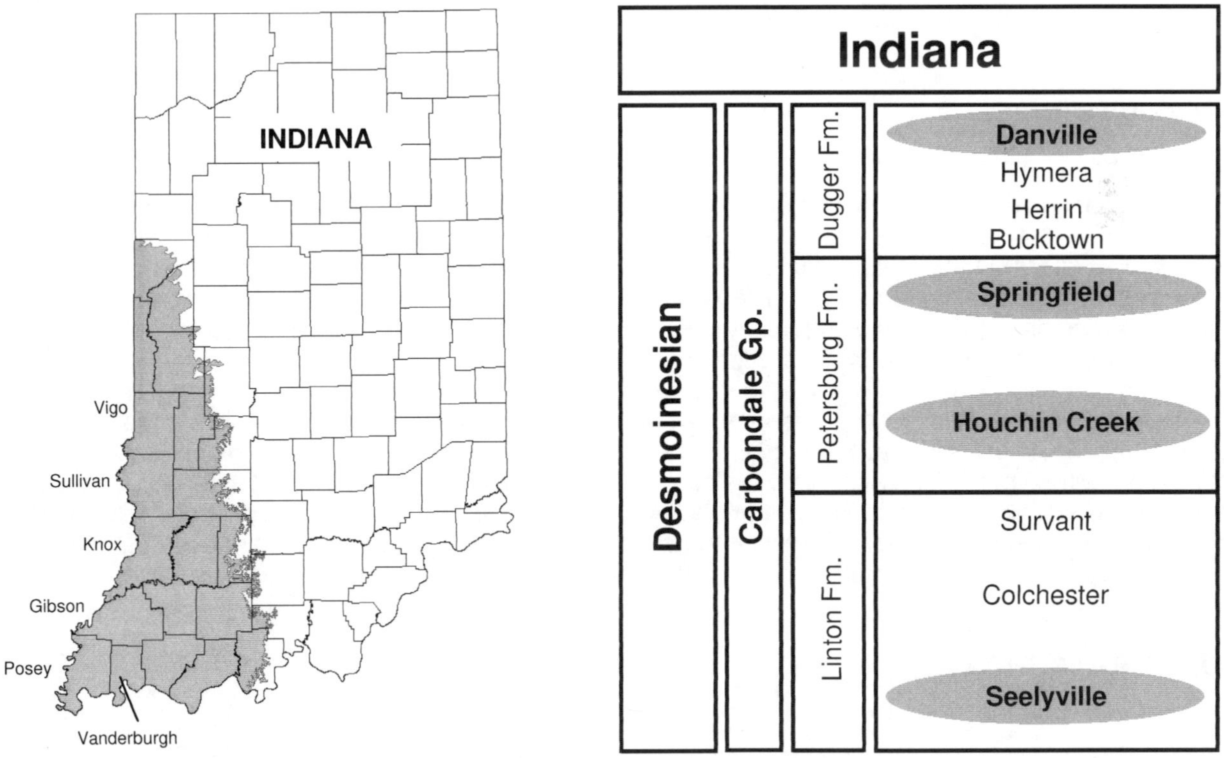

Figure 1. Geographic and stratigraphic distribution of Pennsylvanian coals of Desmoinesian age (Carbondale Group) in Indiana. Circled in gray and highligted are the coal seams evaluated in this study.

Pennsylvanian coal-bearing strata in Indiana are subdivided into three main intervals that span from the Morrowan to the Desmoinesian Epochs (Raccoon Creek, Carbondale, and McLeansboro Groups). In the lower interval (Raccoon Creek Group), there are six coal seams, some of which are considered of economic importance due to their low sulfur contents. However, none of them are thick and their extent is somewhat limited (Mastalerz and Harper 1998).

The Carbondale Group (Figure 1) contains coals that account for most of the mining operations in Indiana (Mastalerz and Harper 1998). This group is divided into three Formations (Linton, Petersburg, and Dugger). The sulfur content of Carbondale coals is greater than that of the Raccoon Creek coals. However, greater thickness, widespread distribution, and closer proximity to the surface makes them more attractive for mining. The McLeansboro Group overlies the Carbondale group and is considered a barren sequence because of the absence of thick and continuous layers of coal (Mastalerz and Harper 1998).

The formations comprising the lower, middle, and upper Desmoinesian groups have been defined based on the intervals between mappable units such as limestone, coal beds, and shales. However, there are still some correlation problems among units because of the discontinuous nature of most strata, or alternatively, due to other problems such as coal seams that may be split into various benches. Correlations across Pennsylvanian age strata in the Illinois Basin are a difficult task that requires detailed subsurface information in addition to palynological data. Estimating the continuity of coal-bearing strata in Indiana has been traditionally done by correlating marker beds from both limestones and shales.

Upper and middle Pennsylvanian strata can be consistently identified on traditional oil and gas geophysical well logs due to their repetitive patterns and therefore predictable facies succession. Lower Pennsylvanian strata are not as consistent as in the Desmoinesian and younger units and they lack thick and widespread marker beds. The identification of parasequences, as well as the use of paleosols recognized from continuous core data proved to be excellent tools in correlating intervals within the lower Pennsylvanian (Kvale, Mastalerz et al. 2004).

PREVIOUS STUDIES

Investigations on coal distribution and occurrence in western Indiana are available elsewhere (Harper 1985, 1988, 1994; Eggert 1994; Harper and Eggert 1995; Mastalerz, Ames et al. 2003; Drobniak, Mastalerz et al. 2004; Mastalerz, Drobniak et al. 2004). These works include data from surface and underground coal mining, geophysical logs from oil and gas drillings, and stratigraphic test holes. Although abundant information is available on geometry and structure of coal beds, the information on the gas content and gas distribution trends is limited.

Coalbed gas determinations have been carried out by some investigators (Harper 1991; Mastalerz and Kvale 1998; Mastalerz and Kvale 2000; Drobniak, Mastalerz et al. 2004). The gas content shows a direct correlation with depth in Daviess and Gibson counties (Mastalerz and Kvale 1998). However, such correlation is not so evident in Posey County, where an erratic distribution of gas content is observed (Mastalerz and Kvale 2000). Gas contents measured with canister desorption techniques in Indiana vary anywhere from as little as 13 to as much as 149 standard cubic feet per ton (scf/ton, raw basis) .

COAL DISTRIBUTION

Depth and thickness maps have been created for the Danville, Springfield, Houchin Creek, and Seelyville coal members of the Pennsylvanian Carbondale Group. Figures 2 – 5 show the depths at which the coal could be expected, and the average thickness values for each seam. The thickness of a coal seam is necessary information to determine the suitability of a coal seam for CBM well completions.

The depth to the Danville coal in ft below the surface is charted geographically by contour lines in Figure 2a. Most Danville coals at depths shallower than 200 ft are discarded for analysis from the CBM exploration point of view. The Danville coal seam is generally very shallow, and it is one of the most extensively mined intervals due to its proximity to the surface. The thickness map for the Danville coal (Figure 2b) shows very limited areas where this coal is thicker than 66 inches and relatively small areas where the coal is between 44 and 66 inches, particularly where this coal occurs at depths greater than 200 ft. Coal thickness distribution in this coal seam is very patchy, particularly in Gibson and Sullivan counties, where this resource has been extensively mined and abundant information is available (Figure 2b). Its fairly constant distribution in Knox, Posey, and Vanderburgh counties is more an artifact of the sample distribution, which is significantly more sparse than in the rest of the mapped area. Assessing the suitability of this seam for CBM in Indiana requires additional information including gas collection for determining its origin. Although gas contents for the Danville coal have been measured from canister desorption techniques, there are no data about the isotopic composition of the gases associated with this seam. This isotopic information is critical in defining the exploration strategy for coal-bed methane. Additional information is required to properly estimate the gas potential from this seam.

Similarly to the Danville coal seam, the depth to the Springfield coal member was mapped (Figure 3a) showing in detail the occurrence of the Galatia channel in Gibson County, Indiana. The geographic thickness distribution of this seam (Figure 3b) indicates an extensive and continuous coal deposit characterized by an overall thickness greater than 42 inches, with few areas in Knox and Sullivan counties where its thickness is characteristically below 28 inches. Canister desorption data as well as isotopic gas compositions have been recently collected making this seam ideal for understanding the occurrence of CBM and therefore allowing better predictions during exploratory phases.

The Houchin Creek coal seam was mapped for depth (Figure 4a) and thickness (Figure 4b) following the same methodology as for Danville and Springfield coals. Available information is significantly less detailed than that for the two overlying seams, but correlations made from geophysical logs from petroleum exploration indicate that this unit is one of the most continuous throughout southwestern Indiana, making this unit an ideal marker bed for regional correlations. Thickness of this unit is very variable, with significant contrasts in Gibson, Knox, and Sullivan counties. Very little is known regarding the gas content and origin of this gas.

The Seelyville coal seam was represented for depth and thickness based on available information (Figures 5a and 5b). This unit is considered to be a primary target under investigation for CBM in Indiana because it features relatively large depth, thickness, and a wide regional distribution. Mapped thicknesses show great variability, particularly in Knox and Posey counties (Figure 5b), where the adjacent lithologies are also characterized by the presence of discontinuous lenses of sandstone. The associated changes in adjacent lithology seem to play an important role in the occurrence of gas associated with this coal seam, and more detailed studies are required to understand the relationship between gas content and lithologic distribution.

GAS CONTENT

Direct gas measurements using canister desorption data have been used in Indiana to estimate the potential of Indiana's CBM resource as an alternative energy source. Available canister desorption data from 58 wells drilled in Gibson, Knox, Posey, Sullivan, Vanderburgh, and Warrick counties in southwestern Indiana were used to determine the maximum and minimum values for depth and gas contents of the studied coal seams. Table 1 shows the minimum and maximum depth and measured gas content for the Danville, Springfield, Houchin Creek, and Seelyville coals.

Table 1. Minimum and maximum depths and gas contents measured from canister desorption on Indiana coal samples.	Coal Seam	Depth (ft)		Gas Content (scf/ton, raw basis)	
		Min	Max	Min	Max
	Danville	145	504	29	70
	Springfield	217	669	25	96
	Houchin Creek	582	772	38	73
	Seelyville	429	943	13	149

The canister desorption data were plotted in terms of 'depth vs. gas content' to evaluate any correlation between these two parameters. The plot of all available data (Figure 6) indicates no significant correlation, probably due to the heterogeneity of facies above coals in some areas and the wide geographic distribution of data. The data for Sullivan a Gibson counties were plotted on Figure 7 indicating an overall increase in gas content with increasing depth. The facies distribution above the coals in these two counties is more homogeneous.

To corroborate preliminary findings about the observed increase in gas content with increasing depth, Figure 8 shows the distribution of available data for the Springfield coal in Gibson County. This diagram shows a consistent increase in gas content with increasing depth with the exception of a data point near 300 ft showing an anomalously high gas content value of 96 scf/ton. A linear regression of all data indicates a poor correlation with $R^2 = 0.29$. However, eliminating the anomalous data point improves the correlation to $R^2 = 0.93$. The anomalously high CBM content of 96 scf/ton may be due to bacterially generated methane at shallow depths. Figure 9 shows a similar diagram as Figure 8 but the available data are from the Seelyville coal in Vanderburgh county. The Seelyville coal in Vanderburgh County is slightly more mature (higher vitrinite reflectance values R_o) than the Springfield coal in Gibson County, and it occurs at greater depths. The data show a fairly good correlation between depth and gas contents (Figure 9). The relatively low gas contents are attributed to mainly thermogenic gas generation with minor contributions of biogenically generated methane in this portion of the basin.

GAS ORIGIN

Coal-bed gases in the Illinois Basin are generally of mixed thermogenic and biogenic origin. The biogenic end-member is generated via the microbial CO_2-reduction pathway, while thermogenic gas results from thermal cracking of organic compounds during burial. Most of the Illinois Basin coal-bed gases plot either within the isotopic field for microbial CO_2-reduction or the field for mixed-origin on the isotopic methane classification graph shown in Figure 10 (Whiticar 1999). Only the most mature gas sample from the Springfield coal of Kentucky indicates a well-defined thermogenic origin (Figure 10).

Table 2. Carbon and hydrogen isotope ratios of CBM from the Illinois Basin.	Coal Seam	Number of CBM samples	Depth (ft)	R_o (%)	$\delta^{13}C$ ($^o/_{oo}$ vs. VPDB)	δD ($^o/_{oo}$ vs. VSMOW)	Biogenic contribution (% of total)
	Springfield (IL)	1	533	0.52	-63.3	-200	100
	Springfield (KY)	4	1100	0.75	-49.5	-180	0
	Springfield (IN)	5	321	0.58	-69.9	-196	60
	Houchin Creek (IN)	2	366	0.54	-58.6	-196	22
	Seelyville (IN)	11	483	0.62	-58.2	-195	54

The contribution of the thermogenic end-member to the bulk gas varies depending on the coal maturity (R_o), which ranges from 0.52 to 0.75% (Table 2) Within this range of maturity, the microbial methane endmember may range from 0 to almost 100% of the coalbed gas (Table 2), according to a rough estimate based on Katz et al.'s method (Katz, Narimanov et al. 2002). Such a wide range of mixtures of microbial and thermogenic gas in Illinois Basis coals is expected because these coals are at or above the threshold for thermogenic gas generation. This threshold, usually considered at R_o = 0.5 - 0.6%, may be shifted up or down due to differences in maceral composition.

CONCLUSIONS

1. Indiana could be attractive for CBM development because of its abundant shallow coal resources that exhibit low reservoir pressure and high permeability. Indiana coals, being of high volatile bituminous rank, did not generate much thermogenic gas. Instead, exploration and plans for economic CBM development should be based on the expectation that CBM occurs as mixtures of thermogenic and biogenic gas.
2. Our CBM isotopic and compositional data from the Illinois Basin indeed indicate that CBM is composed of a variable mixture from two endmember methane sources, namely biogenic methane and thermogenic methane.
3. Because of the mixed thermogenic and biogenic origin of coalbed gas in Indiana, exploration strategies must include elements of thermogenic and biogenic play to successfully predict and identify areas of highest gas contents.

REFERENCES

Archer, P. L. and J. N. Kirr (1984). Pennsylvanian Geology, Coal, and Coalbed Methane Resources of the Illinois Basin - Illinois, Indiana, and Kentucky. Coalbed Methane Resources of the United States. C. T. Rightmire, Eddy, G.E., and Kirr, J.N. **AAPG Studies in Geology 17:** 105-134.

Archer, P. L. and J. N. Kirr (1984). Pennsylvanian Geology, Coal, and Coalbed Methane Resources of the Illinois Basin - Illinois, Indiana, and Kentucky. Coalbed Methane Resources of the United States. C. T. Rightmire, Eddy, G.E., and Kirr, J.N., AAPG Studies in Geology No.17**:** 105-134.

Drobniak, A., M. Mastalerz, et al. (2004). Evaluation of coalbed gas potential of the Seelyville Coal Member, Indiana, USA. International Journal of Coal Geology **57**: 265-282.

Eggert, D. L. (1994). Coal Resources of Gibson County, Indiana. Bloomington, IN, Indiana Geological Survey. **Special Report 50:** 36p.

Gas Research Institute (1999). North American Coalbed Methane Resource Map. 1999 Gas Research Institute, Chicago, IL.

Harper, D. (1985). Coal Mining in Vigo County, Indiana. Bloomington, IN, Indiana Geological Survey. **Special Report 34:** 67p.

Harper, D. (1988). Coal Mining in Sullivan County, Indiana. Bloomington, IN, Indiana Geological Survey. **Special Report 43:** 47p.

Harper, D. (1988). Underground mines in the Survant Coal Member (Pennsylvanian) of Indiana. Bloomington, IN, Indiana Geological Survey. **Special report 41:** 19p.

Harper, D. (1991). Coalbed Methane in Indiana. Bloomington, IN, Indiana Geological Survey. **Occasional Paper 56:** 18p.

Harper, D. (1994). Underground mines in the Hymera coal member (Pennsylvanian) of Indiana. Bloomington, IN, Indiana Geological Survey. **Occasional Paper 62:** 12p.

Harper, D. and D. L. Eggert (1995). Coal Mining in Knox County, Indiana. Bloomington, IN, Indiana Geological Survey. **Special Report 54:** 23p.

Katz, B. J., A. Narimanov, et al. (2002). Significance of microbial processes in gases of the South Caspian Basin. Marine and Petroleum Geology **19**: 783-796.

Kvale, E. P., M. Mastalerz, et al. (2004). Atokan and early Desmoinesian coal-bearing parasequences in Indiana, U.S.A. Sequence stratigraphy, paleoclimate, and tectonics of coal-bearing strata. J. C. Pashin, and Gastaldo, R.A. **AAPG Studies in Geology 51:** 71-88.

Mastalerz, M., P. R. Ames, et al. (2003). Coals of the Brazil Formation (Pennsylvanian) in Indiana: observations of correlation inconsistecies and their implications. International Journal of Coal Geology **54**: 209-222.

Mastalerz, M., A. Drobniak, et al. (2004). Characterization of Indiana's coal resource: Availability of the reserves, physical and chemical properties of the coal, and the present and potential uses. Bloomington, IN, Indiana Geological Survey. **Open-file Study 04-02:** 245p.

Mastalerz, M. and D. Harper (1998). Coal in Indiana: A Geologic Overview. Bloomington, IN, Indiana Geological Survey. **Special Report 60:** 45p.

Mastalerz, M. and E. P. Kvale (1998). Coal-bed gas potential in Daviess County, Indiana. Bloomington, IN, Indiana Geological Survey. **Open-File Report 98-7:** 53p.

Mastalerz, M. and E. P. Kvale (1998). Coal-bed gas potential in Gibson County, Indiana. Bloomington, IN, Indiana Geological Survey. **Open-File Report 99-17:** 39p.

Mastalerz, M. and E. P. Kvale (2000). Coal quality variation and coalbed gas content in boreholes SDH-383 and SDH-384 in Posey County, Indiana. Bloomington, IN, Indiana Geological Survey. **Open-File Study 00-5:** 30p.

Whiticar, M. J. (1999). Carbon and hydrogen isotope systematics of bacterial formation and oxidation of methane. Chemical Geology **161**: 291-314.

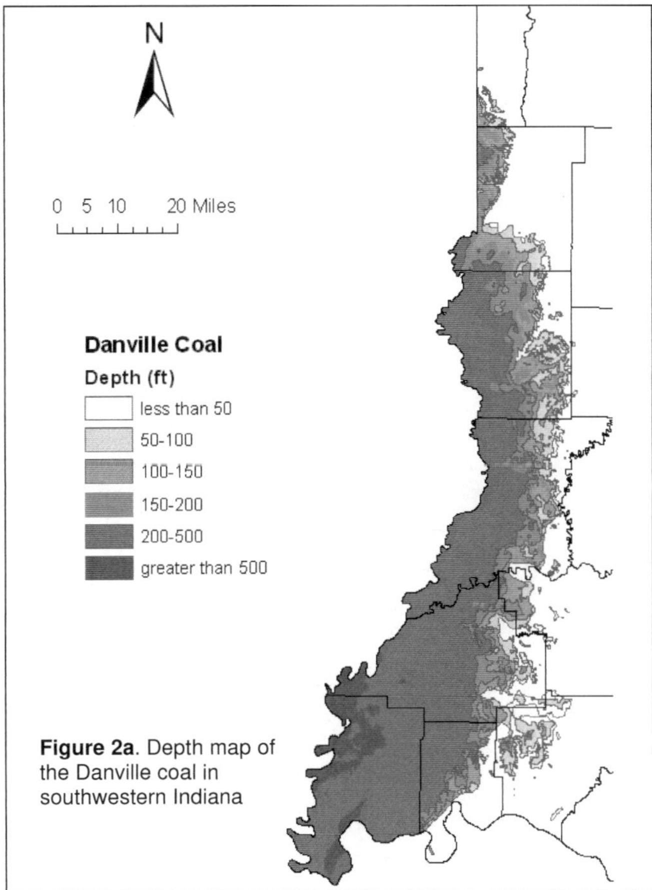

Figure 2a. Depth map of the Danville coal in southwestern Indiana

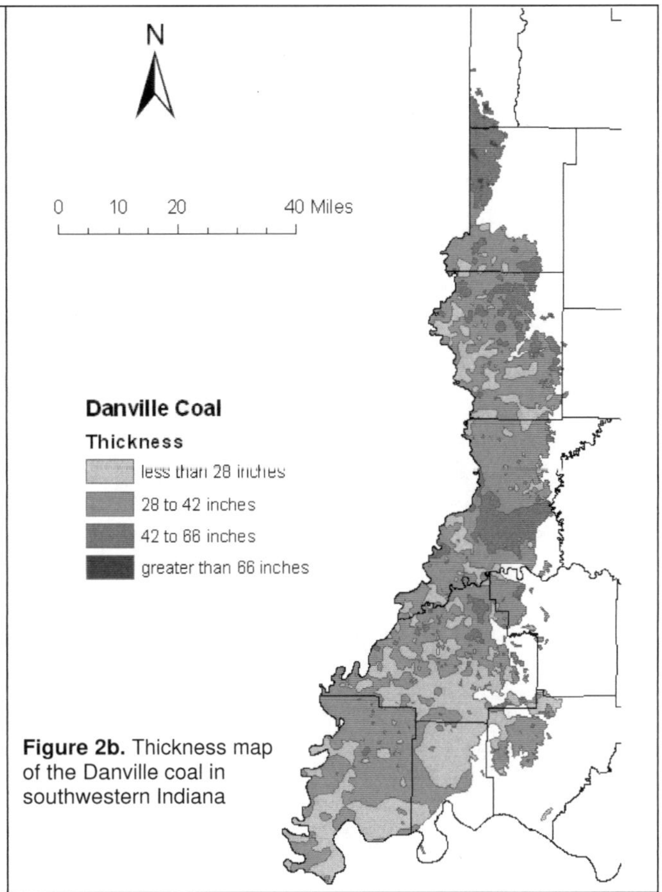

Figure 2b. Thickness map of the Danville coal in southwestern Indiana

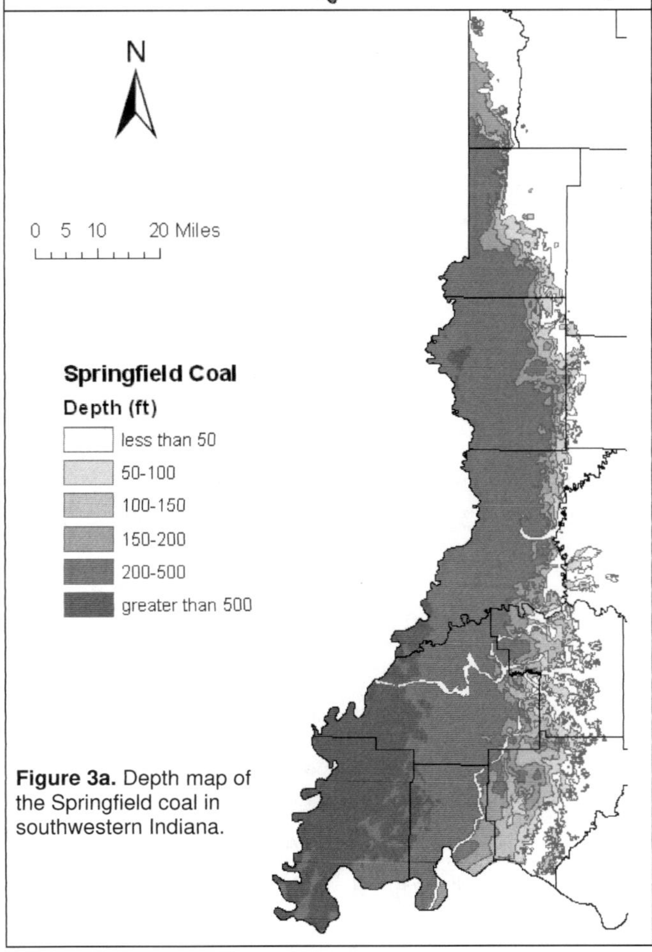

Figure 3a. Depth map of the Springfield coal in southwestern Indiana.

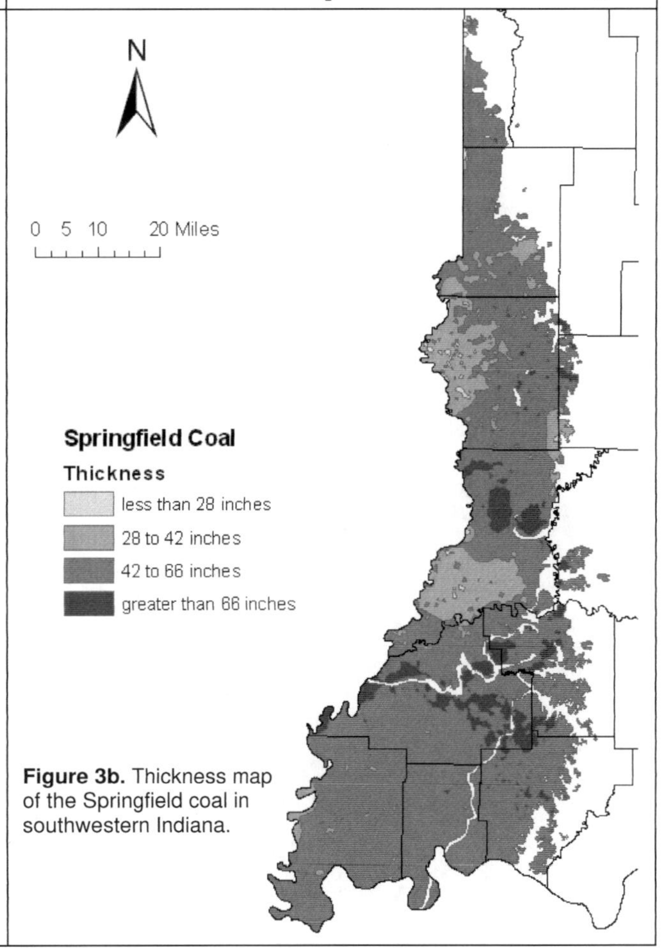

Figure 3b. Thickness map of the Springfield coal in southwestern Indiana.

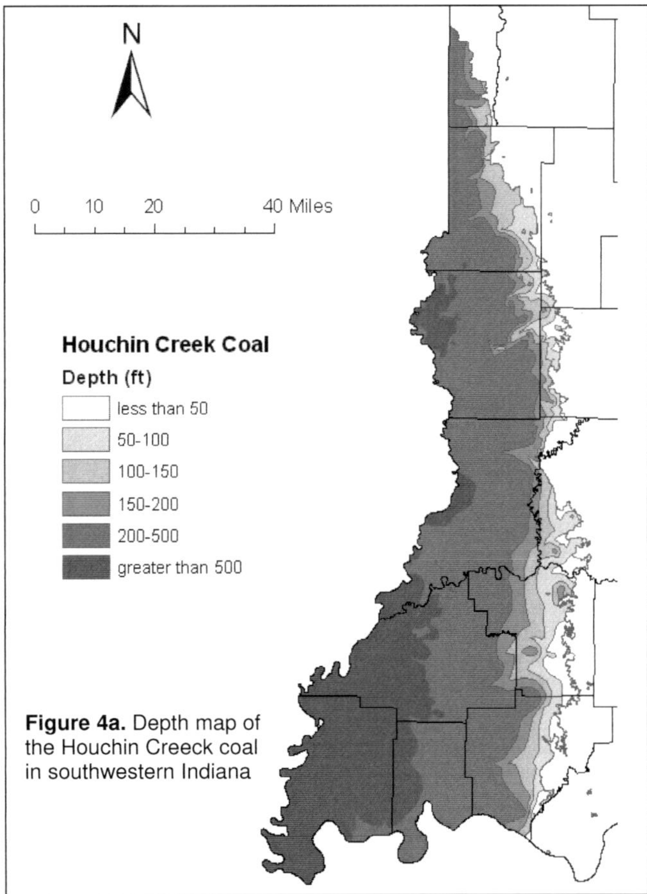

Figure 4a. Depth map of the Houchin Creeck coal in southwestern Indiana

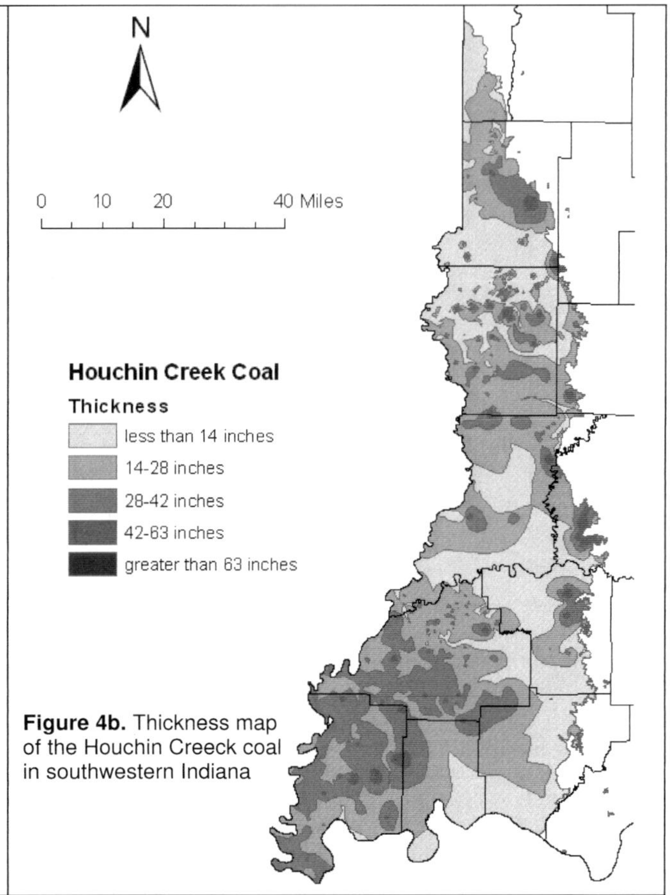

Figure 4b. Thickness map of the Houchin Creeck coal in southwestern Indiana

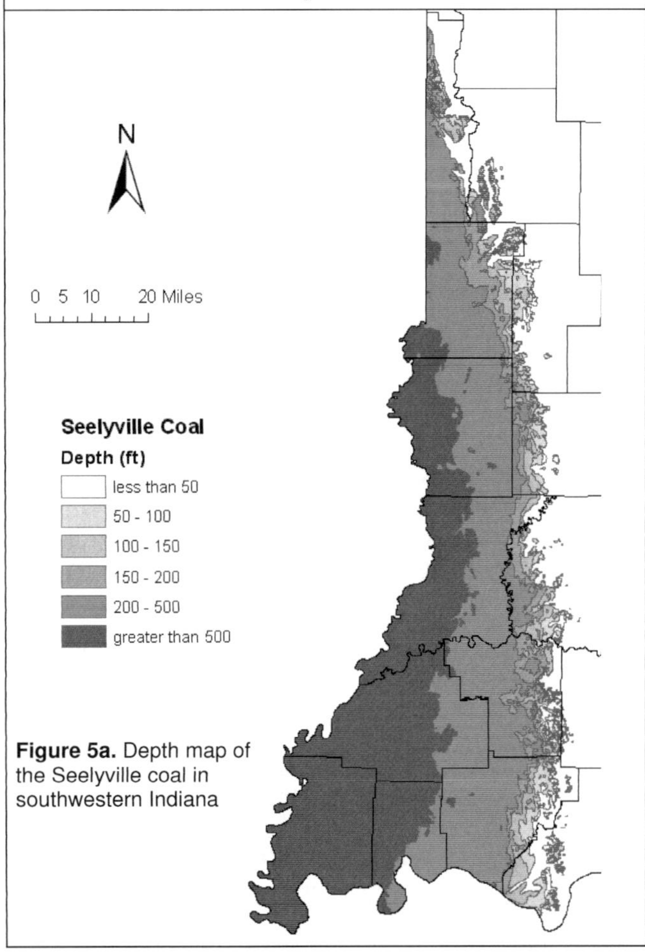

Figure 5a. Depth map of the Seelyville coal in southwestern Indiana

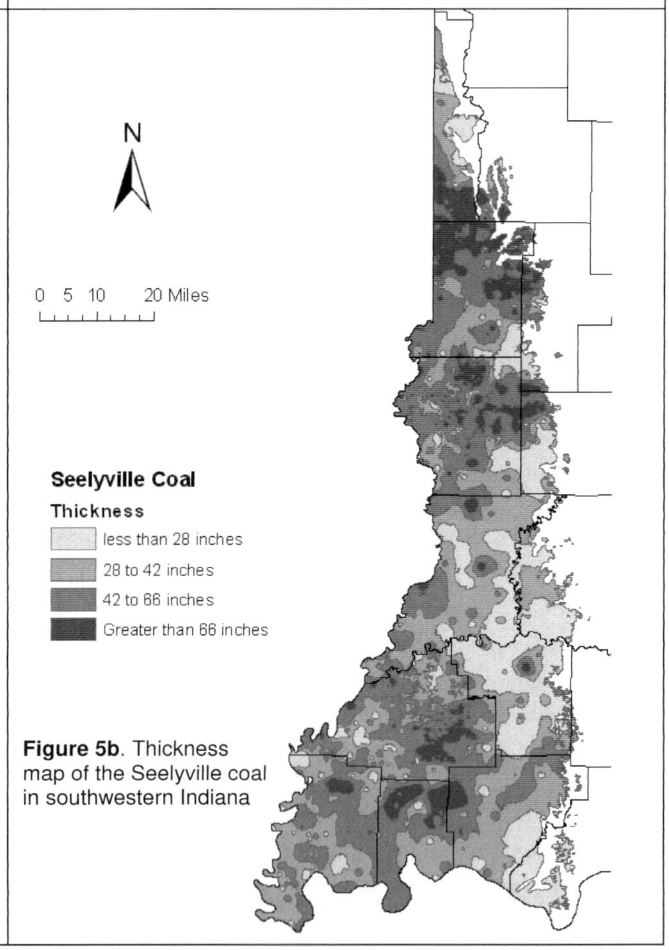

Figure 5b. Thickness map of the Seelyville coal in southwestern Indiana

Figure 6. All Data Available

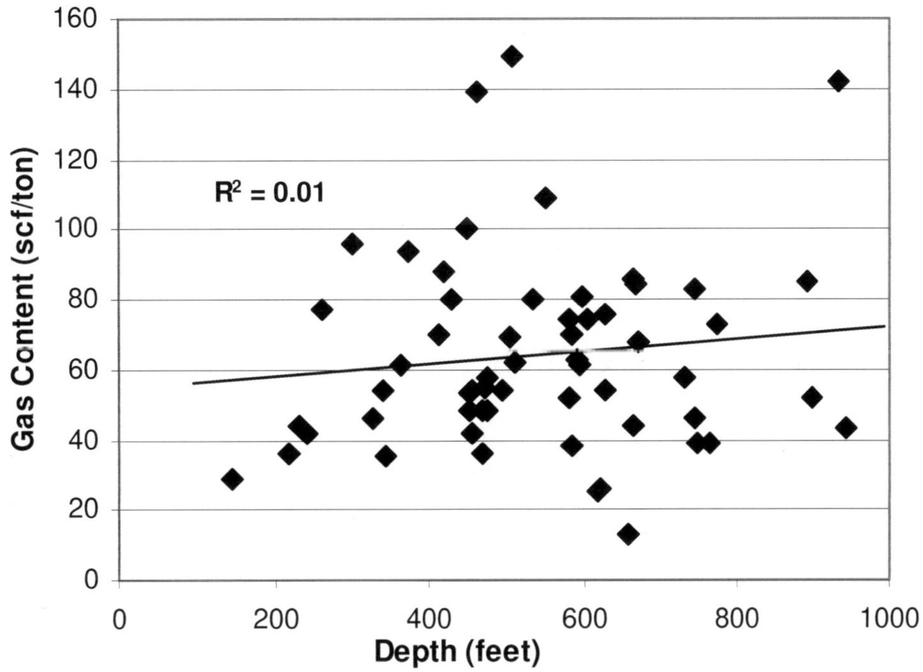

Figure 7. Data for Sullivan and Gibson counties

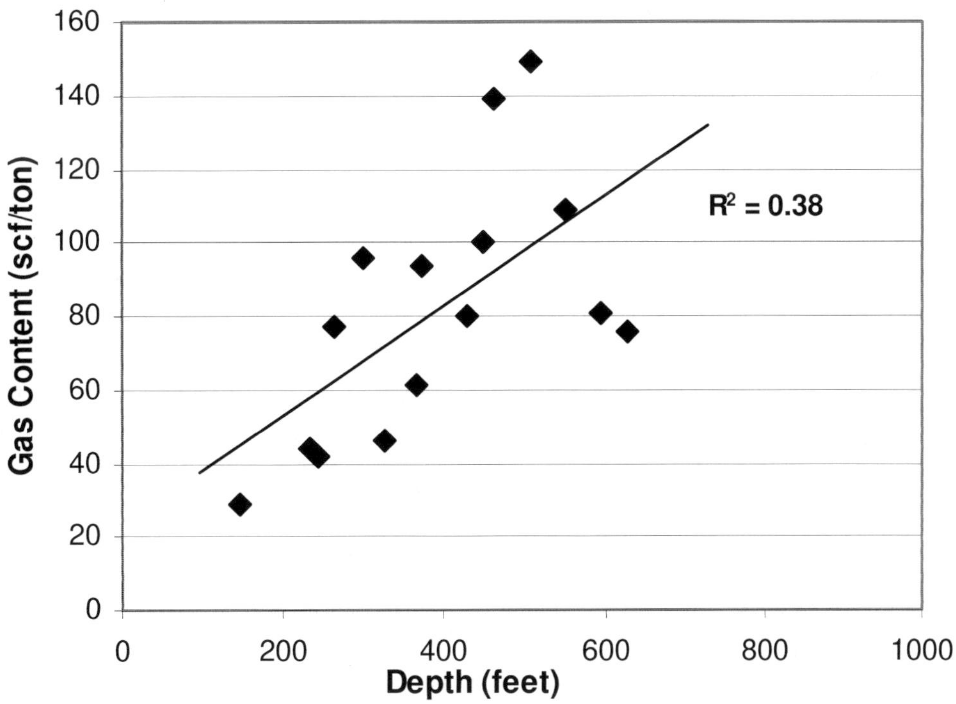

Figure 8. Springfield Coal (Gibson County)

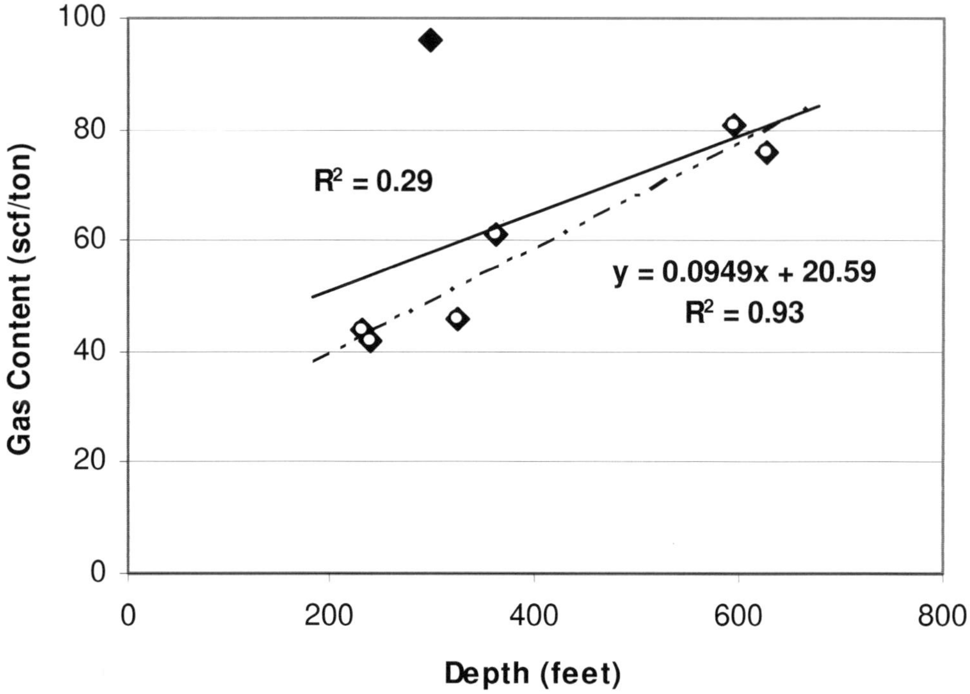

Figure 9. Seelyville Coal (Vanderburgh County)

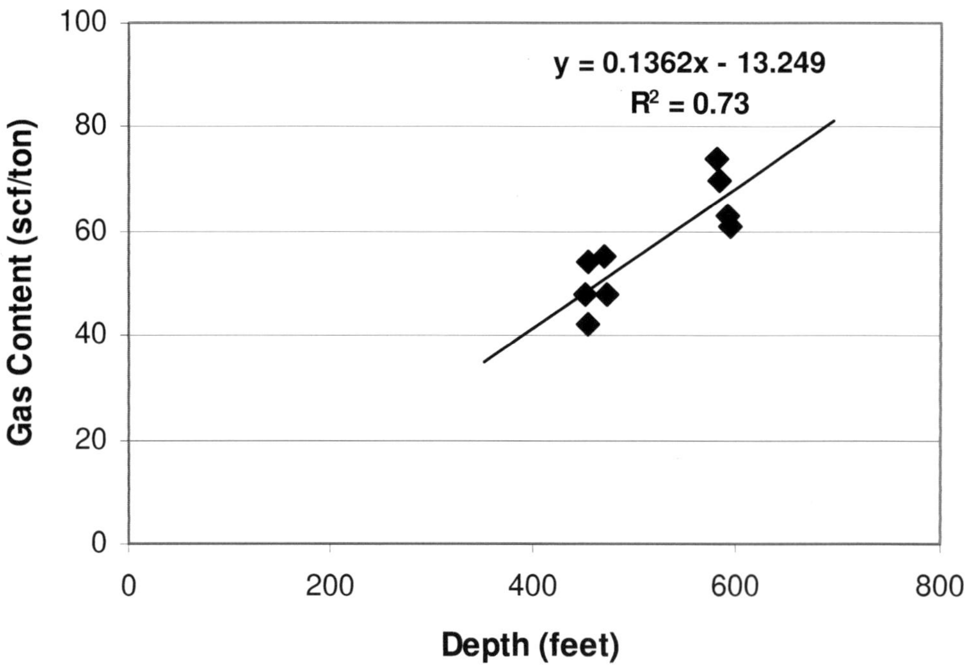

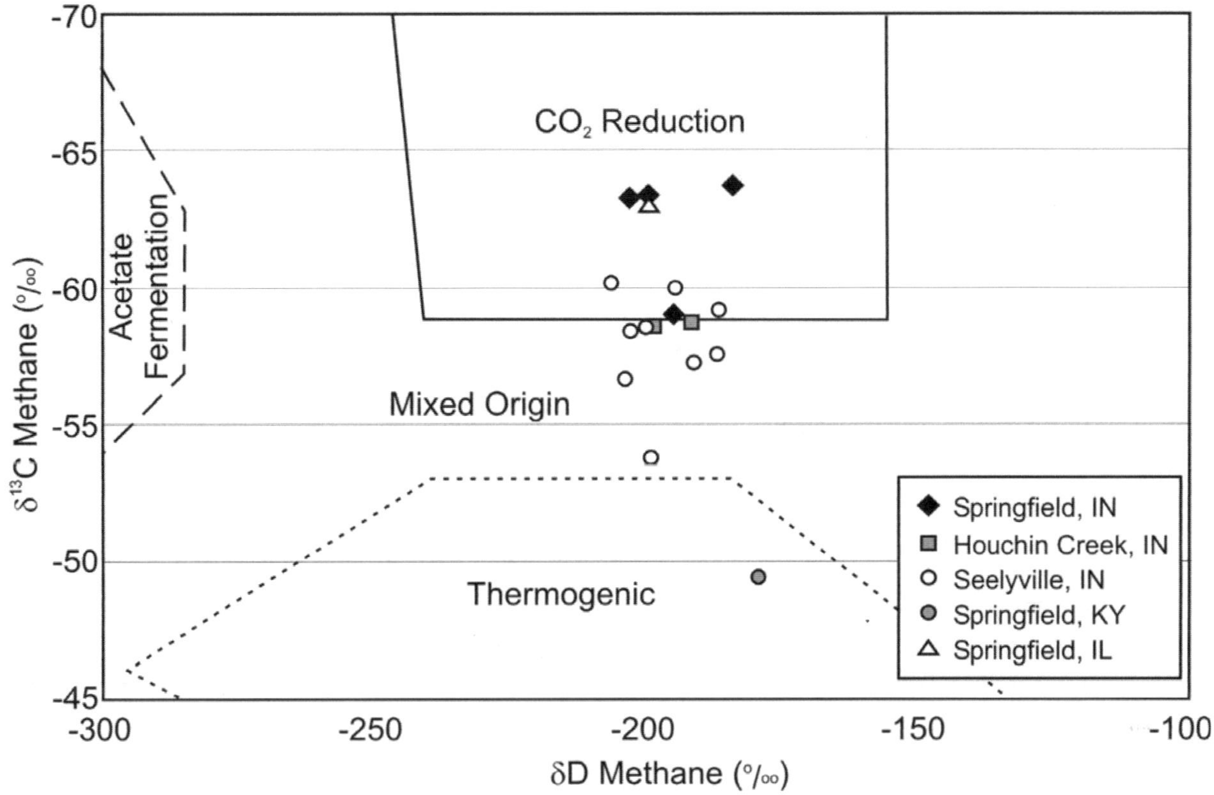

Figure 10. Carbon and hydrogen isotopes for available gas samples from coals in Indiana, Illinois, and Kentucky (isotopic ranges after Whiticar, 1999).

0530

Same Day Downhole Critical Gas Content without the Core

Robert Lamarre, WellDog Inc., Laramie, WY 82072, phone: 307.721.8875, blamarre@welldog.com

ABSTRACT

Gas desorption techniques to analyze coal gas content require expensive sample retrieval methods and extended desorption and analysis times.

WellDog has adapted its new downhole Raman spectrometer to measure gas contents in coalbed methane wells without core sampling and in a matter of hours.

This technology, which has been successfully calibrated in three independent laboratories using state-of-the art Boyle's Law isotherm apparatus, accurately measures *in-situ* gas content in wells containing one or more coal seams.

This presentation will show the methods behind the analysis, and share laboratory calibration and field case study results. It will also review how same-day analysis of gas content is leading to new, more efficient development methods for companies exploring for coalbed methane.

NOTE TO REVIEWER: *Final paper not available for inclusion in the 2005 ICBM Symposium CD. Contact author for more information.*

0503

Key Issues in China Coalbed Methane Exploration & Development and Corresponding Settlements

Liu Honglin, Wang Hongyan, Wang Yibing, Liu Hongjian, Li Yanxiang

Natural Gas institute, Research institute of Petroleum Exp. & Dev., PetroChina Company limited, Langfang, China

ABSTRACT

The potential coalbed methane (CBM) resource in china is about 22.5 trillion cubic meters (Tcm) based on the recently resource evaluation, and mostly distributed in three major CBM provinces: Northwestern, South, and North China (Huabei) basins. CBM pilot testing was mainly located in three areas in the past twenty years. Over 200 wells have been drilled throughout the country, eleven pilot areas have established in china. Up to now, Different type tectonic style coal basin have been studied and carried out exploration , some pilot wells began to produce commercial gas flow, more and more corporation and local governments began to pay more attention to the CBM. The first commercial CBM field had been established in Qinshui basin. However, because of different geological setting, coal basin evolvement, immature exploring method and other factors influenced, operators in china encountered some technology problems. It can be surely forecasted that CBM would play an important role in china future energy industry if these technology problems solved.

INTRODUCTION

During the past two decades, China government has promoted its CBM Industry in its policy aspect[1], and many corporations & joints had interest in CBM exploration and strived to obtain profit from CBM, Over 200 CBM wells have been drilled throughout China and commercial gas flow rate also has been gotten in some CBM wells. Over 150 CBM wells have been drilled in coal basins throughout eastern China. Many foreign corporations such as Arco, Phillips, and Texaco signed CBM licenses and initiated ambitious testing programs within their newly approved exploration blocks. Chinese local government groups, PetroChina and China Union Coalbed methane company limited (CUCBM) are now testing new areas using updated technology. However, most early efforts clustered around gas-prone but low-permeability coal mines were not successful except for some areas in Qinshui and Ordos Coal-bearing Basin.

CHINA CBM GEOLOGY SETTING AND RESOURCE

With the exception of the Ordos basin, Chinese coal basins are structurally more complex than the commercial U.S.A CBM areas. Faults, some active, compartmentalize coal reservoirs into isolated blocks that are difficult to develop CBM. In addition, erosion of the coal section and re-burial during Quaternary time had affected gas saturation in East China coal basins. For studying more easily, five CBM provinces were divided in terms of coal bearing strata, coal basin type and tectonic geology. These are North China (Huabei) CBM Province, South China (Huainan) CBM province, Northeastern

China (Dongbei) CBM province, Northwestern China (Xibei) CBM province and Dianzang CBM Province. North China is the most important CBM Province in china and has most of promising CBM targets. PetroChina recently has calculated China CBM resource about 22.5 Tcm shallower than 2000 m. 11 pilot testing well groups(Jinshi-1 well group, Dacheng, panzhuang, zhaoyuan, shouyang, yangjiaping, linjiaping, gikou, linxing, huaibei, fuxin) had been established all over the country, shown in figure 1.

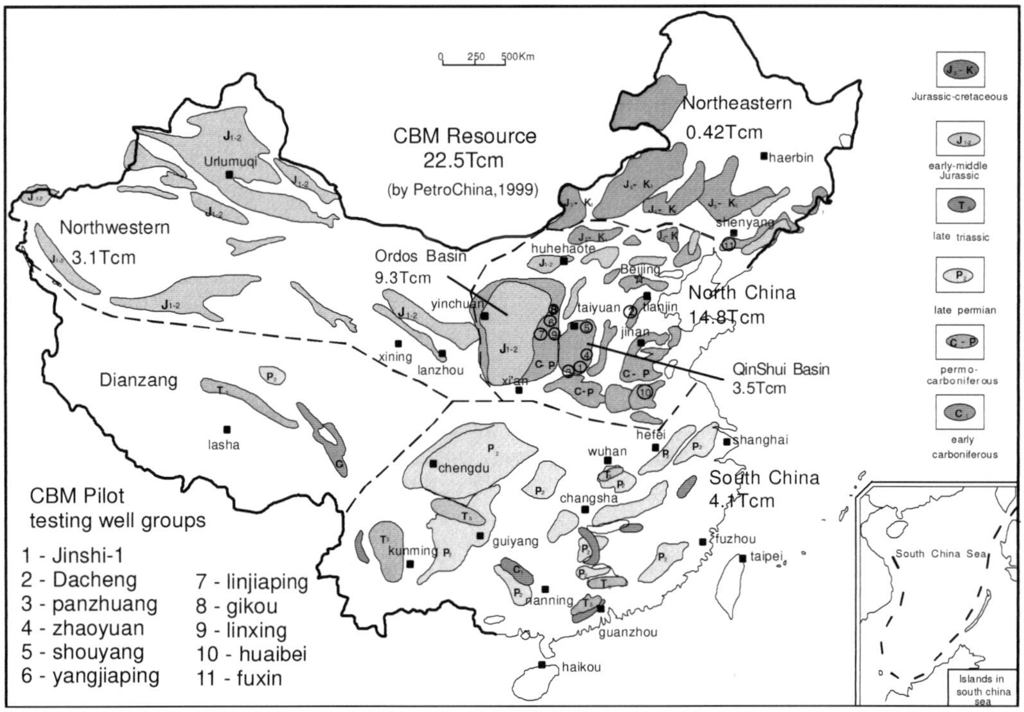

Figure 1 China CBM Resource Distribution and Pilot testing well Groups Location map

The principal coal seam targets in North China occur within the Permo-Carboniferous age Shihezi, Shanxi, and Taiyuan formations[2]. The total CBM resource in North China is around 14.8 trillion m^3, accounting for 66 percent of all. These units comprise a regressive depositional sequence that was laid down over much of northeastern China. Towards the north, individual coal seams occur lower in the section, and are fewer but thicker. Typically, a total of 10 to 30 m of coal occurs within a half dozen potential seams. The coal is generally bright, rich in vitrain, and well cleated, unlike duller Gondwana coals of like age. Gas content varies with coal rank and depth, from about 6 m^3/ton in sub-bituminous coals to over 15 m^3/ton in anthracite. Yet CBM saturation mostly ranges from 30 to 60 percent and 42 percent in average. A water-bearing carbonate interval interbedded with the coal seams often complicated well completion and hydraulic stimulation.

Additionally, lower coal rank CBM targets lie in the Jurassic Yan'an formation in north china and equivalents in Northwest CBM Province, Qsadam basin, Zhungaer basin and Tuha basin as examples, has 3.1 Tcm CBM resource. But seams in these units tend to be thicker and stratigraphically more dispersed, as well as lower in rank (sub-bituminous). Some exceptional Jurassic areas have 100 m thick individual coal seams with gas contents of 12 m^3/ton, so CBM saturation was higher in some targets than that in North China CBM Province. Recent commercial success in the low-rank Uinta and

Powder River basins of the western U.S.A suggest that China's Jurassic coal deposits should be deserved greater attention. In the past two years, CBM wells have been drilled in Jurassic coal bearing strata in Ordos and Tuha basin by PetroChina. In Tuha basin, PetroChina had gotten higher gas saturation of Jurassic coal seam and a new gas bearing area was found. Now new geological appraising activity in Jurassic formation is going on in Tuha basin.

In South China CBM Province, most important targets lie in the late Permian Longtan formation with 10 to 50 layers and 5 to 55 m thick in total, CBM resource is 4.1Tcm. Liupanshui basin is the biggest coal-bearing basin and has a lot of favorable CBM targets. Some targets have higher permeability and gas content because of lately intense tectogenesis. Till now very few CBM wild cat well were drilled in this coal basin for disadvantageous downstream condition and complicate geology setting.

In Northeast china province which coal-bearing basins dispersed, the thick coal seams in cretaceous strata are main CBM target seam. And CBM resource is small, around 0.42 Tcm. Though coal bearing basin and its resources are small, its CBM concentration is higher. Some favorable target is often called "small but rich".

CHINA CBM EXPORATION AND PILOT TESTING ACTIVITIES

China CBM Exp.& Dev. History

From 50's of last century to now, China and foreign companies had evaluated a number of diverse areas and helped to define the high-potential plays, although a few of these projects attained commercial status. Through out the whole country, over 214 coalbed methane wells which widely distributed in four CBM province, had been drilled till Nov. 11th ,2002, including 153 production wells , 59 wild cat wells, 5 gob wells, shown in table 1.

Table 1 Pilot CBM testing wells and location in china

CBM province	Location of pilot wells	operator	Number of Wells
North china	gikou, linxing, shouyang, yangjiaping, linjiaping,shizhuang, panzhuang, jinshi1,huaibei,dacheng,	Arco, Armoco, Taxco, Phillips, Lowell, PetroChina, CUCBM	Over 150
Northeastern china	fuxin	Fushun Mine Bureau	6
South china	Pingxiang, Liupanshui	PetroChina, Sinopec	4
Northwestern china	Tuha basin	PetroChina	1

In virtue of the CBM exploration character and utilization, the history of china CBM Exp.& Dev. can be divided into three stages:

The first stage (from 50's to 60's): the vertical and horizontal gob well and mine ventilation systems for reduction of the coal gas damage, in most cases mine operators discharge gas recovered from gob well directly to atmosphere only for mine safety concern.

The second stage (from 70's to 90's): Early pilot exploration and development. Because of lacking energy, CBM was prospected as potential resource and many CBM favorable targets were appraised.

The third stage (from 90's to now): CBM Exp. &Dev. activities conducting all over the country. Modern CBM drilling and testing techniques was adapted to local geologic conditions as well as well site services such as wireline core retrieval, injection/falloff testing, and methane desorption are of higher quality which more widely used. Coalbed exploration was reaching a crest in 2000. Foreign operators, Molopo, Texaco, Arco, and Phillips have kicked off ambitious testing and evaluation programs in their respective blocks. CUCBM and PetroChina were testing several of its own areas in Qinshui and Ordos basin.

Main Coal Bed Methane Exploring Area

1.Qinshui coal basin

After CBM well Jinshi-1 drilled, PetroChina is testing a six spot CBM pilot in the high-rank Jincheng mining areas of the southern Qinshui basin, close to CUCBM's pilot. The coal seam is about 6 m thick, and not deeply buried. The cost of drilling is lower and economical benefit is higher. The CBM production rate is around 9050 m^3/d and water is 3-5 m^3/d. Up to now, eleven CBM wells have been drilled in Jincheng area including six appraisal wells. Gas in place is about 406.6 billion m^3 and 35.2 billion m^3 CBM is proved by PetroChina. CUCBM CBM project which lied in the northern to PetroChina Project, had drilled over 16 production wells, and gas was compressed into tanker and sold to end user in Zhengzhou city. This area would most probably developed into the first china CBM producing area in next ten years for its favorable downstream condition and high production rate[3].

2.Ordos coal basin

ARCO, Texaco, and CUCBM are currently assessing the CBM potential of Hedong Project located in the eastern of Ordos basin. Arco's 5,000 sq km CBM contract in the Ordos basin was approved on Oct. 1, 1998, comprising three separate blocks. Arco drilled four appraisal wells in 1997, and Texaco joined the project in1998 and participated in four pilot well drilled in sanjiao block, Previously, Enron Exploration had drilled eight test wells in this area, reporting favorable permeability in the Shanxi and Taiyuan formation coal reservoirs. Since assuming control of the license from Enron in 1996, Arco has drilled and hydraulically stimulated nine production wells, production evaluation suggested that this project was successful for high gas production rate[4] .

Phillips China Inc. is testing the Hedong CBM Contract area-Linxing block just north of Arco's lease in the Ordos basin. Following acquisition of high-frequency seismic data during 1997, Phillips has tested gas content and permeability at seven wells within its 3,120 sq km block. Coal rank and gas content are somewhat lower than in the southern Ordos, but the coal reservoirs here are thicker and permeability is higher. Phillips' block is located about 100 km south of the newly completed 36 in., Ordos-Beijing natural gas pipeline, nearby West-to-east gas pipeline and 120 km west of Taiyuan, a major industrial city. Tens of CBM well groups have been drilled from 1997 to 2000, including a 5-point CBM wells in Xing County[5].

3.Huainan and Huaibei coal basin

The Permo-Carboniferous coal-bearing strata of the Huainan coalfield contain about 32 to 40 closely spaced coal seams with a total thickness of 42 m. The minable coal seams are distributed in the middle lower part of the Permian strata, which contained 9-18 target coal seams with a total thickness of 18-34 m. Data control is relatively good, with hundreds of coal exploration coreholes drilled throughout the area. Given its East China setting, geologic settings are fairly complex, but potential gas markets in this industrial area are attractive. Texaco was the first to sign a CBM contract with CUCBM in Huaibei area, commencing on Mar. 1, 1998. Texaco has committed to drill three wells within its 2,663 sq km Huaibei block in northern Anhui Province[6]. The principal coal seams at Huaibei average 9 to 15 m thick with gas contents of about 12 m^3/ton at a depth of 600 m. Three new wells had been drilled till now.

4.Other basins and areas

Dacheng CBM area lies in North China nearby Tianjing city, PETROCHINA have drilled 6 wells since 1994 including 5-point pilot wells. And 8 layer coal seams were encountered, of all coal seams, No.3 and 6 coal seam was the thicker coal seam about 3-6 meter. The highest gas content was 16.7 m^3/t and 8.6 m^3/t in average. The average gas saturation was about 30%. After water fracturing, the gas production rate ranged from 1500 to 2500 m^3, the water production rate was higher about 100 m^3/d.

Liupanshui CBM basin lies in the western of Guizhou Province with area of 1579 km^2, gas in place is about 1.4 Tcm. The main coal-bearing formation is upper Permian Longtan and Changxing formation with 5-11 layers in total. The accumulative thickness is about 9.23-53.44 m; the thickest single coal seam is about 6 m. Coal rank ranges from sub-bituminous to anthracite, and the average gas content is above 15 m^3/t. Some targets like Panguan syncline, Tucheng syncline, Liuzhi syncline and Gemudi syncline are favorable in geology condition. Sinopec had drilled two CBM well in 2001 in Liupanshui.

TECHNOLOGICAL PROBLEM ENCOUNTERED IN EXPLORE ACTIVITIES

After tens of year development, elementary coalbed methane exploring and developing technology has formed including geology appraisal, well drilling, casing, completion, fracture, well group production. More and more special institute and operators had constructed, and a series of technology criterion & standard have been enacted. Many advantage policy has brought into action one after another.

Only 7 percent of the total coalbed methane wells obtained commercial gas flow while exploration & developing activities spreading out the whole country. Most early efforts clustered around coalmine with low-permeability were not very successful. Challenges still face CBM developers in China, including complex reservoirs, restricted permeability, and few gas pipelines[7]. Well testing in eastern China indicates restricted permeability widespread, generally below 1.0 md for complex geologic setting. In some coal basin such as Qinshui and Ordos basin, the coalbed methane exploration hot spot are getting warmer and warmer, and new coalbed methane hotspot in other places are coming forth one after another.

Although vast progress made, feature technology formed, and exploration and development experience attained, the following technological difficulties are still existing in our coalbed methane exploring and developing activity, such as special appraising for complex coal basin and CBM target selecting method, multi-branch well drilling, horizontal well drilling, coal seam stimulation, well completion and so on.

Firstly, it is necessarily to establish a geological appraisal method and favorable CBM area selection method. During the past, prospective effect could not attained by the way of directly using the foreign typical technology and methods in china, because of more complicate geological condition, different coal bearing basin, multiplicity gas controlling parameters and so on. If we have an available geological appraisal method, we would manage China's complex geologic setting through a program of reservoir mapping and characterization. An integrated approach combining remote imagery, seismic, and core hole data analysis can identify areas of favorable gas saturation and permeability, reducing exploration risk.

Secondly, unclearly cognition about china coalbed methane enrichment and reservoir, lack of coalbed methane exploration database, which due to low exploration successful rate, also have a great disadvantageous effect on china coalbed methane industry.

Thirdly, well drilling technology including under balance well drilling, multi-branch well drilling, horizontal well drilling and high cost of drilling and testing coalbed methane are another important factors affected china coalbed methane exploration. While Coal seam was easer to damaged by drilling mud than conventional gas reservoir, we should develop the under balance well drilling technique. In order to enhancing the gas production rate, we should develop multi-branch and horizontal well drilling technique.

In technology of enhancing coalbed methane production aspect, whether to and How to exert hydraulic fracture technology in coal bed also is a problem in China CBM exploration. Although hydraulic fracture can increase the production, it can not used to every wells. In some CBM testing place such as an-1, an-2, da-1, da1-6, hydraulic fracture had communicated the water-bearing strata, which leads to high water production rate and low gas production rate. Whether to fracture was decided by coal mechanical characteristic, and blindly hydraulic fracturing would arise to damage to coal seam permeability.

ACKNOWLEDGEMENTS

The authors thank the technical staff and management of China United Coalbed Methane Co., China National Petroleum Co., Arco, Phillips, and Texaco for numerous contributions to this article. And special thanks for staff in CBM center of Lang fang-branch RIPED, PetroChina.

REFERENCE

1. Li, H.X., The development strategy and supportive policies for coalbed methane industry in China, United Nations Investment Promotion Seminar on Coalbed Methane Development in China, November 1998.

2. Han, D.X., and Yang, Q., Coal geology of China, Coal Industry Press, 1979.

3. Jin, A.X., Li, G.F., Wang, F.M., Liu, G., and Tian, Y.T., Elementary inquiring of transforming permeability of anthracite coal seam in Jincheng by hydraulic fracturing method, United Nations International Conference on Coal Bed Methane Development and Utilization, Conference Proceedings, Beijing, Oct. 17-21, 1995, pp. 356-361.

4. Li, J.L., The potential for coalbed methane development in Anyang, China, Proceedings of the 1997 International Coalbed Methane Symposium, University of Alabama, Tuscaloosa, May 12-17, 1997, pp. 351-355.

5. Quan, Y.K., Wang, L., Hou, H.B., Su, F.Y., and Cai, Y.F., Simulation technique to evaluate

coalbed methane reservoir of Yangjiaping pilot, Liulin area, United Nations International Conference on Coal Bed Methane Development and Utilization, Conference Proceedings, Beijing, Oct. 17-21, 1995, pp. 102-108.

6. Derickson, J.P., Horne, J.S., Fisher, R.D., and Stevens, S.H., Huaibei coalbed methane project, Anhui Province, People's Republic of China, Society of Petroleum Engineers, 1998 SPE International Conference and Exhibition in China, Beijing, China, Nov. 2-6, 1998, pp. 517-530.

7. Reeves, S., et al., Designing practical reservoir evaluation programs for frontier international coalbed methane plays, Proceedings of the 1993 International Coalbed Methane Symposium, University of Alabama, Tuscaloosa, May 17-21, 1993, pp. 511-517.

Hartman, R.C., Pratt, T.J.: "A Preliminary Study of the Effect of Moisture on the Carbon Dioxide Storage Capacity in Coal", paper 0533, *Proceedings 2005 International Coalbed Methane Symposium,* University of Alabama, Tuscaloosa (May 2005)

0533

A Preliminary Study of the Effect of Moisture on the Carbon Dioxide Storage Capacity in Coal

Chad Hartman, Timothy J. Pratt , TICORA Geosciences, Inc.

ABSTRACT

An investigation was conducted to determine the effect of excess moisture on the carbon dioxide sorption of a Ft. Union sub-bituminous coal sample. A series of representative splits were prepared and brought to moisture equilibration. Additional moisture was added to the series of splits, increasing incrementally. Volumetric isotherms were conducted at 21.1 °C up to pressure in excess of 700 psia. Results indicate that excess moisture (surface moisture) increases carbon dioxide's storage capacity, unlike methane isotherms which have previously been shown to be unaffected by moisture content beyond in-situ moisture. The data presented demonstrate that extraneous moisture and/or moisture in coal matrix porosity can behave as a potential solvent of CO_2, implying that volumetric adsorption isotherm analysis is not a measure of the carbon dioxide *sorption* behavior, but rather a measure of the *total gas storage* behavior. The results help to explain previously reported data that show the CH_4:CO_2 storage capacity ratio increases with increasing coal rank.

INTRODUCTION

Accurate determination of gas storage capacity in coal and shale gas reservoirs is a critical parameter in the suite of data required to ultimately predict gas saturation and production behavior. Storage capacity is a measurement of the volume of gas a sample can hold at a given reservoir temperature and pressure. Gas storage capacity in coal is typically assumed to be dominated by the sorption process because the free gas in the primary matrix and secondary fracture porosity systems is negligible as both pore systems are typically saturated with water. Dissolved gas in the two porosity systems is also thought to be negligible as the hydrostatic pressure and total porosity is insufficient for significant *methane* dissolution in coal gas reservoirs typically targeted for development. Storage capacity is determined in the laboratory using comprehensive sample preparation and analysis protocols, which are specific to the thermal maturity of the sample rock [1].

The gas *sorption* behavior and the maximum volume of gas *adsorbed* within the organic matrix of coal and shale gas reservoirs are conventionally evaluated through *sorption isotherm analysis*. The analysis is performed on a representative, non-oxidized, crushed particle sample that is typically equilibrated to the in-situ (i.e., native-state) moisture concentration. The sample is placed into a high pressure vessel and subjected to gas invasion at reservoir temperature and increasing pressure scenarios. When performing "volumetric" isotherm analysis, the adsorbed storage capacity is not directly measured, instead temperature and pressure data are measured and gas storage capacity is computed [2].

Gas sorption capacity is strongly dependent upon sample moisture content. One of the greatest sources of error for in-situ gas storage capacity estimates is the measurement of isotherm data on samples with moisture contents that deviate from in-situ conditions. Errors will also result if sample moisture content varies during the measurements. Joubert [3,4] and Yee et. al. [5] have previously reported that increasing moisture content in coal decreases the methane sorption capacity until in-situ moisture content is obtained, at which point no further decrease in storage capacity is observed, regardless of the quantity of extraneous moisture. Extraneous moisture content in coal does not adversely affect the methane sorption isotherm analysis because methane's solubility in water is negligible over the typical pressure range of investigation. The experimental isotherms discussed in this report indicate that, unlike methane, excessive moisture can have a strong impact on the storage capacity of CO_2 in coals.

Clarkson and Bustin [6] have shown that the storage capacity of methane and carbon dioxide increase for both wet and dry coals as a function of micropore volume, wherein monolayer sorption is assumed to occur. The IUPAC pore distribution classification [7] states that low rank coals (carbon content < 75%) contain mostly macroporosity (30-2960 nm), coals in the 76-84% fixed carbon content range contain mainly micro (0.4-1.2 nm) and meso porosity (1.2-30 nm), and high ranking coals (carbon content > 85%) contain mostly microporosity. Clarkson and Bustin also observed that coals containing a higher vitrinite content adsorb more than those with a low vitrinite content. Furthermore, Figures 1 and 2 respectively show that storage capacity is inversely proportional to moisture content, and similarly, moisture content is inversely proportional to rank (figures 1 and 2 were generated from TICORA Geosciences' large data base of experimental results).

To summarize, the increase in adsorption capacity of both methane and carbon dioxide that is observed with increasing coal rank is generally attributed to an increase in microporosity, the corresponding decrease in moisture content, and an increase in thermal maturity [5,6,8]. However, the rate of adsorption capacity increase as rank increases is not proportional between the two gases (Figure 3). At lower coal rank (subbituminous C) the ratio between methane and carbon dioxide storage capacity is about 1:10, while at higher ranks (high volatile A bituminous) the ratio is about 1:2 [8]. This indicates that there is one or more parameters directly proportional to coal rank that is responsible for the change in the storage capacity ratio. We propose that the primary parameter is specifically moisture content and that water saturated mesoporosity and macroporosity provide dissolution sites for carbon dioxide molecules. This implies that volumetric carbon dioxide sorption isotherm analysis, unlike the volumetric analysis of methane, is an evaluation of the total gas storage behavior (i.e., sorbed gas plus dissolved gas).

EXPERIMENTAL

A Ft. Union subbituminous C coal sample, the properties of which are listed in Table 1, was stage ground to a mesh size of -60 mesh and moisture equilibrated at reservoir temperature following protocol outlined by Testa and Pratt [1]. Four representative splits were prepared from the sample and the moisture contents were determined by taking two gram aliquots from each split (in triplicate) and drying the aliquots in an oven for 1.5 hours at 110 °C. All but one of the splits was then further wetted with deionized water to produce the desired incremental excess moisture weight fractions using a misting spray bottle and thoroughly mixing after each individual spray. The splits were then placed in laminate foil bags under a positive pressure of helium and manually tumbled for five minutes and then allowed to sit at approximately 29.4 °C for 48 hours to ensure the added moisture was evenly distributed and diffused throughout the coal particles. The "post-wetting" moisture contents were then determined. Table 2 provides the pre and post "wetting" moisture contents of each split.

Using a volumetric isotherm apparatus, CO_2 isotherms were performed on each split of the subbituminous Ft. Union coal sample at 21.1 °C and at pressures ranging from 50 to 790 psia. Details of the isotherm apparatus and related analytical techniques and procedures have been described by Mavor et. al. [2]. The results of the isotherm analyses are presented in Table 3 and Figure 4.

RESULTS AND DISCUSSION

Joubert [3,4] has previously conducted methane isotherms which show that increasing moisture content in coal decreases storage capacity until in-situ moisture content is obtained, at which point no further decrease in storage capacity is observed, regardless of the quantity of excess moisture. The CO_2 storage capacities calculated from the isotherms presented herein reveal that not only does carbon dioxide's storage capacity not decline with increasing excess moisture, but continues to increase, never approaching a sorption potential "wall". This phenomenon can be rationally explained by the disparity between the solubility of methane and carbon dioxide. Methane has a solubility of 3.122×10^{-5} mole fraction while carbon dioxide has a solubility of 8.21×10^{-4} mole fraction [9] at 1 atmosphere and 15 °C, a factor of 26.3 times greater. The data shown in Figure 5 was extrapolated from Whitson and Brule's study of phase behavior [10] and shows the solubility of CO_2 and CH_4 at 37.8 °C from 0 – 5000 psia. Carbon dioxide's solubility shows a sharp linear increase up through 1000 psia, increasing ~28 scf/STB per 100 psia. On the other hand, methane's solubility shows a gentle increase up through 1000 psia, increasing ~0.9 scf/STB per 100 psia.

As previously mentioned, all four of the Ft. Union coal splits prepared were initially brought to in-situ moisture content, therefore the increasing storage capacity with increasing moisture content can only be attributed to dissolution of carbon dioxide within the excess surface moisture. To explore the direct effects of the added excess moisture to the resulting isotherm curves, the storage capacity for each split was calculated using the Langmuir model at 700 psia and plotted to determine what relationship if any existed between the two parameters (Figure 6). Surprisingly there is a very good correlation, revealing the excess moisture content of each split is directly proportional to the observed changes in storage capacity and resulting Langmuir parameters. The results of the excess moisture isotherms presented in this report demonstrate that the solubility of carbon dioxide in water is sufficient enough to cause a measurable increase in calculated storage capacity by dissolution in surface moisture alone, suggesting that inherent moisture contained in meso and macroporosity can similarly store CO_2 via dissolution.

Equation 1 defines the number of molecules sorbed at the stabilized sample cell pressure and temperature at the end of given isotherm pressure step [2].

$$n_2 = n_1 \frac{\rho_s - \rho_{f1}}{\rho_s - \rho_{f2}} + \left(1 - \frac{\rho_{f2}}{\rho_s}\right)^{-1} \left[V_r \left(\frac{p_{r1}}{z_{r1} R T_{r1}} - \frac{p_{r2}}{z_{r2} R T_{r2}} \right) + V_{v0} \left(\frac{p_{s1}}{z_{s1} R T_{s1}} - \frac{p_{s2}}{z_{s2} R T_{s2}} \right) \right] \quad (1)$$

The number of sorbed molecules is equal to the number of molecules that flowed out of the reference cell less the increase in the number of molecules in the void volume of the sample cell. The void volume in the sample cell is reduced by the volume of the sorbed phase. Therefore, the void volume at the start and end of the step differ and must be computed with Equations 2 and 3 [2]. The free gas density in the void volume is computed with Equation 4 at the stabilized sample cell end point pressure and temperature conditions. The sorbed phase density is generally assumed to be equal to the liquid density of the gas of interest at the atmospheric pressure boiling point [5].

$$V_{v1} = V_{v0} - \frac{n_1 \hat{M}}{\rho_s} \quad (2) \qquad V_{v2} = V_{v0} - \frac{n_2 \hat{M}}{\rho_s} \quad (3) \qquad \rho_f = \frac{\hat{M} p}{z R T} \quad (4)$$

Based on equations 1 – 4 the calculated storage capacity (sorbed number of molecules) does not distinguish between the phenomena of sorption and dissolution, rather it is solely dependent upon changes in free gas density between the beginning and end pressures of the step. The only parameter that *does* effect accurate calculation of storage capacity is the sorbed phase density which comes into play to appropriately correct for a reduction in void volume from pressure step to pressure step. Because the calculations assume the void volume is that volume of matrix porosity available for sorption (i.e. free of moisture) the portion of CO_2 that has potential to dissolve in water saturated meso and macro pores rather than sorb in available microporosity is not appropriately accounted for. This could potentially lead to the overestimation of the gas storage capacity of CO_2 in coals with high moisture contents by over estimating the sorbed phase volume (i.e. decrease in void volume).

Further studies need to be conducted to properly characterize the interaction of carbon dioxide with available microporosity for sorption and dissolution within water filled porosity. This will require investigations into matrix pore size distributions and the probable distribution of moisture content (wt. %) within available porosity. Other factors requiring investigation include the complex interaction of carbon dioxide with coal, due to carbon dioxides quadrupole moment, potential multi-layer adsorption, coal swelling during adsorption [11], and the formation of carbonic acid in water.

CONCLUSIONS

Based upon the data and discussion presented, we conclude that carbon dioxides solubility in water at elevated pressures is substantial enough to contribute to the calculated storage capacities of coals with large in-situ moisture contents. The idea that water within the coal matrix porosity has the potential to behave as a solvent for CO_2 provides some insight into the mechanism that causes a carbon dioxide storage capacity at 10:1 over methane in low ranking coals (subbituminous C) as compared to higher ranking coals (high volatile bituminous A) which are usually found to have a 2:1 storage capacity for carbon dioxide when compared to methane.

Carbon dioxide isotherm analysis conducted on samples with moisture contents both below and above in-situ moisture (i.e., native state moisture condition) can lead to potential overestimation of carbon dioxide storage capacity due to an increase in porosity available for sorption (dry) and the solubility of carbon dioxide at elevated pressure (excessively wet).

Future research should include carbon dioxide and methane isotherms run on excessively wet and excessively dry representative coal splits to investigate the impact of moisture within the coal fabric and apparent storage capacity. Figure 7 models the predicted storage capacity behavior for a coal sample on which said isotherm analyses are carried out. If such experimentation were conducted over a range of coal ranks the results would help provide understanding into how important inherent moisture content is compared to coal petrology, chemistry, and rank when determining the ratio between the gas storage capacity of methane and carbon dioxide.

NOMENCLATURE

M	*molecular weight*
n	*number of moles*
p_{r1}	*reference cell start pressure with sorbing gas, psia*
p_{r2}	*reference cell end pressure with sorbing gas, psia*
p_{s1}	*sample cell start pressure with sorbing gas, psia*
p_{s2}	*sample cell end pressure with sorbing gas, psia*
R	*gas constant, 10.7316 psia-ft^3/lbmole-^{o}R*
T_{r1}	*reference cell start temperature with sorbing gas, degrees Rankine*
T_{r2}	*reference cell end temperature with sorbing gas, degrees Rankine*
T_{s1}	*sample cell start temperature with sorbing gas, degrees Rankine*
T_{s2}	*sample cell end temperature with sorbing gas, degrees Rankine*
V_r	*reference cell volume, ft^3*
V_{vo}	*void volume determined with helium calibration, ft^3*
z_{r1}	*reference cell start real gas deviation factor with sorbing gas, dimensionless*
z_{r2}	*reference cell end real gas deviation factor with sorbing gas, dimensionless*
z_{s1}	*sample cell start real gas deviation factor with sorbing gas, dimensionless*
z_{s2}	*sample cell end real gas deviation factor with sorbing gas, dimensionless*
ρ_{f1}	*sample free gas density at start conditions, lbm/ft^3*
ρ_{f2}	*sample free gas density at end conditions, lbm/ft^3*
ρ_s	*sorbed gas density, lbm/ft^3*

ACKNOWLEDGMENTS

We appreciate the assistance in sample preparation and isotherm analysis that Mr. Matt Winkler provided as well as editorial contributions provided by Mr. Dave Young.

REFERENCES

1. Testa S.M. and Pratt T.J., 2003: "Sample Preparation for Coal and Shale Gas Resource Assessment", Paper SPE 356, Proceedings of the 2003 International Coalbed Methane Symposium, University of Alabama, Tuscaloosa
2. Mavor M.J., Hartman C., Pratt T.J., 2004: "Uncertainty in Sorption Isotherm Measurements", Paper SPE 411, Proceedings of the 2004 International Coalbed Methane Symposium, University of Alabama, Tuscaloosa
3. Joubert J.I., Grein C.T. and Bienstock D., 1973: "Sorption of Methane in Moist Coal" ; Fuel, V. 52, p. 181-185
4. Joubert J.I., Grein C.T. and Bienstock D., 1974: "Effect of Moisture on the Methane Capacity of American Coals" ; Fuel, V. 53, p. 186-191
5. Yee D., Seidle J.P., and Hanson W.B., "Gas Sorption on Coal and Measurement of Gas Content," Law B.E. and Rice D.D., eds., 1993: Hydrocarbons From Coal, AAPG Studies in Geology #38: Tulsa, OK, AAPG
6. Clarkson C.R., Bustin R.M., 1999: "The Effect of Pore Structure and Gas Pressure Upon the Transport Properties of Coal: a Laboratory and Modeling Study" ; Fuel, V. 78, p. 1333-1344
7. International Union of Pure and Applied Chemistry, 1994 ; Pure Applied Chemistry, V. 66, p. 1739
8. Pratt T.J., Baez L.R.G., 2003: "Critical Data Requirements for Coal and Gas Shale Resource Assessment", Paper 0367, Proceeding of the 2003 International Coalbed Methane Symposium, University of Alabama, Tuscaloosa
9. Lide D.R., ed., 1996: CRC Handbook of Chemistry and Physics, 77th Edition, New York, CRC Press
10. Whitson C.H., Brule M.R., eds., 2000: Phase Behavior, Monograph Volume 20 SPE Henry L. Doherty Series, Richardson, TX, Society of Petroleum Engineers Inc.
11. Gregg S.J., Sing K.S.W., eds., 1982: Adsorption surface Area and Porosity, 2nd Edition, New York, Academic Press

Moisture holding capacity	wt. fraction	0.2261
Ash content (in-situ)	wt. fraction	0.0949
Volatile matter content (dry, mineral-matter free)	wt. fraction	0.3197
Fixed carbon content (dry, mineral-matter free)	wt. fraction	0.3593
Carbon content (in-situ)	wt. fraction	0.5253
Hydrogen content (in-situ)	wt. fraction	0.0375
Total Sulfur content (in-situ)	wt. fraction	0.0058
Nitrogen content (in-situ)	wt. fraction	0.0080
Oxygen content (in-situ)	wt. fraction	0.1023
Helium density (in-situ)	g/cc	1.34
Gross Calorific Value (moist mineral-matter free)	BTU/lbm	9,091
Sulfur Trioxide in ash (Dry Basis)	wt. fraction	0.03944
Mean Maximum Vitrinite reflectance	%	0.49
Coal Rank Classification (ASTM D 388)	-	Sub-bituminous C

Table 1. Ft. Union Coal Characteristics

Sample	Pre-Wetting Moisture Content	Post-Wetting Moisture Content
	wt. %	wt. %
Ft. Union 1	21.83	21.65
Ft. Union 2	21.81	27.53
Ft. Union 3	22.16	36.47
Ft. Union 4	21.94	50.927

Table 2. Ft. Union Coal Moisture Contents (wt. %)

Ft. Union 1					
Step	Pressure	Temperature	Storage Capacity	Langmuir Volume	Langmuir Pressure
	psia	°C	scf/ton	scf/ton	psia
1	55.11	21.15	86.70	1112.93	686.83
2	177.90	21.18	224.34		
3	335.80	21.19	355.05		
4	504.52	21.18	466.85		
5	650.63	21.20	548.12		
6	770.17	21.20	591.69		
Ft. Union 2					
Step	Pressure	Temperature	Storage Capacity	Langmuir Volume	Langmuir Pressure
	psia	°C	scf/ton	scf/ton	psia
1	59.80	21.21	88.00	1301.69	844.17
2	179.95	21.26	224.54		
3	337.15	21.28	366.67		
4	490.81	21.26	479.84		
5	653.09	21.31	578.79		
6	761.29	21.33	610.84		
Ft. Union 3					
Step	Pressure	Temperature	Storage Capacity	Langmuir Volume	Langmuir Pressure
	psia	°C	scf/ton	scf/ton	psia
1	56.30	20.70	90.15	1368.93	805.33
2	174.95	20.90	239.64		
3	331.83	21.09	396.56		
4	489.48	22.19	538.24		
5	641.03	21.88	594.71		
6	796.99	21.68	683.19		
Ft. Union 4					
Step	Pressure	Temperature	Storage Capacity	Langmuir Volume	Langmuir Pressure
	psia	°C	scf/ton	scf/ton	psia
1	56.81	21.32	90.52	1820.65	1101.42
2	159.19	21.36	230.61		
3	299.83	21.34	384.58		
4	446.20	21.33	512.72		
5	593.26	21.36	652.75		
6	735.90	21.33	727.44		

Table 3. CO2 Isotherm Results of Four Representative Splits of a Ft. Union Coal Sample

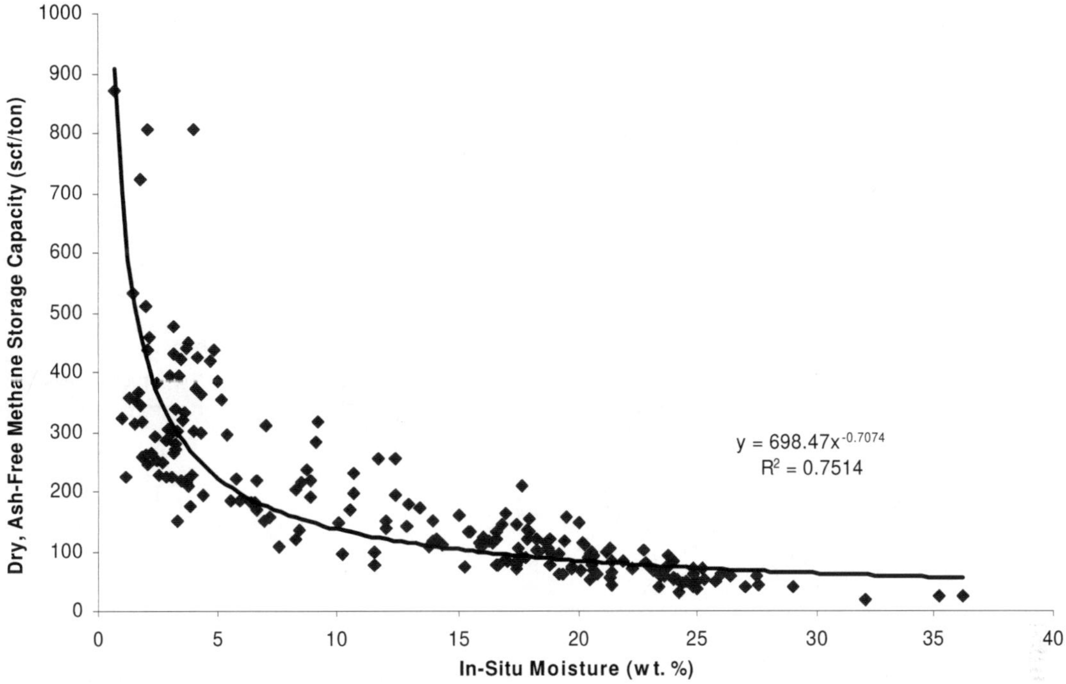

Figure 1. Methane Storage Capacity (scf/ton) vs. Moisture Content (wt. %) for a Large Series of Coal Samples Analyzed by TICORA Geosciences, Inc.

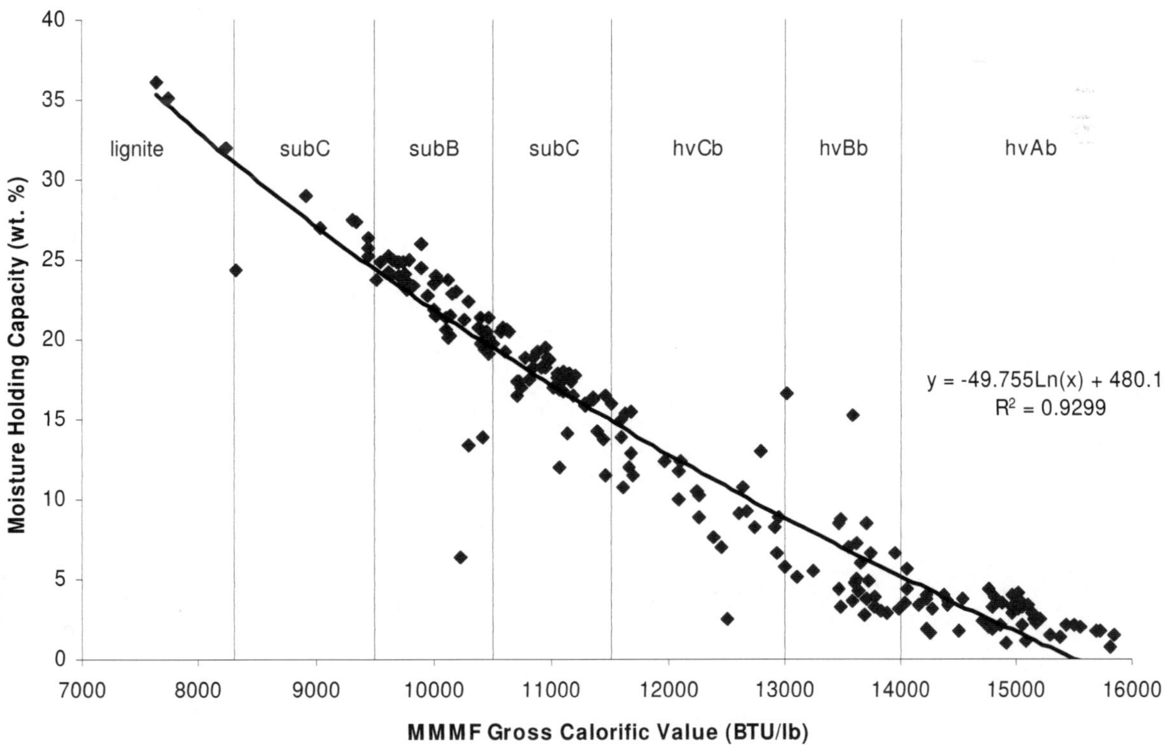

Figure 2. Moist Mineral Matter Free Gross Calorific Value (BTU/lb) vs. Moisture Content (wt. %) for a Large Series of Coal Samples Analyzed by TICORA Geosciences, Inc.

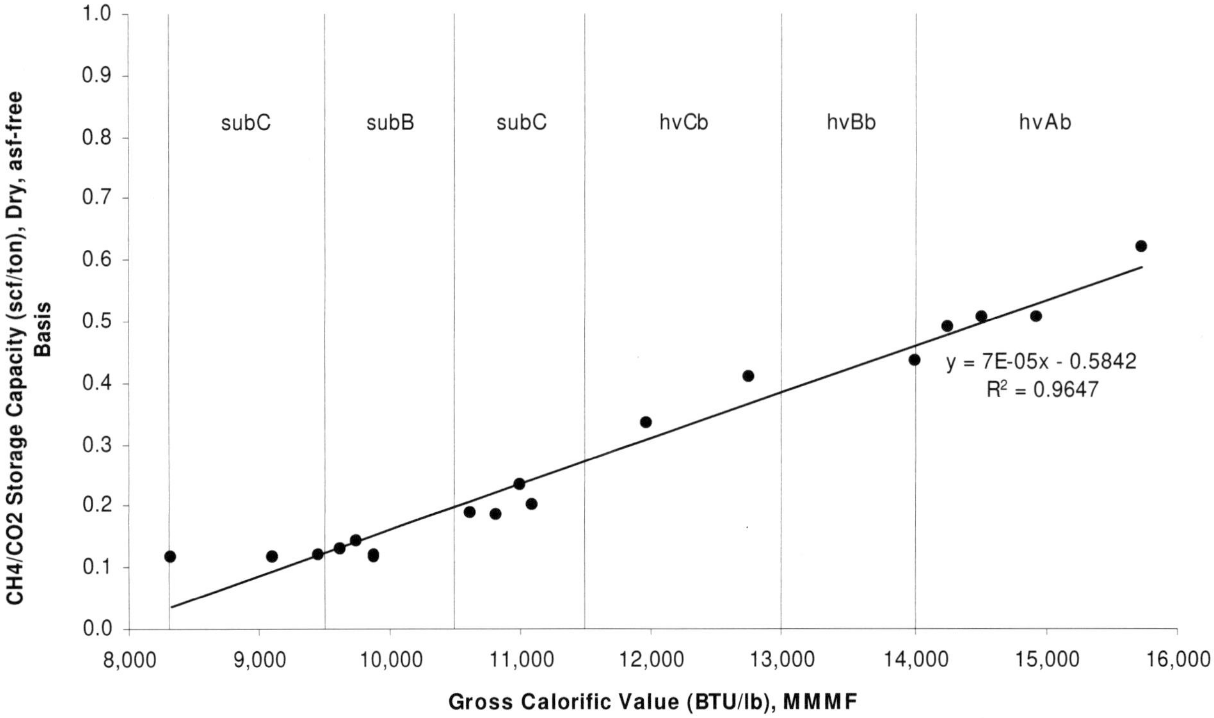

Figure 3. Dry, Ash-Free CH4/CO2 Storage Capacity (scf/ton) Ratio vs. Moist Mineral Matter Free Gross Calorific Value (BTU/lb) for a Series of Coals from the Powder River, San Juan, Hanna, and Denver Basins

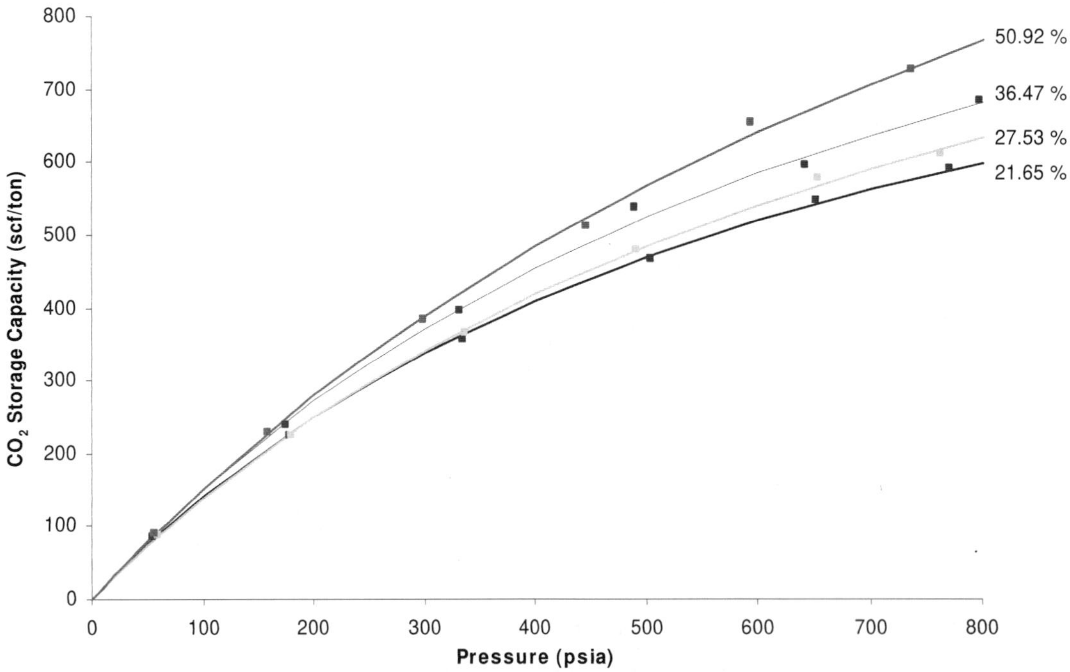

Figure 4. Carbon Dioxide Adsorption Isotherms of Representative Splits of a Sub-bituminous Ft. Union Coal at 21.1 °C and moisture contents of 21.65, 27.53, 36.47, and 50.92 weight percent.

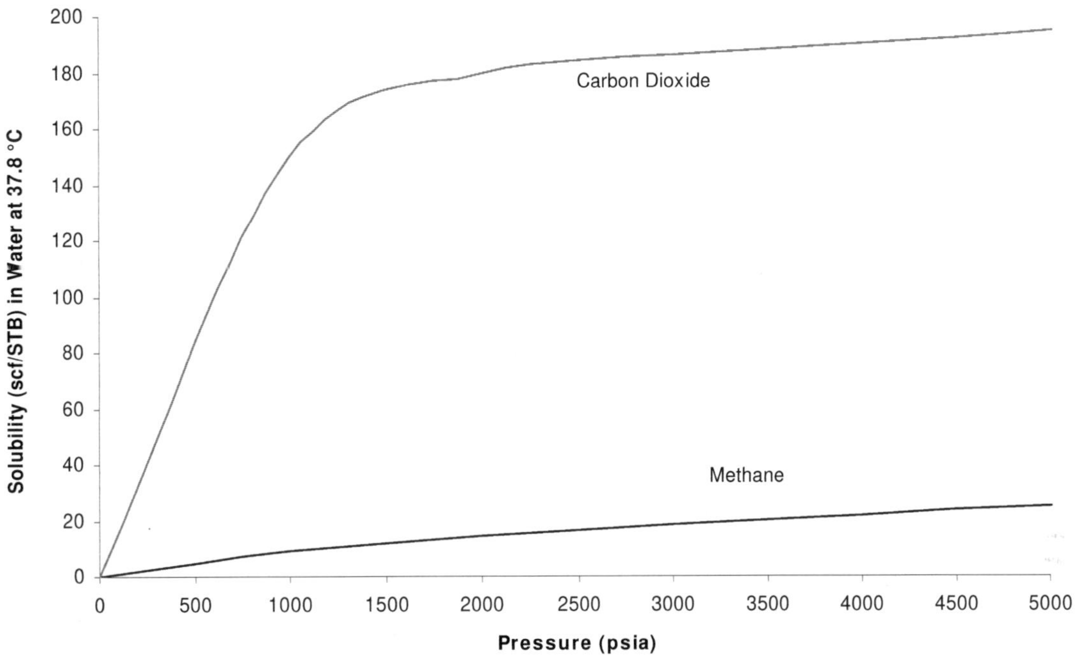

Figure 5. The Solubility of Carbon Dioxide and Methane in Water @ 37.8 °C

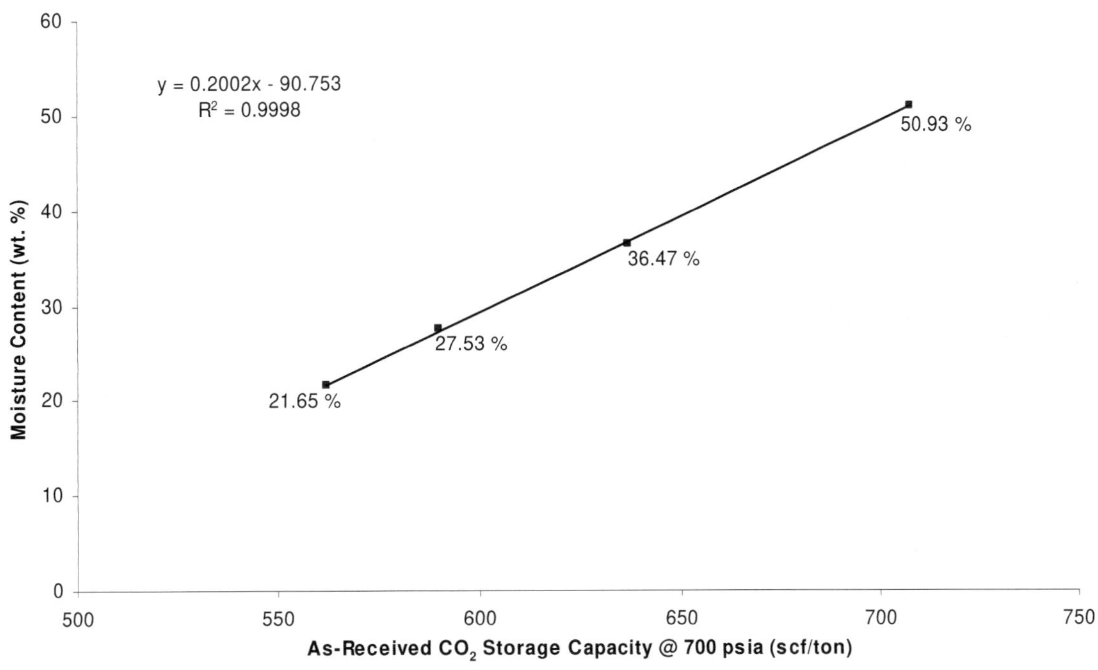

Figure 6. The Calculated As-Received Carbon Dioxide Storage Capacity (scf/ton), at 700 psia, for 4 Ft. Union Coal Representative Splits with Varying Moisture Contents (wt. %)

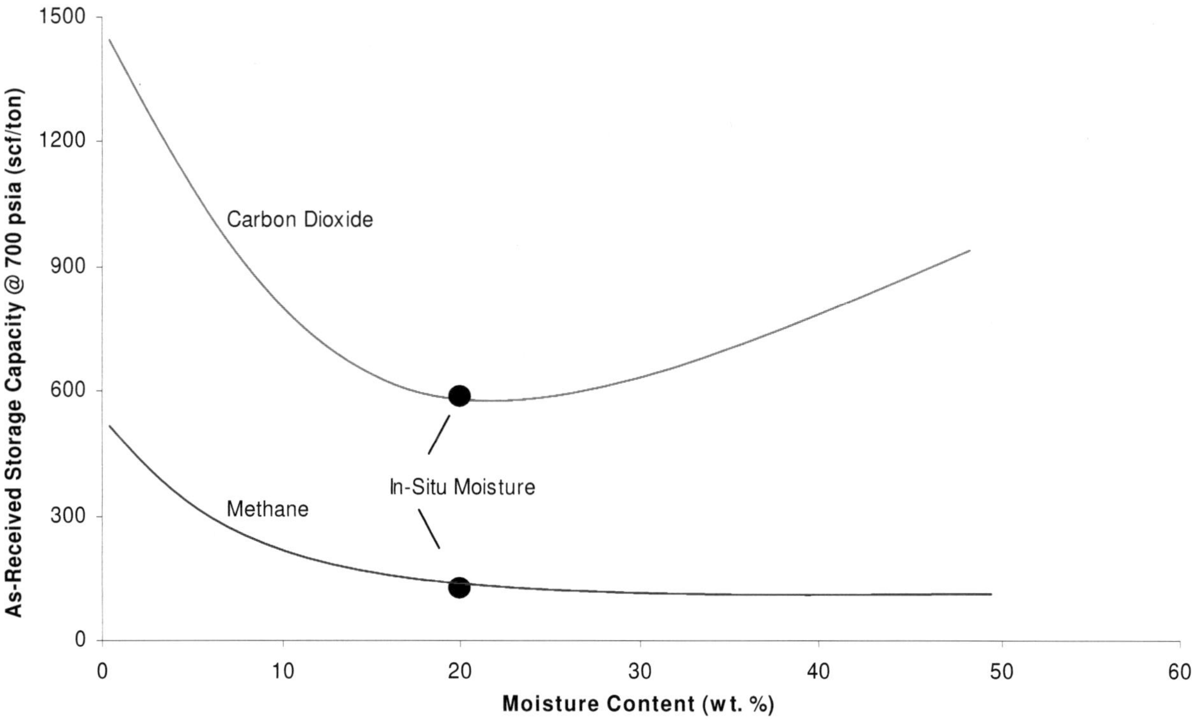

Figure 7. The Predicted Carbon Dioxide and Methane Storage Capacity (scf/ton), at 700 psia, for Representative Splits with Moisture Contents (wt. %) that Vary from Excessively Wet to Excessively Dry

0505

Numerical Analysis of the Impact of Longwall Panel Width on Methane Emissions and Performance of Gob Gas Ventholes

C.O. Karacan; W.P. Diamond; G.S. Esterhuizen and S.J. Schatzel

National Institute for Occupational Safety and Health (NIOSH), Pittsburgh Research Laboratory

ABSTRACT

In coal mining, longwall mining is a preferred method to maximize production by extracting large blocks of coal that have been outlined with a set of development entries. In U.S. mines, longwall panels are typically 305 m (1,000ft) wide (with a continual trend towards even wider panels), and usually over 3,050 m (10,000ft) in length. The increasing size of longwall panels, while helping to increase coal production, may also increase methane emissions due to the exposure of the mining environment to a larger area of fractured, gas-bearing strata. Thus, understanding the impact of increased panel widths on methane emissions, and designing gob gas ventholes and bleeder systems accordingly, can enhance the safety of the underground workforce by reducing their exposure to potentially explosive accumulations of methane/air mixtures.

As part of its mine safety research program, NIOSH's Pittsburgh Research Laboratory has initiated a reservoir modeling effort to better understand the interaction of the various geotechnical factors influencing gas flow within and to the underground longwall mining environment. A focus of this modeling effort has been 1) the prediction of the incremental amount of methane emissions to be expected due to increasing longwall panel widths, and 2) optimizing gob gas venthole completion practices to capture more of the gas in the subsided strata above longwall panels before it can enter the ventilation system of the underground workplace. The history matching phase of the study has been carried out on a 381-m (1,250ft) wide panel in the Pittsburgh Coalbed. Simulations have also been completed for a 442 m (1,450ft) wide panel to estimate the incremental increase in methane emissions and the performance of current gob gas venthole placement configurations. Additional venthole placement configurations were simulated to investigate options for optimizing methane capture from the subsided strata above the wider longwall panel.

INTRODUCTION

Longwall mining is a high-volume coal extraction method where a rectangular "panel" in the coalbed has been outlined with a set of development entries. However, longwall mining creates large-scale disturbances around the longwall face and in the overlying strata. The immediate consequences of these disturbances are caving of the unsupported immediate roof strata into the void left by the progressive extraction of the coalbed. The height of caving is dependent on the mining height as well as the strength and stratigraphy of the roof strata, generally extending upwards 3 to 6 times the thickness of the mined coalbed [1-3]. Caving of the roof causes an area of relieved stress in the overlying strata where rocks are fractured vertically and horizontally along bedding planes. The thickness of the fractured zone can vary up to 100 times the height of the mined coalbed [2], depending on the rock layers, thickness of the overburden, and the size of the panel (Figure 1).

The occurrence of such an extensive area of stress relief and the resultant rock damage changes the gas transport characteristics, particularly permeability, in the overlying strata. Any gas that is contained in this area of relieved stress will be released (coalbed gas initially by desorption and diffusion flow mechanics), as the permeability of these zones is dramatically increased and new permeability pathways are created by the mining-induced fracturing. Relaxation of the roof and floor strata and the associated fracture connectivity allows gas to flow from all of the surrounding gas sources towards the lower-pressure sink of the underground workings. These gas sources include overlying (and in some cases underlying) coalbeds, carbonaceous shales, and to some extent other more traditional gas reservoirs such as sandstones [4]. In their study, Diamond et al. [4] found that as much as 91% of the longwall gob gas originated in the overlying coalbeds, with those as high as 61 m (200ft) above the mined coalbed contributing gas to the longwall gob. The magnitude of the gas flow depends on the degree of strata relaxation and on the type and strength of the rocks in the adjacent strata [5], as well as the volume of gas available in the affected strata. Gas-bearing strata can be directly exposed to the caved zone which is in contact with the mine ventilation system, or connected to it by fractures.

In mines where the released gas from subsided and relaxed strata places a burden on the mine ventilation system, gob gas ventholes are commonly drilled from the surface above the panel to capture the gas before it enters the underground working environment. As mining advances towards the venthole, the surrounding gas-bearing strata will fracture and establish preferential pathways for the released gas to flow towards the pressure sink of the venthole once it has been mined under (Figure 2). Gas production may vary over the life of the venthole, generally with high concentrations of methane (> 80%) at early stages, and becoming a mixture of methane and mine ventilation air later in mining, depending on the proximity of the venthole to the caved zone, the gas content of the overlying strata, and the interaction of the venthole with the ventilation system [6].

Methane control using gob gas ventholes in the longwall mining environment has become a standard practice in many U.S. longwall coal mines, and has proven to be useful in reducing the amount of methane that enters the working environment from relaxed and fractured strata. However, there are various factors to be considered in designing gob gas ventholes for specific mining conditions. For instance, the number of ventholes used and the well spacing on a particular longwall panel depends on the rate of mining and the gas content of the strata above and below the mined coalbed. The location of the ventholes on the panel is also important. Diamond et al. [7] showed that placing the gob gas ventholes on the margin of the panels instead of in the traditional centerline location could increase gas production substantially (almost 80% in the site-specific study).

In recognition of the importance of gob gas ventholes for methane control, Mills and Stevenson [8] reported that approximately 30 to 40 percent of the total gas liberated from several mines operating in the Mary Lee-Blue Creek coal interval, Alabama, was produced by gob gas ventholes. Dixon [9] stated that the mines could not operate at economic levels without gob gas drainage and also reported that the methane level in mine bleeder entries can be reduced as much as 80% using gob gas ventholes.

As a proven technology for today's longwall operations, gob gas ventholes offer an extra margin of safety. The competitive nature of the domestic and international coal industry is constantly pushing mining companies to increase the size of their longwall panels, especially the panel width. In addition to the larger volumes of gas released from the surrounding strata, increasing panel widths can result in increased methane emissions due to the greater coal surface area exposed on the longwall face, and the larger amounts of coal being transported on the pan line and conveyor belt systems.

Diamond and Garcia [10] conducted a study in mines operating in Pocahontas No. 3 Coalbed in Virginia to investigate the consequences of increasing panel width on face emissions in gassy coalbeds. Continuous longwall face emission monitoring studies were conducted at two adjacent mines where the face length was to be extended from 229 to 305 m (750 to 1,000ft). A regression analysis of the face emission data predicted that a 75-m (250-ft) longer face would produce 7%-13% more gas. In a more recent study to estimate the face emissions with increased longwall face lengths in the Pittsburgh Coalbed in Pennsylvania, Schatzel et al. [11] used both a graphical regression analysis and a more detailed linear optimization technique on the same set of measured face emission and ventilation data to differentiate methane increases and contributions from different sources. They concluded that methane emission rates would increase by 35% to 48% when comparing a longwall face length of 305 m (1,000ft) to 488 m (1,600ft).

The larger volume of surrounding gas-bearing strata being affected by mining-related disturbances due to increased longwall panel widths potentially results in higher gas emissions into the underground workplace, if the gas is not captured effectively by gob gas ventholes. Conventional venthole design strategies that are practiced with current panel widths may not be sufficient to remove the expected additional methane volumes from wider panels. Several approaches may be utilized to deal with the potential for these increased volumes of released methane. Rather then drilling additional gob gas ventholes based on trial and error, a theoretical reservoir modeling approach should be used to better understand the complexities of the interaction between the mine and the associated gas reservoirs. This approach would help to plan gob gas venthole strategies in advance of extending the panel width to more efficiently respond to changing gas emission conditions.

OBJECTIVES OF THE STUDY

Advanced numerical models can simulate the rock mass and associated reservoir and gas flow responses to longwall mining, thus allowing for the simulation of methane flow and emissions without costly trial and error field experiments. This paper presents the development and application of geomechanical and reservoir models to investigate the effect of increased longwall panel widths on rock response, permeability changes, methane emissions, and the design and performance of gob gas ventholes.

To meet the study objectives, a two-stage approach was taken. The first stage was the use of FLAC2D (Fast Lagrangian Analysis of Continua 2D, [12]) finite difference code to simulate the geomechanical response of rock layers to longwall mining in terms of stress, strain, and fracturing, and the use of this output to calculate permeability changes based on empirical relationships. The calculated permeability fields were then used to update the initial permeabilities in a dynamic reservoir model developed using Computer Modeling Group's GEM software [13] to simulate gas flows and emissions associated with longwall mining. Finally, the reservoir model was used to simulate the performance of various configurations of gob gas ventholes for a panel 61 m (200ft) wider than the 381-m (1,250-ft) base case for which actual gob gas venthole production data were measured over the life of the panel.

GEOLOGY AND MINE LAYOUT

The study mine operates in the Pittsburgh Coalbed in Greene County, Pennsylvania. Overburden depths range between 152 and 274 m (500 and 900ft). The study area is in a new longwall mining district for the mine, and is not surrounded by other mined-out panels. Longwall panels in this area are super-critical, i.e., the panel width is greater than the height of the overburden, which results in a more complete and immediate caving of the overburden strata into the mine void. The first two panels in this area are 381 m (1,250ft) wide and the widths are planned to be increased to 442 m (1,450ft) starting with the third panel in the district. The panel lengths are generally 3,350 to 3,960 m (11,000-13,000ft).

A generalized stratigraphic section of the strata above the Pittsburgh Coalbed in the study area was obtained from cores and is shown in Figure 3. The Sewickley Coalbed as well as rider coals directly above the Pittsburgh Coalbed are believed to be the primary source of strata gas in the area. Within this interval, the thickest coalbed is the Sewickley Coalbed. Gas released from the Pittsburgh Rider coals located in the caved zone is expected to migrate to the mine ventilation system, while gas in the Sewickley Coalbed, as well as any other gas-bearing horizons above the caved zone, will migrate to the operating gob gas ventholes placed in the fractured zone, as shown in Figure 2. The Waynesburg Coalbeds, approximately 95 m (310ft) above the Pittsburgh Coalbed, are generally characterized by low gas contents, and are not expected to contribute significant volumes of gas to the mining environment.

A sandstone paleochannel complex is present in parts of the study area. This sandstone channel is located immediately over the Pittsburgh Coalbed and replaces part, or in some areas all, of the overlying shale unit. The thicknesses of the sandstone and shale section vary considerably throughout the study area, but the combined thickness of this unit above the Pittsburgh Coalbed is about 12 m (40ft).

METHANE CONTROL PRACTICES

Longwall methane control at the study mine includes the bleeder ventilation system, horizontal methane drainage holes in the Pittsburgh Coalbed, and gob gas ventholes. The bleeder ventilation system includes the peripheral bleeder entries surrounding the panel and the associated bleeder fan shaft. The bleeder fan in the study area is at the top of a 1.8-m (6-ft) diameter air shaft.

Gob gas ventholes ideally are generally drilled to within 12 or 13 m (40 or 45ft) of the top of the Pittsburgh Coalbed (into the fractured zone, but not into the caved zone, Figure 2), and are completed with 17.8-cm (7-in) casing and 61 m (200ft) of slotted pipe on the bottom, as shown in Figure 4. Usually, a vacuum is also applied to these ventholes by installing pumps (commonly referred to as exhausters or blowers) on the surface well head. The purpose of this completion strategy is to create a pressure sink across several potential gas-bearing horizons and to capture the released gas before it can enter the underground workplace. For these gob gas ventholes to work effectively, they must be drilled close enough to the mine to capture methane from the fractured zone but far enough above the caved zone to minimize the amount of ventilation air that is drawn into the ventholes.

The gob gas ventholes in the study area are instrumented by NIOSH in cooperation with the coal company to continuously measure gas production rates and methane concentrations after the wells become operational, and monitoring continues as mining progresses over time. These gas flow data are an integral part of the database necessary to conduct the reservoir modeling studies presented in this paper.

DESCRIPTION OF GEOMECHANICAL MODEL FOR ANALYSIS OF STRATA RESPONSE AND PERMEABILITY CHANGES DURING LONGWALL MINING

Geomechanical models were used to evaluate the effects of longwall mining on the surrounding rock mass and to calculate mining-related permeability changes. The FLAC model simulated mining in increments, starting from the left side of the grid and advancing to the right (Figure 5). The rock mass was modeled with considerable detail, capturing the layered nature of the coal measure rocks. Each layer was provided with appropriate strength and elastic properties based on laboratory and field test results [14]. The model runs captured the strength of the intact material as well as the strength of the bedding planes by using a ubiquitous joint approach in which bedding planes are assumed to be uniformly distributed throughout the rock mass. Rock failure was based on the Coulomb failure criterion, with strain softening and dilation occurring once the rock has failed. The model was run to equilibrium at each step so that rock failure and stress redistribution would occur. The model runs included a sufficient lateral extent of mining so that full subsidence of the overburden strata would occur over the simulated mined area.

Calculation of permeability changes was carried out using the final stress and rock failure distributions. Initial permeabilities in the rock mass were based on field scale data from mines in the Eastern United States, as well as other published sources [15-17]. Table 1 summarizes the initial horizontal and vertical permeabilities used in the FLAC models.

The effect of both stress changes and rock fracturing on permeability was determined from the FLAC model runs. An exponential relationship was used to compute the effect of stress changes on rock mass permeability after the work of Ren and Edwards [18] and Lowndes et al. [19]. Permeabilities were calculated independently for the horizontal and vertical directions as follows:

$$k_h = k_{h0} e^{-0.25(\sigma_{yy} - \sigma_{yy0})} \quad (1)$$

$$k_v = k_{v0} e^{-0.25(\sigma_{xx} - \sigma_{xx0})} \quad (2)$$

The parameters for these equations are given in the nomenclature section.

In the model, the parameters were set so that a stress change of 10 MPa would result in a change in permeability of about one order of magnitude. In addition, fractured rock was assumed to experience an increase in permeability of 100 md above the current permeability in both the horizontal and vertical directions, regardless of rock type. Similarly, bedding shear failure was assumed to increase the permeability by 50 md in the bedding (horizontal) direction.

Figure 5 shows an example of the permeability distribution around an advancing longwall face calculated by the FLAC models. The figure shows a low permeability zone ahead of the face where stresses are high as well as a higher permeability zone behind the face where stress relief has taken place. The fractured zone, location of bedding shearing, and the typical location of a gob gas venthole within the fractured zone is also shown as indicated on the figure. The details of the permeablllty calculation and geomechanical analysis are given in Esterhuizen and Karacan [20].

DESCRIPTION OF RESERVOIR MODEL FOR ANALYSIS OF METHANE EMISSIONS DURING LONGWALL MINING AND GOB WELL PERFORMANCE

Grid Model of the Area

For this study, a three-dimensional grid model of the mine area was created from mine maps using Cartesian grids. The dimension of the grid model was based on actual panel dimensions and was constructed in such a way that the width of the grid model will be about 5 times the current panel width of 381 m (1,250ft) as shown in Figure 6. The total width of the development mining on either side of the panel, including the gate roads and the pillars, was determined from the mine maps to be about 61 m (200ft). Panel 1 was located in the middle of the grid system, which was fine grided to enable increasing panel widths for the modeling exercise.

The number of vertical layers and their thicknesses were based on the generalized stratigraphic log of the area shown in Figure 3. In the grid model, the bottom-most layer was dedicated to the Pittsburgh Coalbed, or mining layer. The other lithologies shown in the log were represented in the grid model based on their thicknesses. Although this log presents only major layers in the mine area, it was convenient for determining the number of vertical layers and their thicknesses for generating the modeling grids in the vertical direction.

For convenience in generating the grids, the layers shown in Figure 3 were assumed to be uniform in thickness and continuous throughout the simulated mine area, except for the sandstone paleochannel complex and associated shale unit overlying the Pittsburgh Coalbed. For these two layers, non-uniform grids were used based on the spatial thicknesses determined from isopach contour maps. In generating the model elevations, the formation top map with respect to sea level for the mining layer (Pittsburgh Coalbed) was digitized as the reference elevation, and the depths of other layers were determined based on their thicknesses as determined from stratigraphic logs.

In the grid model, the Pittsburgh Coalbed layer was constructed differently from the other layers in the area to include the details of the longwall mining operation, the related mining environment, and the lower part of the caved zone. This layer was constructed in such a way that it would host both the mined and unmined Pittsburgh Coalbed, and the gateroads that surround the panel. To simplify the ventilation pathways, the mine entries were represented by a single grid rather than a detailed gridding for each gateroad on the tailgate and headgate sides of the panel (Figure 6).

Figure 7 shows the 3-D grid reservoir model that was constructed for this study. In this figure, only the primary layers representing the Waynesburg, Sewickley, and Pittsburgh Coalbeds and the sandstone paleochannel complex were retained, while the other intervening rock layers have been removed for visualization purposes. The figure also shows the actual gob gas ventholes from the first longwall panel in the district and the wellbores that are used to represent the basic elements of the ventilation system in the model.

Gob Gas Ventholes and Pseudo-Ventilation System

Four gob gas ventholes with 17.8-cm (7-inch) casing and 61 m (200ft) of slotted pipe at the bottom were drilled on the first panel in this new mining district. However, the ventholes were drilled to varying proximities [14, 11, 9 , and 12 m (47, 35, 30, and 40ft)] above the top of the Pittsburgh Coalbed as opposed to the preferred ~12 m (~40ft) distance, which places the bottom of the venthole above or close to the top of the caved zone. The ventholes are located about 100 m (300ft) from the tailgate side of the panel, and are 166, 830, 1,635, and 2,458 m (544, 2,720, 5,362, and 8,060ft), from the start-up end of the panel. The ventholes are modeled based on these actual completion parameters.

A simplified version of the ventilation system was also incorporated into the model to simulate ventilation airflows during longwall mining. Since the reservoir model does not have the inherent capability to simulate the geometries of a mine ventilation system, wells are used to simulate the injection (intake) and removal (return) of ventilation airflows, including the bleeder fan shaft (Figure 7). A single well injecting air into the entries with a rate constraint of 1,700 m^3/min (60,000 cfm) to simulate the typical ventilation airflow rates at the mine site was used to model the air intake part of the ventilation system. The bleeder fan was modeled with a large-diameter vertical well on the tailgate side of the panel, with the completion interval being equal to the height of the entries (thickness of the Pittsburgh Coalbed). During simulations, the bleeder fan was operated with a bottom-hole pressure constraint of 1.36 KPa (0.2 psia) suction pressure. Another vertical well was employed to model the return air side of the ventilation system. The vertical wells representing the ventilation intake and return were also used to control the ventilation pressures at the mining layer at the assigned bottom hole pressure constraint.

Reservoir Description and Reservoir Parameters

In the mining industry, it is not common practice to perform extensive data gathering for gas reservoir analysis. Thus, most of the data required by the reservoir simulator for this unique mining application was gathered from previous NIOSH publications that addressed general reservoir and gob gas venthole performance issues in this same general study area, from external reports, and from personal communications with the operating mining company to obtain reasonable assumptions for setting the initial, pre-mining reservoir properties to the coal and non-coal units.

In the current study, coal/sand sequences and coalbeds interspersed with conventional gas reservoirs and non-productive shales were handled by turning off the coalbed related special features by model layer and forcing the simulator to behave as a conventional dual-porosity model in those layers. In the aerial sense, lithology and flow-related parameter changes (especially in the mining layer and in the caved and subsided zones as the face was advancing) were modeled by assigning different rock regions in which the reservoir parameters, including the fracture properties, were varied.

Table 2 gives representative reservoir parameters used in the model for coal and non-coal lithologies that were calculated, obtained from the literature, and/or estimated from history matching techniques. These values, particularly permeabilities, represent the values prior to mining and thus prior to any disturbance on the strata. In order to make geomechanical calculations to determine the response of the overlying strata to mining-induced stresses and to calculate the permeability changes, the previously described FLAC software was used. The changes in competent rock properties in the strata overlying the panel during and after mining disturbance were coupled with the reservoir model during simulation of the longwall face advance.

The most important coal-related reservoir parameters are the nature of the cleats, their spacing, gas content, and adsorption data. Law [21] reported cleat spacing of 0.5 to 9.7 cm (0.016 to 0.32ft) in the Northern Appalachian Basin. In this study, cleat spacing was estimated at 3.0 cm (0.1ft), which was a value based on the reported mean cleat spacing values of 2.4 cm (0.08ft) and 3.2 cm (0.11ft) from outcrops of the Pittsburgh and Sewickley Coalbeds, respectively [21]. The gas content and adsorption related data for the Pittsburgh Coalbed and for the other major coalbeds in the area were obtained from the results of laboratory methane adsorption isotherm tests and direct method gas content determination tests [22] on various coal samples. The fracture permeabilities for the Pittsburgh, Sewickley and Waynesburg Coalbeds were estimated to be 4 md in the face cleat direction and 1 md in the butt cleat direction as estimated by history matching techniques. These values were taken to be uniform throughout the layers.

The permeability of the caved zone in the longwall gob (Figures 1 and 2) is not easily predictable since it mostly depends on the type of rock units, the extent of their fragmentation and packing in the void space, and caving height. Therefore, it is natural to assume that this permeability value may vary widely for different locations and for different roof rocks when they cave. Brunner [23] investigated the problem of the migration of ventilation air through longwall gobs using a mine ventilation model and a laminar resistance grid constructed over the caved zone. He assigned a resistance value to each resistance element in the grid based on idealized gob conditions. The gob permeability values he used and tested in his model ranged between 10^{-5} and 10^{-7} m^2 (10^{10} and 10^8 md). Based on a fractured porous medium vs. permeability classification, these values fall into the heavily fractured rock category. In this study, the permeability of the caved zone was estimated using history matching techniques. The detailed approach and analysis of this parameter for the study resulted in a permeability of about 10^{-8} m^2 (10^7 md) in the horizontal direction and 10^{-9} m^2 (10^6 md) in the vertical direction, due to compaction of the gob. These values fall into the open-jointed to heavily fractured rock classification in the horizontal direction and into jointed rock classification in the vertical direction.

In the model constructed for this study, the open entries surrounding the longwall panel serve as main paths for ventilation airflow. The permeabilities for the area including the three entries and intervening coal pillars were assigned high values [9×10^{-7} m^2 (10^8 md)] within the allowed limits of the simulator for minimum resistance. The associated fracture porosity and fracture spacing for these areas were calculated from the mine maps as 40% and 30 m (100ft) respectively.

Simulation Strategy for Longwall Mining

During longwall mining, the strata disturbances described previously move along with the face, and gas production from the ventholes starts at the time each successive venthole location is intercepted by the mining-induced disturbances. This leads to a moving-boundary problem for modeling this type of process. In this study, the problem was addressed with "restart" models, each of which represents a time interval for the mining advance and related strata disturbance. Each model restart run was performed so that it would progress up to either the next venthole location or to a defined location on the panel for the distance and time characterizing the intervening face movement. In this process, the simulation outputs from the previous model run were written in a "restart" file, then used by the next model run, which updated the longwall face and reservoir changes for the next face position.

RESULTS AND DISCUSSION

Model Calibration and History Matching of Actual Production Data

History matching is a well-accepted method in reservoir engineering for calibrating models and estimating unknown parameters. There is always a valid concern as to the degree of correctness of calibrations and the range for unknown parameters determined as a result of history matching. However, it has been shown and widely accepted that with careful analysis of reservoir conditions and a good sense of the reasonable value ranges for the unknowns and their expected behaviors, history matching is a valid and valuable technique. This is especially true in a situation where every parameter cannot be determined by practical means and many unknowns are present, therefore, history matching is probably the only choice.

In this study, performing a dynamic simulation approach by restart models to simulate the advance of a longwall face (described previously) was an integral part of the calibration and history matching phase of the simulation, as well as for modeling venthole performance scenarios for different panel widths. The major matching parameters that were changed at each restart were disturbed/undisturbed permeability values for the different horizons, desorption time constants for coal, relative permeabilities, and permeability of the caved zone. During calibration and prediction runs, the fractured rock permeabilities predicted by FLAC computations, fracture spacing changes, coal desorption time changes, relative permeabilities for subsided and fractured rock, and transmissibilities in the vertical direction were incorporated into the zones above the mined out section of the panel as the face was advanced to various locations between restarts. Since modifying porosity between restarts for different regions causes material balance instabilities, this effect was addressed with fracture spacings. During dynamic runs, all known and fixed values and the initial estimates of the unknowns were used, and field measurements of gob gas venthole production values were compared with model predictions. Based on this comparison for the entire 268-day simulation, adjustments were made to the unknown parameters, and additional calibration runs were completed until an acceptable agreement was obtained.

Longwall gob gas ventholes are normally not water producers, except for occasional condensation of water vapor at the well head. Most of the strata water, especially within the caved zone, flows to the mine void unless this zone has been desaturated or drained previously by other means. Although there might be regional differences, Hunt and Steele [24] report that the Central and Northern Appalachian Basin is depressurized and desaturated due to its geological history, extensive coal mining, and the drilling of numerous oil and gas wells through the coalbeds to deeper horizons. No water production data were available for this panel; however, no major water influx into the caved zone was reported by the operating mining company. Therefore, although the model has been set up as a two-phase model for flow, water flow was turned off by setting the connate water saturation to initial water saturation values in the relative permeability curves assigned to each layer.

The study panel has four gob gas ventholes that were continuously monitored by NIOSH for gas production rates and volumes (corrected to STP conditions), and methane concentration in the produced gas stream. The measurement of flowing bottom-hole pressures is not commonly done in these types of wells due to the difficulty in instrumenting the ventholes for this purpose. These ventholes are produced using exhausters that create a negative pressure against the mine ventilation pressure to help extract the gob gas. The pressure sink created by the producing gob gas ventholes also helps to induce gas flow towards the ventholes, which helps to reduce the volume of methane that can enter the underground workplace.

Some of the produced methane is used to fuel the engines that operate the exhausters. While using produced methane to fuel the engines is an operational advantage for remote locations, operationally (and for modeling purposes) this sometimes presents a problem. If the methane concentration in the produced gas stream falls below the minimum requirement to run the engines, or when mechanical or other operational problems result in the wells going off-line or running at less than optimum conditions, changes occur in the flowing bottom-hole pressures of the ventholes. In fact, pressure measurements taken at the well head below the flame arrestors indicate that well pressures are not constant, but fluctuate depending on various factors. Thus, in the absence of reliable flowing bottom-hole pressure data, the alternative approach that was taken for this modeling study was to use gas flow rates as the constraints in the model at the expense of computational time and convenience, and to match the flow rates as well as concentration and pressure values to field data and expectations.

Figures 8 and 9 show the actual and simulated gas production rates and methane concentrations for the four gob gas ventholes monitored on the first longwall panel in the study area. The production rate and methane concentration data match quite well for most of the field measurements for the entire 268-day period when mining was active on this panel. The shifts on the time scale for each well to become productive are caused by, and depend on, when the well location was intercepted by the longwall face.

The calculated bottom-hole pressures for the four gob gas ventholes are given in Figure 10. This figure shows that when the venthole location is intercepted by the longwall face on the panel, the pressure sharply drops to atmospheric pressures or below due to their completion depths being in close proximity to the mine ventilation pressures (less than atmospheric pressure since exhausting ventilation system is used) by way of the mining induced fractures. This was the criterion for accepting the calculated bottom-hole pressures and was basically maintained by the mining-induced fracture permeabilities assigned to each layer. Even during the periods when the ventholes are shut-in, they stay close to or under atmospheric pressure conditions as expected. The bottom-hole pressures calculated during this step were used later as the operating constraints for the different gob gas venthole placement scenarios, presented in the following discussion.

Analysis of the Effect of Longwall Panel Width on Methane Emissions and Gob Gas Venthole Performance

The mine operating in the study area currently has a single completed longwall panel in the new mining district with a width of 381 m (1,250ft) and a length of approximately 3,350 m (11,000ft). The mine plan includes one additional longwall panel of the current dimension, then increasing the panel width by 61 m (200ft) to 442 m (1,450ft). Since the purpose of this study is to investigate the effect of increased panel widths on gas emissions, the primary question to be answered from a mine safety and methane control perspective is whether the current number and configurations of ventholes will adequately control the projected increase in gob gas on the larger longwall panels. To answer this question, the venthole production from the first panel in the district has been analyzed with the assumptions and limitations stated previously, and then compared to the simulated venthole production and the methane emissions for the increased panel width.

Different numerical model runs for well placement and operating scenarios were performed for the increased panel width to investigate the performance of different options to optimize the extraction of gob gas by the ventholes and to thus minimize the migration of gob gas to the ventilation system. In these scenarios, the four actual ventholes from the first panel were placed and produced based on their measured production histories as discussed in the previous sections, then operated according to the bottom-hole pressure history constraint calculated in the calibration/history matching phase of this study. Additional optimally completed and produced ventholes - i.e., ventholes completed in the fractured zone ~12 m (~40ft) above the top of the Pittsburgh Coalbed layer with 61-m (200-ft) of slotted casing and operated with -0.2 atm (-2.7 psi) suction pressure at all times after they become productive were then added as infill ventholes. Unless specified differently, these additional ventholes also used 17.8-cm (7-in) diameter casing. In two of the case scenarios, only optimal ventholes were used with different configurations in order to produce the same amount of methane as in the actual and optimal well combinations to minimize the number of ventholes. Scenarios were also simulated where the well casing diameter was increased to 20.3 cm (8 in) for the 442-m (1,450-ft) wide panel, and alternatively decreased to 12.7 (5 in). The various venthole placement scenarios and completions (cases) are presented schematically in Figure 11.

Analysis of Expanding Panel Width from 381 m (1,250ft) to 442 m (1,450ft) with only Four Actual Ventholes in Operation

For this analysis, the results of simulation runs for the 381-m (1,250-ft) wide panel using the four actual gob gas ventholes were compared with the results when the panel width was increased to 442 m (1,450ft) using the same venthole placement, completions, and production history. In order to expand the panel size, the grids representing entries and the panel on the headgate side were moved 61 m (200ft) outward within the fine-grid section in the model, and the reservoir properties in this and overlying layers were assigned according to the new geometry. The locations of the ventholes that were 91 m (300ft) from the tailgate side of the panel, the completion data, and their operating constraints were kept constant between two cases.

Figure 12 presents the simulated methane concentrations and methane production rates for the two panel sizes using the same venthole performance criteria. As can be seen from these figures, the gob gas venthole performance predicted for changing panel sizes are very similar with the same set of operating ventholes in place. This behavior is to be expected once the wellbore flow model (Eq. 3) is analyzed, and its parameters are used by the reservoir simulators. The wellbore model equation is basically a summation of flow, as defined by the product of productivity index (Eq. 4) by total mobility (Eq.

5) and difference between grid block and the well bottom hole pressures from individual layers for a multilayer completion.

$$Q_J = \sum_l PI_l \lambda_{T,l} (P_g - P_{wbh}) \tag{3}$$

$$PI = 2 f_{comp} kh \frac{w_{frac}}{\ln\left(\frac{r_e}{r_w}\right) + S} \tag{4}$$

$$\lambda_T = \sum_j \frac{k_{rj}}{\mu_j} \tag{5}$$

The parameters used in these equations are listed in the nomenclature section at the end of this paper.

It should be noted that there is no term directly related to panel size in Eqs. 3-5. The most likely parameter that may be influenced as a consequence of subsidence variations for the increased panel width is the permeability associated with the productivity index term (Eq. 4). However, for supercritical panels (as are both of these base cases), subsidence should be complete and thus stress conditions should be similar irrespective of the panel width. In fact, the FLAC computations confirmed this behavior. This behavior also suggests that differences in the wellbore conditions should not be expected between the two cases. Geomechanical calculations showed that there is not a major change in vertical and horizontal permeabilities in the subsided regions as a result of changing the panel width, as shown in Figure 13. This figure shows the permeabilities for 366-m (1,200-ft) and 488-m (1,600-ft) panel widths; we can expect similar permeability changes for the 381-m (1,250-ft) and 442-m (1,450-ft) cases.

The gob gas venthole performance in terms of methane production for the two base cases was also similar as shown in Figure 12. The cumulative methane production at the end of simulations of 268 days of mining is only about 0.95 MMSCF (~1%) less for the wider panel base case. This may be due to the increased distance methane released on the headgate side of the panel must travel to the ventholes that are located on the tailgate side of the panel. In the event that all of the gas released from the surrounding strata on the headgate side of the panel does not efficiently migrate to the ventholes on the tailgate side, additional methane must be handled by the mine ventilation system.

Although the performance of gob gas ventholes is not influenced significantly with an increased panel width of 61 m (200ft), the simulation results indicate that an additional 137 MMSCF (355 cfm) of methane will be released from the overlying strata as a result of mining on the larger panel over the 268 days of simulation. The additional methane liberated from the overlying strata must be handled by the mine ventilation system when only the four actual venthole are operated on the 442-m (1,450-ft) wide panel.

Alternative Gob Gas Venthole Placement Scenarios and Performance Analysis for a 442 m (1,450ft) Wide Panel

The following discussions present alternative gob gas venthole placement and completion scenarios to investigate potential options to maximize methane capture to minimize the methane loading of the ventilation system on the 442-m (1,450-ft) wide panel in the Pittsburgh Coalbed. It should be noted that these simulations are based on site-specific reservoir simulations and are not intended to be a general prediction of gob gas venthole performance for other coalbeds.

The alternative gob gas venthole placement and completion scenarios (cases), as shown schematically in Figure 11, are: (A) moving the existing four actual ventholes to locations 152 m (500ft) from tailgate side, i.e., 61 m (200ft) closer to the centerline of the panel than on the first panel; (B) adding four optimal infill ventholes located between each actual well; (C) adding four optimal ventholes located

90 m (300ft) from the gateroads on the headgate side of the panel, positioned diagonally to the actual ventholes; and (D) adding four optimal ventholes located 90 m (300ft) from the gateroads on the headgate side and positioned directly opposite the actual wells on the tailgate.

In addition to the venthole placement scenarios (Cases A-D), the cased diameters of the ventholes were changed for two different cases as shown on Figure 11. For Case E, the diameters of four actual ventholes were increased to 20.3 cm (8 in) without any optimal ventholes in the model, and for Case F, the diameters of the ventholes in Case B were decreased to 12.7 cm (5 in). The results of each of these scenarios were compared to the actual production obtained from the four ventholes on the 381-m (1,250-ft) panel. To determine the minimum number of optimal ventholes that will produce the same maximum amount of methane as in the highest-producer configuration of actual and optimal ventholes, two final scenarios where simulated. In Case G, six optimal ventholes were used along the tailgate side of the panel, and in Case H, two of the optimal ventholes were placed on the headgate side opposite the first two ventholes on the tailgate side Figure 11.

In order to compare the production performance of optimal ventholes that are included in the scenarios with the performance history of the four actual ventholes on the first panel, and also to emphasize the importance of completing and operating the wells properly, simulation runs were conducted to replace the four simulated actual ventholes with four optimal ventholes on the 442-m (1,450-ft) wide panel. As shown on Figure 14, optimal ventholes produce gas that is 85-90% methane through the entire mining period, whereas the methane concentration of the gas produced by the actual wells in the study area decline significantly with time, with the methane concentration being in the 65-70% range for most of the mining period. The difference in methane concentrations between the two scenarios is due to the fact that some of the actual ventholes were drilled into or very close to the caved zone, and thus their produced methane was diluted with mine ventilation air.

The total cumulative methane production is about 98.6 MMSCF for the four actual ventholes over the 268 days simulated for the 442-m (1,450-ft) wide panel, whereas the four optimal ventholes produce about 141 MMSCF of methane, or about 50% more than the actual wells produced on the first panel in the study area. This basic analysis points out the importance of implementing proper completion practices and paying close attention to gob gas ventholes' operation, i.e., keeping them running, to optimize their methane control capability.

Figures 15 and 16 show methane concentrations and cumulative methane produced, respectively, from gob gas ventholes located with the different configurations (A-F) shown on Figure 11. Methane concentration data presented in Figure 15 show that the highest and lowest methane concentrations result when the diameters of the wells are reduced (Case F) and increased (Case E), respectively. The decrease in methane concentration in the produced gas from the four simulated larger diameter ventholes (Case E) may be due to the increased capture of mine air from the four actual ventholes, several of which were drilled into the caved zone.

The next configurations to produce gas of lower methane concentration are those only using the four actual ventholes [the 381-m (1,250-ft) base case and Case A with the four actual 17.8-cm (7-in) ventholes placed at 152 m (500ft) from the panel margin instead of at 90 m (300ft)]. Again, the reason for the lower methane concentration in the produced gas is attributed to the increased capture of mine air from the simulated actual ventholes, several of which were drilled into the caved zone. Gob gas venthole configurations, B, C, and D, on the other hand, produce similarly high methane concentrations. This is due to the infill optimal ventholes and their associated completion and performance parameters, as discussed previously.

Figure 16 and the accompanying data summarized in Table 3 compare the cumulative methane production volumes from each of the gob gas venthole configuration scenarios (Cases A-F) shown in Figure 11. The lowest cumulative methane production (Case A) is obtained when the four ventholes with the actual completions are located 152 m (500ft) from the tailgate entry of the 442-m (1,450-ft) panel. This configuration produces 13.2 MMSCF less methane compared to the base case production from the

381-m (1,250-ft panel), thus illustrating the importance of near-margin well placements where enhanced fracture permeabilities exists for a longer time [7].

The highest cumulative methane productions are achieved when additional optimal infill ventholes are used. Simulation results show that Case D, with the four actual ventholes on the tailgate side of the panel and 4 additional optimal ventholes located on the headgate side of the panel (directly opposite the tailgate ventholes), produces 138 MMSCF more methane compared to the four actual ventholes on the 381-m (1,250-ft) panel. Case C, with four actual ventholes on the tailgate side of the panel and 4 additional optimal ventholes located on the headgate side of the panel (but diagonal to the tailgate ventholes) is the next highest incremental producer of methane at 119 MMSCF. While Cases C and D are similar, Case D is probably the higher producer because the tailgate and headgate ventholes are intercepted at the same time, resulting in a quicker overlap of their drainage radius which enhances gas desorption from the overlying coalbeds. Also, when headgate and tailgate ventholes are intercepted at the same time, the headgate ventholes start producing earlier and stay on production longer compared to Case C. The third highest incremental methane producer is Case B (four actual and four optimal infill ventholes on the tailgate side of the panel) with 113 MMSCF more methane than the four actual ventholes on the 381-m (1,250-ft) panel.

This analysis shows that drilling additional gob gas ventholes increases gas production considerably, provided that they are drilled and operated properly. Also, placing optimal wells on the headgate side of the panel may improve production by a marginal amount by capturing additional gas that is released from the overlying strata on that side of the panel. This may become an important methane control issue as panel widths continue to increase, because at some critical panel width it is possible that gob gas originating in the headgate area may not be able to efficiently migrate to the pressure sink of the wells placed on the tailgate side of the panel. Under these conditions, the additional gob gas may require additional gob gas ventholes to be placed in this area.

The cases where the diameters of four actual gob gas ventholes were increased by one inch (Case E) and the diameters of all gob gas ventholes, including the infill ventholes, were decreased by two inches (Case F) produce 31 and 38 MMSCF more methane, respectively, compared to the 381-m (1,250-ft) base case. However, this incremental amount is much less compared to the cumulative methane captured by 7-in ventholes (Cases B, C, and D).

One of the main considerations in gob gas venthole design and operation should be to locate and drill the ventholes optimally and operate them continuously. This will minimize the number of ventholes that need to be drilled to produce the same amount of methane that can be achieved with a larger number of less than optimally drilled and operated ventholes. In scenarios G and H (Figure 11), the performance of two configurations using six optimal gob gas ventholes were simulated and compared with the performance of Case D, the highest methane producer (138 MMSCF).

Figure 17 and the accompanying data summarized in Table 3 compare the cumulative methane production from gob gas venthole configuration Cases G and H with that of Case D. The data shows that Case G, where six optimal ventholes were located along the tailgate side of the panel, produced about 86 MMSCF more methane than the 381-m (1,250-ft) base case, whereas Case H with two of the optimal wells on the headgate side produced 134 MMSCF more methane. Thus, the six optimal ventholes of Case H produced almost as much methane as the eight (four actual and four optimal) ventholes of Case D. The performance difference between Case G and H are due to the location of the ventholes and to the length of time they stay on production.

Analysis of the Amount of Methane Released on the Longwall Face as a result of Expanding Panel Width from 381 m (1,250ft) to 442 m (1,450ft)

The approximate amount of methane present in the coal at reservoir conditions which may report to the ventilation system as face emissions can be calculated using adsorption isotherm analysis performed on Pittsburgh coal samples. The Langmuir volume and pressure for Pittsburgh coal samples were 490 scf/ton and 350 psi, respectively, on a moisture equilibrated basis. Thus, the amount of methane that can theoretically be released when the in situ reservoir pressure (~ 90 psi) is reduced to atmospheric pressure is about 80 scf/ton. For the 581,500 tons of incremental coal that will be mined when the panel size is increased by 61 m (200ft), the cumulative methane volume will be about 47 MMSCF. This suggests that any additional amount of methane greater than 47 MMSCF in the simulations will come either from the immediate roof or from overlying strata, and may migrate to the ventilation system if not captured by the gob gas ventholes.

It should be noted that the actual mined panel at the study site was partially degasified with horizontal methane drainage boreholes. This volume of produced methane should be subtracted from the 47 MMSCF of predicted total methane given above for the additional 61 m (200ft) of mined panel width. However, no methane production data is available for those horizontal boreholes; therefore, this unknown volume of actual produced methane has not been subtracted from the predicted increase in longwall face emissions.

Impact of Gob Gas Venthole Configurations on Controlling Ventilation System Methane on a Wider Face

From a mine safety perspective, it is also important to evaluate the impact of gob gas venthole configurations on controlling the migration of methane to the ventilation system on wider longwall panels. As discussed previously, modeling an increase in panel width from 381 m (1,250ft) to 442 m (1,450ft) results in about 47 MMSCF of additional methane liberation from the coal mined on the longwall face and 137 MMSCF from the overlying disturbed strata, i.e., gob gas, for a total of 184 MMSCF of additional methane. The 47 MMSCF of additional methane liberation on the face will report to the ventilation system unless additional methane drainage provisions such as horizontal boreholes are used. Alternatively, ventilation airflow can potentially be increased to dilute the additional gas. The additional 137 MMSCF of methane released from the overlying strata may report to the ventilation system if no extra methane control measures (such as additional gob wells or different configurations) are implemented. This volume of cumulative methane represents the potential of an average of about 355 cfm of additional methane entering the ventilation system over the 268 days of the simulated mining for this study. The following discussion will address the additional 137 MMSCF of gob gas that may enter the ventilation system and the effectiveness of the various configurations of simulated ventholes to remove the additional gob gas resulting from the mining of the 442-m (1,450-ft) wide panel.

Table 4 presents the amount of uncaptured methane that will be available for flow to the ventilation system on a 442-m (1,450-ft) wide panel with the various gob gas venthole configurations that have been simulated relative to the gob gas volume for a 381-m (1,250ft) panel. The greatest amount of uncaptured methane for potential flow from the overlying disturbed strata to the ventilation system (148 MMSCF or 380 cfm) is Case A, where the gob gas ventholes are placed an additional 61 m (200ft) away from tailgate side towards the center of the panel. The minimum amounts of uncaptured methane reporting to the ventilation system on the wider panel occurs with Case B (25 MMSCF or 65 cfm), Case C (17 MMSCF or 44 cfm), and Case D (-1 MMSCF) when optimal ventholes are used. In particular, Case D (where the additional optimal ventholes are operated on the headgate side of the panel directly opposite the existing actual ventholes) is very effective in reducing the potential volume of methane reporting to the ventilation system. The minimal uncaptured methane flow into the ventilation system for Case D is comparable to the 3 MMSCF uncaptured methane for Case H where six optimal ventholes are used (Figure 11) instead of eigth ventholes (four actual and four optimal).

SUMMARY AND CONCLUSIONS

Increasing the size of longwall panels may also increase methane emissions due to the exposure of the mining environment to a larger area of fractured, gas-bearing strata. Thus, understanding the impact of increased panel width on methane emissions, and designing gob gas ventholes and bleeder systems accordingly, can enhance the safety of the underground workforce.

In this study, a 3-D dynamic reservoir model was developed to simulate longwall operations for a mine operating in Pittsburgh Coalbed. Permeability data for the mining-induced disturbed rock mass were computed by a geomechanical model as input to the reservoir model. The reservoir model was calibrated by matching historic gas production rates and methane concentration data from four actual gob gas ventholes while keeping the well pressures at expected levels for the 381-m (1,250-ft) wide panel in the study area to determine unknown parameters and constraints. This base case was then used as a comparison for the simulation predictions performed for the 442-m (1,450-ft) wide panel. Reservoir simulations for various gob gas venthole configurations were used to investigate the effect of mining a panel of 61 m (200ft) of additional face length on venthole performance. The effectiveness of these gob gas venthole configurations to drain gas from subsided strata on the wider panel, thus minimizing methane emissions to the ventilation system, was also addressed in the simulations.

The general conclusions from the simulations conducted for this study can be summarized as follows:

1. Increasing the longwall panel width by 61 m (200ft) resulted in an additional 184 MMSCF of cumulative methane emissions (47 MMSCF from the longwall face, and 137 MMSCF from the overlying disturbed strata) to the ventilation system if not captured by an optimized methane control system.

2. Production from individual gob gas ventholes does not change appreciably when the longwall panel width is increased, provided that permeability and saturation fields are not affected. Therefore, given the increased volume of gas available for flow to the ventilation system on the wider panel, an optimized methane control system is probably required to keep methane concentrations in the ventilation system within safe, statutory limits.

3. The performance of optimally completed and operated gob gas ventholes are far superior to the actual wells that were drilled and operated on the first panel at the study mine site. Optimal gob wells produce about 50% more methane for the same period of time with very high methane concentrations (85%-90%). This shows the importance of operating the gob gas ventholes continuously and monitoring them constantly to minimize the amout of time they are off-line.

4. Moving the gob gas ventholes 61 m (200ft) away from their actual location on the tailgate side of the panel towards the centerline resulted in a 12 MMSCF decrease in their simulated gas drainage capability.

5. Simulating an increase in the diameters of the four actual wells by one inch on a wider panel increases their methane production by about 38 MMSCF compared to the same wells with 7-in diameter casings on the 381-m (1,250-ft) panel. However, this is not enough of an increase in gas drainage to eliminate the increase in methane emissions resulting from mining the larger panel. The same is true in the case of employing four optimal infill ventholes along with the four actual ventholes, while decreasing their diameters to 5 inches in an attempt to reduce drilling costs.

6. The optimum configuration simulated to increase gob gas venthole methane recovery on the wider panel was six optimally completed and produced ventholes (four on the tailgate and two on the headgate) and four actual ventholes on the tailgate and four additional optimal ventholes located on the headgate side of the panel. Placing eight ventholes on the tailgate side of the panel (four actual and four optimal infill ventholes) was a slightly less productive configuration.

7. Increasing the face length from 381 to 442 m (1,250 to 1,450ft) on longwall panels in the Pittsburgh Coalbed will require that additional methane control measures be implemented, including increasing the number of gob gas ventholes completed in an optimal configuration with the ventholes operated in a manner to reduce downtime, thus effectively draining a larger volume of the additional gas expected to be encountered on the wider panels.

NOMENCLATURE

f_{comp} Fraction of grid block penetrated by the well

h	Thickness
k	Permeability
k_h	Horizontal permeability
k_v	Vertical permeability
k_{h0}	Horizontal permeability (initial)
k_{v0}	Vertical permeability (initial)
$k_{r,j}$	Relative permeability to the phase
w_{frac}	Fraction of well
r_e	Effective radius of the well
r_w	Wellbore radius
Q_J	Total flow rate of the phase
PI_l	Productivity index from layer l
P_g	Grid block pressure
P_{wbh}	Well bottom-hole pressure
S	Wellbore skin
$\lambda_{T,l}$	Total mobility of the phase
μ_j	Viscosity of the phase
σ_{yy}	Vertical stress
σ_{xx}	Horizontal stress

SI METRIC CONVERSION FACTORS

ft x 0.3048	m
md x 10^{-15}	m^2
ft^3 x 28316.85	cm^3
psi x 6894.757	Pa
inch x 2.54	cm
cfm x 0.00047195	m^3/s

ACKNOWLEDGMENTS

The valuable assistance of Fred Garcia (NIOSH) in designing and conducting the gas flow monitoring field studies used in this paper is gratefully acknowledged. The interest and assistance of the management and engineering staff of the cooperating mine that provided access and data for the successful completion of this study is gratefully acknowledged. It is through cooperative research efforts of this kind that important mine safety and health innovations can be made readily available to the industry as a whole.

REFERENCES

1. Singh, M.M and Kendorski, F.S. 1981. "Strata Disturbance Prediction for Mining Beneath Surface Water and Waste Impoundments", Proc. 1st Conference on Ground Control in Mining, pp. 76-89.

2. Palchik, V., 2003. "Formation of Fractured Zones in Overburden due to Longwall Mining", Environmental Geology, 44, 28-38.

3. Mucho, T., 2003. Personal Communications.

4. Diamond, W. P., Ulery, J. P., and Kravitz, S. J., 1992. "Determining the Source of Longwall Gob Gas: Lower Kittanning Coalbed, Cambria County, PA", Bureau of Mines, Information Circular No: 9430.

5. Lunarzewski, L. 1998. "Gas Emission Prediction and Recovery in Underground Coal Mines", International Journal of Coal Geology, 35, pp. 117-145.

6. Mucho, T.P., Diamond, W.P., Garcia, F., Byars, J. D., and Cario, S.L., 2000. "Implications of Recent NIOSH Tracer Gas Studies on Bleeder and Gob Gas Ventilation Design", Society of Mining Engineers Annual Meeting, Feb. 28-Mar. 1, Salt Lake City.

7. Diamond, W. P., Jeran, P. W., and Trevits, M. A., 1994. "Evaluation of Alternative Placement of Longwall Gob Gas Ventholes for Optimum Performance", Bureau of Mines, Information Circular No: 9500.

8. Mills, R. A., and Stevenson, J. W., 1991. "History of Methane Drainage at Jim Walter Resources, Inc." Paper 9133, Proceedings of the 1991 Coalbed Methane Symposium, Tuscaloosa, AL, May 13-17, pp. 143-151.

9. Dixon, C. A., 1987. " Coalbed Methane: A Miner's Viewpoint", Paper 8704, Proceedings of the 1987 Coalbed Methane Symposium, Tuscaloosa, AL, Nov 16-19, pp. 7-10.

10. Diamond, W.P and Garcia, F. 1999. "Prediction of Longwall Methane Emissions: An Evaluation of the Influence of Mining Practices on Gas Emissions and Methane Control Systems", NIOSH, Report of Investigations No: 9649.

11. Schatzel, S., Krog, R.B., Trackemas, J., Garcia, F., and Marshall, J. 2005. "Face Methane Liberation Evaluation for Wider Longwall Faces", presented at the SME Annual Meeting and Exhibition, Salt Lake City, UT, Feb 28-March 2.

12. Itasca Consulting Group, Fast Lagrangian Analysis of Continua, 2nd Edition, Minnesota, 2000.

13. Computer Modeling Group Ltd. 2003. "Generalized Equation of State Model-GEM", User's Guide, Calgary, Alberta, Canada.

14. Rusnak, J.A. and Mark, C. 1999. "Using the Point Load Test to Determine the Uniaxial Compressive Strength of Coal Measure Rock", Proc. 19th Int. Conference on Ground Control in Mining, pp. 362-371.

15. Hoek E and Bray JW. Rock Slope Engineering, 3rd edition, Inst. Min. Metall. London, 1981.

16. Booth C.J. A Numerical Model of Groundwater Flow Associated with an Underground Coal Mine in the Appalachian Plateau, Pennsylvania. PhD. Thesis, University of Pennsylvania. 1984.

17. Franklin J.A. and Dusseault M.B. Rock Engineering, McGraw-Hill, 1989.

18. Ren TX, and Edwards JS. Goaf gas modeling techniques to maximize methane capture from surface gob wells. Mine Ventilation (2002): 279-286.

19. Lowndes, I.S., Reddish, D.J., Ren, T.X., Whittles, D.N., and Hargreaves, D.M. 2002. Improved Modeling to Support the Prediction of Gas Migration and Emission from Active Longwall Panels, Mine Ventilation, ed. Euler De Souza, Balkema. pp 267-272.

20. Esterhuizen, G. and Karacan, C.O., 2005. "Development of Numerical Models to Investigate Permeability Changes and Gas Emission Around Longwall Mining Panels", Proc. AlaskaRocks 2005, 40th US Symposium on Rock Mechanics, Anchorage, Alaska, 25-26 June.

21. Law, B. E. 1993. "The Relationship between Coal Rank and Cleat Spacing: Implications for the Prediction of Permeability in Coal", Proceedings of the 1993 Coalbed Methane Symposium, Tuscaloosa, AL, May 17-21, pp. 435-441.

22. Diamond, W. P., La Scola, J. C., and Hyman, D. M., 1986. "Results of Direct-Method Determination of the Gas Content of the US Coalbeds", Bureau of Mines, Information Circular No: 9067.

23. Brunner, D. J., 1985. "Ventilation Models for Longwall Gob Leakage Simulation", Proceedings of 2nd US Mine Ventilation Symposium, Reno, Nevada, September 23-25, pp. 655-663.

24. Hunt, A. M. and Steele, D. J., 1991. "Coalbed Methane Development in the Appalachian Basin", Quarterly Review of the Methane from Coal Seams Technology, 1 (4), 10-19.

Table 1. Initial horizontal and vertical permeabilities used in the FLAC models.

Rock Class	Rock types	Horizontal Permeability, md	Vertical Permeability, md
Soil	Clay bands	0.1	0.1
Very low permeability	Black Shale	0.2	0.1
Low permeability	Gray shale	1.0	0.5
Coal-face cleat direction	Coal	4.0	0.1
Coal-butt cleat direction	Coal	1.0	0.1
Moderate permeability	Limestone	2.0	2.0
High permeability	Sandstone	10.0	10.0

Table 2. Representative examples of prior-to-mining basic reservoir-rock properties (for fractures) used in the study.

Parameter	Pittsburgh Coalbed	Sandstone	Limestone	Shale	Entries
Permeability-x, md	4.0	10	2.0	0.2	$9x10^7$
Permeability-z, md	0.1	10	2.0	0.1	$9x10^7$
Effective Porosity, %	4.0	10	2.0	1.0	40
Effective Fracture Spacing, m/(ft)	0.03/(0.01)	15/(50)	60/(200)	60/(200)	30/(100)
Langmuir Pressure, MPa/(psi)	2.38/(350)	-	-	-	-
Langmuir Volume, cc/g (scf/ton)	15.4/(490)	-	-	-	-
Desorption Time, days	20	-	-	-	-
Coal Density, g/cc	1.35	-	-	-	-

Table 3. Cumulative predicted methane production differences from gob gas ventholes on a 442 m (1,450-ft) wide panel relative to the production from the 381-m (1,250-ft) wide panel base case. Case configurations are shown schematically in Figure 11.

Case ID, Gob Gas Venthole Configuration TG = Tailgate HG = Headgate	Cumulative Methane Production Difference, MMSCF
4 actual, 90 m (300ft) to TG	- 0.9
A, 4 actual, 152 m (500ft) from TG	- 13
B, 4 actual +4 optimal infills, all 90 m (300ft) from TG	113
C, 4 actual 90 m (300ft) from TG +4 optimal, 90 m (300ft) from HG, diagonal to TG ventholes	119
D, 4 actual 90 m (300ft) from TG +4 optimal, 90 m (300ft) from HG, directly across from TG ventholes	138
E, 4 actual, 90 m (300ft) from TG, 8-in diameter	38
F, 4 actual, 90 m (300ft) +4 optimal infills, all 90 (300ft) from TG, 5-in diameter	31
G, 6 optimal, all 90 m (300ft) from TG	86
H, 4 optimal 90 m (300ft) from TG +2 optimal 90 m (300ft) from HG, directly across from TG ventholes	134

Table 4. Cumulative total predicted uncaptured methane volume potentially available for flow from the overlying disturbed strata to the ventilation system for the simulated gob gas venthole configurations on the 442-m (1,450-ft) panel as compared to the 1,250-ft face actual base case.

Case ID, Gob Gas Venthole Configuration TG = Tailgate HG = Headgate	Cumulative Uncaptured Methane Volume, MMSCF (cfm)
4 actual, 90 m (300ft) to TG	137 (355)
A, 4 actual, 152 m (500ft) from TG	148 (380)
B, 4 actual +4 optimal infills, all 90 m (300ft) from TG	25 (65)
C, 4 actual 90 m (300ft) from TG +4 optimal, 90 m (300ft) from HG, diagonal to TG ventholes	17 (44)
D, 4 actual 90 m (300ft) from TG +4 optimal, 90 m (300ft) from HG, directly across from TG ventholes	-1
E, 4 actual, 90 m (300ft) from TG, 8-in diameter	99 (255)
F, 4 actual, 90 m (300ft) +4 optimal infills, all 90 (300ft) from TG, 5-in diameter	104 (270)
G, 6 optimal, all 90 m (300ft) from TG	50 (130)
H, 4 optimal 90 m (300ft) from TG +2 optimal 90 m (300ft) from HG, directly across from TG ventholes	3 (10)

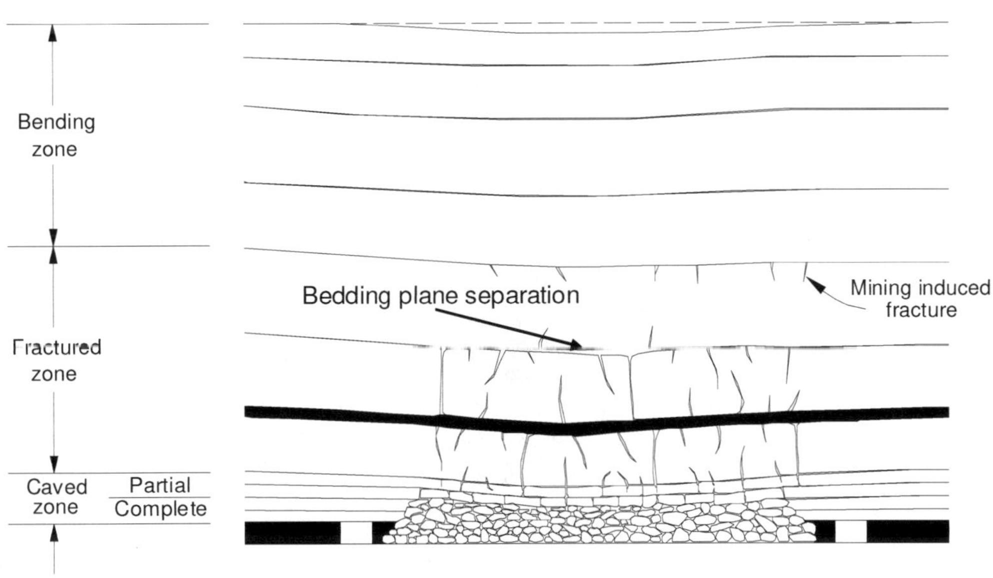

Figure 1. Schematic of strata response to longwall mining (modified from [1])

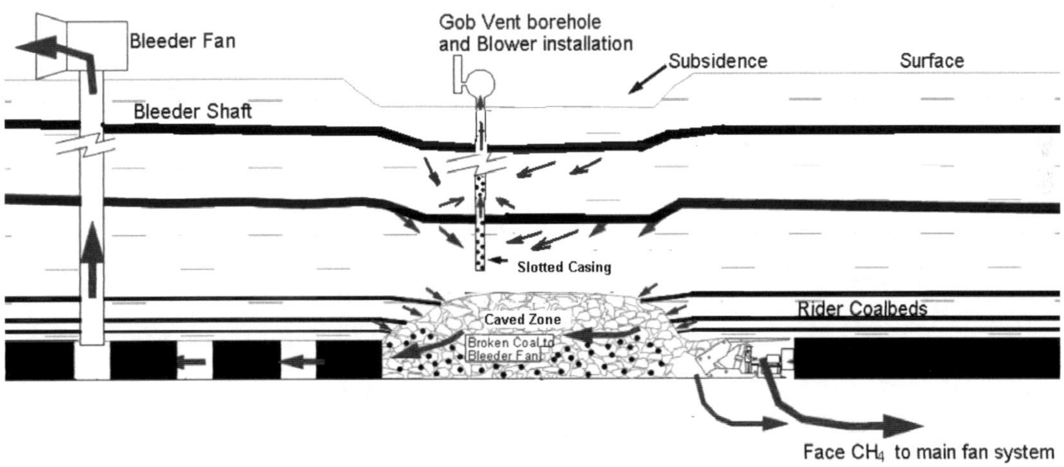

Figure 2. Schematic of a supercritical panel and associated gas migration paths.

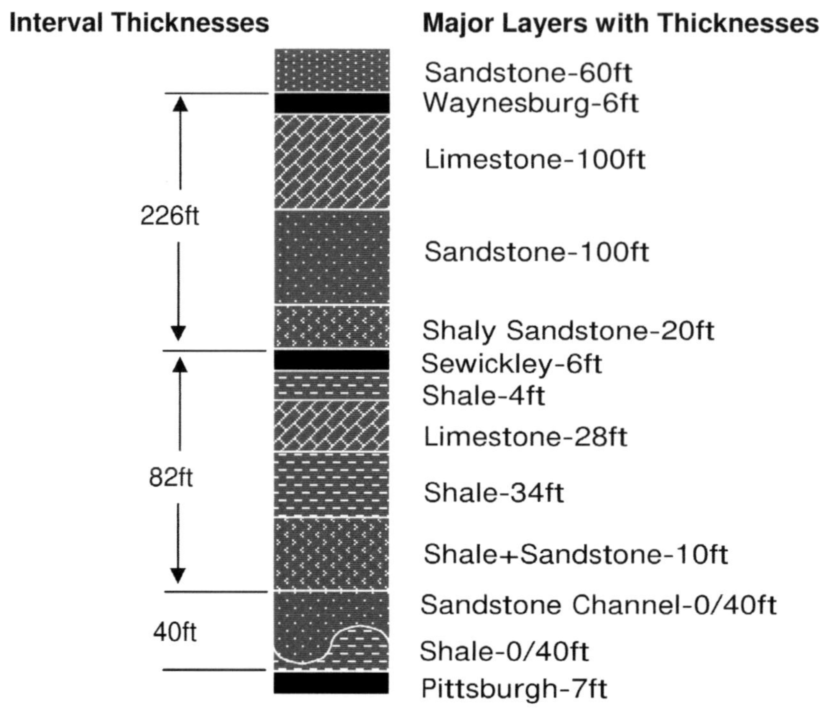

Figure 3. Generalized stratigraphic section of the study area (not to scale) showing major rock units and their thicknesses used in the model.

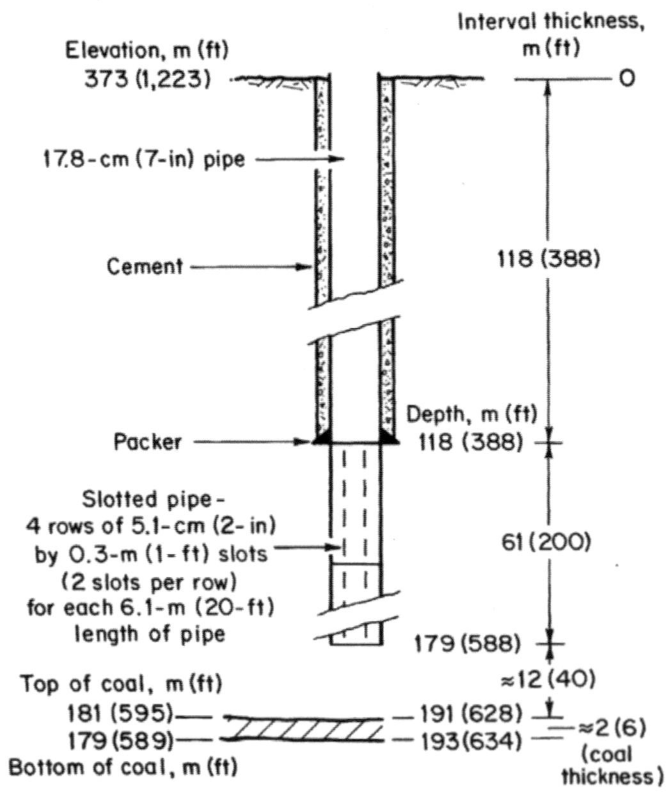

Figure 4. Typical gob gas venthole completion used in the study mine.

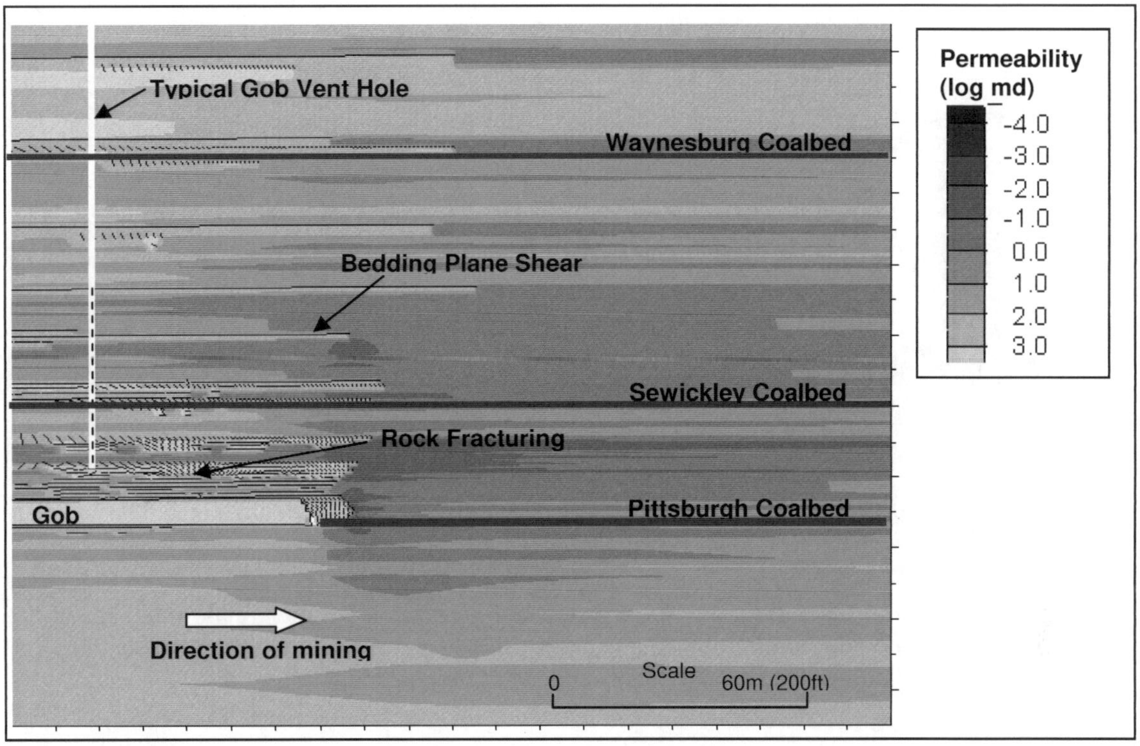

Figure 5. FLAC model results of a vertical longitudinal section through an advancing longwall face showing horizontal permeability contours, rock fracturing and bedding plane shear.

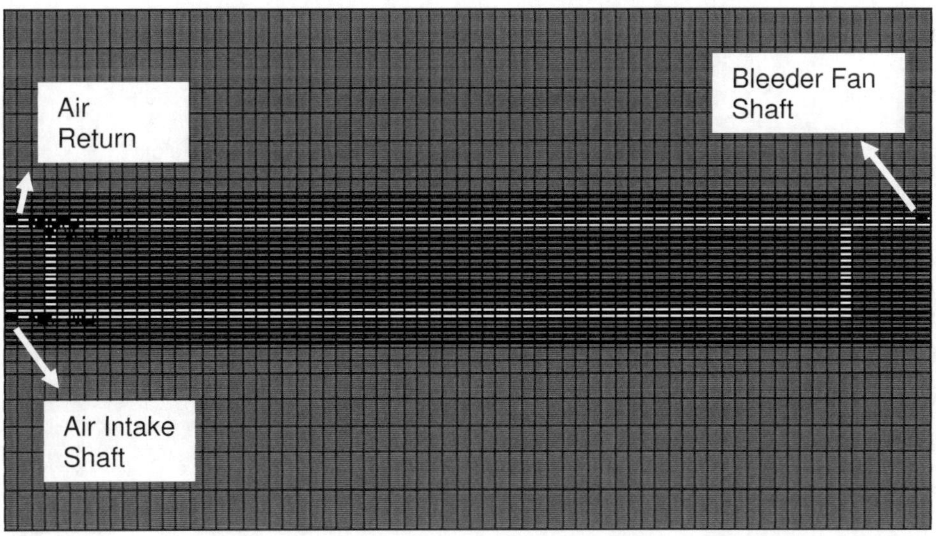

Figure 6. Pittsburgh Coalbed model layer grid showing simulated longwall panel, development entries, and "wells" used to model the ventilation system.

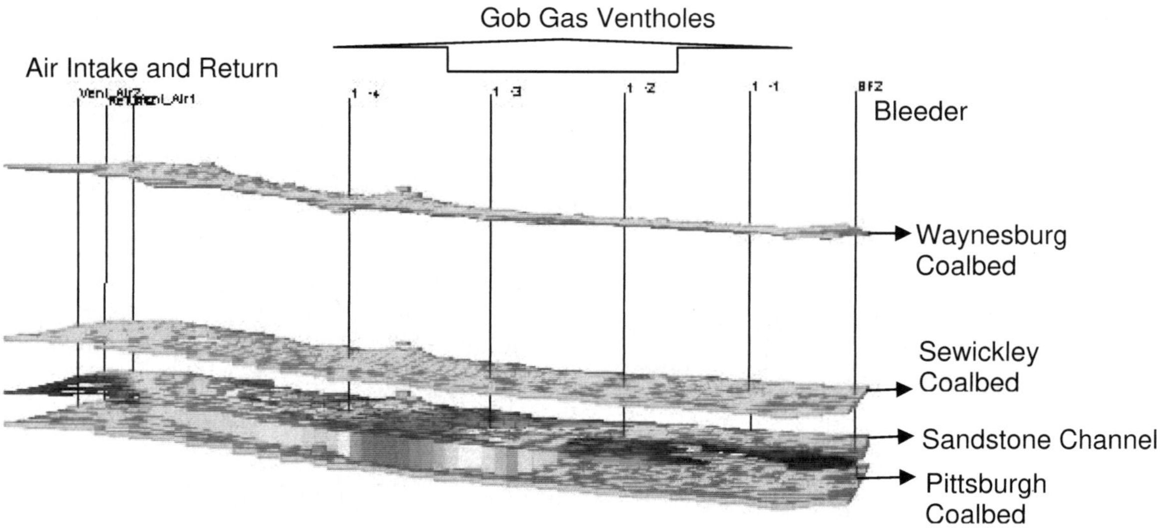

Figure 7. 3-D grid model of the study area (inner layers removed) showing the major coalbeds and the sandstone paleochannel. Figure also shows actual wells used in the model. Colors represent different thicknesses (red= 35-40ft, green= 5-15ft, blue < 5ft).

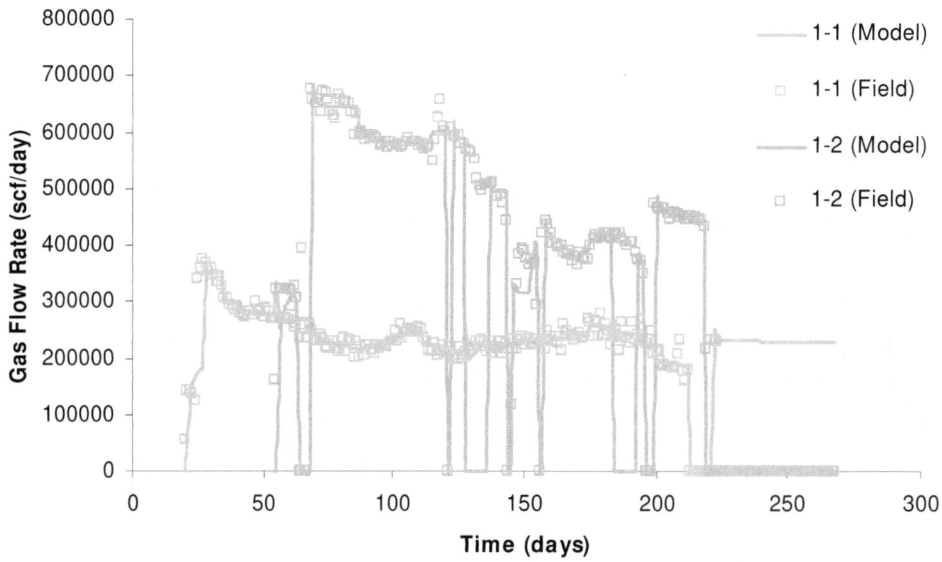

Figure 8-a. Graphs showing the history matches for gas flow rate from actual field measurements versus the model simulations for gob gas ventholes 1-1 and 1-2.

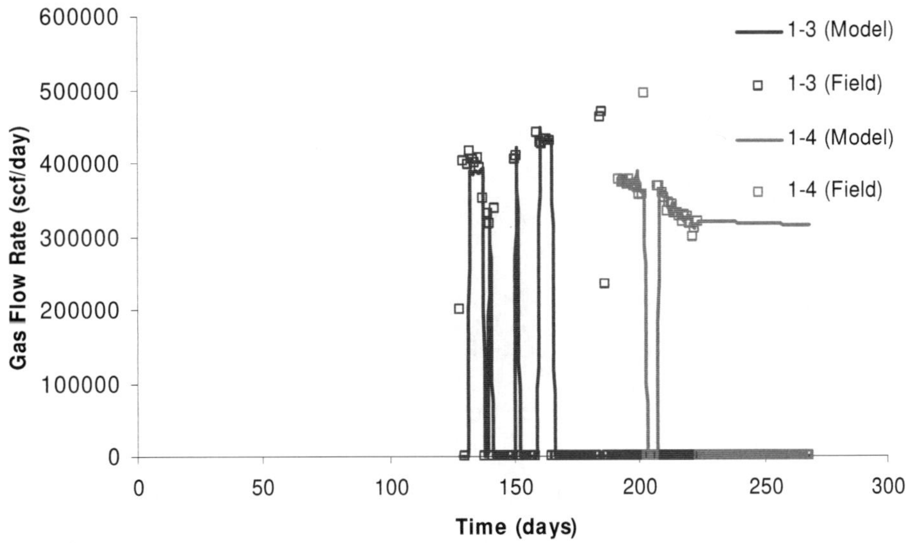

Figure 8-b. Graphs showing the history matches for gas flow rate from actual field measurements versus the model simulations for gob gas ventholes 1-3 and 1-4.

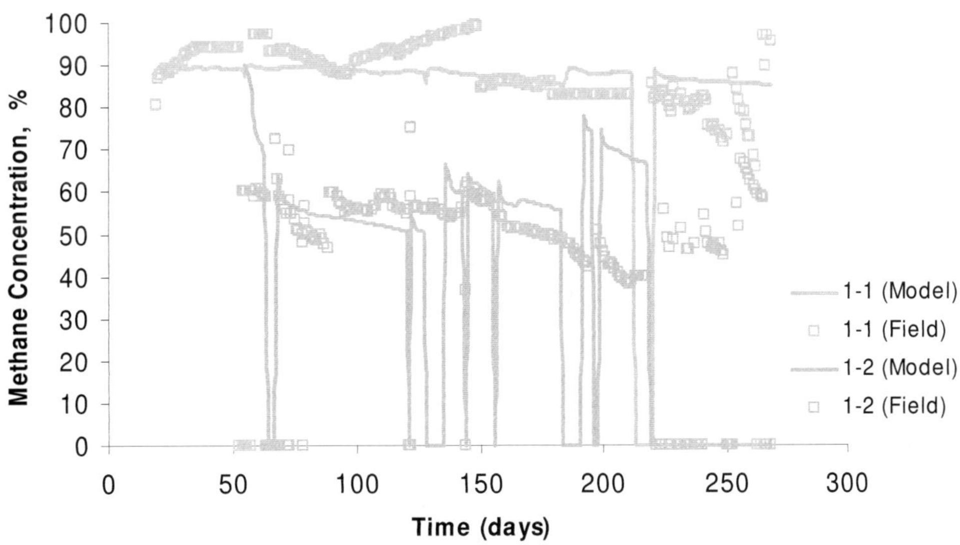

Figure 9-a. Graphs showing the matches for methane concentration in produced gas between field measurements and model runs for gob gas ventholes 1-1 and 1-2.

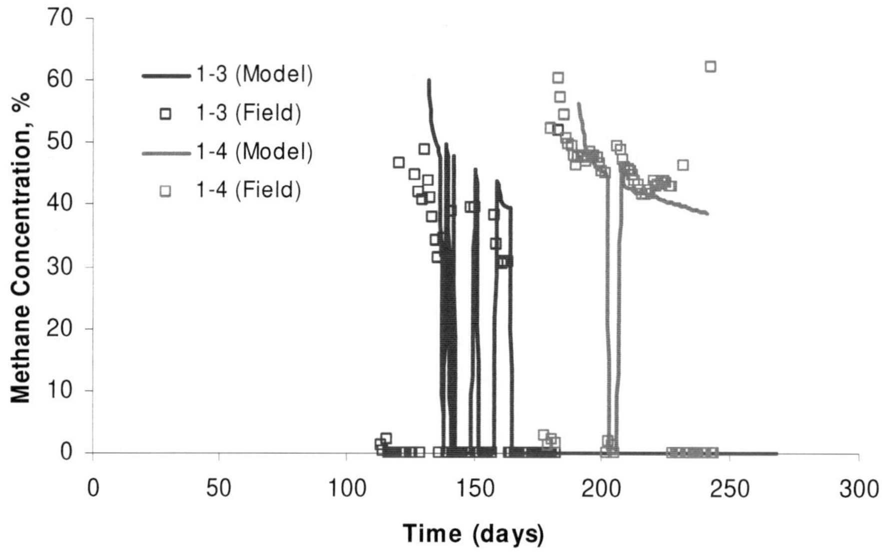

Figure 9-b. Graphs showing the matches for methane concentration in produced gas between field measurements and model runs for gob gas ventholes 1-3 and 1-4.

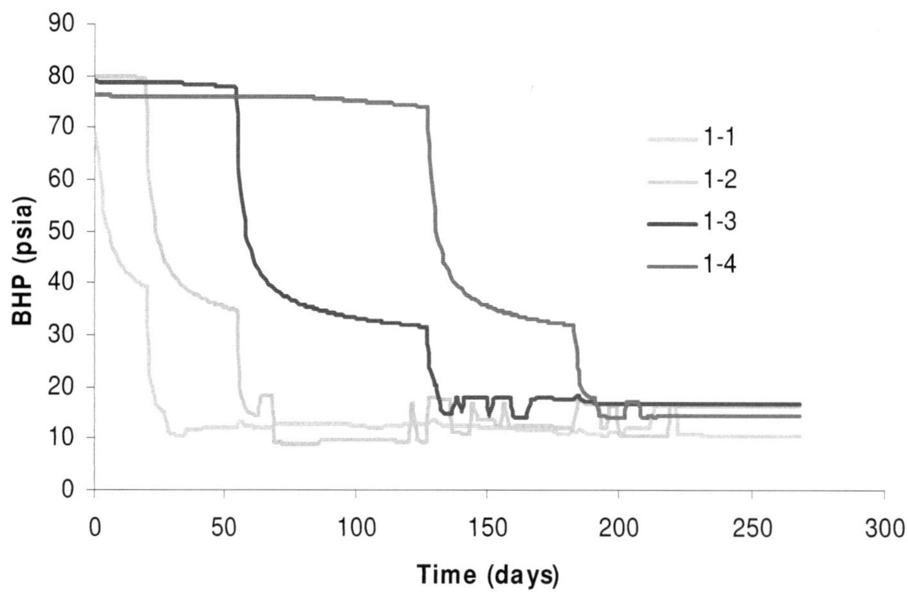

Figure 10. Predictions for bottom-hole pressures for the actual gob gas ventholes.

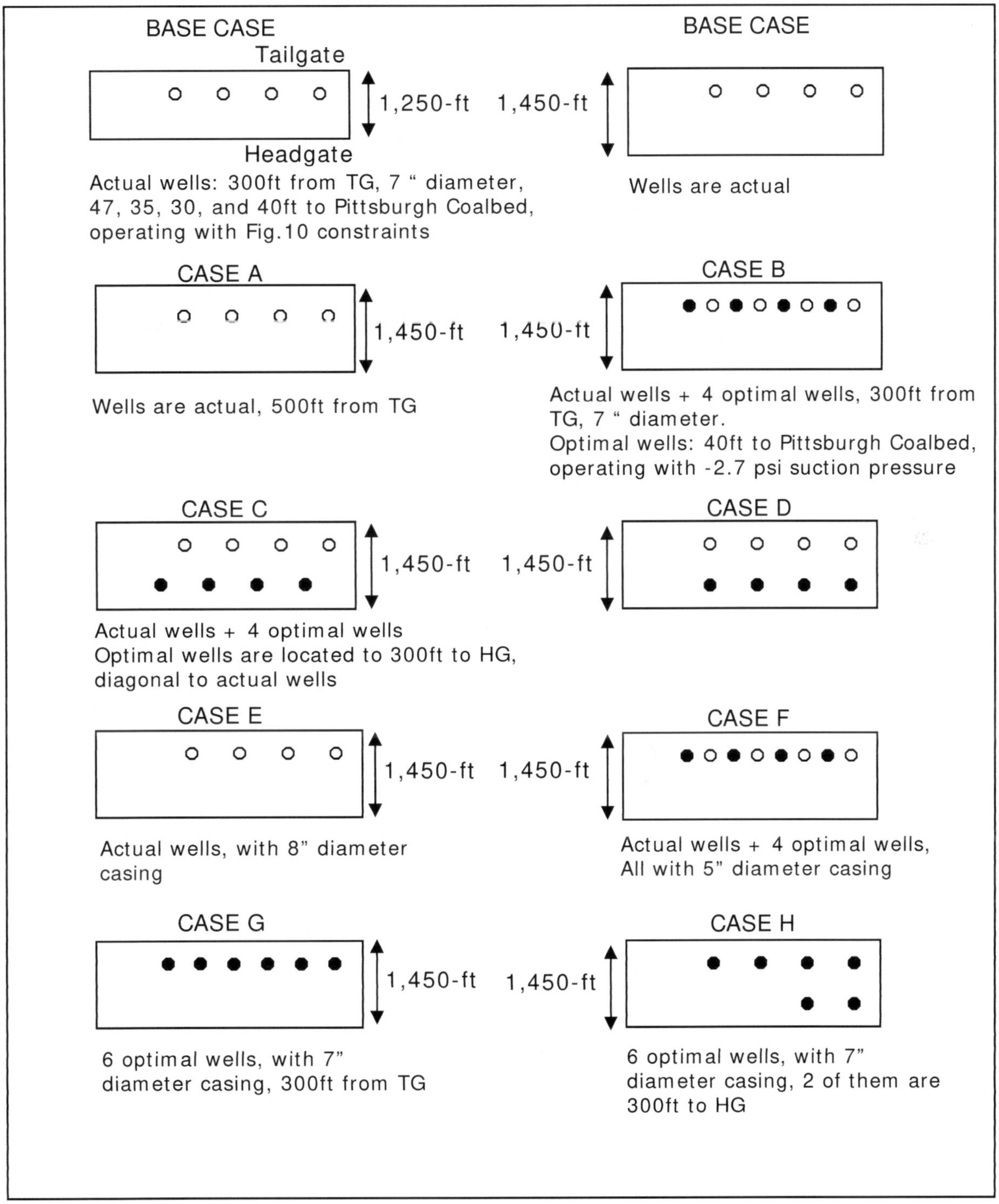

Figure 11. A schematic representation of well configurations and locations for the cases tested.

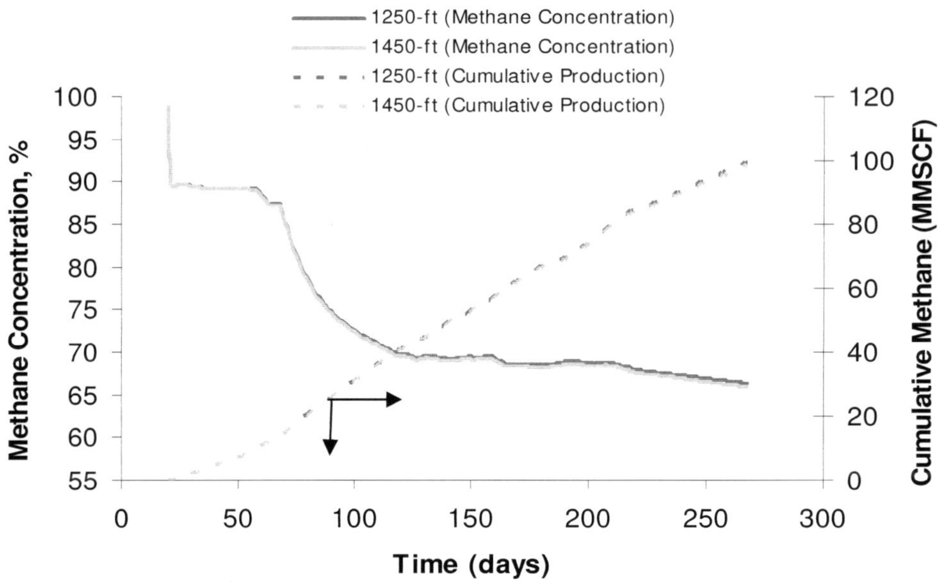

Figure 12. Comparison of cumulative methane production and methane concentration for gob gas venthole productions for different panel sizes.

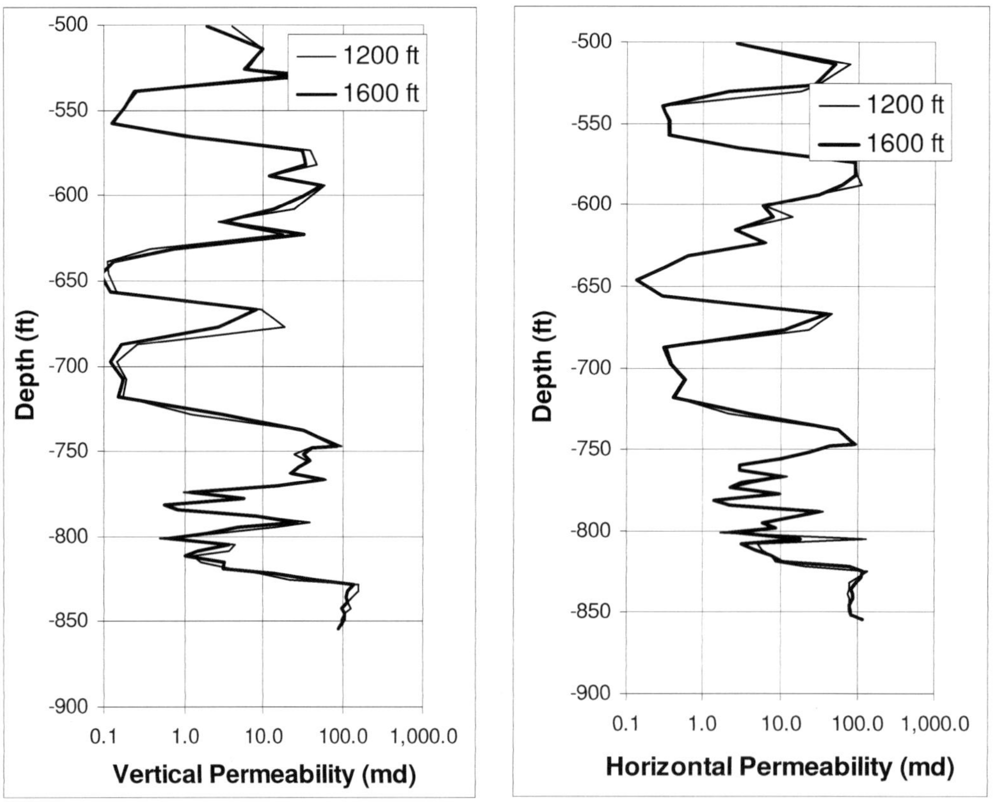

Figure 13. Comparison of horizontal and vertical permeability changes for 1,200-ft and 1,600-ft panel widths.

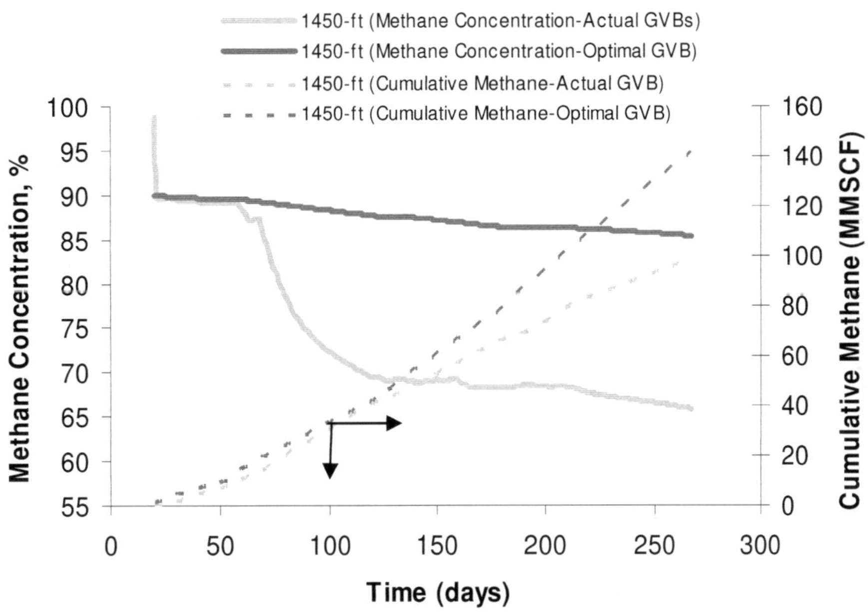

Figure 14. Comparison of the performance of actual GVBs (gob gas ventholes) at the study site with the optimal GVBs for the 1,450-ft wide panel.

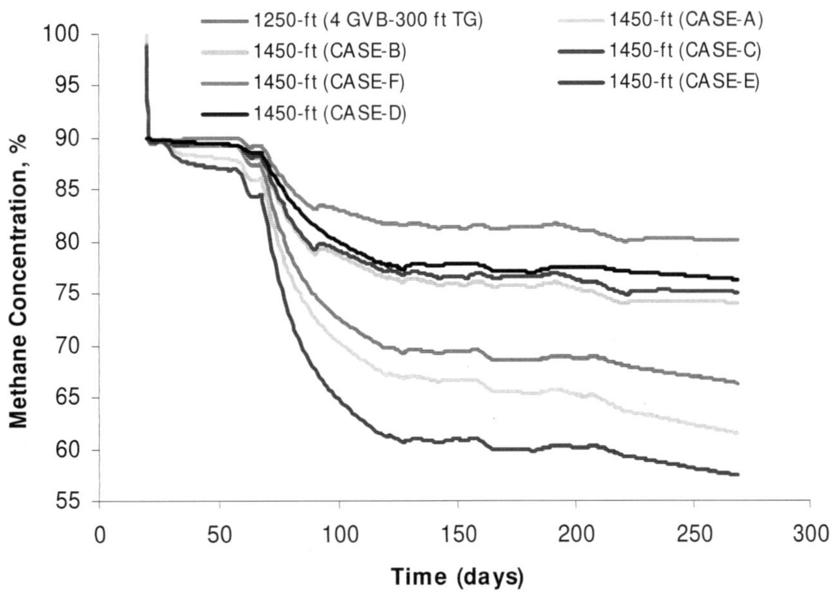

Figure 15. Comparison of methane concentrations in the produced gas from gob gas ventholes on the 1,250-ft and 1,450-ft panels with various well configurations as presented in Figure 11.

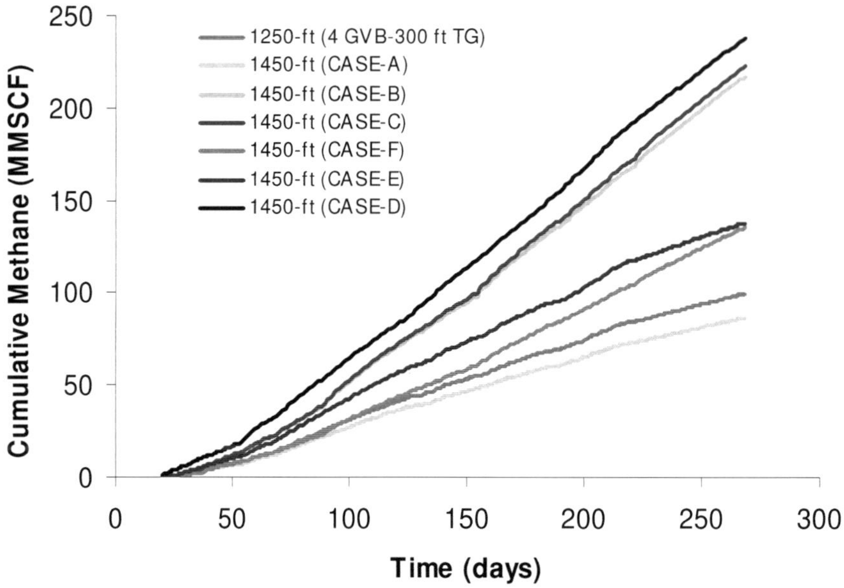

Figure 16. Comparison of cumulative gob gas venthole methane production from the 1,250-ft and 1,450-ft panels with various well configurations (A-F) presented in Figure 11.

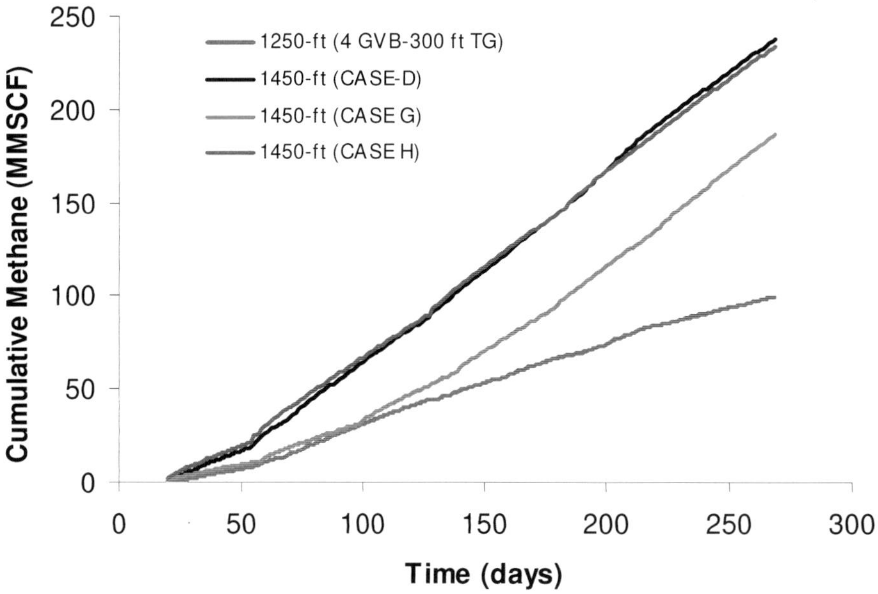

Figure 17. Comparison of cumulative gob gas venthole methane production from the 1,250-ft and 1,450-ft panel using CASE-D with configurations (G-H) that use only optimal wells presented in Figure 11.

0520

Rapid Determination of Equilibrium Gas Isotherms in Wet Coalbeds from Non-equilibrium State Measurements

Hossein Jahediesfanjani[1] and Faruk Civan,
the University of Oklahoma, Mewbourne School of Petroleum and Geological Engineering, Norman, Oklahoma, 73019

Abstract

Reliable estimation of gas reserves in coalbed methane (CBM) reservoirs requires accurate determination of adsorption isotherms relating the ability of coal to adsorb gas in the presence of water. Gas diffuses through the water, enters the matrix structure, and then adsorbs on the coal surface. This process is relatively slow and therefore, the time and effort required for constructing equilibrium isotherms are usually very demanding. This paper presents a rapid approach for determining the equilibrium gas isotherms by means of the measurements conducted with the coal samples taken from the wet coalbed methane (CBM) reservoirs containing a system of gas, water, and coal. A new data interpretation method is developed in order to reduce the time required to construct the equilibrium gas isotherms using the experimental data obtained by means of non-equilibrium measurements. This is accomplished with a theoretical approach, implementing non-equilibrium thermodynamics of gas and water adsorption on the coal surface, using well-established isotherms along with unsteady-state description of mass transfer between the cleat and matrix of the coal. The applicability of this approach is verified by analyzing a variety of non-equilibrium experimental data.

Introduction

The process of the coalification starts off with peat. After a considerable amount of time, burial heat and pressures metamorphose from peat to lignite, bituminous, and finally anthracite coal. The degree of coalification, affect the rank and composition and the ability to contain different materials like water and methane, hence the water or gas contents of coal vary [1]. The secondary biogenic gases formed in many coal-bearing basins indicate that the origin of the coalbed methane is not thermogenic [2]. Almost all coals have some weak affinity for water [3]. In general, the CBM gas reservoirs are composed of coal, water and gas at thermodynamic equilibrium at the initial prevailing in-situ conditions. Therefore, the chemical potential of the various species for different phases are equal. The majority of the gas (more than 98%) is adsorbed over the internal coal surfaces. However, a small percentage of the methane remains as a free gas inside the cleats and fractures, where water is the dominant phase. Moreover, water usually diffuses into and adsorbs over the coal surface. This complicates the adsorption process significantly. The adsorption phenomena can be studied by constructing an isotherm in the laboratory. An isotherm for any given system of coal and gas represents the adsorbed gas volume as a function of pressure at a constant temperature. However, because the gas and water adsorption are usually slow processes, constructing an equilibrium isotherm is very time-

[1] T-301 Sarkeys energy Center, 100 Boyds st, Norman, OK, 730119-1003; electronic Email: Hossein@ou.edu

demanding. To reduce the time required to obtain an equilibrium isotherm, several techniques have been used, like crushing the coal samples, increasing the temperature, and adding some extra materials to the coal-gas system. Any of these methods has its own shortcomings, so that the final obtained isotherm can be far beyond the reality. Moreover, when an isotherm is represented based on the equilibrium data points, it does not give enough information about the non-equilibrium intermediate stages.

In this paper, a theoretical and experimental approach is introduced to obtain both equilibrium and non-equilibrium isotherms as a function of pressure and temperature. Applications are demonstrated for pure methane, gas mixture, and wet-gas adsorption on the coal and several other carbonaceous materials. An experimental set-up is designed to perform the necessary experiments to verify the developed equations and validate the data interpretation procedure. The experimental set-up allows the volume of the total system containing gas, water, and coal to vary at prescribed pressure and temperature condition. A computer code has been prepared to solve the gas and water non-equilibrium adsorption equations and proper coal swelling correlations based on the obtained experimental data to calculate the adsorbed volume of various components as a function of time. This procedure enables obtaining gas and water contents of the CBM reservoir as a function of time, pressure, and equilibrium and non-equilibrium adsorption isotherms rapidly with greater accuracy. For example, instead of spending weeks to obtain the equilibrium adsorption data points, the same isotherm can be developed using only one or two sets of non-equilibrium data points in a much shorter time (less than about a day) and with better quality without crushing the coal samples. The following brief review of the literature provides the background for the present work.

1. Equilibrium Adsorption

Suttoo and Davies [4] developed an equilibrium-based isotherm to examine the adsorption of methane in the coal surface. Their data fit the Freundlich equation. Choudhary and Mayadevi [5] obtained the isotherm of adsorption of methane on silicalite-I by the gravimetric adsorption. Their equilibrium data fit the Langmuir isotherm. Choi et al. [6] obtained the equilibrium data for adsorption of methane on activated carbon and fitted with the Langmuir-Freunlich equations. Clarkson [7] fitted the equilibrium adsorption data using the vacancy solution and Dubinin-Polani theories. This approach provided accurate results for binary gas/coal systems.

2. Water Adsorption

Presence of water in the wet coal is a result of the chemical and physical bonding of coal and water, and influences the properties of the coal [3]. Gosiewska et al. [8] show that the mineralogical nature of coal largely influences the wettability of the coal surface with water. At macroscopic scale, carbon surface is hydrophilic resulting on a little water adsorption [8]. However, if H-bonding sites are present at the surface, forming strong bonds between the water and the coal surface may change the wettability and enhance water adsorption. Nordon and Bainbridge [9] reported that particularly in lower rank of bitominous coals, the self-heating phenomenon may cause the releasing of some heat when water is adsorbed on the coal. Kross et al. [10] reported that moisture-equilibrated coals showed lower methane adsorption capacity by 20-25% with respect to the dry coal. Allardice and Evans [11] modified the BET equation to fit the obtained equilibrium water adsorption data. Stamm [12] used the steady-state diffusion measurements to represent the diffusion of water into uncoated cellophane_II. Monazam [13] presented a model to predict the coal moisture content at any time after exposing the coal to the moist air. They developed a model based on the Fick's law and material balance concept. Clarkson and Bustin [14] examined the effect of the moisture on binary gas adsorption/desorption isotherms. They concluded that the Dubinin and Astakhov (D-A) and Dubinin-Radushkevich (D-R) isotherms better fit the equilibrium adsorption data for both dry and moisture-equilibrated coals.

3. Non-Equilibrium Adsorption

Models proposed for adsorption are commonly based on the thermodynamic and kinetic properties of the materials in contact with each other. Non-equilibrium adsorption models have been developed for gas-liquid [15], solid-gas, and solid-liquid solutions [16]. The activated adsorption theory and Elovich equation are the two major models used in the early adsorption kinetic studies [17]. Ward and Findlay [18] pointed out that the activated adsorption theory eventually yields in an expression containing several parameters which are difficult to measure independently, where as the Elovich equation correlates the adsorption data better. They represented an expression for the non-equilibrium coverage based on the Elovich equation and Langmuir isotherm. Lorenz and Engler [19] developed a procedure based on the statistical thermodynamic to study the adsorption phenomenon. They pointed out that the adsorption energy or heat of adsorption is coverage dependent and changes with the amount of the adsorbed gas changes. Hamilton and Goodstein [20] represented a thermodymiac approach for the adsorption of methane on the graphite surface. Do and Rice [21] developed a model that considers the partial surface diffusion in the non-equilibrium adsorption rates. Payne and Kreuzer [22] developed a non-equilibrium adsorption/ desorption model using the quasichemical approximation and the sticking coefficient concept. Langmuir assumed that adsorption of a molecule into a mobile second layer in which migration of gas molecules can take place until an empty site on the solid surface occurs. In the contrast, King [23] observed the presence of an intermediate state on desorption kinetics. King concluded that ignoring the presence of the intermediate state can lead to serious errors in the derived values of the activated energy. Kreuzer and Payne [16] systematically formulated a set of phenomenological rate equations that control the time evolution of a two-phase adsorbate. They treated the 2-dimensional (2D) gas as an ideal phase and described the 2D condensate phase as modeled by Einstein's equation. Ward and Elmoselhi [24] applied the statistical rate theory (SRT) to predict the net rate of the molecular adsorption at a gas-solid interface. They developed an expression for non-equilibrium adsorption using the SRT approach. The parameters of this model are considered to be material properties of the gas-solid surface condition. Therefore, if such parameters are determined they can be used in the SRT approach to predict the rate of adsorption in circumstances that are independent of the experimental conditions at which they were measured. This approach has been used to model various adsorption systems. This approach provides better results and predicts the adsorption phenomenon with higher accuracy as concluded in several papers [25, 26]. Rudzinsky and Panczyk [27] adopted the SRT with the Langmuir isotherm and developed non-equilibrium adsorption models to describe the volume-dominated, solid-dominated, and equilibrium-dominated cases. They [28] also introduced the Statistical Rate Theory of Interfacial Transport (SRTIT) approach and developed the rate equation using the D-A isotherm.

4. Coal Swelling or Shrinkage

The early works show that the percentage of linear swelling of a charcoal block during adsorption is directly proportional to the change in the free energy of the block [29]. Maggs [30] concluded that the charcoal swelling obeys the Hook's law. This means that the dimensional change is proportional to the pressure of the films formed at its inner surfaces, and thus the stress-strain relationships are complicated. However, the coal swelled in a manner very similar to the charcoal in the presence of adsorbed films. Seidle and Huitt [31] reported that water content of the coal and amount of the adsorbed gas (especially methane in CBM reservoirs) have significant swelling effects in coal. They reported several swelling coefficients for matrix, methane, and carbon dioxide.

Formulation

An experimental and interpretation procedure is developed to monitor the changes in coal, water, and gas systems with respect to time. According to **Figure 1** the equipment is composed of a container located inside a bigger vessel filled with water to adjust and fix the temperature at a constant value. The pressure above the container can be fixed at a constant value and the volume is allowed to change while the coal, water, and gas are in contact with each other. The next step is to increase the pressure to a new level and monitor the volume changes of the system at different times until the equilibrium is established and the volume is fixed at a constant value. For each time interval the volume change is measured. This volume change is expressed as:

$$\Delta V(t) = \left[V_g(t) + V_w(t) + V_c(t)\right] - V_i \quad (1)$$

$$V_w(t) = V_{ww}(t) + V_{gw}(t) \quad (2)$$

$$V_g(t) = V_{gg}(t) + V_{wg}(g) \quad (3)$$

where $\Delta V(t)$ is the total vessel volume change at different time steps, *cc*, and *V* is volume in cc. The first subscript is a phase indicator and the second one is the specific component at that phase. For instance V_{gw} refers to the volume of water component in the methane phase. According to Maggs [36] it is assumed that the coal swelling is proportional to the amount of adsorbed water and methane. Thus,

$$V_c(t) = (a+1)\left[V_{wc}(t) + V_{gc}(t)\right] \quad (4)$$

The mole balance equations for methane and water phases are given respectively by:

$$n_w(t) = n_{gw}(t) + n_{cw}(t) + n_{ww}(t) \quad (5)$$

$$n_{wg}(t) = n_{wg}(t) + n_{cg}(t) + n_{wg}(t) \quad (6)$$

The relationships between the volume of each phase and its moles can be defined by:

$$V_{gw}(t) = \frac{n_{gw}(t)\ Mw_g}{\rho_{gw}} \quad (7)$$

$$V_{ww}(t) = \frac{n_{ww}(t)\ Mw_w}{\rho_{ww}} \quad (8)$$

$$V_{wg}(t) = \frac{n_{wg}(t)\ Mw_g}{\rho_{wg}} \quad (9)$$

$$V_{cg}(t) = \frac{n_{cg}(t)\ Mw_g}{\rho_{cg}} \quad (10)$$

$$V_{cw}(t) = \frac{n_{cw}(t)\ Mw_w}{\rho_{cw}} \quad (11)$$

The average density of the gas phase is given by:

$$\rho_{gg}(t) = y_{gg}\rho_g(t) + y_{gw}\rho_w(t) \quad (12)$$

where the mole fractions are defined as:

$$y_{gg} = \frac{n_{gg}(t)}{n_{gg}(t) + n_{gw}(t)} \quad (13)$$

$$y_{gw} = \frac{n_{gw}(t)}{n_{gw}(t) + n_{ww}(t)} \quad (14)$$

Using the equation of state for real gas we can write:

$$\rho_{gg}(t)=\frac{n_{gg}(t)}{n_{gw}(t)+n_{gg}(t)}\frac{Z_gRT}{P}+\frac{n_{gw}(t)}{n_{gw}(t)+n_{ww}(t)}\frac{Z_wRT}{P} \tag{15}$$

$$\rho_{gg}(t)=\rho_{wg}(t) \tag{16}$$

$$\rho_{ww}(t)=\frac{n_{wg}(t)}{n_{wg}(t)+n_{ww}(t)}\frac{Z_gRT}{P}+\frac{n_{ww}(t)}{n_{wg}(t)+n_{ww}(t)}\rho_w \tag{17}$$

The next step is to define the relationship necessary to calculate the density of the adsorbed phase inside the coal. The results of the analysis [32] of some fluids tested on many different carbonaceous solids show that the adsorbed phase density is very close to that the liquid phase density at high pressure. Thus, the following expression for the density of the adsorbed phase can be written:

$$\rho_{cg}(t)=b'\rho_{Lg}(t)=b\frac{Z_{Lg}RT}{P} \tag{18}$$

$$\rho_{cw}(t)=c'\rho_{cw}(t)=c\rho_{ww}(P,T) \tag{19}$$

where b' and c' are the fraction of the adsorbed phase densities to that of the liquid phase. The subscript L refers to the liquid phase. **Figure 2** shows water, coal, and gas in contact with each other in the CBM reservoir condition. For the control volume indicated in this figure for the equilibrium between gas, water and coal system, the chemical potentials are equal to each other. Thus,

$$\mu_{cg}=\mu_{wg}=\mu_{gg} \tag{20}$$

$$\mu_{cw}=\mu_{gw}=\mu_{ww} \tag{21}$$

For the gas phase behavior, Ward and Elmoselhi [24] proposed to use the Born-Oppenhiemier approximation, the chemical potential per molecule of an ideal, asymmetric, and diatomic gas is given by:

$$\mu_{gg}=kT\,Ln(P_g\phi_g) \tag{22}$$

$$\phi_g\;\frac{\left[1-\exp\left(\frac{h\omega_g}{kT}\right)\right]\;\exp\left(-\frac{D_o}{kT}\right)h^5(2\pi)^{3/2}}{(kT)^{7/2}\,2m_1m_2r_e^2(m_1+m_2)^{1/2}} \tag{23}$$

The above equation is for pure gas. For the present case (coal, water, and gas), we can write:

$$P_g=y_gP \tag{24}$$

Therefore, Equation 22 becomes:

$$\mu_{gg}=kT\,Ln(Py_g\phi_g) \tag{25}$$

Similarly for water we have:

$$\mu_{wg}=kT\,Ln\left[P(1-y_g)\phi_g\right] \tag{26}$$

Rudzinsky and Panczyk [28] proposed the following procedure to develop an equation for the adsorption kinetics.

$$J_{gc}=R_a-R_d \tag{27}$$

According to the SRIIT approach [28] the following expressions are expressed for the rate of adsorption and desorption.

$$R_a = K'_{gs} \exp\left(\frac{\mu_{gc} - \mu_{gg}}{kT}\right) \quad (28)$$

$$R_d = K'_{gs} \exp\left(\frac{\mu_{gg} - \mu_{gc}}{kT}\right) \quad (29)$$

where J_{gc} is the simultaneous net change of gas between coal and gas phases.
Dubinin and Astakhov (D-A) (1971) used Polany's theory of adsorption (1932) that was based on physical and chemical concepts, and developed an isotherm for adsorption of vapors on microporos adsorbents [7]. As described earlier the D-R and D-A isotherms have recently been emphasized as the semi-empirical isotherms for adsorption of various gases and water vapors on coal surfaces. These isotherms can be expressed as:

$$\theta_i = \exp\left\{-\left[-D_i \, Ln\left(\frac{P_i}{P_o}\right)\right]^r\right\} \quad (30)$$

Where r and D_i are the empirical values to be determined using the experimental data points. For r=2 Equation 30 is called D-R and for other values of r it is called D-A. The value of r is usually varied between 1 and 2 for carbons with large micropores [33]. However; the values less than 1 also have been reported for adsorption of gases on the activated carbons [28]. Using the D-A isotherm and substituting equations 25, 28, 29, and 30 into Equation 26 and integrating using the initial condition of $\theta(t=0)=0$, we have:

$$\theta_i(t) = \exp\left\{-\left[-D_i \; a \; Ln\frac{P_i}{P_o} \; \tanh\left(2PK_{gsi}t\right)\right]^r\right\} \quad (31)$$

where a is an other empirical constant to be determined by fitting the experimental data points.
Similarly for the gas-water case we can develop the following equations:

$$J_{gw} = R_{ds} - R_s \quad (32)$$

$$R_{ds} = K_{gw} \exp\left(\frac{\mu_{gg} - \mu_{gw}}{kT}\right) \quad (33)$$

$$R_s = K_{gw} \exp\left(\frac{\mu_{gw} - \mu_{gg}}{kT}\right) \quad (34)$$

$$\mu_{gg} = kT \, Ln\left(Py_{gg}\right) \quad (35)$$

$$\mu_{gw} = kT \, Ln\left(H_{gw}x_{gw}\right) \quad (36)$$

Eventually, by substituting equations 33, 34, 35, and 36 in Equation 32 we have:

$$\frac{dy_{gg}}{dt} = K_{gw}\left(\frac{Py_{gg}}{H_{gw}x_{gw}} - \frac{H_{gw}x_{gw}}{Py_{gw}}\right) \quad (37)$$

$$\frac{dx_{gw}}{dt} = K_{gw}\left(\frac{H_{gw}x_{gw}}{Py_{gw}} - \frac{Py_{gg}}{H_{gw}x_{gw}}\right) \quad (38)$$

Equations 37 and 38 are subject to the following initial conditions and the auxiliary equations:

$$x_{gw}(t=0)=0 \ \ and \ \ y_{wg}(t=0)=0$$

$$\sum_{i=1}^{n} y_i = 1 \ \ and \ \ \sum_{i=1}^{n} x_i = 1 \qquad (39)$$

This procedure can be expanded for the multi-component gas and water adsorption on the coal surface. Multi-component adsorption is the case that the gas phase is composed of more than one component. Therefore, each component will adsorb in the coal separately. If an ideal gas mixture is assumed, we have:

$$P_i = Py_i \qquad (40)$$

where y_i is the mole fraction of component i in gas phase at each time.
According to Clarkson, for the adsorbed phase we have:

$$\theta_i = \gamma_i \ X_i^s \ \theta \qquad (41)$$

where γ_i is the activity coefficient of each component at the adsorbed phase that can be calculated using the Laar equation for binary system [34]:

$$\gamma_i = \exp\left[\frac{A'_{i,j} \ \left(A'_{j,i}\right)^2 \ X_j^2}{\left(A'_{i,j} \ X_i^s + A'_{j,i} \ X_j^s\right)^2}\right] \qquad (42)$$

where *i,j* can be either 1,2 or 2,1.

Calculation Procedure

A computer code is developed to solve the above equations and calculate the amount of each gas component at liquid, gas and coal phase. This computer code is based on the following procedure.

1. Input the measured non-equilibrium data points for two different pressures for each component and water
2. Calculate the values of P_o for various gas components using the method proposed with Kapoor [35]
3. Plot $Ln(V_i)$ versus $X = \left\{-Ln\left[\frac{P_i}{P_o}\tanh\left(2P_iK_{i,gs}\right)\right]\right\}^r$ according to Equation 31

which is a convenient form of Equation 43, and fit the best curve for the data points so that the parameters of *D, r,* K_{gs} and V_m for each component can be obtained.

$$Ln(V_i) = Ln(V_{mi}) - D_i^r\left\{-Ln\left[\frac{P_i}{P_o} \ \tanh\left(2P_iK_{gsi}t\right)\right]\right\}^r \qquad (43)$$

4. The parameters c and d in Equation 44 that is a convenient form of Equation 30 can be obtained by fitting two equilibrium points which have already been measured for each component.

$$Ln(\theta) = d - cD^r\left[-Ln\left(\frac{P_i}{P_o}\right)\right]^r \qquad (44)$$

5. For a fixed pressure the volume change of the vessel is determined. The obtained volume change value is a combination of adsorbed volume, exchanged volume between water and gas phase, and also coal swelling.
6. Initially, the adsorbed volume of each component is guessed and the parameter x_i^s are calculated.
7. The activity coefficient for each component is calculated using Equation 42.

8. Equations 37 and 38 are simultaneously solved to calculate the estimated values of mole fraction of water and gas components in water and gas phases.
9. The total adsorbed volume in the coal is calculated. The coal swelling coefficient is assumed and the volume of coal swelling is calculated.
10. Using the values obtained in step 2 and by combining Equations 41 and 31, the new adsorbed volume of each component are calculated.
11. Equations 1-19 are solved to obtain the overall mass balance for the particular time period.
12. If the overall mass balance is satisfied, the assumed values are considered adequate and the next time step is carried out. Otherwise, using the estimated values in step 9 steps 7-11 is repeated.

Applications

The applicability of the model is tested and verified using the reported data for the gaseous phase adsorption on the carbonaceous materials. Following cases show the applicability of the proposed model.

Case 1. Propane and propylene adsorption on Chemviron and Wesvaco activated carbon.

Figure 3 shows a set of measured non-equilibrium adsorption data points for adsorption of propane and propylene on commercial activated carbons [33]. The developed computer code uses these data and regenerates the equilibrium adsorption isotherms for each component. **Figure 4** represents difference between measured and predicted isotherm. According to this figure the average error between measured and predicted points is less than 2%.

Case 2. Various gas adsorption on carbon molecular sieves.

Figure 2 shows a control volume composed of a matrix structure surrounding by cleats in a CBM reservoir that is in contact with water and gas. The gas phase can be a mixture of various components like, methane, ethane, carbon dioxide, and/or nitrogen. This model is applicable to obtain equilibrium, and non-equilibrium, isotherms for the systems mentioned above.

<u>2.1. Pure methane adsorption on carbon molecular sieves.</u> The methane and other gas adsorption data on carbon molecular sieves are reported in the literature [36]. **Figure 5** shows the equilibrium adsorption isotherms for various gases on the carbon molecular sieve surface. **Figure 6** represents a set of the non-equilibrium adsorption data points for the same gas and molecular sieve system at pressure of 2.3 psia. Following the steps 2-4 the equilibrium and non-equilibrium isotherm parameters are determined for each gas component. **Table 1** summarizes the values obtained from steps 2-4. **Figure 7** shows a plot of Equation 43 required to predict the mentioned parameters for methane. Similarly, the same set of plots can be developed for other gases. This figure is constructed for different values of r in a fixed value of K_{gs}. To determine the value of r we take into account that this value is usually between 0.1-2.0 for adsorption of different gases on carbonaceous materials. Therefore, any value of r in this range that results in a better straight line is chosen. To improve the value of R^2, parameter K_{gs} is varied until the best possible fit for the reported data is obtained. The slope of the straight line is equal to D^r which is used to estimate the value of D. The intercept is equal to $Ln(V_m)$ which is used to estimate the value of V_m. Now that the non-equilibrium isotherm is established, step 4 is applied to estimate the values of c and d to construct the equilibrium isotherm as well. **Figure 8** represents the non-equilibrium methane adsorption isotherms on the carbon molecular sieves at different pressures. **Figure 9** compares the predicted and measured equilibrium data points. According to this figure, the points are very close to each other and the error value is less than 8%.
<u>2.2. Mixture of methane and carbon dioxide adsorbed on carbon molecular sieves.</u> The gas mixture contains 95% CH_4, 3% CO_2 and 2% C_2H_6. Reported data for equilibrium and non-equilibrium adsorption of each individual gas are expressed in **Figures 3** and **4.** Using these equations and the parameters expressed in **Table 1**, we are the adsorbed volume of each

component at different times can be presented for each pressure. For the vessel size of 160 cc at the initial condition 130 cc of the gas mixture is injected. The vessel temperature is fixed at 75 $^{\circ}$F. Following the above mentioned procedure, using the estimated parameters reported in **Table 1**, and the compressibility values for various gases [37], the equilibrium and non-equilibrium isotherms can be constructed for various pressures. **Figures 10**, **11** and **12** represent the non-equilibrium isotherms for the projected high pressures using the reported low pressure values for methane, carbon dioxide, and ethane, respectively. **Figure 13** shows the relationship between methane, carbon dioxide, and ethane mole fraction in gas phase and the adsorbed volume in coal phase. Ethane adsorbs faster than carbon dioxide and methane. **Figure 14** compares the adsorbed mole fraction of each component at different pressures. Even though the adsorbability of carbon dioxide in carbon molecular sieves is higher than methane, because the product of activity to mole fraction of methane component of the gas mixture in gaseous phase is higher than that of the carbon dioxide, the adsorbed mole fraction of methane is higher than the other components. **Figure 15** shows the relationship between the amount of adsorbed moles and pressure. The adsorbed moles of methane are much higher than carbon dioxide. Moreover, as pressure increases the difference between the adsorbed moles increases.

Case 3. Comparison of equilibrium and non-equilibrium adsorption rates.

After production starts in coalbed methane reservoirs, the state of the equilibrium that initially exists among the three phases of liquid, gas and coal is disturbed. Depending upon the rate of the production and also the reservoir characteristics, time required for establishing a new equilibrium state at the new reservoir pressure and temperature is different. Most of the available coalbed methane reservoir simulators have ignored this fact and used the equilibrium adsorption isotherms to simulate reservoir gas production. **Figure 16** shows that neglecting this fact can result in significant prediction error. This figure compares the equilibrium and non-equilibrium methane adsorbed volume on carbon molecular sieves at different pressures. For instance, at pressure 10 psia, the methane equilibrium adsorbed volume is 220 scf/ton, whereas, the minimum non-equilibrium adsorbed volume is about 140 Scf/ton. Therefore, neglecting this fact will result in about 60% error in the reservoir reserve estimation.

Conclusions

This work represented a model to obtain equilibrium and non-equilibrium isotherms using non-equilibrium data points. Therefore, without spending considerable amount of time to establish an equilibrium state, the non-equilibrium data points which are taken over a very short period of time can be extrapolated to the equilibrium state. The reported data for adsorption of pure methane and gas mixture on carbon molecular sieves show the applicability of this method. The results confirmed the applicability of the developed model. The advantages of this method are: 1. The unnecessary simplifications can be eliminated and the laboratory condition can be a better representative of the real reservoir conditions by running the adsorption experiments in the presence of various gases, water and coal. 2. Because the non-equilibrium data points are used to generate the equilibrium isotherm, the time to obtain the isotherm will be significantly reduced. 3. The obtained isotherm can be better used in reservoir simulation due to its ability to cover the transition time between two successive equilibrium data points.

References

[1] Francis, W., "Coal, Its Formation and Composition", Edward Arnold (1954), 15-35.
[2] Aravena, R., Harrison, S.M., Baker, J.F., Abercrombie, H., and Rudolph, D: "Origin of Methane in the Elk Valley Coalified, Southeastern British Columbia, Canada", Journal of Chemical Geology, 2003 (195) issue 1-4 (219-227).
[3] Snyder, G.T., Riese, W.C.R., Franks, S., Pelzmann, W.L., Gorody, A.W., and Moran, J.E.: "Origin and History of Waters Associated with Coalbed Methane: ^{129}I, ^{36}Cl, and Stable Isotope

Resulting from the Fruitland Formation, CO and NM", Journal of Geochimica et Cosmochimica Acta, 67 (23) 2003, 4529-4544.
[4] Sutton, J.B., and Davis, E.C.H.: "The Adsorption of Methane by Coal", Fuel, 57, 1935 (1785-1787).
[5] Chaudhary, V.R., and Mayadevi, S.: "Adsorption of Methane, Ethane, Ethylene, and Carbon dioxide on Silicalite-I", Zeolites, 17, 1996, 501-507.
[6] Choi, B., Choi, D., Lee, Y.W., and Lee, B.K.: "Adsorption Equilibria of Methane, Ethane, Ethylene, Nitrogen, and Hydrogen onto Activated Carbon", Journal of Chemical Data, 48, 2003, 603-607.
[7] Clarkson, C.R.: "Application of a New Multi-component Gas Adsorption Model to Coal Gas Adsorption Systems" Society of Petroleum Engineers Journal, Sep. 2003, 236-251.
[8] Gosiewska, A., Drelich, J., Laskowski, J.S., and Pawlik, M.: "Mineral Matter Distribution on Coal Surface and Its Effect on Coal Wettability", Journal of Colloid and Interface Science, 247, 2002, 107-116.
[9] Nordon, P., and Bainbridge, N.W.: "Heat of Wetting of a Bituminous Coal", Fuel, 62, 1983, 619-621.
[10] Krooss, B.M., Bergen, F.V., Gensterblum, Y., Siemons, N., Pagnier, H.J.M., and David, P.: "High-Pressure Methane and Carbon Dioxide Adsorption on Dry and Moisture-Equilibrated Pennsylvania Coals", International of Coal Geology, 51, 2002, 69-92.
[11] Allardice, D.J., and Evans, D.G.: "The brown Coal/Water System: Part 2. Water Sorption Isotherm on Bed-Moist Yallourn Brown Coal", Fuel, 50(3), 1971, 236-53.
[12] Stamm, A.J.: "Diffusion of Water into Uncoated Cellophane. I. From Rates of Water Vapor Adsorption, and Liquid Water Absorption", Fuel, 60, 1955, 76-82.
[13] Monazam, E.R.: "Water adsorption and Desorption by Coals and Chars", Energy and Fuels, 12, 1998, 1299-1304.
[14] Clarkson, C.R., and Bustin, R.M.: "The Effect of Pore Structure and Gas Presence upon the Transport Properties of Coal: a laboratory and Modeling Study." Fuel, 78 (1999) 1345-1362.
[15] Ward, C.A.: "The Rate of Gas Adsorption at a Liquid Interface", the Journal of Chemical Physics, 67(1) 1977, 229-235.
[16] Kreuzr, H.J., Payne, S.H.: "Non-Equilibrium Thermodynamics of a Two-Phase Adsorbate", Surface Science, 198, 1988, 235-262.
[17] Low, M.J.D.: "Kinetics of Chemisorption of Gases on Solids", Chemical Reviews, 60, 1960, 276-312.
[18] Ward, C.A., and Findlay, R.D.: "Statistical Rate Theory of Interfacial Transport. III. Predicted Rate of Non-dissociative Adsorption", Journal of Chemical Physics, 76(11), 1982, 5615-5623.
[19] Lorenz, W., and Engler, C.: "Chemisorption with Partial Charge Transfer on Semiconductor Surfaces: Adsorption Equilibrium and Adsorption Isotherm", Surface Science, 114, 1984, 607-613.
[20] Hamilton, J.J., Goodstein, D.L.: "Thermodynamic Study of Methane Multilayer Adsorbed on Graphite", Journal of Physical Review B, 28 (7), 1983, 3838-3848.
[21] Do, D.D., and Rice, R.G.: "On the Relative Importance of Pore and Surface Diffusion in Non-Equilibrium Adsorption Rate Processes", Journal of Chemical Engineering Science, 42(10), 1987, 2269-2284.
[22] Payne, S.H., and Kreuzer, H.j.: "Analysis of Thermal Desorption Data", Surface Science, 222, 1989, 404-429.
[23] King, D.A.: "The Influence of Weakly Bound Intermediate States on thermal Desorption Kinetics", Surface Science, 64, 1977, 431-51.
[24] Ward, C.A., and Elmoselhi, M.: "Molecular Adsorption at a Well Defined Gas-Solid Interphase: Statistical Rate Theory approach", Surface Science, 176, 1986, 457-475.
[25] Elliot, J.A.W., and Ward, C.A.: "Statistical Rate Theory Desorption of Beam-Dosing Adsorption Kinetics", Journal of Chemical Physics, 106(13), 1997, 5667-5677.
[26] Elliot, J.A.W., and Ward, C.A.: "Temperature Programmed Desorption: A Statistical Rate Theory Approach", Journal of Chemical Physics, 106 (13), 1997, 5678-5684.
[27] Rudzinski, W., and Panczyk, T.: "The Langmuirian Adsorption Kinetics Revised: A Farewell to the XXth Century Theories?" Fuel, 2000, 1963-1974.

[28] Rudzinski, W., and Panczyk, T.: "Kinetics of Gas Adsorption in Activated Carbons, Studied by Applying the Statistical Rate Theory of Interfacial Transport", Journal of Physical Chemistry B, 105, 2001, 6858-6866.
[29] Sutton, J.B., and Davies C.H.: "The Adsorption of Methane by Coal", Fuel. Vol. 57, Oct.1935, 1785-1788
[30] Maggs, F.A.P.: "The Adsorption-Swelling of Several Carbonaceous Solids", Transactions of the Faraday Society, 42B, 1946, 284-8.
[31] Seidle, J.P., and Huitt, L.G.: "Experimental Measurements of Coal Matrix Shrinkage Due to Gas Desorption and Implications for Cleat Permeability Increases", Paper SPE 30010 Proceedings from the International Meeting on Petroleum engineering held in Beijing, PR, China, 14-17 November, 1995.
[32] Apol, M.E.F., Amadei, A., and Berendsen, H.J.C.: "Application of Quasi-Guassian Entropy Theory to the Calculation of Thermodynamic Properties of Water and Methane in the Liquid and Gas Phase", Journal of Physical Chemistry 104(17), 1996, 6666-6678.
[33] Mofarahi, M., Sadrameli, M., and Towfighi, J.: "Characterization of Activated Carbon by Propane and Propylene Adsorption", Journal of Chemical and Engineering Data, 48(5), 2003, 1256-1261.
[34] Sandler, S.I.: "Models for Thermodynamic and Phase Equilibria Calculations", Marcel Dekker, INC, 1993, 10-35.
[35] Kapoor, A., Ritter, J.A., and Yang, R.T.: "On the Dubinin-Radushkevich Equation for Adsorption in Microporous Solids in the Henry's Law Region", Langmuir, 5, 1989, 1118-1121.
[36] Vyas, S.N., Patwardhan, S.R., Vijayalakshimi, S., and Ganesh, K.S.: "Adsorption of Gases on Carbon Molecular Sieves", Journal of Colloid and Interface Science, 168, 1994, 275-280.
[37] Arnaud, M., Bose, T.K., Okambawa, R., and Ingrain, D.: "Precise Determination of the Compressibility Factor by Using Dielectric Constant Measurements", Journal of Fluid Phase equilibria, 88, 1993, 137-149.

Nomenclature

a	:	Coal swelling coefficient	dimensionless
b', c'	:	Ratio of the liquid to the adsorbed phase density	dimensionless
c, d	:	Equilibrium isotherm fitting parameters	dimensionless
D	:	D-A isotherm parameter	dimensionless
D_o	:	The dissociation energy of the gas molecule	$[M][L][T]^{-2}[Mol]^{-1}$
H	:	Henry's constant	$[M][T]^{-2}[L]^{-1}$
j	:	Net adsorption or solution rate	$[T]^{-1}$
K	:	Boltzman constant, 1.38×10^{-23}	$\frac{m^2 Kg}{T^2 K}$
K'_{gs}	:	Non-Equilibrium constant for gas-water system	$[T]^{-1}$
K_{gw}	:	Non-Equilibrium constant for gas-water system	$[T]^{-1}$
Mw	:	Molecular Weight	$[M][Mol]^{-1}$
m_1, m_2	:	Masses of the two component atoms	$[M]$
P	:	Pressure	$[M][T]^{-2}[L]^{-1}$
r	:	D-A isotherm exponent	dimensionless
R	:	Universal gas constant	$[M][L][T]^{-2}[Mol]^{-1}[K]$
R_a	:	Adsorption rate	$[L]^{3}[T]$
R_d	:	Desorption rate	$[L]^{3}[T]$
R_s	:	Solution rate	$[Mol][T]^{-1}$

R_{ds}	:	Dissolution rate	$[Mol][T]^{-1}$
V	:	Volume	$[L]^3$
x	:	Mole fraction of the component in liquid phase	dimensionless
x^s	:	Mole fraction of the component in liquid phase	dimensionless
y	:	Mole fraction of the component in gas phase	dimensionless
z	:	Gas compressibility factor	dimensionless
ϕ	:	Function of temperature and gas phase properties	dimensionless
θ	:	Ratio of the adsorbed volume to the maximum volume	dimensionless
ρ	:	Density	$[M][L]^3$
γ	:	Activity coefficient	Dimensionless
ω	:	Characteristic vibration frequency of a gas molecule	$[rad][T]^{-1}$

Subscripts

g	:	Gas
w	:	Water
c	:	Coal
i	:	Component

Table 1. Non-Equilibrium isotherm parameters for various gases, using Arnoud et al. [36] data

Parameter Gas Comp.	*r*	*D*	V_m	*c*	*d*
CH_4	0.3	675	682	2.6	24.14
C_2H_6	0.3	625	906	1.747	10.171
CO_2	1.3	0.17	3719	4.14	5.51
C_3H_8	0.2	623	1630	1.21	4.34
N_2	0.2	857	7380	4.6	3.23

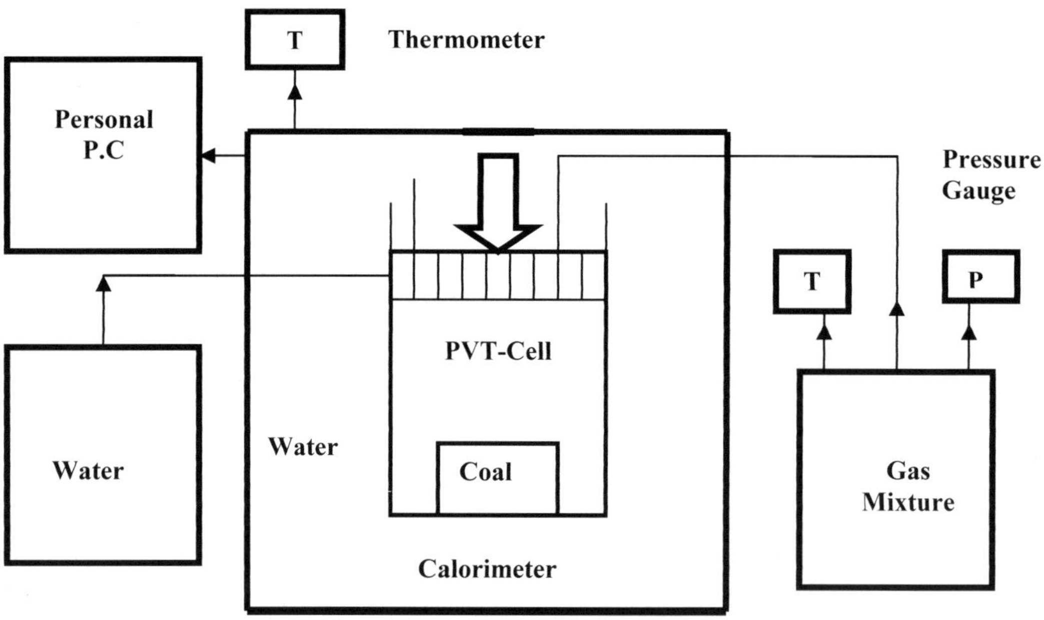

Figure 1. The Schematic Scheme of the Designed Apparatus

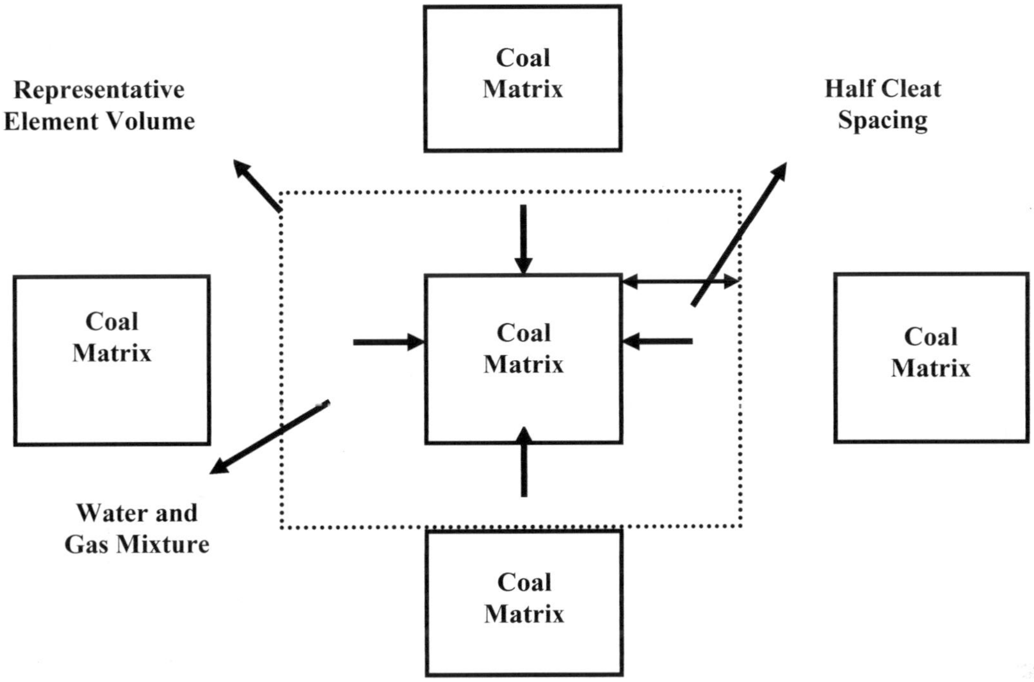

Figure 2. A System of Coal, Water, and Gas in Contact with Each Other

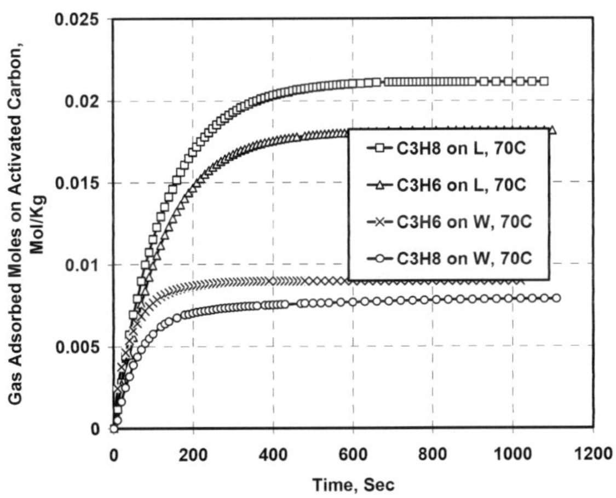

Figure 3. Non-equilibrium isotherm for adsorption of propane and propylene on activated carbons at 14.6 psia [33]

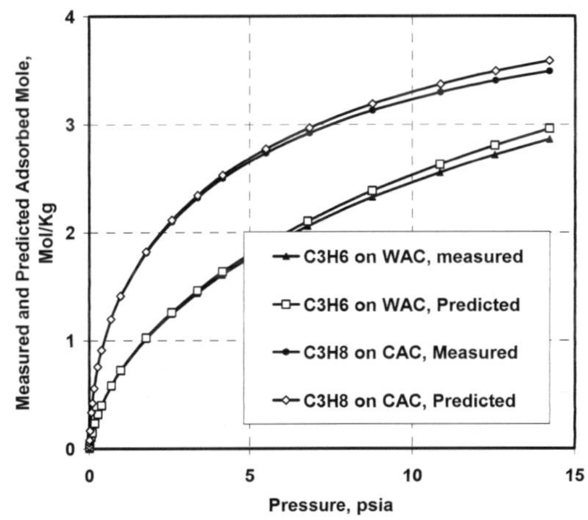

Figure 4. Comparison of predicted and measured [33] equilibrium isotherms for propane and propylene adsorption on activated carbons

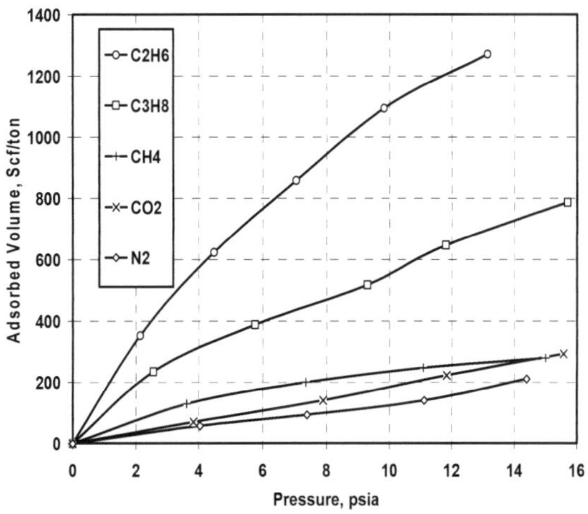

Figure 5. Equilibrium adsorption isotherm of methane on the molecular sieve surface [36]

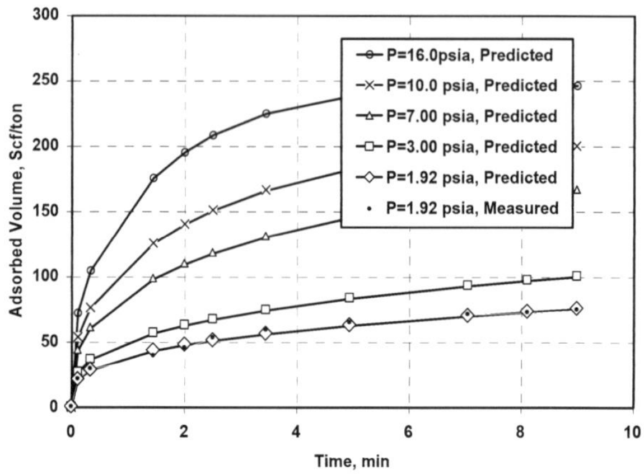

Figure 8. Non-equilibrium isotherm for adsorption of methane on molecular sieve surface at various pressures

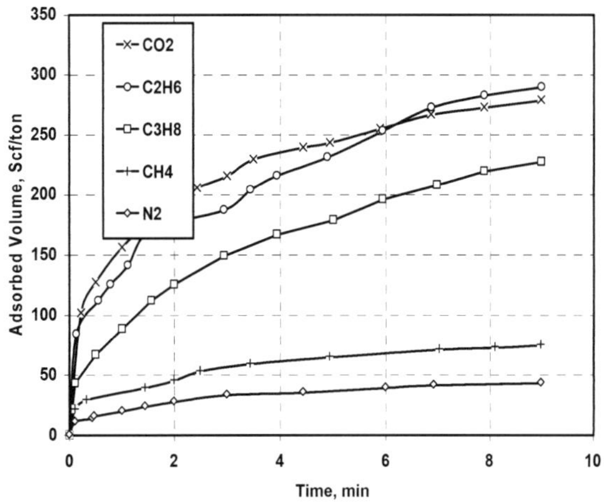

Figure 6. Non-equilibrium isotherm for adsorption of various gases on carbon molecular sieves at 2.3 psia [36]

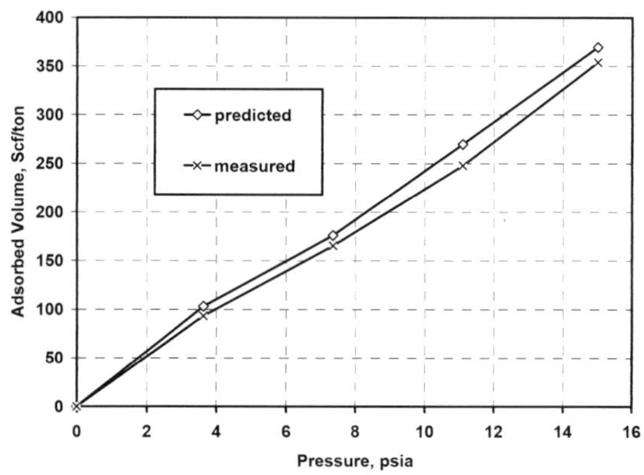

Figure 9. Comparison of the experimental equilibrium data with the estimated values using the non-equilibrium isotherm parameters

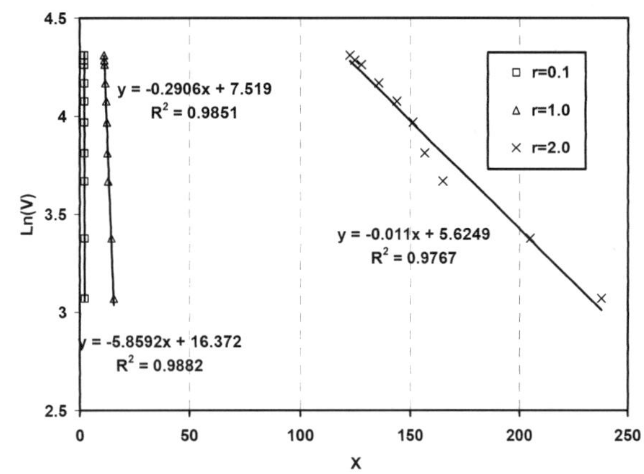

Figure 7. Adsorbed methane volume vs. X explained in step 3

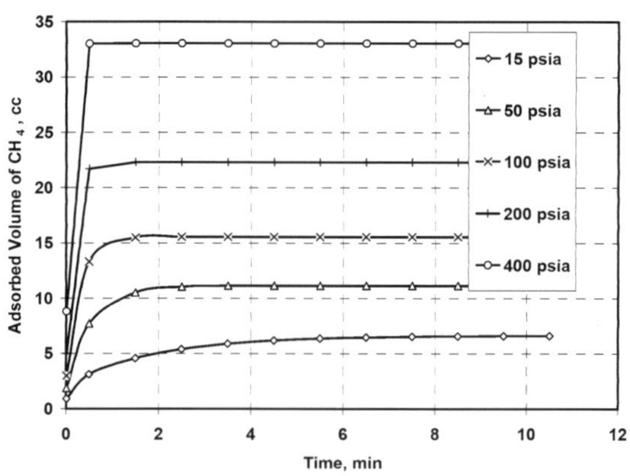

10. Projection of the CH_4 non-equilibrium adsorption isotherms for the higher pressures

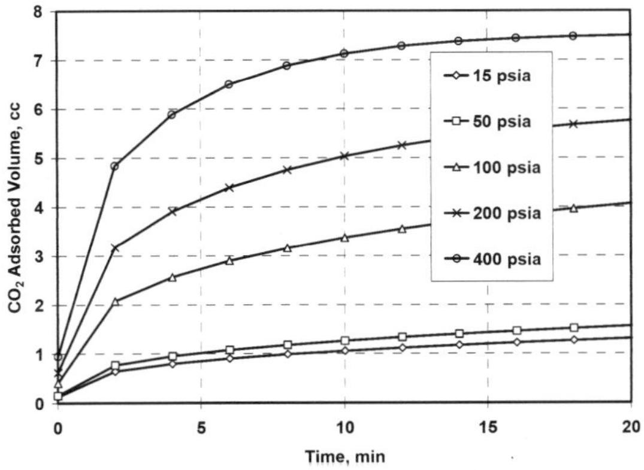

Figure 11. Projection of the CO_2 non-equilibrium adsorption isotherms for the higher pressures

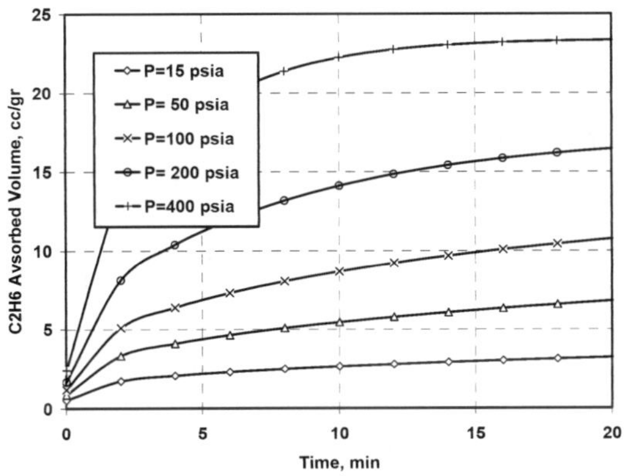

Figure 12. Projection of the C_2H_6 non-equilibrium adsorption isotherms for the higher pressures

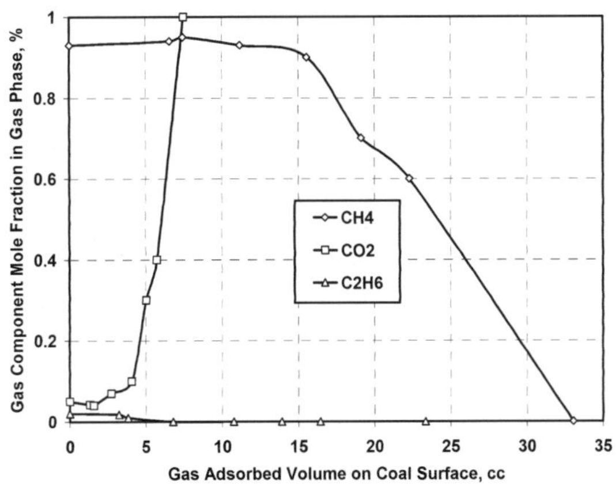

Figure 13. The relationship between gas mole fractions in gas phase vs. the gas adsorbed volume in coal

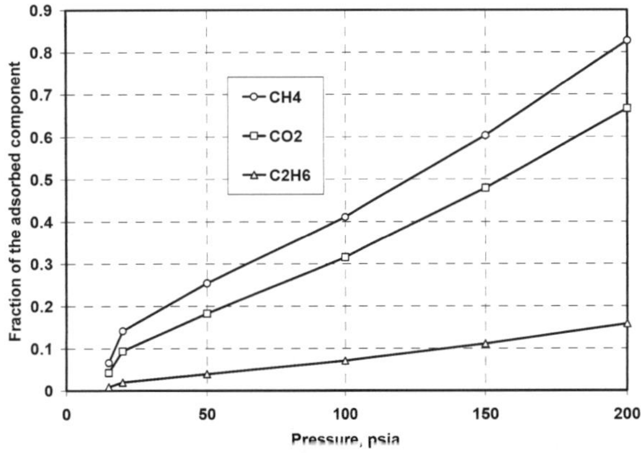

Figure 14. Comparison of the adsorbed fraction of methane and carbon dioxide at equilibrium pressures

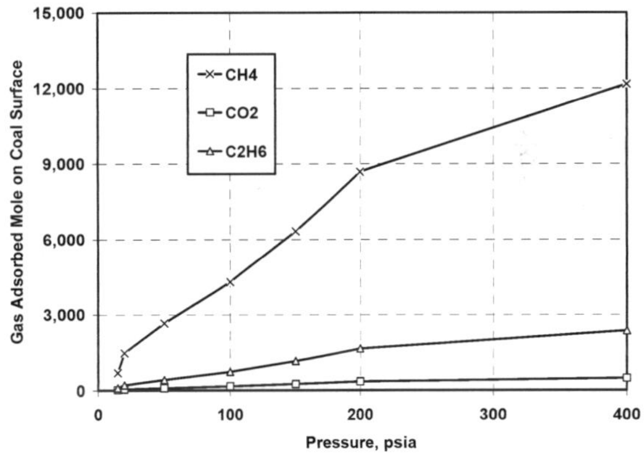

Figure 15. Comparison of the adsorbed moles of methane and carbon dioxide at equilibrium pressure

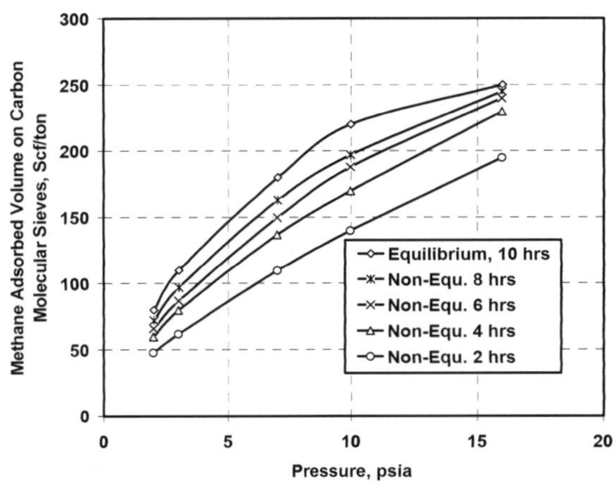

Figure 16. Comparison of Equilibrium and Non-equilibrium methane adsorbed volume on carbon molecular sieves.

0527

Measurement and Prediction of Single- and Multi-Component Methane, Carbon Dioxide and Nitrogen Isotherms for U.S. Coals

Scott Reeves, Reinaldo Gonzalez
Advanced Resources International, Inc.

Khaled A. M. Gasem, James E. Fitzgerald, Zhejun Pan, Mahmud Sudibandriyo, Robert L. Robinson Jr.
Oklahoma State University

ABSTRACT

In October, 2000, the U.S. Department of Energy, through contractor Advanced Resources International, launched a multi-year government-industry R&D collaboration called the Coal-Seq project. The Coal-Seq project investigated the feasibility of CO_2 sequestration in deep, unmineable coalseams by performing detailed reservoir studies of two enhanced coalbed methane (ECBM) recovery field projects in the San Juan basin. As part of the overall project, single-component CH_4, CO_2 and N_2 isotherms were measured (or obtained) on coal samples from the Black Warrior, Cherokee, Forest City, Gulf Coast, Illinois, Northern Appalachian, Piceance, Powder River and San Juan basins. In addition, binary- and ternary-component isotherms were measured on the San Juan basin sample to compare actual multi-component sorption behavior to predictions based on single-component data.

Uni-variate analysis of the single-component results indicates a reasonable correlation of CH_4 and N_2 storage capacity (i.e., Langmuir volume) with coal rank (i.e., vitrinite reflectance) exists, but no such (linear) trend appears to exist for CO_2. No uni-variate trends were found to predict Langmuir pressure. The CO_2/CH_4 and N_2/CH_4 ratios both showed reasonable correlation to coal rank. Multiple linear regression of the data yielded acceptable models to predict Langmuir volume and pressure for the three gases based on coal characterization data.

Comparison of actual multi-component isotherm data to predictions based on single-component data showed poor predictive capability for all the theories evaluated. New and more robust multi-component sorption theories are therefore required to predict multi-component sorption behavior from single-component data.

INTRODUCTION

In October, 2000, the U.S. Department of Energy (DOE), through contractor Advanced Resources International (ARI), launched a multi-year government-industry R&D collaboration called the Coal-Seq project[1]. The Coal-Seq project investigated the feasibility of CO_2 sequestration in deep, unmineable coalseams by performing detailed reservoir studies of two enhanced coalbed methane (ECBM) recovery field projects in the San Juan basin. As part of the project, single-component CH_4, CO_2 and N_2 isotherms were measured (or obtained) on coal samples from the Black Warrior, Cherokee, Forest City, Gulf Coast, Illinois, Northern Appalachian, Piceance, Powder River and San Juan basins. Maceral composition, vitrinite reflectance and proximate analyses were also measured on each coal sample to aid in their characterization. One of the objectives of the data collection effort was to establish the beginnings of a public-domain database of such information to serve as a useful reference for companies and/or other organizations considering enhanced coalbed methane and/or carbon sequestration projects.

In addition to the single-component data, binary- and ternary-component isotherms were measured on the San Juan basin sample to compare actual multi-component sorption behavior to predictions based upon various theories using single-component data (e.g., extended Langmuir, loading ratio correlation, equations -of -state, etc.). No such analysis has been published previously, yet modeling of ECBM and carbon sequestration is founded on the assumption that such predictions are reliable. The objective of this work was to test that assumption.

DESCRIPTION OF SAMPLES, MEASUREMENT RESULTS

From 2003 through 2004, a total of 11 coal samples were collected from different basins throughout the United States. Those basins included the Black Warrior, Cherokee, Forest City, Gulf Coast, Illinois, Northern Appalachian, Piceance, Powder River and San Juan basins. All but one sample was in the form of a core from a well drilled from the surface, the exception being a mine sample from the Black Warrior basin. Each sample was identified by location (county, state), and coal seam (name, depth), and was characterized by coal rank (vitrinite reflectance), maceral composition, and chemical composition (proximate analysis). Following characterization, moisture-equilibrated single-component N_2, CH_4 and CO_2 isotherms were measured at the temperature corresponding to in-situ conditions. The results were reported on an as-received, dry, and dry-ash-free basis.

Coal characterization data for the samples are provided in Table 1. The isotherm results, presented in the form of Langmuir isotherm constants (V_L and P_L), are presented in Table 2. Table 2 also provides the conditions at which the isotherms were measured. Plots of the individual isotherms for N_2, CH_4 and CO_2 are presented in Figures 1-3. Initial examination of the results suggests that all of the data appear as one would expect - gas storage capacity generally increases with coal rank, and for a given sample the gas storage capacity is greatest for CO_2, followed by CH_4, and is the least for N_2.

SINGLE-COMPONENT ISOTHERM PREDICTIONS FROM COAL CHARACTERIZATION DATA

While the publication of this dataset in and of itself was an important objective of this effort, an additional objective was to identify the existence of any relationships in the data that would enable one to estimate single-component isotherm properties from coal characterization data alone. Such a capability would enable one to reasonably estimate sorptive properties of a coal in the absence of actual isotherm measurements.

The first approach employed was simple uni-variate analysis. Since coal rank is usually an important factor influencing coal sorptive capacity, vitrinite reflectance, a proxy for coal rank, was selected as the independent variable. Plots of Langmuir volume and Langmuir pressure versus vitrinite reflectance for each gas and for each sample were created; the results are presented in Figures 4 and 5.

The results indicate that CH_4 and N_2 sorptive capacity increases with coal rank in a statistically meaningful fashion. However, no such trend exists for CO_2. If a relationship does exist between coal rank and CO_2 sorptive capacity, it must be influenced by other factors not accounted for in this simple model. Figure 5 suggests there is no meaningful relationship between Langmuir pressure and vitrinite reflectance for any of the gases. Again, other factors may be influencing the result.

The ratio of N_2/CH_4 and CO_2/CH_4 sorptive capacities were also examined as function of vitrinite reflectance. Both Figures 6 and 7 suggest that some relationship does exist in this regard, with the ratio increasing with increasing coal rank for N_2, and decreasing with increasing coal rank for CO_2. Figure 8 illustrates both of these ratios and a function of pressure for the Piceance basin sample. Similar trends existed for the other samples.

Principle component analysis together with multiple-linear regression and analysis of covariance were then employed to determine if a more sophisticated mathematical model could be constructed to predict single-component isotherm properties (Langmuir constants) based upon coal characterization data alone. Six individual models were constructed (one for each Langmuir constant for each gas), each with seven input variables. The variables used for each model, their relative impact, and the predictive accuracy of each model are presented in Table 3.

Actual versus predicted Langmuir constants for each model are also provided in Figures 9 & 10. Each model appears to predict the objective Langmuir parameter with excellent accuracy. Note that the corrected R^2 values in Table 3 are different from those in Figures 9 & 10 because the corrected R^2 measures the combined predictive capability of each input parameter in multi-dimensional space, and this is the real indicator of the efficiency of the model[2]. While the predictive accuracy is excellent, the few number of samples limits the widespread applicability and robustness of the model at this time.

In all cases, predictions of CO_2 isotherms were generally more difficult than for N_2 or CH_4, and predictions of Langmuir pressure more difficult than Langmuir volume. Nevertheless, acceptable models were created for all parameters, which indicates that predictions of single-component isotherm properties are possible from coal characterization data alone, even with a limited dataset as was the case here.

MULTI-COMPONENT ISOTHERM PREDICTIONS FROM SINGLE-COMPONENT DATA

In addition to investigating relationships between coal characterization data and single-component isotherm properties, the validity of multi-component sorption predictions based on single-component data was also examined. Two isotherm models were specifically evaluated for the purposes of predicting multi-component isotherm behavior from single component data. They were:

<u>Extended Langmuir/Loading Ratio Correlation (LRC).</u>

Historically, simple models have been used to represent the behavior of pure and mixed gas adsorption on coal. The extended Langmuir model is used almost exclusively in literature studies[3], although the Ideal Adsorbed Solution (IAS) model has also been employed[4,5]. Both of these models work well for essentially ideal adsorbed solutions, but neither is capable of handling nonidealities in the adsorbed phase with any accuracy. The extended Langmuir model is shown below as an illustration of the simple modeling approach used in most previous studies. For mixtures, it takes the form:

$$\theta_i = \frac{\omega_i}{L_i} = \frac{B_i P y_i}{1 + \sum_j B_j P y_j} \quad (1)$$

where ωi is the amount of component "i" adsorbed (standard cubic feet of "i" adsorbed per ton of coal), L_i and B_i are Langmuir constants for "i", P is pressure, and y_i is the mole fraction of "i" in the gas phase. This relation allows mixture adsorption to be predicted from pure-component data, since values of L_i and B_i may be determined from the pure-component form of Equation (1). For pure-fluid adsorption, Equation (1) reduces to the Langmuir model:

$$\theta = \frac{\omega}{L} = \frac{BP}{1 + BP} \quad (2)$$

The combined Langmuir-Freundlick adsorption isotherm, expressed in terms of ωi, yields the loading ratio correlation (LRC) for mixtures:

$$\theta_i = \frac{\omega_i}{L_i} = \frac{(B_i P y_i)^{\eta_i}}{1 + \sum_j (B_j P y_j)^{\eta_j}} \quad (3)$$

The additional parameter (η_i) in the LRC lends the Langmuir model more flexibility. Although the simplicity of Langmuir models is attractive, data show that they are inadequate to represent the behavior of mixtures of the gases CO_2, methane, and nitrogen. In fact, errors greater than 100% were found when the extended Langmuir model was applied to data on the adsorption of nitrogen from nitrogen and CO_2 mixtures[6].

Zhou-Gasem-Robinson Equation-of-State (ZGR-EOS).

Simulations of coalbed gas recovery and CO_2 sequestering require reliable, yet simple, analytic models beyond Langmuir-type correlations. Equation-of-state (EOS) frameworks offer an attractive potential for such requirements. A general two-dimensional EOS can be written as follows[7]:

$$\left[A\pi + \frac{\alpha\omega^2}{1 + U\beta\omega + W(\beta\omega)^2}\right]\left[1 - (\beta\omega)^m\right] = \omega RT \quad (4)$$

where A is the specific surface area, π is the spreading pressure, ω is the specific amount adsorbed, and α, β and m are model parameters. The model coefficients, U, W, and m must be specified to obtain a specific form of the 2-D EOS for application. For example, an analog of the van der Waals (VDW) EOS is obtained by setting m = 1 and U = W = 0; similarly for the Soave-Redlich-Kwong (SRK) (m = U = 1 and W = 0); the Peng-Robinson (PR) (m = 1, U = 2, and W = -1); and the Eyring (m = 1/2 and U = W = 0) EOS.

This general 2-D EOS can be used to investigate EOS behaviors by specifying various combinations of model coefficients. Selection of the model coefficient m is the most important among the EOS model coefficients, because it has a significant effect on the shape of the pure adsorption isotherm. If U and W are equal to zero, then by setting m to values of ∞, 1, and 1/2, we obtain the 2-D ideal gas law, the VDW EOS, and the Eyring EOS, respectively. Actually, the pure gas isotherms vary considerably in shape and we have found that it is sometimes desirable to select an m value even smaller than 1/2 to describe pure isotherms. We have determined that an equation with m = 1/3 and U = W = 0 (the ZGR EOS) is promising. The 2-D EOS can be applied to adsorbed phases containing mixtures by utilizing the traditional mixing rules (where x is the mole fraction in the adsorbed phase):

$$\alpha = \sum_i \sum_j x_i x_j \alpha_{ij} \quad (5)$$

$$\beta = \sum_i \sum_j x_i x_j \beta_{ij} \quad (6)$$

along with the non-traditional combination rules,

$$\begin{aligned} \alpha_{ij} &= (1 - C_{ij})(\alpha_i + \alpha_j)/2 \\ \beta_{ij} &= (1 + D_{ij})\sqrt{\beta_i \beta_j} \end{aligned} \quad (7)$$

where C_{ij} and D_{ij} are the EOS binary interaction parameters.

Modeling Results

On the San Juan basin sample, in addition to the three single-component isotherms, three binary isotherms, and one ternary isotherm were measured[8]. The feed-gas compositions of the multi-component isotherms were:

- 50% CH_4, 50% N_2
- 40% CH_4, 60% CO_2
- 20% N_2, 80% CO_2
- 40% CH_4, 10% N_2, 50% CO_2

The three single component isotherms for this sample are presented in Figure 11. The three binary isotherms for the coal sample are presented in Figures 12 -14, and the ternary isotherm in Figure 15. Note that in each plot the data points represent the actual measured data, with error bars indicating the experimental error associated with each data point. The lines represent model predictions of the data based on single-component isotherms.

One can readily observe from Figures 10-13 that the ability of either model (LRC or ZGR-EOS) to predict multi-component isotherm behavior from single-component data is frequently well outside the bounds of experimental error. This is an expected outcome since neither model accurately accounts for the physics of adsorption on wet coal. The modeling results are also presented in Table 4, where the LRC results are shown for the case of using an overall optimum fixed value of $\eta = 0.9$. In this table, the absolute average deviation for the model predictions from the actual isotherm data for each gas in each multi-component system is provided. Note the large number of cases where the error is greater than the experimental error (noted with asterisks). This seems to indicate that two commonly-used models for predicting multi-component sorption behavior based on single-component data are inadequate, which could lead to misleading predictions of ECBM and sequestration project performance. Work is currently underway to understand the causes for the errors and develop improved multi-component isotherm models.

CONCLUSIONS

Based on the results of this work, the following conclusions have been drawn:

- Uni-variate analysis of the single-component results indicates a reasonable trend of CH_4 and N_2 storage capacity (i.e., Langmuir volume) with coal rank (i.e., vitrinite reflectance) exists, but no such (linear) trend appears to exist for CO_2. No uni-variate trends were found to predict Langmuir pressure. The CO_2/CH_4 and N_2/CH_4 sorption capacity ratios also showed identifiable trends with coal rank. Multiple linear regression of the data yielded reasonable models to predict Langmuir volume and pressure for the three gases based on coal characterization data.

- Comparison of actual multi-component isotherm data to predictions based on single-component data showed inadequate predictive capability for all the theories evaluated. New and more robust multi-component sorption theories are therefore required to predict multi-component sorption behavior from single-component data if, in fact, such calculations are feasible.

ACKNOWLEDGEMENTS

This work was funded by the U.S. Department of Energy, National Energy Technology Laboratory, contract number DE-FC26-00NT40924. The DOE project manager was Charlie Byrer. The authors also wish to thank the various companies and organizations that provided coal samples on which to perform the coal characterization and isotherm measurements (or that provided existing data), which included BP, ConocoPhillips, CONSOL, Illinois State Geological Survey, Geological Survey of Alabama, Harvest Gas, Kansas Geological Survey, and Tom Brown (now EnCana). All isotherm measurements were performed by Terra Tek with the exception of the San Juan basin sample, which were performed by Oklahoma State University.

REFERENCES

1. Reeves, S.R.: "Geologic Sequestration of CO_2 in Deep Unmineable Coalbeds: An Integrated Research and Commercial-Scale Field Demonstration Project", SPE 71749, presented at the SPE Annual Technical Conference and Exhibition, New Orleans, September 30-October 3, 2001.

2. Johnson, D.E.: "Applied Multivariate Methods for Data Analysts", Brooks Cole Publishing Company, 1998.

3. Arri, L. E., and Yee, D., *Modeling Coalbed Methane Production with Binary Gas Sorption*, SPE Paper 24363, presented at the SPE Rocky Mountain Regional Meeting, Casper, Wyoming, May 18-21, 1992.

4. Myers, A. L., and Prausnitz, J. M., *Thermodynamics of Mixed-Gas Adsorption*, AIChE J. <u>11</u> 121-129 (1965).

5. Stevenson, M. D., Pinczewski, W. V., Somers, M. L., and Bagio, S. E., *Adsorption/Desorption of Multicomponent Gas Mixtures on Coal at In-Seam Conditions*, SPE Paper 23026, presented at the SPE Asia-Pacific Conference, Perth, Western Australia, November 4-7, 1991.

6. Hall, F. E., *Adsorption of Pure and Multicomponent Gases on Wet Fruitland Coal*, M. S. Thesis, Oklahoma State University, December 1993.

7. Zhou, C., Gasem, K. A. M., and Robinson, Jr., R. L., *Predicting Gas Adsorption Using Two-Dimensional Equations of State*, I&EC Research <u>33</u> 1280-1289 (1994).

8. Gasem, K.A.M., Robinson, R.L., Jr., Reeves, S.R., :"Adsorption of Pure Methane, Nitrogen and Carbon Dioxide and Their Mixtures on San Juan Basin Coal", Topical Report, DOE Contract No. DE-FC26-00NT40924, May, 2002.

Table 1: Coal Characterization Data

Basin	County & State	Seam Name	Sample Depth (ft)	Coal Rank (% Vro)	Coal Rank Description	Proximate Analysis (as received): Moisture (%)	Ash (%)	Volatile Matter (%)	Fixed Carbon (%)	Heating Value (BTU/lb)	Petrographic Analysis (w/o minerals): Vitrinite (%)	Liptinite (%)	Inertinite (%)
Powder River	Campbell, WY	Big George	1225	0.36	Sub	22.6%	2.6%	35.2%	39.6%	9,659	90.2%	5.8%	0.4%
Gulf Coast	Catahoula, LA	Wilcox	4898	0.41	Sub	10.8%	9.0%	38.7%	41.6%	10,948	57.9%	8.4%	33.7%
Forest City Basin	Franklin, KS	Mineral	780	0.54	HVB	2.1%	15.2%	42.5%	40.2%	12,564	89.8%	4.8%	5.4%
Illinois	Clark, IL	Herrin (No. 6)	604	0.59	HVB	13.1%	11.0%	29.6%	46.3%	11,064	83.8%	7.2%	8.9%
N. Appalachian	Marshall, WV	Pittsburgh	714	0.62	HVB	1.8%	8.0%	38.7%	51.5%	13,567	85.1%	2.7%	12.2%
Cherokee Basin	Montgomery, KS	Mineral	772	0.69	HVB	1.7%	13.0%	36.7%	48.6%	13,128	86.4%	6.9%	6.7%
N. Appalachian	Marshall, WV	Lower Freeport	1361	0.76	HVA	1.3%	11.6%	36.2%	50.9%	13,167	70.4%	12.1%	17.5%
Piceance	Rio Blanco, CO	Cameo	6700	0.80	HVA	1.3%	5.1%	35.6%	58.0%	14,020	92.1%	1.5%	6.3%
Warrior	Walker, AL	Mary Lee	0	0.90	HVA	2.6%	13.7%	30.1%	53.6%	12,345	87.4%	4.6%	8.0%
San Juan	La Plata, CO	Fruitland	3100	1.33	MV	n/a	n/a	30.0%	70.0%	n/a	96.0%	0.8%	3.2%
Warrior	Tuscaloosa, AL	Newcastle	1948	1.40	MV	1.6%	11.5%	21.3%	65.6%	13,600	90.6%	1.0%	8.4%

Table 2: Isotherm Results

			Isotherm Information (moisture-equilibrated, daf/dmmf)								
			Measurement	Equilibrium	Ash	Nitrogen		Methane		Carbon Dioxide	
Basin	County & State	Seam Name	Temperature	Moisture	Content	VL	PL	VL	PL	VL	PL
			(deg F)	(%)	(%)	(scf/ton)	(psi)	(scf/ton)	(psi)	(scf/ton)	(psi)
Powder River	Campbell, WY	Big George	75	23.9%	2.9%	81	1344	217	820	2331	1172
Gulf Coast	Catahoula, LA	Wilcox	140	12.0%	9.2%	154	2315	379	1010	1109	885
Forest City Basin	Franklin, KS	Mineral	70	2.6%	14.0%	218	744	465	406	881	185
Illinois	Clark, IL	Herrin (No. 6)	65	13.2%	11.3%	208	1125	432	458	1248	303
N. Appalachian	Marshall, WV	Pittsburgh	64	2.3%	7.7%	427	2919	569	537	987	240
Cherokee Basin	Montgomery, KS	Mineral	75	2.1%	13.1%	249	819	477	357	853	171
N. Appalachian	Marshall, WV	Lower Freeport	69	1.6%	11.2%	393	2478	547	560	898	261
Piceance	Rio Blanco, CO	Cameo	155	1.8%	4.7%	411	2101	616	871	911	340
Warrior	Walker, AL	Mary Lee	86	4.7%	14.0%	256	1023	612	845	1061	292
San Juan	La Plata, CO	Fruitland	130	n/a	n/a	521	1587	848	663	1153	223
Warrior	Tuscaloosa, AL	Newcastle	86	2.3%	18.6%	770	1245	1081	559	1308	218

Table 3: Results of Multiple-Linear Regression Models

Objective Function	Parameter Ranking												Goodness of Fit	
	Vro	Moist.	Ash	VM	FC	HV	Vitr.	Lipt.	Inert.	Temp.	E.Moist.	Ash Cont.	% AAD*	Corr. R^2
VL(N_2)	4	3	n/a	n/a	7	1	2	5	n/a	6	n/a	n/a	4.81%	0.954
PL (N_2)	n/a	3	1	n/a	n/a	5	7	6	4	2	n/a	n/a	7.65%	0.884
VL(CH_4)	1	7	n/a	n/a	5	4	2	3	n/a	6	n/a	n/a	2.80%	0.977
PL (CH_4)	n/a	n/a	2	n/a	3	1	n/a	6	5	7	4	n/a	1.99%	0.982
VL(CO_2)	2	n/a	n/a	3	5	7	n/a	6	n/a	4	1	n/a	1.59%	0.991
PL (CO_2)	6	n/a	n/a	1	3	4	5	n/a	n/a	n/a	2	7	7.91%	0.973

* AAD = average absolute deviation

Table 4: Accuracy of Multi-Component Isotherm Predictions from Single-Component Data

Mixture, (Feed Mole %)	Langmuir % AAD	LRC (n=0.9) % AAD	ZGR-EOS % AAD	Estimated Experimental Error % AAD
$CH_4 - N_2$:				
CH_4 (50%)	15.8*	12.0*	11.9*	7.0
N_2 (50%)	6.2	9.3	10.0	17.0
Total	12.2*	8.2*	11.5*	7.0
$CH_4 - CO_2$:				
CH_4 (40%)	25.9*	21.0*	27.0*	7.0
CO_2 (60%)	9.0*	10.5*	10.4*	6.0
Total	1.2	2.2	1.4	4.0
$N_2 - CO_2$:				
N_2 (20%)	44.9*	37.3*	48.7*	29.0
CO_2 (80%)	5.2	5.7	4.9	6.0
Total	3.5	3.8	3.5	5.0
$N_2 - CH_4 - CO_2$:				
N_2 (10%)	47.8*	44.5*	55.9*	14.0
CH_4 (40%)	20.7	5.2	21.6	27.0
CO_2 (50%)	13.2*	15.8*	17.6*	5.0
Total	2.9	5.4*	4.3	5.0

* Greater than experimental error

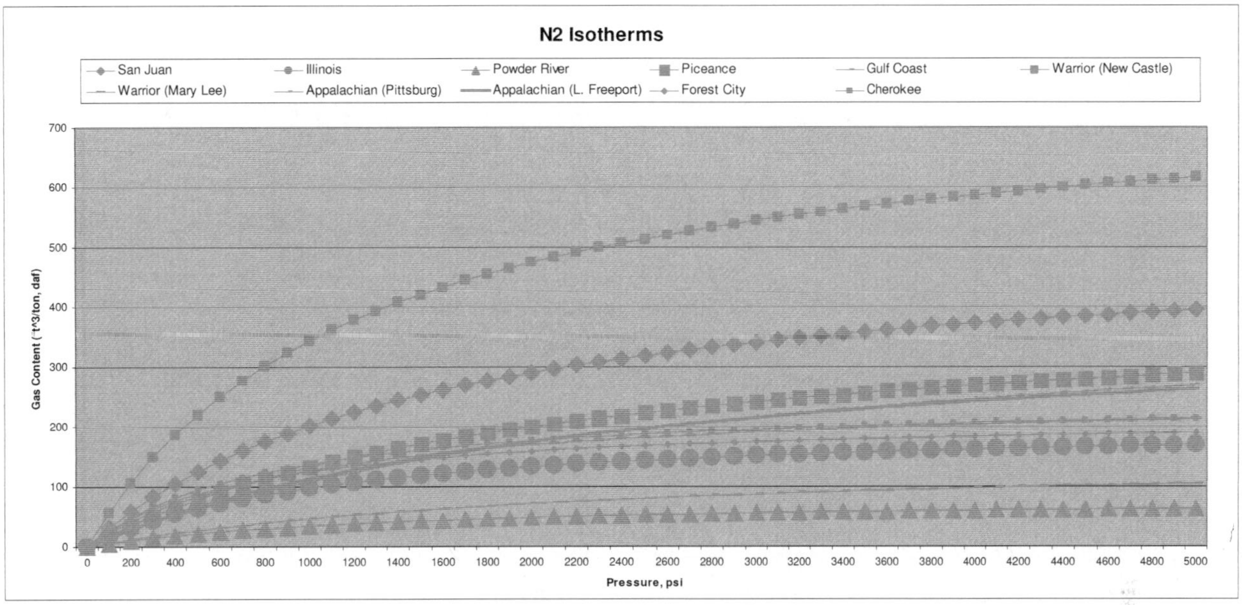

Figure 1: Langmuir Nitrogen Isotherms

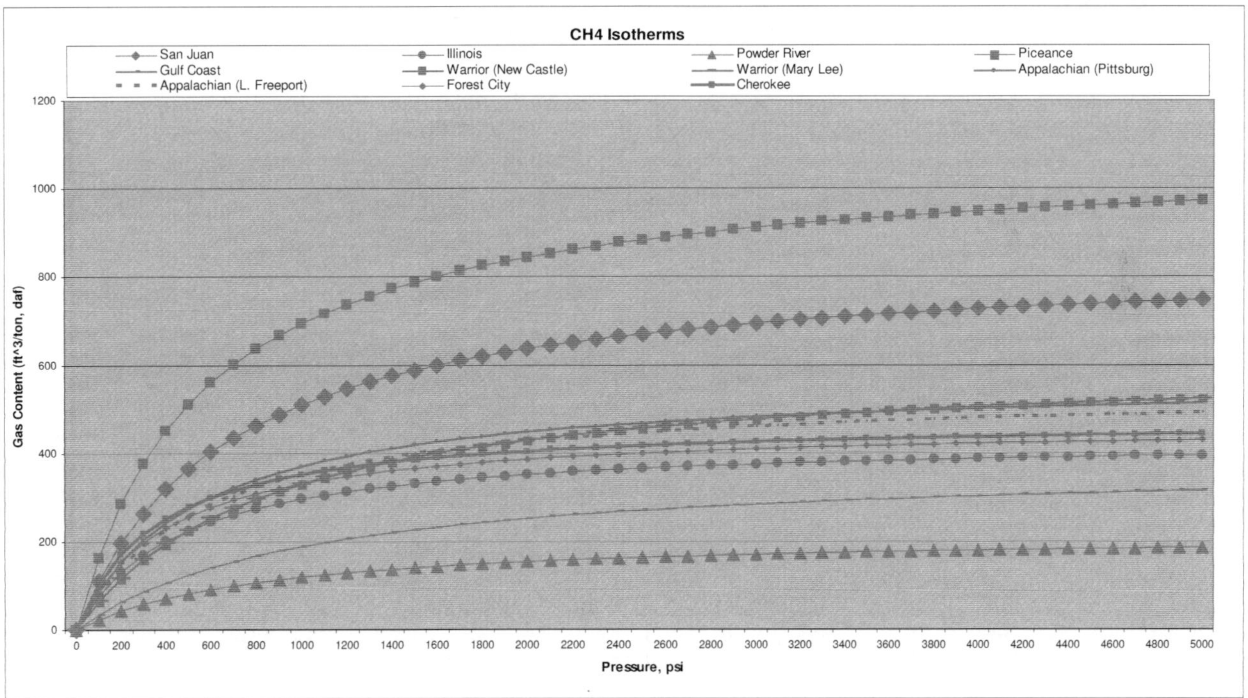

Figure 2: Langmuir Methane Isotherms

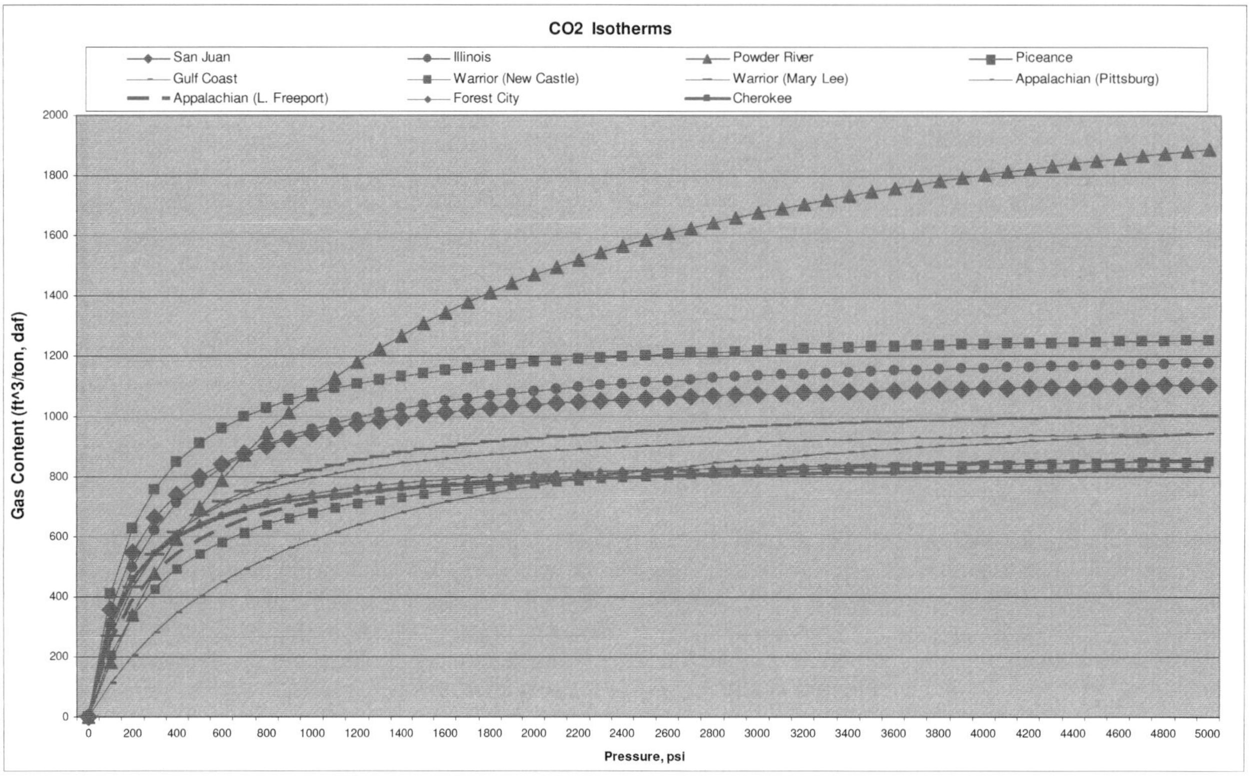

Figure 3: Langmuir Carbon Dioxide Isotherms

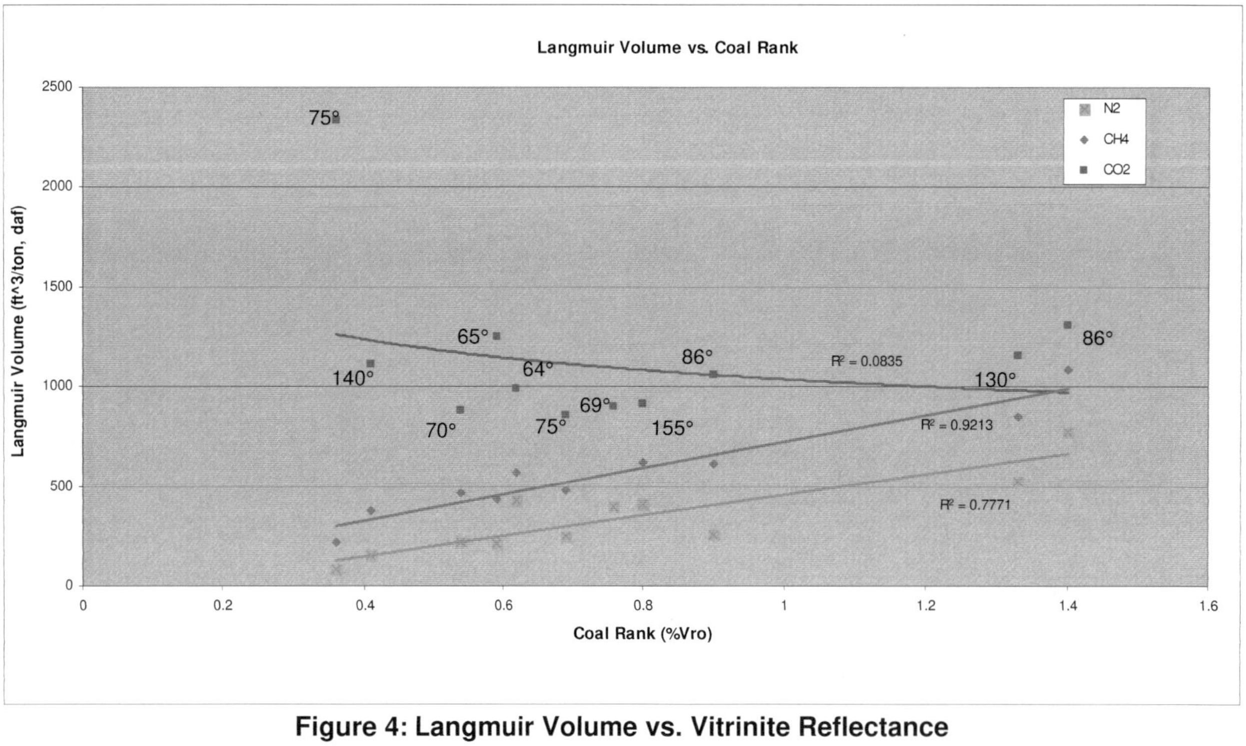

Figure 4: Langmuir Volume vs. Vitrinite Reflectance

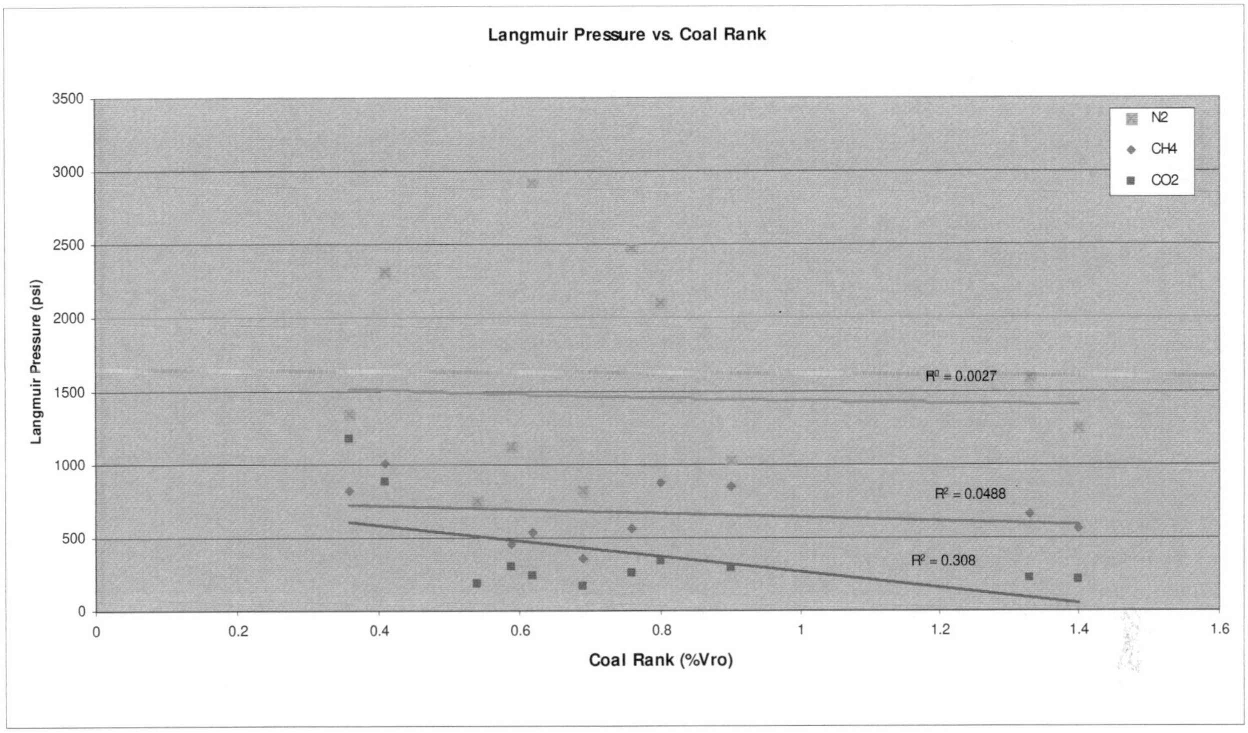

Figure 5: Langmuir Pressure vs. Vitrinite Reflectance

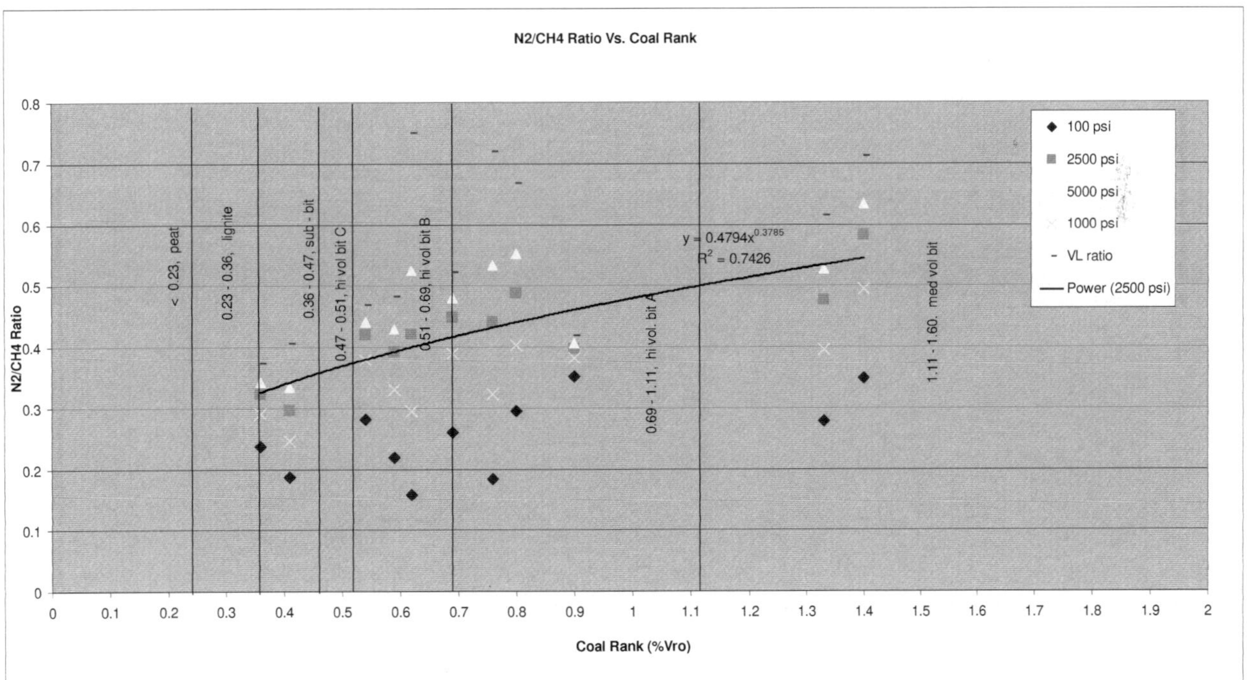

Figure 6: N_2/CH_4 Sorption Capacity vs. Vitrinite Reflectance

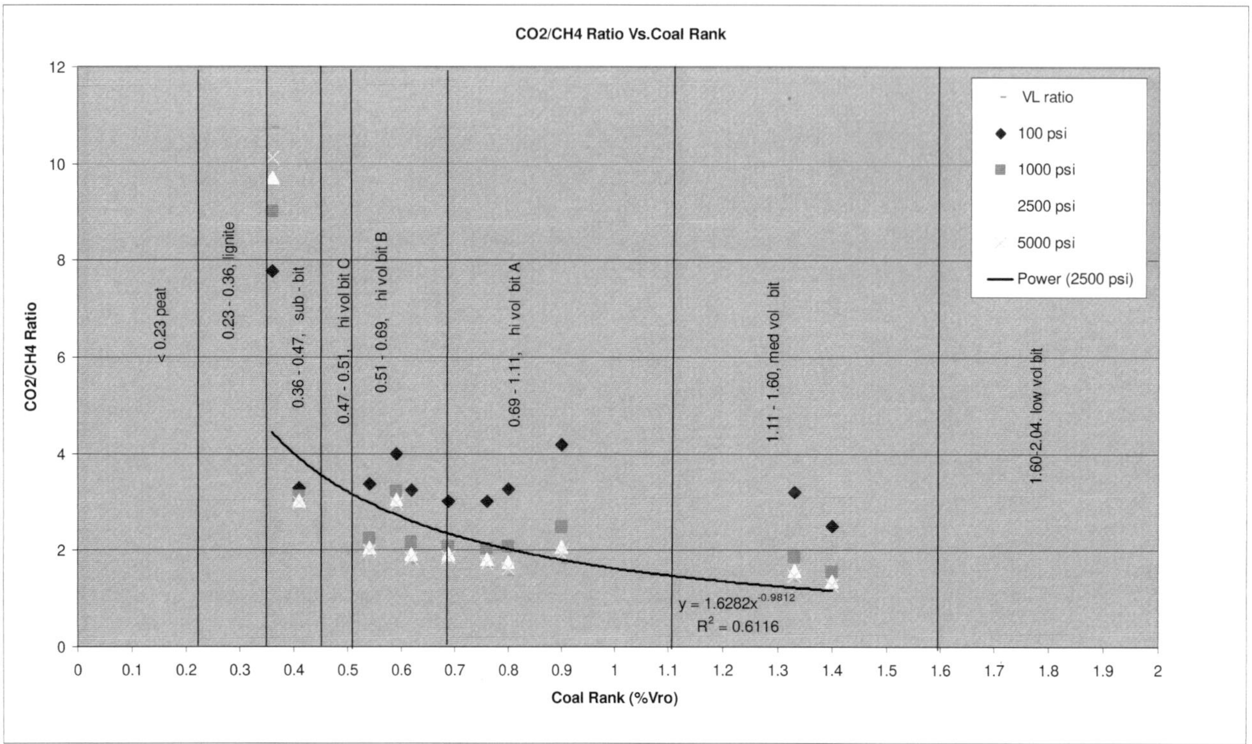

Figure 7: CO_2/CH_4 Sorptive Capacity vs. Vitrinite Reflectance

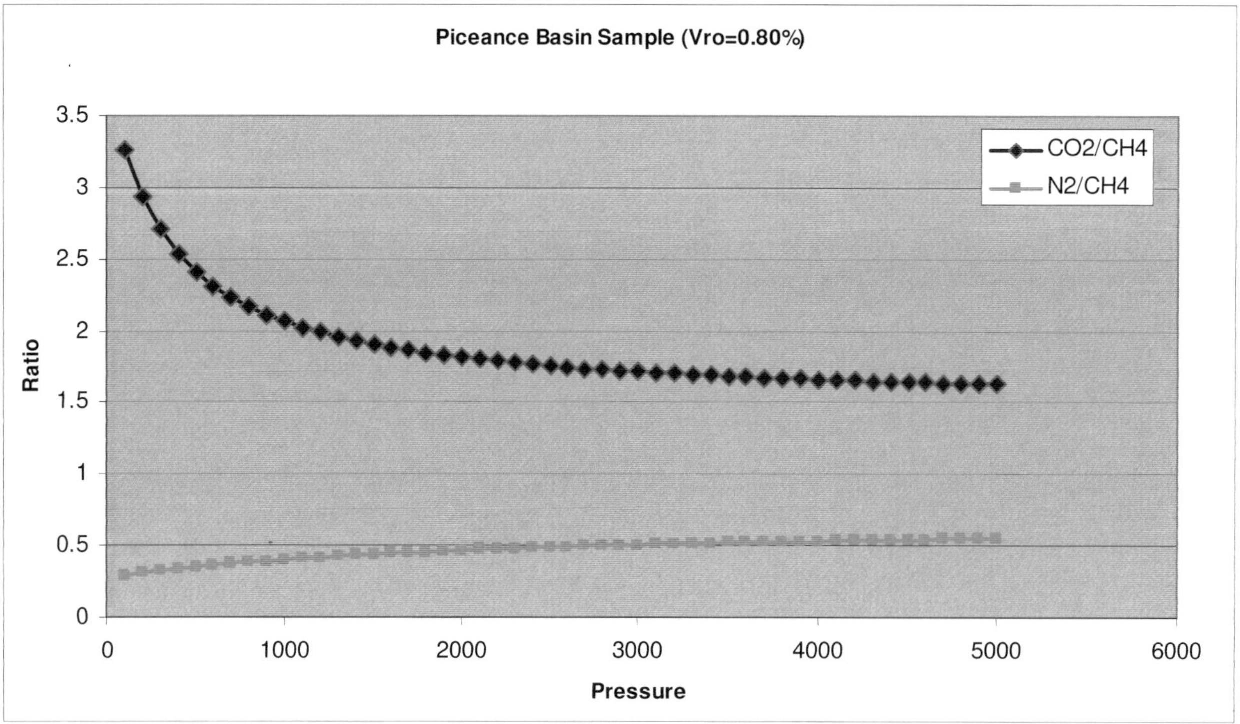

Figure 8: N_2/CH_4 and CO_2/CH_4 Sorptive Capacity Ratios, Piceance Basin

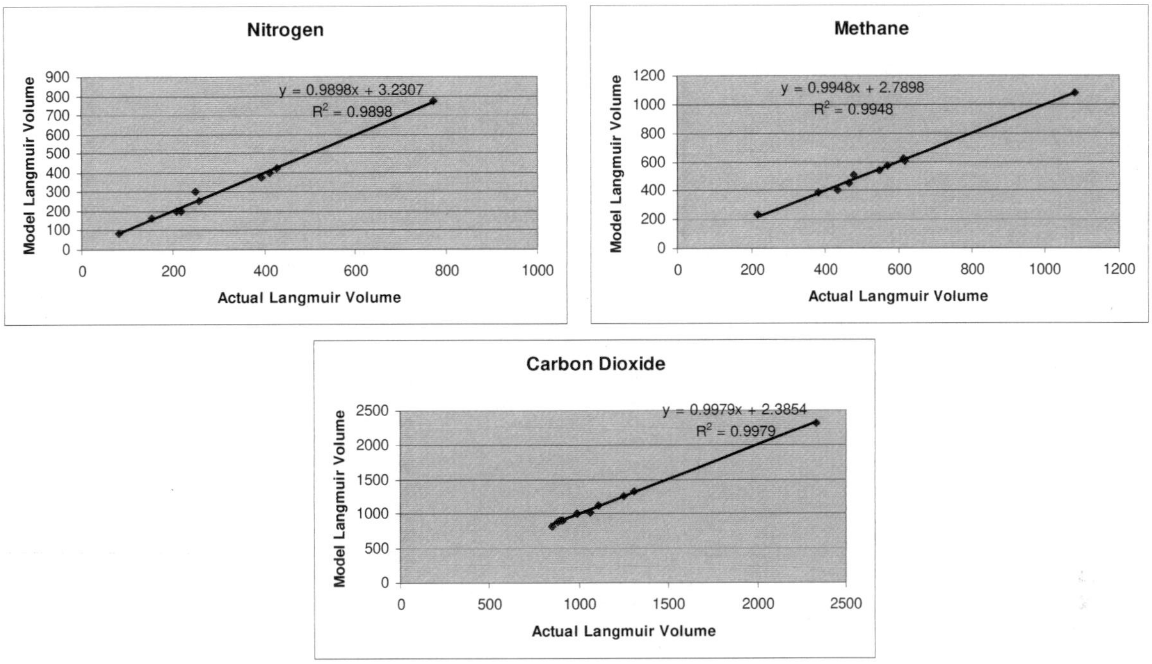

Figure 9: Actual vs. Predicted Langmuir Volumes, Multiple Linear Regression Models

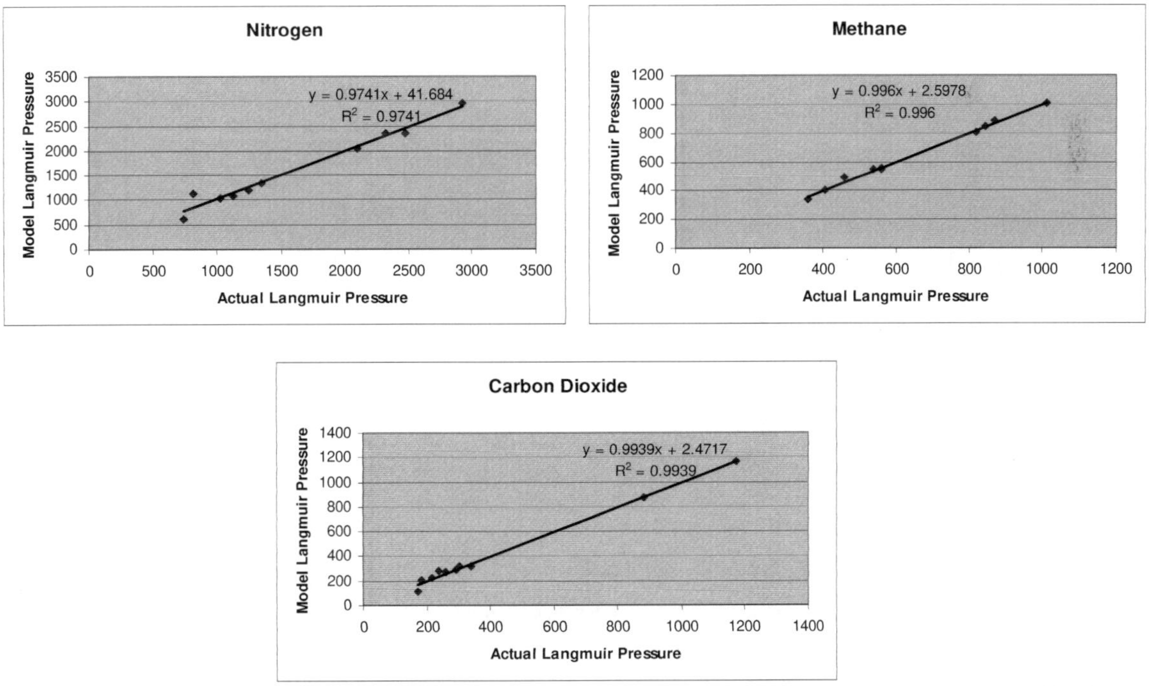

Figure 10: Actual vs. Predicted Langmuir Pressures, Multiple Linear Regression Models

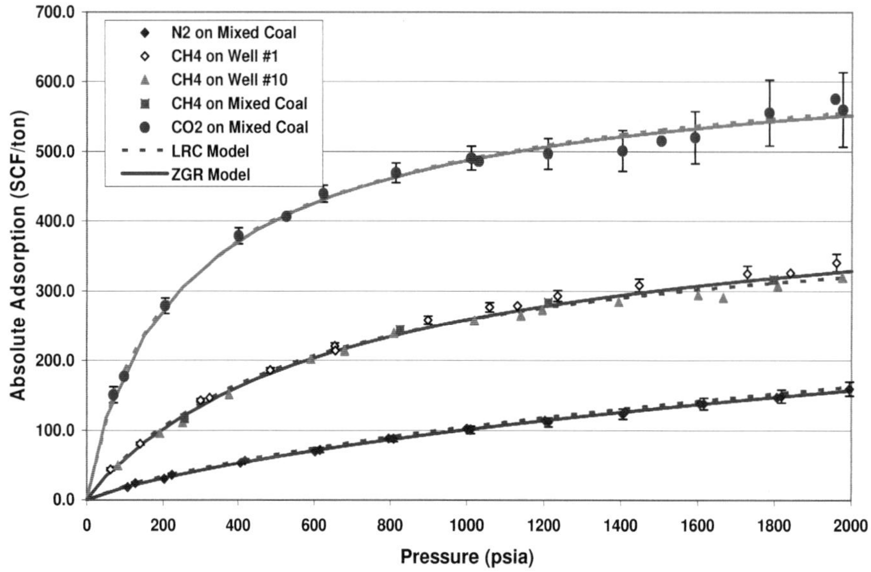

Figure 11: Single-Component Nitrogen, Methane and Carbon Dioxide Isotherms for San Juan Basin Coal at 130ºF

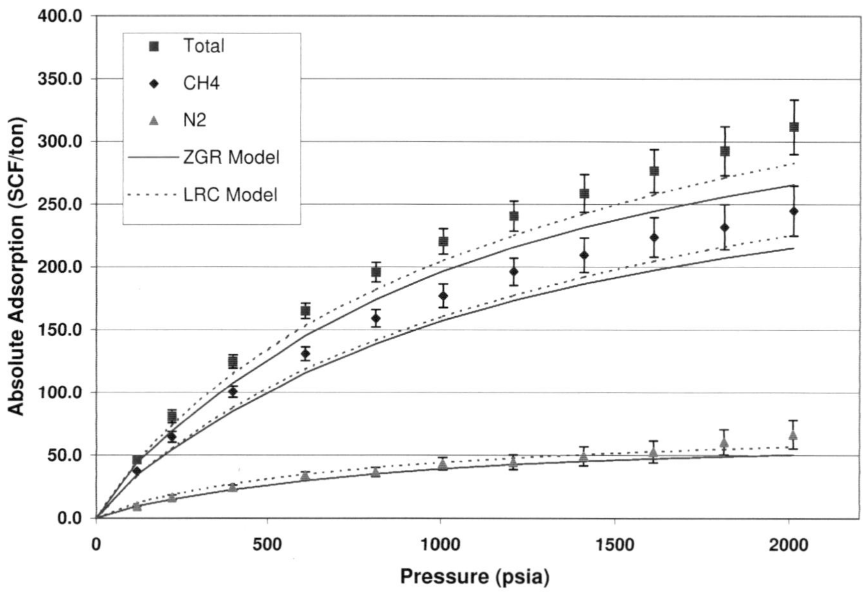

Figure 12: CH_4-N_2 Binary Adsorption Isotherms for San Juan Basin Coal (50% CH_4, 50% N_2) at 130ºF

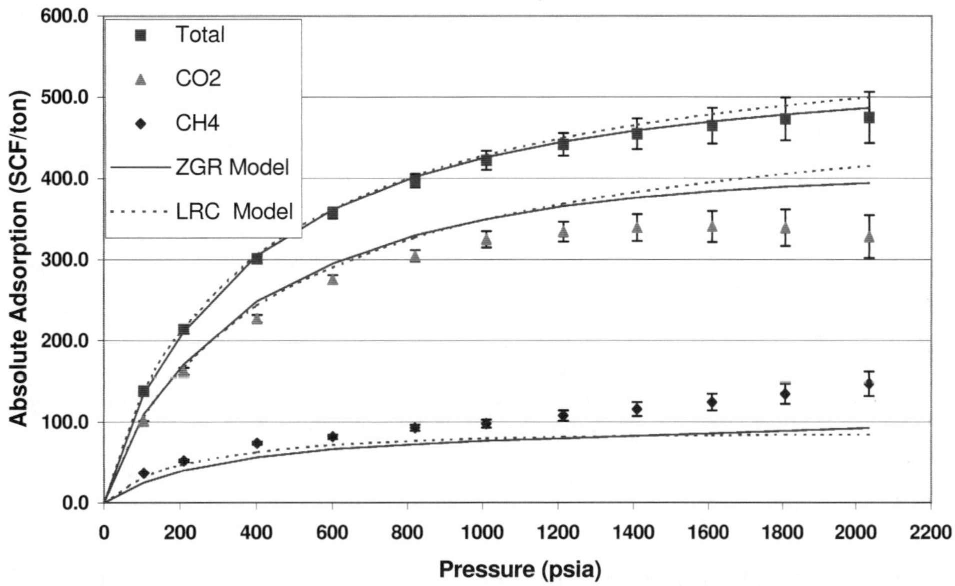

Figure 13: CH_4- CO_2 Binary Adsorption Isotherms for San Juan Basin Coal (40% CH_4, 60% CO_2) at 130ºF

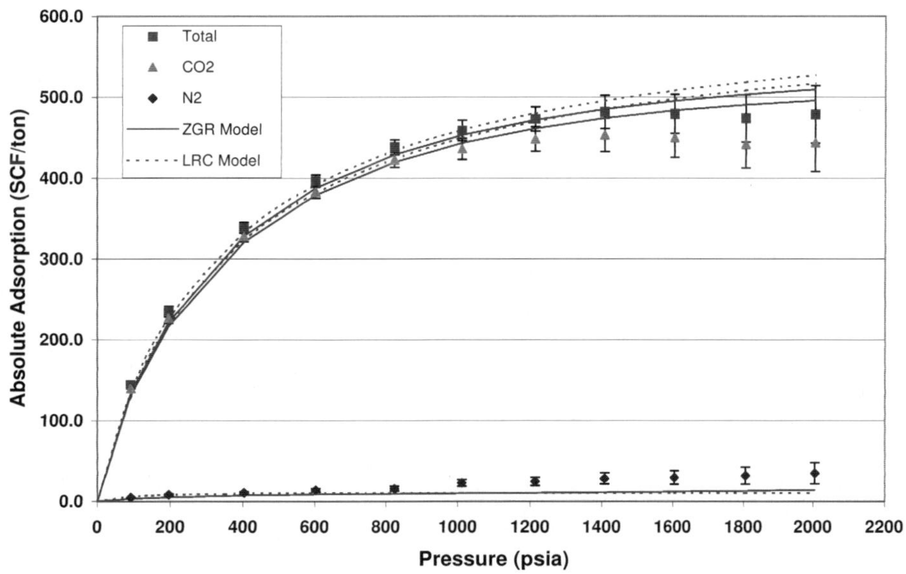

Figure 14: N_2-CO_2 Binary Adsorption Isotherms for San Juan Basin Coal (20% N_2, 80% CO_2) at 130ºF

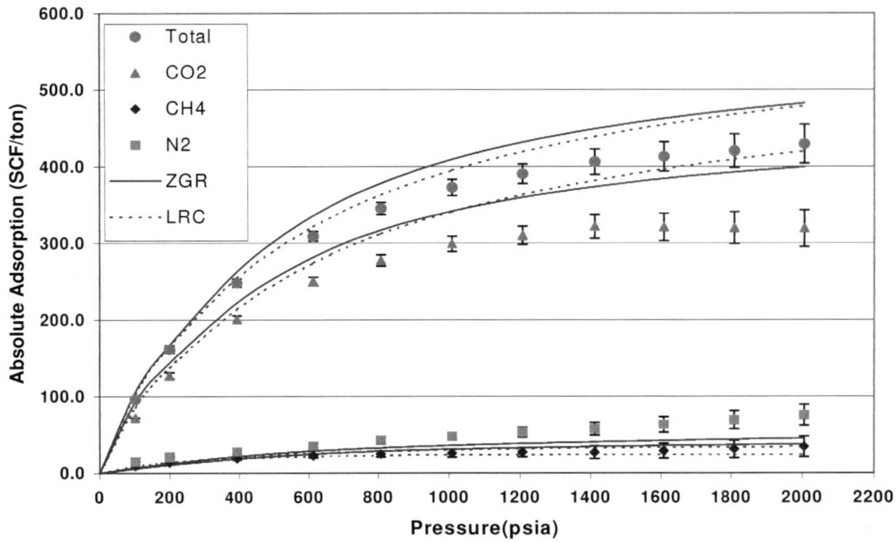

Figure 15: Ternary Adsorption Isotherms for San Juan Basin Coal (40%CH_4, 10% N_2, 50% CO_2) at 130ºF

2005 INTERNATIONAL COALBED METHANE SYMPOSIUM

0529

SIGNIFICANCE OF HYDRAULIC FRACTURING IN MEETING U.S. NATURAL GAS SUPPLY REQUIREMENTS

C.M. Boyer II; J.R. Williamson, Data & Consulting Services, a Division of Schlumberger Technology Corporation, Pittsburgh, Pennsylvania; and J.R. Duda, U.S. Department of Energy, National Energy Technology Laboratory, Morgantown, West Virginia

ABSTRACT

The projected increase in demand for natural gas in the U.S. will require the continued development of conventional and unconventional U.S. gas reservoirs and the increased use of hydraulic fracturing in these formations to achieve the required gas production levels. Well completion and production data were compiled into a database that contained almost 72,000 wells drilled and completed during the period 1997 through 2001. Review of these data indicated that almost 60 percent of these wells were hydraulically fractured with an average of 1.7 hydraulic fracture treatments per well. Assuming the current rate of hydraulic fracturing remains constant, total future hydraulic fracture treatments are projected to remain relatively constant at about 35,000 treatments per year through 2025. It is estimated that 75 to 79 percent of new natural gas production coming on line during the 2003-2025 period will be from wells that have been hydraulically fractured. By identifying such a reliance on the well stimulation technique, researchers – including those at the U.S. Department of Energy – managers, and policy makers can best address the efficiency, cost, and environmental aspects of fracturing.

INTRODUCTION

As discussed in a previous companion paper on this study, hydraulic fracturing is a widely employed method for enhancing the flow of oil and natural gas from underground formations [1]. The exploration and production (E&P) sector of the petroleum industry has relied on hydraulic fracturing for decades to commercially develop natural gas resources that are not economically viable at "natural" flow rates and to improve the return on investment for other natural gas plays. It has been projected that a significant portion of the U.S. future supply of natural gas will be produced from formations that typically require hydraulic fracturing to produce at economical rates, Fig. 1 [2]. In fact, production from unconventional formations is forecast at 8.5 to 9.0 trillion cubic feet (Tcf)/year, contributing significantly more to domestic supply when compared to natural gas from conventional formations. Because of this, the U.S. Department of Energy's National Energy Technology Laboratory (DOE-NETL) initiated a critical and comprehensive assessment of hydraulic fracturing and its role in providing future U.S. gas supply. Results of the natural gas portion of assessment are summarized in this paper.

BACKGROUND

Natural gas consumption in the U.S. has varied from 16 to 23 Tcf during the last 33 years, Fig. 2 [3]. However, by most accounts, natural gas demand in the U.S. is expected to significantly increase during the next 20 years, from 22.0 Tcf in 2003 to 30.7 Tcf in 2025, Fig. 2 [4]. This significant increase in gas consumption – 40 percent or 1.5 percent growth per year – will require a parallel growth in natural gas supplies available in the U.S., either through increased domestic production or imports.

Production of natural gas (dry) in the U.S. during the last 33 years has also varied from 16 to 22 Tcf, Fig. 3 [5]. While demand for natural gas has grown 35 percent during the last 17 years (1986-2003), natural gas production has lagged significantly behind, growing only 21 percent, with the difference being supplied primarily by imported gas. During the next 20 years, it is projected that domestic natural gas production will exhibit some growth to match the anticipated growth in demand, from 19 Tcf in 2003 to 22 Tcf by 2025, Fig. 3 [4].

Closely tied to the projected increases in natural gas production is the availability of new natural gas resources that can be readily converted into proven natural gas reserves. Natural gas resource estimates developed by the U.S. Geological Survey and the U.S. Minerals Management Service have identified the technically recoverable natural gas resources in the U.S. These estimates, Fig. 4, indicate that the resource base is adequate to sustain growing production volumes for many years [6].

Equally important and key to this study is the distribution of the resource base across the various resource categories. As shown in Fig. 4, the largest resource category is unconventional gas at 475 Tcf. New gas production will increasingly rely upon these unconventional sources – primarily tight sands and coalbed methane in the Rocky Mountain region. Successful development of these resources will require the continued and increased use of hydraulic fracturing.

Over the past decade however, hydraulic fracturing has come under fire, especially when applied to coalbed methane (CBM) wells. Many refer to the regulatory controversy and related litigation surrounding hydraulic fracturing of coalbed methane wells as the *LEAF* decision or *LEAF* v. EPA. Puder has provided a thorough and well written accounting of this subject, and interested parties are encouraged to read his article [7]. In short, the question was whether hydraulic fracturing fell within the definition of underground injection, which is regulated under the Safe Drinking Water Act.

In part, because of the continuing controversy, the U.S. Environmental Protection Agency (EPA) initiated a study to assess the potential impacts of hydraulic fracturing of CBM wells on underground sources of drinking water (USDW). That study was completed in 2004 and concluded "that the injection of hydraulic fracturing fluids into CBM wells poses little or no threat to USDWs and does not justify additional study at this time" [8]. The Department of Energy (DOE) maintained (and continues to have) a liaise with EPA and contributed substantially in terms of technical expertise, data, and reviews. Controversy centering on hydraulic fracturing remains and given the Department's role in ensuring natural gas and oil supplies in an environmentally responsible manner, it was prudent for the DOE to examine the well stimulation technique in the context of future natural gas and crude oil production – hence the study from which this paper is sourced [9].

ANALYSIS OF HISTORICAL FRACTURING DATA

As a first step in assessing the potential future impact of hydraulic fracturing on gas supply in the U.S., a historic review of past hydraulic fracturing activities was undertaken. Initially it was believed that sources of data for this assessment (completion and production data) would be readily available from service company files, production company files, and state regulatory agencies. However, it was soon discovered that either 1) data confidentiality limited access to a large data set or 2) the formats used for publicly available data were not consistent or readily useable. Because of this, data for this study were obtained from IHS Energy Group - Petroleum Information/Dwights LLC (IHS).

The completion database used in this study consists of data from all oil, gas, and other wells drilled in the lower-48 U.S. states. Offshore Gulf of Mexico and California were not included because of the inability to separate hydraulic fracturing and gravel-pack/frac-pack operations in the data set. Alaska was also not included because forecasted natural gas supply from this region was not envisioned until after 2020 (this has subsequently been revised to 2015). The original study began in 2003 using historic data for the period 1997 through 2001 as the basis for future projections.

Analysis of these data for the five study years resulted in the selection of all identified gas wells – 71,798 – and oil wells – 45,837 – drilled and completed during the period 1997 through 2001, for a total well count of 117,635 wells. Using these wells, we identified the number of new wells with hydraulic fracture treatments, Fig. 5. As shown, approximately 58 percent of the gas wells and 38 percent of the oil wells completed during the period 1997 through 2001 were hydraulically fractured.

Further analysis of the completion data shows that, on average, approximately 1.7 treatments were performed in each gas well and approximately 1.6 treatments were performed in each oil well. In total, more than 100,000 hydraulic fracture treatments were performed in oil and gas wells in the onshore lower-48 U.S. states during the period 1997 through 2001, for an average of more than 20,000 treatments per year, with the locations of these wells shown in Fig. 6.

The preliminary analysis presented above provides an initial understanding of the order of magnitude of hydraulic fracturing in onshore oil and gas wells in the lower-48 U.S. states. However, more detail about the type of hydraulic fracture treatments performed and the type of reservoirs that are typically hydraulically fractured was required to fully understand the current state of the industry and to assist in the prediction of future hydraulic fracturing activities. Because of space limitations, the results presented in the remainder of this paper will focus only on natural gas wells. A similar assessment was performed on oil wells and these results have been published in a DOE-NETL report [9].

Initially the data were screened on the basis of geographic distribution – states and U.S. Energy Information Agency (EIA) producing regions, Fig. 7. Once this was completed, it became apparent that a more refined distribution of hydraulic fracture treatments was required by this study. To accomplish this, the wells and hydraulic fracture treatments were differentiated by geologic formation and reservoir type:

1. ***Geologic Formation*** **-** The 71,798 gas wells in the database were completed in more than 1,200 identified formations. The number of wells per individual formation ranged from a high of 4,804 (Fort Union formation, Powder River basin) to a low of 1 well in over 350 separate formations. Because we planned to compare the details of hydraulic fracture treatments on a formation basis, it was decided that only those formations with more than 50 wells completed would be included in the analysis. This resulted in the analysis of 55,290 wells in 157 formations, or 77 percent of all gas wells completed in the onshore lower-48 U.S. during the period 1997 through 2001.

2. ***Reservoir Type*** **-** The 157 gas-producing formations identified above were further subdivided by reservoir type. The reservoir type was based upon the EIA designations for unconventional gas – coalbed methane, shale gas, and tight sandstone – and the EIA designations for conventional gas (non-associated).

The analysis of the historic data (1997 through 2001) was performed to provide a basis for projecting future hydraulic fracturing requirements. The percentage of wells hydraulically fractured and the average number of treatments per well were determined for each of the 157 formations discussed above. These data were also determined for each of the reservoir type by EIA geographic area and are summarized in Table 1.

The database also contains a variety of hydraulic fracture treatment characteristics. These variables were further analyzed to determine more detailed characteristics of the hydraulic fracture treatments:

1. ***Fluid type*** - **Gas**, referring to either a gas as the fracturing fluid or a gas/liquid foam or **liquid**, referring to a water- or oil-base liquid (linear gel, cross linked gel, etc.);
2. ***Total proppant volume*** - The total proppant placed into the formation during the treatment, reported in pounds.

Using these two variables, the average volume (pounds) of proppant injected with each treatment and the fracture fluid type percentages were determined for each of the 157 formations studied. In addition, an estimate of the average volume of fluid (barrels) that were pumped with each treatment was determined using the following formulas:

- **Liquid** – *Volume of Proppant x 0.01 = Barrels of Fluid*
- **Gas** – *(Volume of Proppant x 0.01) x 0.25 = Barrels of Fluid.*

These formulae were developed based on analysis of several technical publications identified in a literature review, with the results of this analysis presented in Table 2.

In addition to the assessment of the hydraulic fracture characteristics in gas wells, an assessment of the depth of the hydraulic fracture treatments was also completed. Within the database, each hydraulic fracture treatment record included a variable identified as "Objective Depth." This depth was assumed to represent the depth of the formation that was targeted by hydraulic fracture treatment. To assess the depths of the treatments in the U.S., 1,000-foot increments were developed and populated by the corresponding wells in the database. Fig. 8 shows the depth range for all gas wells with hydraulic fracture treatments during the period 1997 through 2001.

FUTURE REQUIREMENTS FOR HYDRAULIC FRACTURING IN THE U.S. NATURAL GAS INDUSTRY

The previous summary of historic hydraulic fracturing operations in the contiguous (onshore) U.S. provides a "snapshot" of the current state of the hydraulic fracturing industry in the U.S. At the same time, these findings can also provide a basis for forecasting future requirements of hydraulic fracturing. Using these data, we estimated the need for hydraulic fracturing during the next 20 years, specifically as applied to the growing U.S. natural gas industry.

Future Well and Gas Production Projections

The drilling of new natural gas wells during the period 2003 to 2025 has been projected by EIA [10]. These projections of drilling activity provided the basis for estimating future requirements for hydraulic fracturing. EIA projects the number of new natural gas wells using the previously described reservoir types – coalbed methane, shale gas, tight sandstone, and conventional gas. In addition, for this study the forecast number of oil wells with associated dissolved gas were also included. Fig. 9 summarizes the EIA forecasts for new gas wells (and oil wells with associated gas) that will be drilled in the U.S. during the period 2003 to 2025.

In addition to the drilling projections for various well types, EIA also projected the future production from these well types [10]. Fig. 10 summarizes the EIA gas production forecasts for new gas wells (and oil wells with associated gas) that will be drilled in the U.S. during the period 2003 to 2025. These production forecasts provide the basis for forecasting the role that hydraulic fracturing will play in providing the nation with natural gas in the future. It is important to recognize that current EIA forecasts, from the Annual Energy Outlook 2005, for new gas wells and production are influenced by the projected increase in imports, primarily liquefied natural gas (LNG). The importation of LNG is projected to increase from less than 1 Tcf in 2003 to over 6 Tcf by 2025 [10]. If future supplies from this source are less than projected, an increase in the projections of natural gas wells (and the associated hydraulic fracturing operations) may be required.

Future Hydraulic Fracture Treatment Projections

The final step in projecting future hydraulic fracture requirements for gas wells in the U.S. required combining the results of the historic assessment with the EIA well and production projections. The methodology used for the various reservoir and well types and the results obtained are briefly described as follows.

Future Unconventional Gas Hydraulic Treatment Projections - EIA future well and production projections for unconventional formations were provided on a U. S. Geological Survey (USGS) play specific basis. Historic hydraulic fracture and well data were cross-correlated with the USGS plays on a formation-by-formation basis.

As an example, EIA projected three future coalbed methane-producing plays in the State of Alabama. Each of these plays targets a geologically similar coal sequence (Pennsylvanian-age coal with minor to moderate tectonic deformation) with similar reservoir characteristics (thin, multiple, high-rank coal seams spread across a thick stratigraphic interval) in a geographically small area (north-central Alabama). Because of the similarity in characteristics, especially as it relates to the hydraulic fracturing process, we correlated these three EIA plays with the historic hydraulic fracturing data for the coal-bearing formations of the Black Warrior basin. Thus, each of the projected USGS unconventional gas plays was directly correlated with the historic data.

Future Conventional Gas and Associated Dissolved Gas Hydraulic Treatment Projections - Future projections of conventional gas wells and oil wells with associated gas by EIA were not on a play basis but rather by the EIA gas production regions, Fig. 7. To determine the future conventional gas wells and oil wells with associated gas that will be hydraulically fractured we relied upon historic results for these well types condensed on an EIA region basis.

Future Re-Fracturing Projections for Gas Wells - The focus of this study is the role and impact of hydraulic fracturing on future natural gas supply in the U.S. For the projections discussed above, we focused on the new gas wells that will be drilled in the U.S. during the period 2003 through 2025. However, hydraulic fracturing of existing wells (re-fracturing) to improve the recovery of gas is also an important aspect that needs to be addressed.

In the Spears 2000 study [11], it was estimated that of the 305,978 producing gas wells in 2000, 5,889 were re-fractured, or 2 percent. Since the number of wells that were re-fractured was relatively significant, we estimated the number of hydraulic fracture treatments that would be associated with the re-fracturing of existing gas wells in the future using this historic data as a basis.

Summary of Hydraulic Fracture Projections for Gas Wells in the U.S. - With the methodology discussed above, the historic well completions data, and the future projections of well drilling, projections of the future hydraulic fracturing requirements were made. Fig. 11 summarizes the forecasts for the number of hydraulic fracture treatments that will be performed in the U.S. during the period 2003 through 2025. These forecasts provide the basis for determining the role that hydraulic fracturing will play in providing the nation with natural gas in the future, as discussed in the next section.

Impact of Hydraulic Fracturing on Future Gas Production

As shown in Figure 11, the number of hydraulic fracture treatments in gas wells is projected to remain relatively constant during the next 22 years. These data pose the question – *What percent of the future gas production is attributed to wells that will be hydraulically fractured?* To answer this question, the projected future gas production from new wells and the percentage of new wells that are hydraulically fractured were combined. The methodology used to combine these data is succinctly described below:

1. ***Coalbed Methane*** – The percentage of gas production from forecasted new hydraulically fractured wells was estimated to be the same as the historical percentage of wells that were hydraulically fractured, determined on a play basis (for example, historically 97 percent of all coalbed methane wells in the Piceance basin were hydraulically fractured; therefore 97 percent of the projected future production from new coalbed methane wells in the Piceance basin will be from hydraulically fractured wells).

2. ***Shale Gas*** – Same method as Coalbed Methane.

3. ***Tight Sandstone*** – All wells in all future EIA tight sandstone gas plays were assumed to be hydraulically fractured, because the designation of tight sandstone (<0.1 md permeability) implies the need for hydraulic stimulation. Therefore, the projected production for all of the tight sandstone plays will be coming from hydraulically fractured wells.

4. ***Conventional Reservoirs and Associated-Dissolved Gas –*** The percentage of gas production from forecasted new hydraulically fractured wells was estimated to be the same as the historical percentage of wells that were hydraulically fractured, determined by EIA region.

Using the methodology described above, we determined the percentage of production attributed to hydraulically fractured wells for each of the gas production types – unconventional, conventional, and associated gas. Figs. 10 and 12 illustrate the projected future production from new wells drilled after 2003 and the projected future production from new, hydraulically fractured wells. As shown in these figures, 75 to 79 percent of the new gas production in the U.S. during the period 2003 through 2025 will be coming from wells that are hydraulically fractured.

DISCUSSION

Hydraulic fracturing has grown to be an important part of the U.S. oil and gas industry. Since the mid-1960s, the primary method of formation stimulation has been hydraulic fracturing [12,13]. Currently, hydraulic fracturing is the most widely used method for stimulating oil and gas wells. Of the approximately 25,000 new oil and gas wells drilled in the year 2000 in the U.S. (lower-48 onshore), a hydraulic fracturing operation was performed in almost 50 percent (12,612 wells) of these wells.

As discussed above, hydraulic fracturing is especially suited for the lower permeability formations, such as tight sandstone, coal, and fractured organic shale. Recent projections indicate that natural gas production from unconventional sources (tight sands, shale, and coalbed methane, which are generally low permeability) is expected to increase at a faster rate, growing from 6.49 Tcf in 2003 to over 8.5 Tcf in 2025, than production from more conventional sources [14]. As the U.S. natural gas industry increases the development of these low permeability formations, the use of hydraulic fracturing must necessarily increase.

Notwithstanding this expected growth in unconventional gas production, new conventional gas production will continue to be a major source of natural gas for the next 20 years, as shown in Fig. 12. There will be a continued dependence on the use of hydraulic fracturing in these conventional gas formations – currently 67 percent of all conventional gas wells are hydraulically fractured, Table 1.

The current reliance of the U.S. natural gas industry on hydraulic fracturing and the projected increase in this reliance necessitate the continuation of and improvement in the safety and environmental soundness of the operation. Currently, fracture stimulation treatments are pumped in a safe and environmentally responsible fashion in accordance with existing Federal, state, and local regulations. In fact, the largest pressure pumping companies voluntarily entered into a Memorandum of Agreement (MOA) with the EPA to eliminate the use of diesel fuel in fluids used to hydraulically fracture coalbed methane wells [15]. Notwithstanding the progress with respect to fracturing coalbed methane wells, recent actions concerning hydraulic fracturing have raised stakeholder awareness and increased dialogue. Late in 2004, five members of Congress wrote to the EPA seeking an investigation of the regulation of hydraulic fracturing [16].

CONCLUSIONS

The projected increase in demand for natural gas in the U.S. will require the continued development of the natural gas resources of the U.S., especially unconventional gas. This will necessitate the continued use of hydraulic fracturing in these formations to achieve the required production levels. Historically (1997 – 2001), 58 percent of natural gas well completions required hydraulic fracturing for commercial production, amounting to more than 20,000 hydraulic fracture treatments per year. Assuming the historic rate of hydraulic fracturing remains constant, it is projected that about 35,000 hydraulic fracture treatments will be pumped per year through 2025. Importantly, it is projected that 75 to 79 percent of all new gas production from the onshore lower-48 U.S. states will be produced from wells that have been hydraulically fractured. Additional details of this study can be found in the referenced report [9].

America's craving for natural gas continues to grow in spite of prices having risen to very high levels. Considerable efforts are underway to expedite the receipt of additional natural gas supplies, such as liquefied natural gas (LNG), but the ability of industry to responsibly develop domestic resources must be maintained and even broadened. Dialogs with industry, regulators, and myriad other stakeholders remain ongoing by the DOE in order to support sound policy making, guide technology development efforts, and to reduce costs and to conserve our resources. The assessment reported herein is just one example of how the DOE works to ensure adequate oil and natural gas supplies for the nation.

ACKNOWLEDGEMENTS

This paper was prepared with the support of the U.S. Department of Energy (DOE), under P.O. No. DE-AP26-05NT50226. However, any opinions, findings, conclusions, or recommendations expressed herein are those of the authors and do not necessarily reflect the views of the DOE. The authors wish to acknowledge the support of the Strategic Center for Natural Gas and Oil, National Energy Technology Laboratory, and Schlumberger Technology Corporation.

METRIC CONVERSION FACTORS

ft^3 x 2.831 685 E-02 = m^3

ft x 3.048 000 E-01 = m

REFERENCES

1. Duda, J.R., Boyer C.M. II, Delozier, D., Merriam, G.R., Frantz, J. H. Jr., and, M.D.: "Hydraulic Fracturing: The Forgotten Key to Natural Gas Supply," paper SPE 75712 presented at the 2002 SPE Gas Technology Symposium, Calgary, Alberta, Canada, 30 April–2 May.

2. Energy Information Administration, *Annual Energy Outlook 2005*, Energy Information Administration, Office of Integrated Analysis and Forecasting, U.S. Department of Energy, Washington, D.C. (February 2005), DOE/EIA-0383 (2005), 95.

3. Energy Information Administration, *Historical Natural Gas Annual 1930 Through 1999*, Energy Information Administration, Office of Oil and Gas, U.S. Department of Energy, Washington, D.C., (October 2000), DOE/EIA-E-0110 (2001), 10.

4. Energy Information Administration, *Annual Energy Outlook 2005*, Energy Information Administration, Office of Integrated Analysis and Forecasting, U.S. Department of Energy, Washington, D.C. (February 2005), DOE/EIA-0383 (2005), 207.

5. Energy Information Administration, *Historical Natural Gas Annual 1930 Through 1999*, Energy Information Administration, Office of Oil and Gas, U.S. Department of Energy, Washington, D.C., (October 2000), DOE/EIA-E-0110 (2001), 4.

6. Energy Information Agency, "Assumptions for the Annual Energy Outlook 2004 With Projections to 2025," Energy Information Administration, Office of Oil and Gas, U.S. Department of Energy, Washington, D.C., (October 2000), DOE/EIA-E-0554 (2004), 90.

7. Puder, Markus G., "Did The Eleventh Circuit Crack "Frac"? – Hydraulic Fracturing After the Court's Landmark LEAF Decision 1999," Virginia Environmental Law Journal, Volume 18, Number 4, pp 508-554.

8. U.S. Environmental Protection Agency, "Evaluation of Impacts to Underground Sources of Drinking Water by Hydraulic Fracturing of Coalbed Methane Reservoirs," U.S. Environmental Protection Agency, Office of Water, Office of Ground Water and Drinking Water, Drinking Water Protection Division, Prevention Branch, Washington, D.C. (June 2004) EPA 816-R-04-003.

9. U.S. Department of Energy, "Significance of Hydraulic Fracturing in Meeting U.S. Natural Gas Supply Requirements," U.S. Department of Energy, National Energy Technology Laboratory, Morgantown, WV (April 2005).

10. Energy Information Administration, *Annual Energy Outlook 2005*, Energy Information Administration, Office of Integrated Analysis and Forecasting, U.S. Department of Energy, Washington, D.C. (February 2005), DOE/EIA-0383 (2005), 159-160.

11. Spears & Associates, Inc., "The North American Testing and Completion Market – Oilfield Markets 1997," Special Proprietary Report, Gas Research Institute, Chicago, IL, 1997.

12. Spears & Associates, Inc., "The Global Frac Market – Regarding Market Research On: Hydraulic Fracturing and Related Services," Special Proprietary Report, Gas Research Institute, Chicago, IL, November 2000.

13. Martinez, S.J., Steanson, R.E., and Coulter, A.W.: "Chapter 55 - Formation Fracturing," in *Petroleum Engineering Handbook*, Howard B. Bradley, editor, Society of Petroleum Engineers, Richardson, TX, February 1992, 55-1.

14. Energy Information Administration, *Annual Energy Outlook 2005*, Energy Information Administration, Office of Integrated Analysis and Forecasting, U.S. Department of Energy, Washington, D.C. (February 2005), DOE/EIA-0383 (2005), 95.

15. U.S., Environmental Protection Agency, "A Memorandum of Agreement - Elimination of Diesel Fuel in Hydraulic Fracturing Fluids Injected Into Underground Sources of Drinking Water During Hydraulic Fracturing of Coalbed Methane Wells," http://www.epa.gov/safewater/uic/pdfs/moa_uic_hyd-fract.pdf.

16. Editors, "Lawmakers Call for Investigation of Hydraulic Fracturing," Oil and Gas Journal OnLine, October 18, 2004.

Gas Well Type	Wells Hydraulically Fractured (%)	No. of Treatments per Well
Coalbed Methane	32	1.8
Shale Gas	79	1.6
Tight Sandstone	74	1.9
Conventional (non-associated)	67	1.8

Table 1 – Percentage of wells hydraulically fractured and average number of treatments per hydraulically fractured well during the period 1997 through 2001.

Gas Well Type	Average Pounds of Proppant per Treatment	Frac Fluid Type		Average Fluid Volume per Treatment, barrels	
		Gas (%)	Liquid (%)	Gas	Liquid
Coalbed Methane	48,084	15	85	120	481
Shale Gas	51,540	26	74	129	515
Tight Sandstone	101,891	5	95	255	1,019
Conventional (non-associated)	90,249	8	92	226	902

Table 2 – Average pounds of proppant, percent of fluid type, and average fluid volume for single hydraulic fracture treatments by gas well type during the period 1997 through 2001.

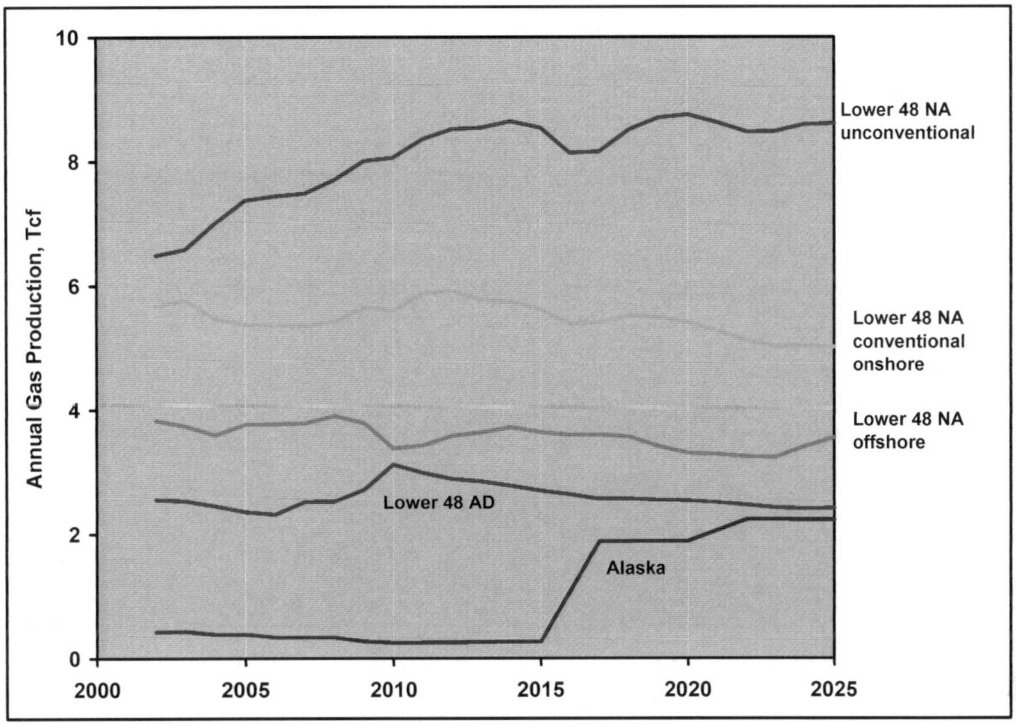

Figure 1 – Projected natural gas production by source in the United States, 2002 to 2025.

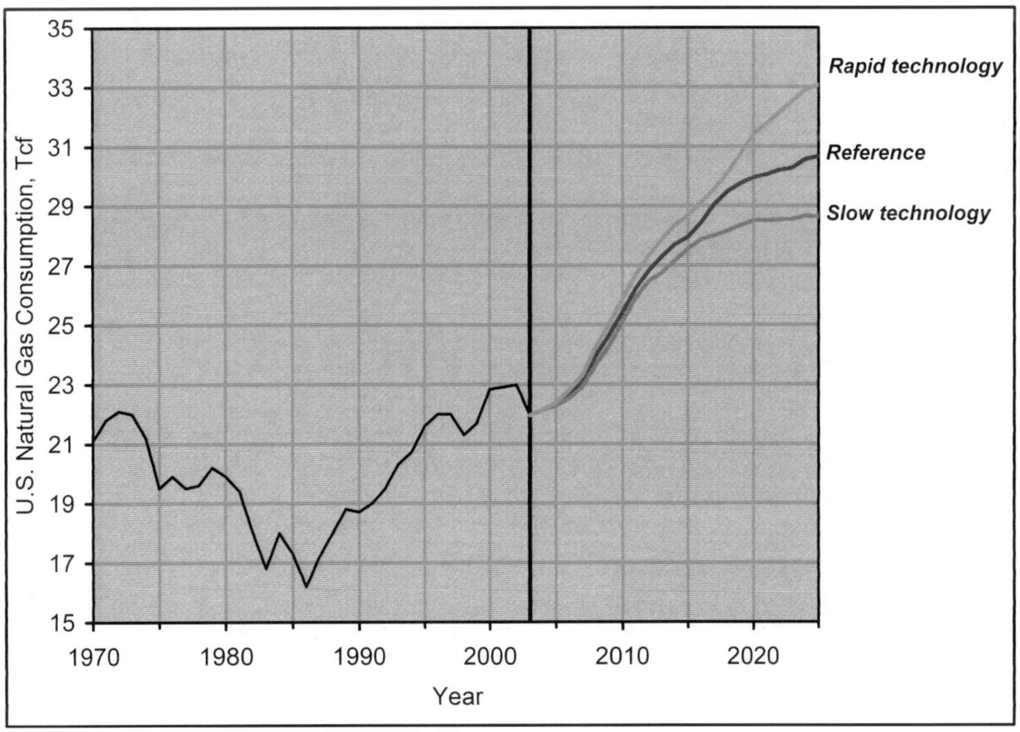

Figure 2 – Historic and projected natural gas consumption in the United States, 1970 to 2025.

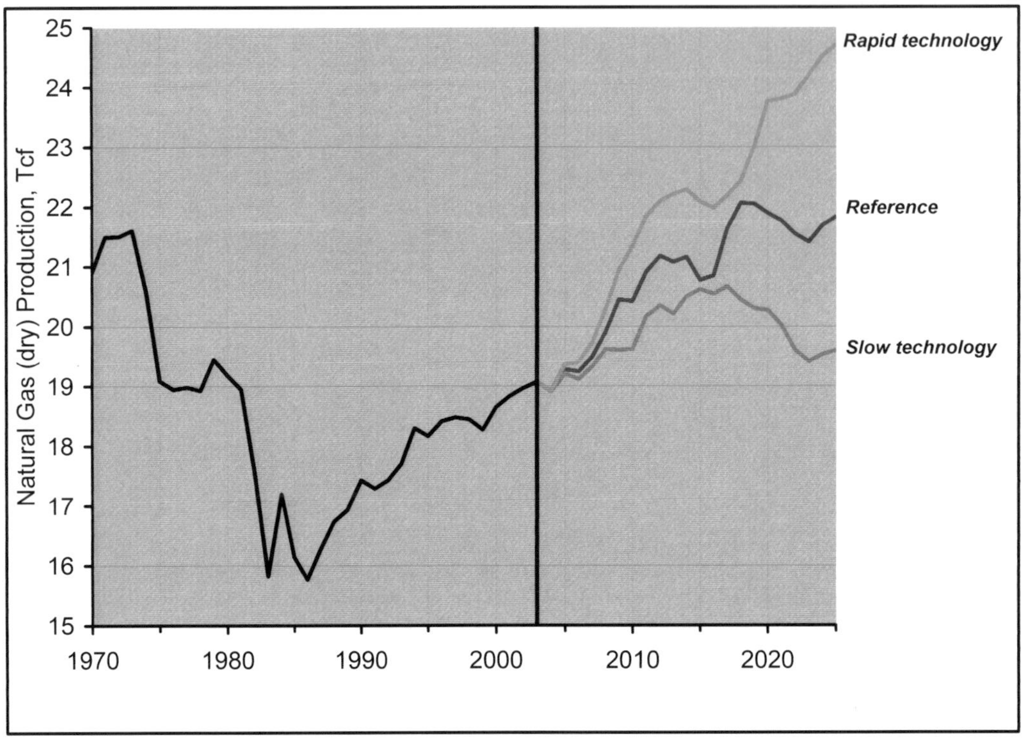

Figure 3 - Historic and projected natural gas (dry) production in the U.S. – 1970 to 2025.

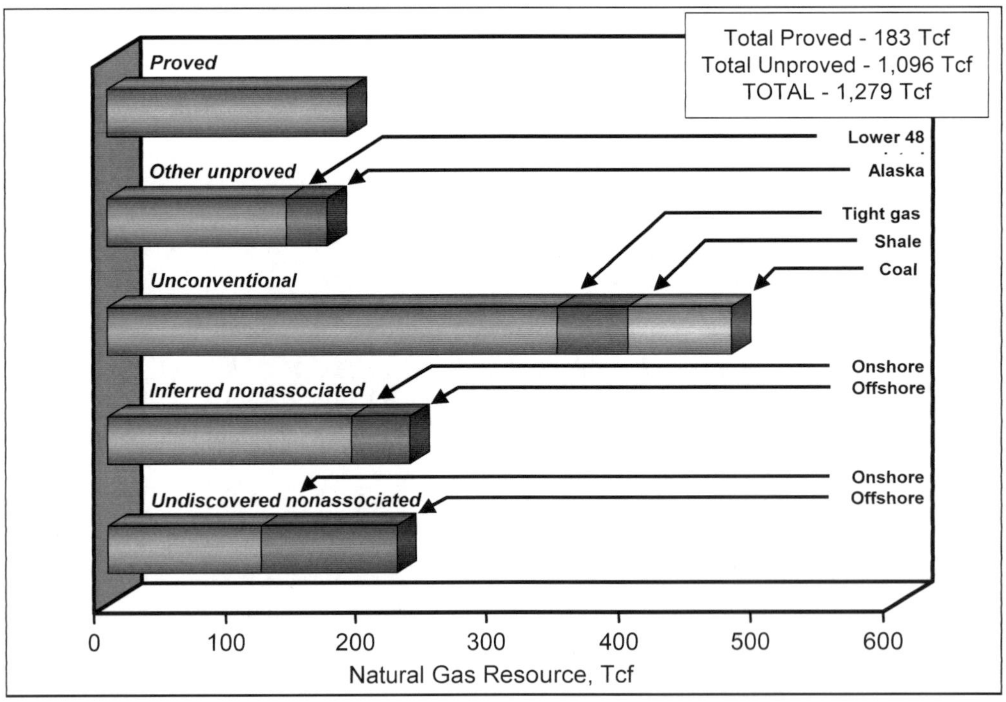

Figure 4 - Technically recoverable U.S. natural gas resources as of January 1, 2002.

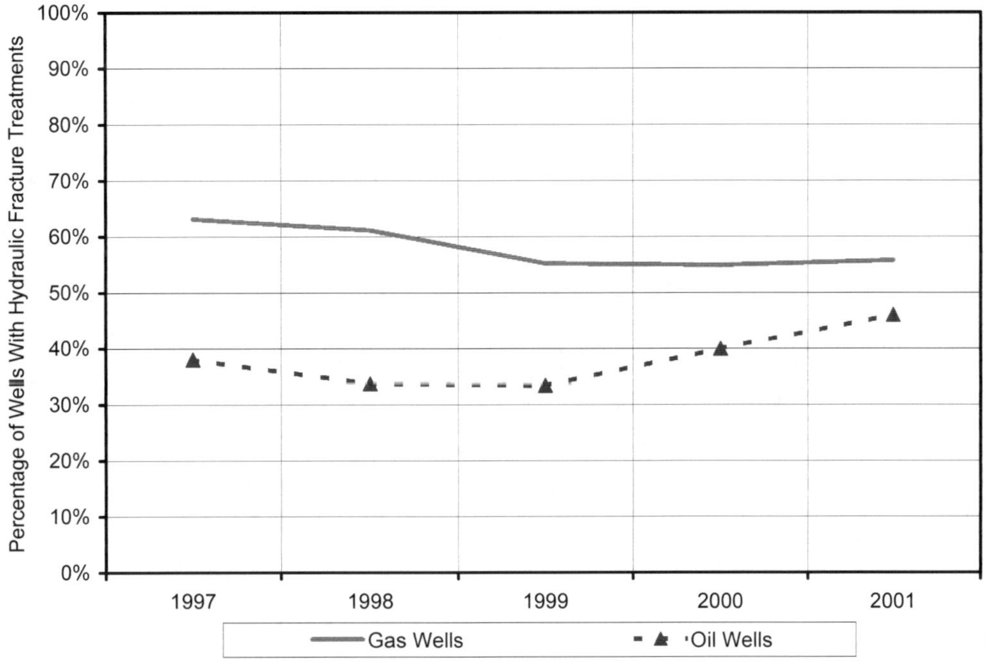

Figure 5 – Percentage of new oil and gas wells hydraulically fractured in the U.S.

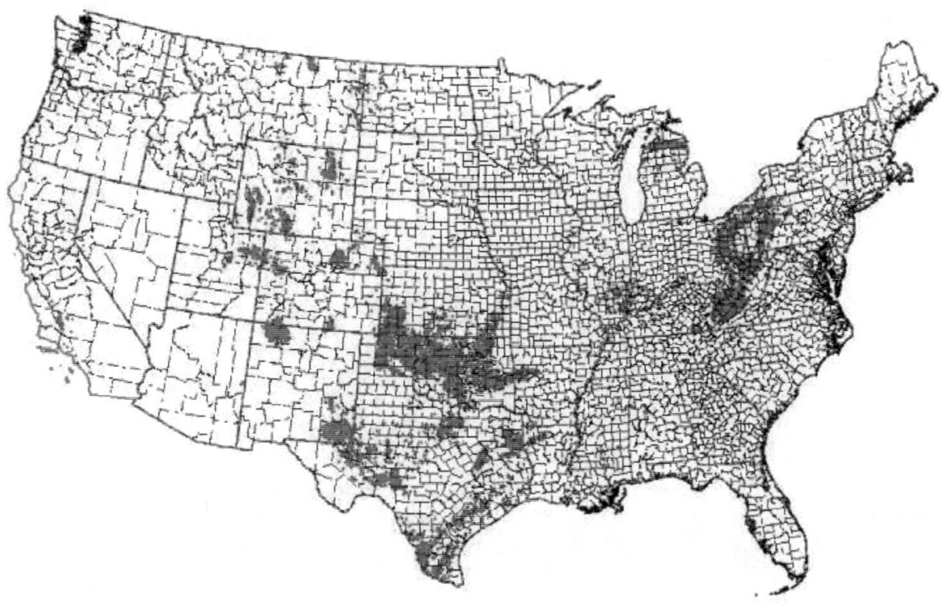

Figure 6 - Location of hydraulically fractured oil and gas wells (shaded red) in the lower-48 U.S. during the period 1997 through 2001.

Figure 7 - Oil and gas supply regions of the lower 48 U.S. used in the EIA Oil and Gas Supply Model (OGSM) and used in this assessment.

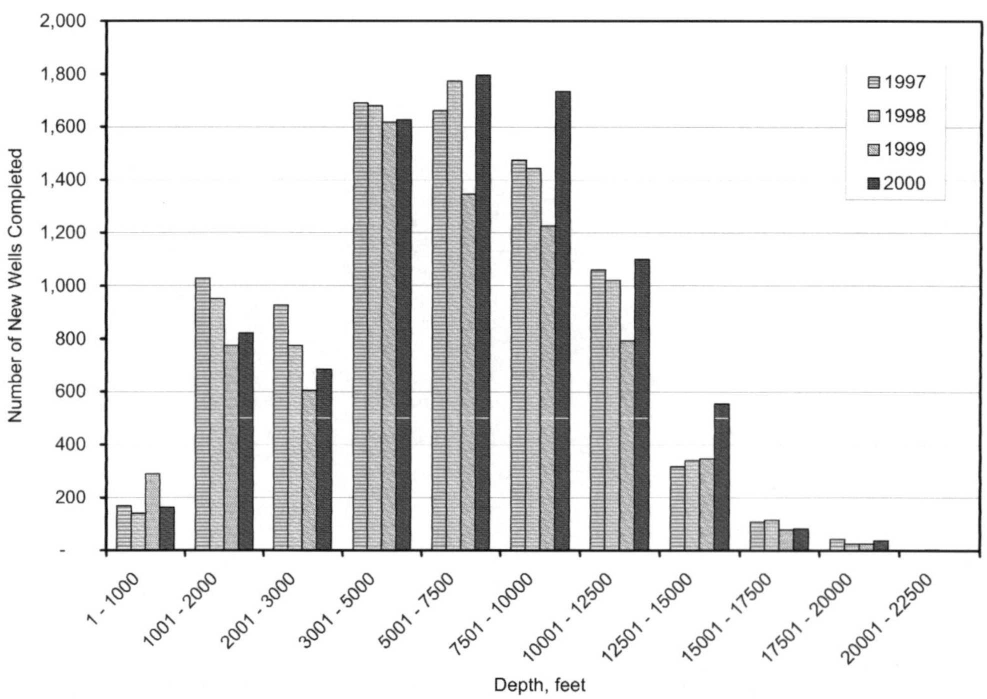

Figure 8 – Depth range of new gas wells with hydraulic fractures in the U.S.

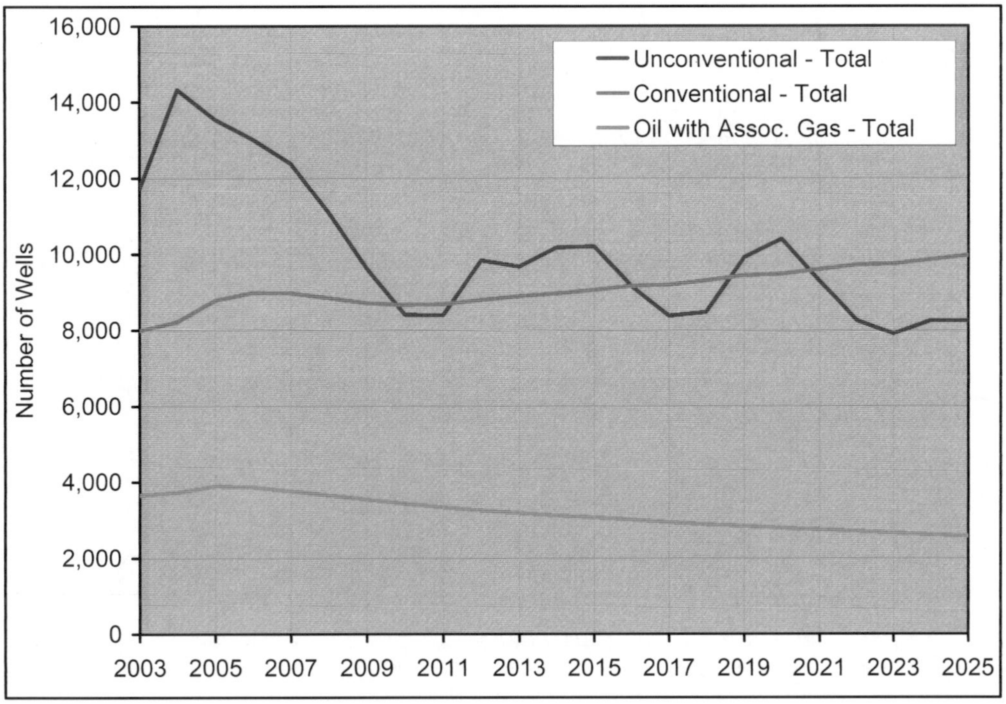

Figure 9 - Projected future conventional gas wells, unconventional gas wells, and oil with associated gas wells in the U.S.

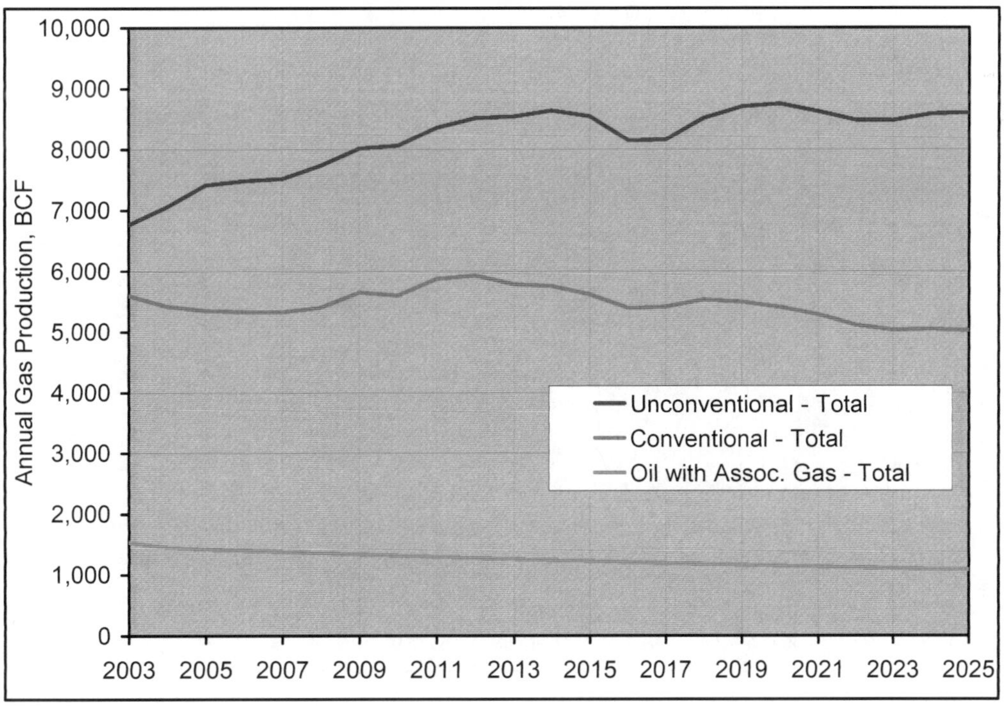

Figure 10 – Projected future gas production from unconventional gas wells, conventional gas wells, and oil wells with associated gas.

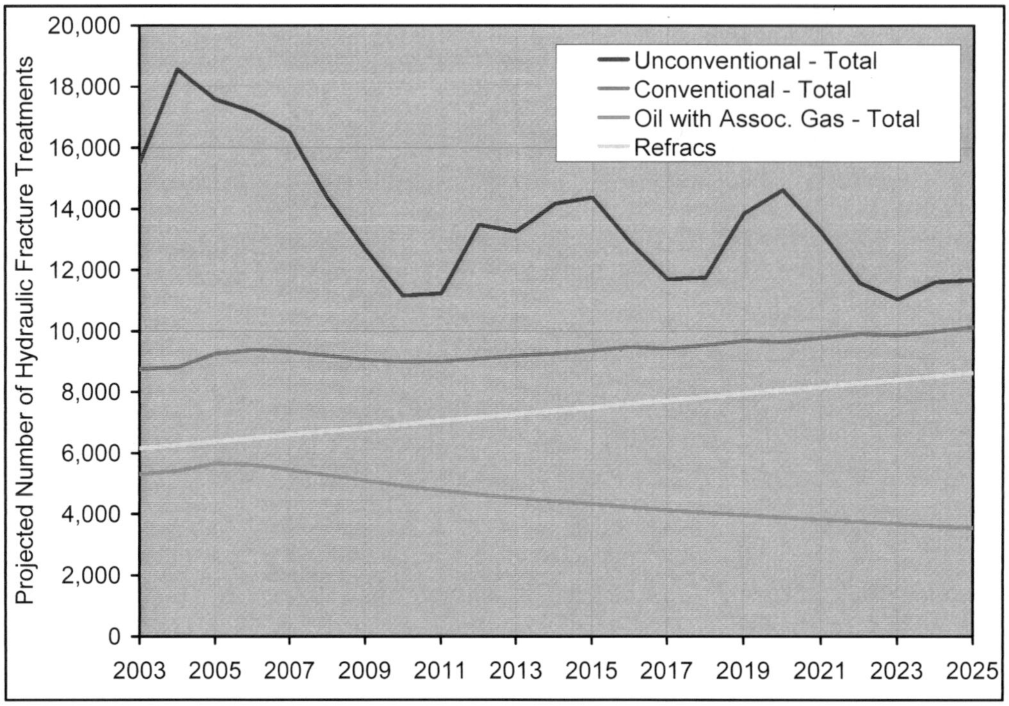

Figure 11 – Projected number of hydraulic stimulation treatments that will be performed in new unconventional and conventional gas wells, new oil wells with associated gas production, and existing producing gas wells.

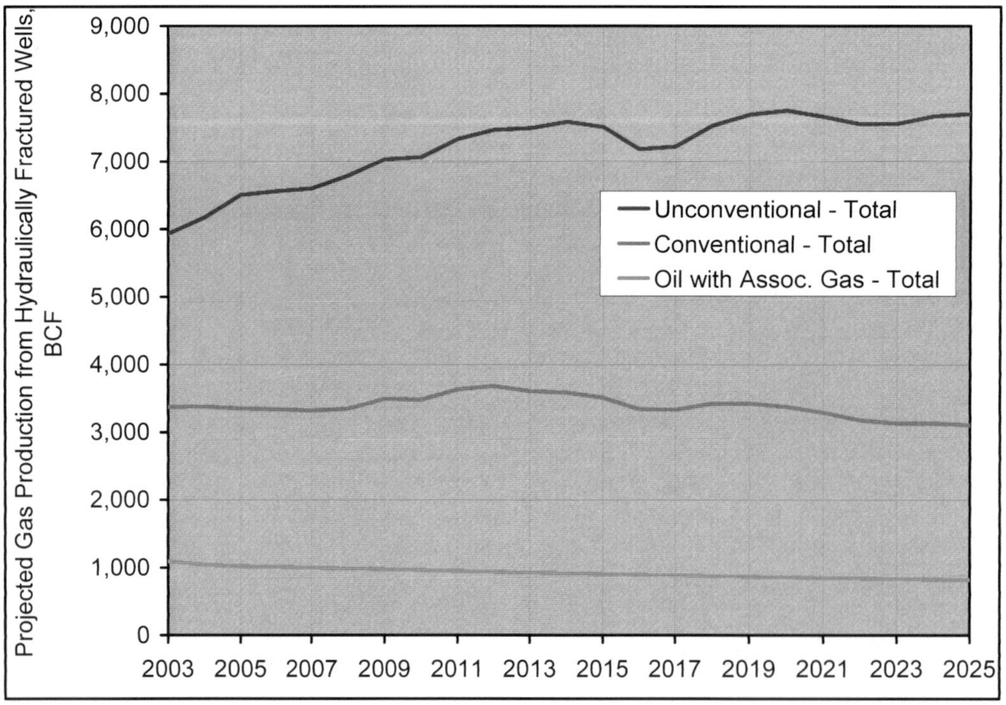

Figure 12 – Projected future gas production from hydraulically fractured unconventional, conventional, and oil with associated gas wells.

2005 INTERNATIONAL COALBED METHANE SYMPOSIUM

0507

Implementing Sand Propped Hydraulic Fracture Stimulation for In-seam Drainage Holes: A Demonstration of an Enhanced Drainage Technology

Rob Jeffrey, CSIRO Petroleum; Christian Boucher, Dartbrook Coal; and Ken Mills, SCT Operations Pty. Ltd.

ABSTRACT

In 2001 and 2002, Dartbrook Coal Mine in the Hunter Valley of NSW, encountered low permeability coal in Longwalls 107, 108 and 109. Standard horizontal in-seam drain holes were found to require extremely long times to drain the coal to a gas content level below 5 cubic meters per tonne. Hydraulic fracture stimulations, using water and sand, were therefore carried out in two boreholes in Longwalls 108 at Dartbrook to assess the effectiveness of sand propped fractures in stimulating gas drainage from in-seam boreholes. The results from this trial were used to develop and carry out hydraulic fracturing treatment in horizontal holes drilled throughout Longwall 109 to accelerate gas drainage in that panel.

A hydraulic fracturing system, that could be deployed and operated in the panel roadway, was assembled for the test. A water-inflated straddle packer system was used to isolate sections of the holes for each treatment. The stimulations produced a significant increase in gas drainage rates from the two boreholes with one increasing by a factor of about 180 while the other hole increased its rate by about 22 times. The stimulated gas rates continuously increased for several weeks and the higher rates were sustained for the entire period the holes were monitored.

Fracture modeling suggests the sand proppant bank may extend to 15m from the borehole. The unpropped portion of the fracture may extend to more than 40m. A purpose-built fracturing system was developed and used at Dartbrook to stimulate holes that covered most of LW109. This full-scale enhancement of gas drainage was successful, producing an average increase in gas drainage rate of 5, which allowed efficient mining of that panel.

Fracture stimulation of horizontal uncased boreholes in coal is limited in application by borehole conditions. Borehole conditions at Dartbrook were exceptionally good, but this is not generally the case at other mines. Packer life is severely affected by borehole failure which results in oversize sections of hole and sharp broken out borehole walls. At the same time, the low and variable fracture initiation pressure along an open borehole in coal limits the applicability of slurry jet induced fracturing, which relies in part on holding the borehole pressure at a value near fracture extension pressure. Methods to overcome the problem posed by borehole stability are being developed.

INTRODUCTION

Hydraulic fracturing is the most common stimulation method applied in the coalbed methane industry to stimulate CBM gas wells. A typical stimulation treatment starts with a clear pad fluid stage injected at rates and pressures sufficient to initiate and extend a hydraulic fracture. Once a large enough fracture has been formed, proppant is added to the fluid and the slurry is pumped into the fracture. The proppant is carried in the slurry into the fracture and, after the injection stops, is left behind to prop the fracture open, forming a permeable channel in the reservoir through which the hydrocarbons can be produced. The propped hydraulic fracture connects the wellbore to a large surface area of the reservoir and provides a conductive pathway to carry the water and gas back to the well. The propped fracture also bypasses any skin damage in the near wellbore region.

Previous experience at other sites

Investigations into using propped fractures to stimulation horizontal in-seam drain holes have been carried out in Australia and overseas. The earliest work that we are aware of was carried out in Queensland by the Department of Mines. Starting in late 1979, the Department conducted a four month long research project to investigate hydraulic fracture stimulation of horizontal drain holes (Croft, 1980). The project's intent was to design and test a system for placing sand propped fractures in horizontal holes. The fracturing equipment was to be located underground next to the coal rib. A number of equipment developmental problems were encountered and no fracture treatments were successfully performed. However, the concept of stimulating horizontal holes from underground is, essentially, the same as what was undertaken in the trial at Dartbrook.

Fracture stimulation work is done in horizontal wells in the petroleum industry on a fairly frequent basis (Walker, Ehrl, and Arasteh, 1993; Weijers, et al., 1992). Such wells are drilled into conventional reservoir materials (sandstones or limestones) and are stimulated by fracturing for the purpose of establishing a better connection with the reservoir.

A long section of a borehole can be treated without isolating each treatment zone but the hydraulic fracture is likely to initiate and grow from the point or points where pre-existing weaknesses in the borehole exist. Once a fracture starts growing, it typically will continue to extend at a lower pressure than required to initiate other fractures along the borehole. Therefore, trying to treat a long section of hole in one injection may, in the worst case, only produce one major hydraulic fracture in the entire open section of the hole.

Diverting agents are used to partially overcome this problem. Diverting agents are materials added to the fluid that act to block the entry to the fractures and are pumped at various times during a treatment in order to divert the fluid and proppant from an established fracture into a new area of the wellbore.

Initiating multiple fractures or extending and propping too few fractures, are problems that can be avoided by limiting the length of the interval being treated, usually by casing the well and then selectively perforating a small zone for each treatment.

In 1993, propped hydraulic fracture stimulation was attempted in horizontal drain holes at the Soldier Canyon Coal Mine, in Utah (Kravits, 1993). Five propped fractures were placed along a 2505 foot-long horizontal borehole. Gas production from the hole increased by 46 percent and then, over a period of about 4 weeks, declined back to pre-stimulation rates. Stress conditions at this site in Utah are suspected to have allowed the hydraulic fractures to grow vertically into overlying and underlying rock, resulting in poor stimulation of the coal seam.

ACARP (Australian Coal Association Research Program) sponsored a project to test propped fractures for stimulation of horizontal in-seam boreholes (Jeffrey, 1999). Field tests were carried out at Central Colliery near Middlemount, Queensland. An openhole straddle packer system was used in the horizontal drain hole. No fracture growth could be achieved at several positions along the borehole because of fluid bypassing the packers. This failure illustrates one serious problem that use of packers in naturally fracture rock can face. This paper describes work carried out at Dartbrook mine that successfully demonstrated the stimulation effect of sand propped fractures (Jeffrey and Boucher, 2004).

IN-SEAM DRAINAGE AT DARTBROOK

Dartbrook coal mine extracts 4 m of the Wynn upper seam by longwall mining methods. The mined section lies within the 28m thick mega seam and a coal floor and roof is left after mining. Figure 1 contains a drawing showing Longwalls 8 and 9. The fan-array holes shown, that are drilled essentially across the panels, are the standard in-seam drainage method used at Dartbrook. The holes drilled for the sand propped hydraulic fracturing work are shown with thicker lines and are aligned mostly along the panel length. Fractures placed in these holes extend east-west across the hole and parallel to the maximum stress direction and to the panel width.

Drainage holes in panels mined before LW7 typically produced gas ranging in rate from 1 to 4 litres per minute per metre of hole. However, in LW7, 8 and 9 this rate was reduced, because of low permeability coal, to 0.1 to 0.3 litres per minute per metre of hole. Coal permeability was estimated to be about 0.02 to 0.03 md while gas content in these three panels was 8 to 9 cubic metres per tonne, composed of 90 percent carbon dioxide and 10 percent methane. The unstimulated drainage rate was too low to effectively reduce the gas content to a below 5 cubic metres per tonne, which is known to be low enough to avoid gas delays during longwall mining.

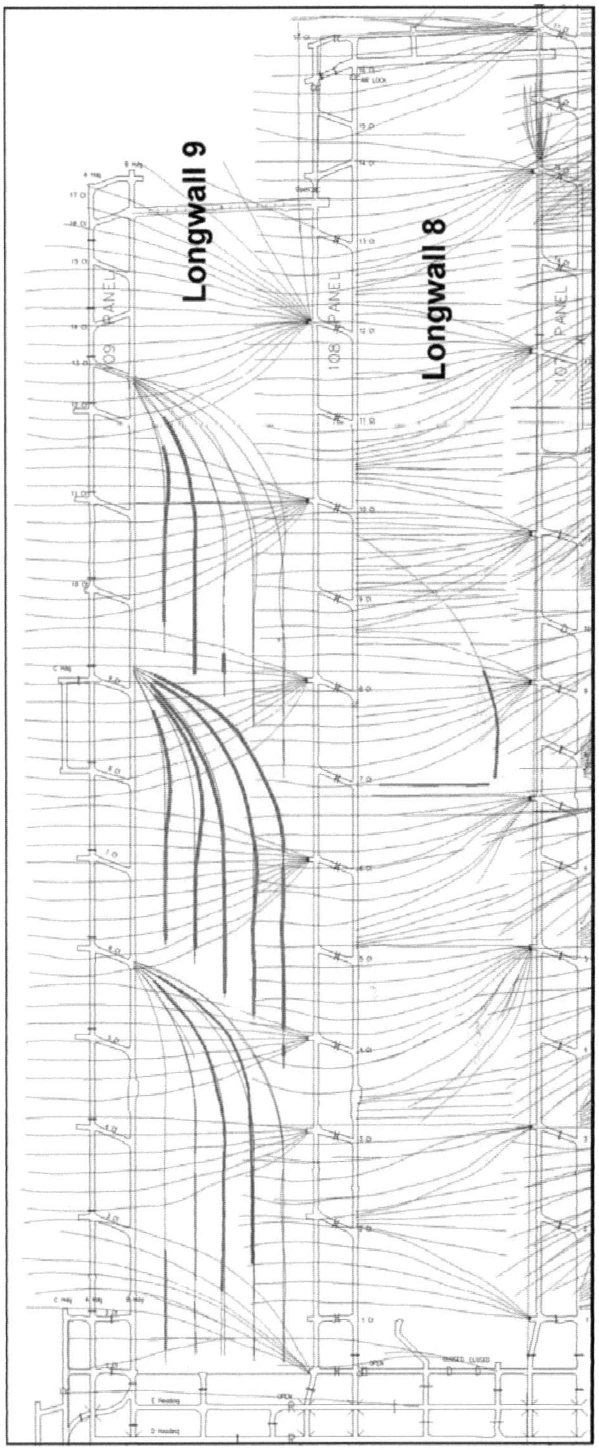

Figure 1: Longwall 9 with hydraulic fractured holes indicated.

Water Only Fractures

The initial fracturing program used small water-only treatments placed at 3m intervals along several boreholes in Longwall 107. Typical treatments consisted of 5 minutes of injection at 160 to 170 litres per minute (800 to 900 litres total volume). The gas rate per metre of borehole was increase by these treatments by a factor ranging from 2 to 3. However, this stimulated gas rate was too low for the coal to be drained in time available and the stimulated gas rate dropped off quickly after the drainage started (Gray, 2002).

Because of the poor stimulation achieved without sand, sand-propped hydraulic fractures next tested in Longwall 8 to determine if the additional and longer lasting fracture conductivity produced by propped fractures would provide the additional degree of stimulation needed.

STIMULATION TRAIL IN LW8

A system of fracturing equipment to undertake the sand-propped fracture trial from the maingate side of LW8, was assembled from existing equipment and mobilized for use underground at Dartbrook. The equipment used was powered by hydraulic motors and the hydraulic power needed to run the equipment was obtained from drill rigs and hydraulic power packs available at the mine.

The major joint system in the coal at Dartbrook is subvertical and strikes at about 90 degrees in the LW8 area (Byrnes, 2002). This joint system typically runs parallel to the maximum horizontal principal stress in the seam, which is also the direction that hydraulic fractures grow. The hydraulic fractures were, therefore, expected to be subvertical, growing along this joint direction.

The first borehole, drilled at cut-through 10, was designed to run perpendicular to the expected fracture direction over its last 100m of length while the second hole, drilled from cut-through 7, was drilled more parallel to the expected fracture direction (see Figure 1). Fracture stimulating these two holes would then provide a comparison of the stimulation effect as a function of borehole orientation. Purpose-drilled boreholes targeted for this type treatment would best be drilled so the fractures form across the borehole axis, but extensive pre-existing drainage holes exist in the seam. A comparison of these two boreholes with different orientations was useful in determining if the existing drainage holes could be successfully stimulated or if new boreholes with the preferred orientation should be drilled to optimize the stimulation process and effect.

The hole at 10 cut-through was fracture stimulated over its last 62m of extent (dark portion in Figure 1) with twenty fractures placed along this hole at a 3 m. spacing. In contrast, the fractures placed in the hole at 7 cut-through were more aligned with the hole. A total of 10 fractures were placed over 88 m of this hole. In both cases, about 100 kg of 30/60 mesh sand proppant was pumped into each fracture formed.

Hydraulic Fracture Growth and Proppant Distribution

A numerical hydraulic fracture model has been used to approximately match the pressure measured during the fracture treatment at 327.7 m in the borehole at cut-through 10. The injection rate used in the treatment was 250 litres per minute. Coal and site properties used for this match consisted of a Young's modulus of 3,500 MPa, Poisson's ratio of 0.34, minimum horizontal stress of 5.4 MPa, and coal permeability of 0.03 md. Figure 2 contains a plot of the pressure match obtained and Figure 3 shows the model-predicted fracture growth with time. Figure 4 shows the model-predicted proppant distribution. The propped fracture is predicted to have a conductivity of 10 md-m with this conductivity extending out to 15 m each side of the borehole.

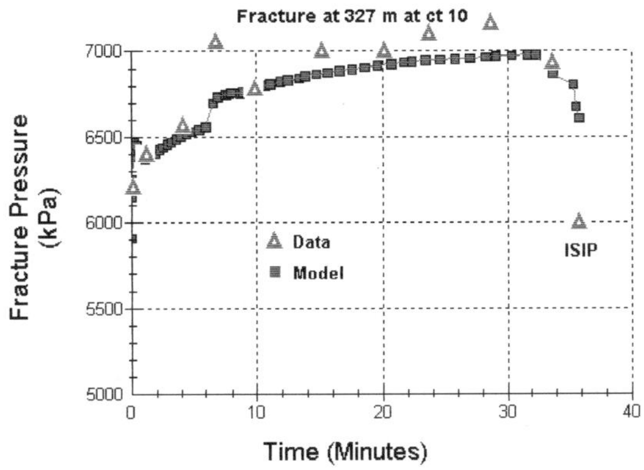

Figure 2: Fracturing pressure versus time for the treatment at 327m at cut-through 10. The red squares are measured pressure.

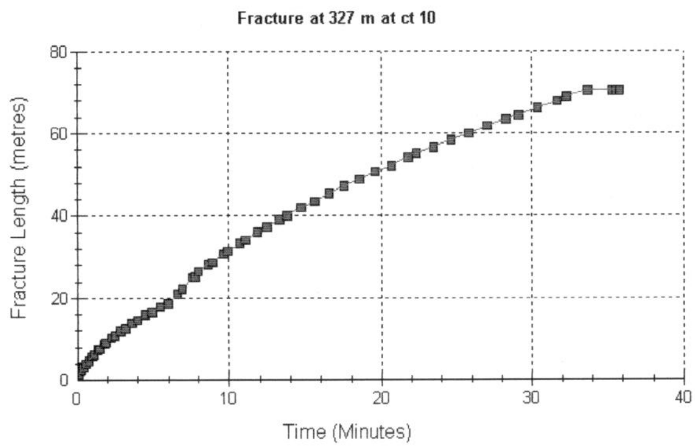

Figure 3: The growth of the fracture with time for the treatment at 327 m at cut-through 10. The fracture grows to 32 m size in 10 minutes which is approximately the growth rate indicated by the observations of water at the rib during the last treatment at 7 cut-through.

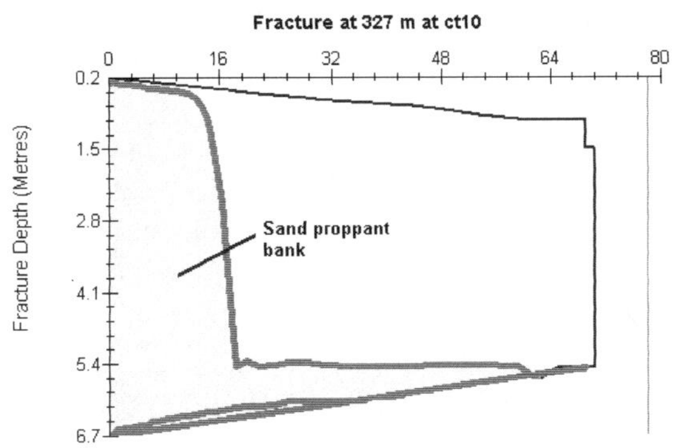

Figure 4: The model-predicted proppant bank for the treatment at 327 m at cut-through 10. 160 kg of sand was placed in this treatment.

Most of the proppant, as indicated in Figure 4, lies within 15 m of the borehole, but the fracture conductivity is enhanced out to 50 m. The model predicts the fractures grow vertically to about 7m in height.

Gas rate from each hole was monitored on a regular basis from the time they were drilled. The results of this trial were striking. Gas rate from both holes increased dramatically, with the hole drilled from 10 cut-through showing the largest increase. Figure 5 shows the specific gas rate history for the holes, expresses as gas rate per metre of hole. Not only did the gas rate increase significantly, but the increased rate was maintained for months. The curves shown are for gas rate per metre based on the entire hole length and for the length of the holes actually stimulated. The rate in the later case increased by a factor of 180 for hole 108-7-IF1 and by a factor of 22 for hole 108-10-10L. Results of this trial justified a drilling and sand-propped fracture stimulation campaign for LW9.

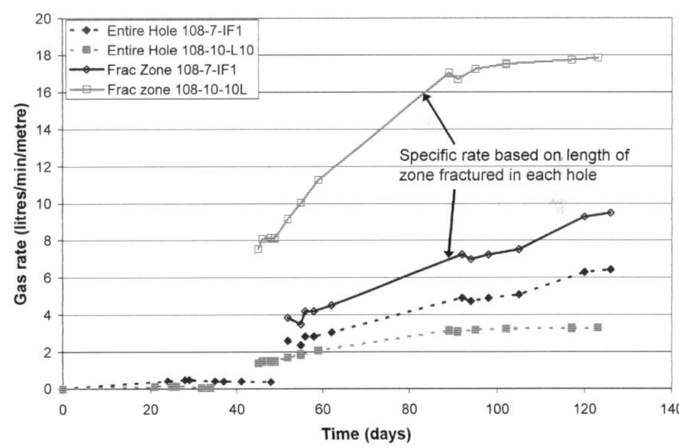

Figure 5: Measured gas rate before and after sand-propped fracture stimulation. The fracturing occurred during the period in the data that is not connected by a line.

STIMULATION AND DRAINAGE OF LW9

As anticipated, hole 108-10-L10 which was drilled along the axis of the panel gave the best stimulation results. In addition, holes with this orientation allow few holes to be drilled to effectively cover the panel with propped hydraulic fractures, since the fracture propagate in the direction of the maximum horizontal stress which runs east-west across the panel. Therefore, the drain holes drilled in LW9, for the fracture treatments, were drilled to align with the long axis of the panel (see Figure 1). The holes were drilled from three locations, at cut-through 13, 9 and 6 with five holes drilled from each site and steered so they were spaced about 35m apart across the panel.

The equipment used in the trail was reconfigured to fit onto a mine trailer. A new small paddle blender was built and refurbished centrifugal slurry pumps were installed to transfer the slurry from the blender to the high-pressure slurry pump. The sand was fed to the blender using a 100mm diameter auger.

The slurry pump was rated at 159 litres per minute at 16 MPa. This delivery rate was boosted by running a water-only high pressure pump in parallel. The water pump delivered 200 lpm at up to 30 MPa. Thus a total injection rate of 360 lpm was obtained. The packers were inflated with water by a small triplex water pump rated to 20 MPa. A valve bank was built that allowed the operator to switch from pumping slurry to clean fluid by moving a single lever. The control for all pumps and motors was located onto one central panel with oil and water pressure gauges available to provide feedback to the operator.

Finally, a special packer inflation rod system was developed that allowed the packer inflation line to be run inside the main injection rod string. This configuration eliminated the problem of an inflation hose being exposed to catching on the packers or hole during packer movement. The packer inflation rods consisted of 3m long flush thick-walled tubes with integral O-ringed thread seals. Packers were run in and out of the holes by first adding a packer inflation rod to the previous one. This resulted in a 3.3m long inflation rod string extending from the last injection rod. Then a injection rod was added by sliding it over the exposed inflation rods. By alternating in this way any length of injection line could be run. An adaptor was added to the injection rod before each fracture that allowed the packer inflation rod to extend out through an O-ring seal and allowed the slurry line to be connected at a side port.

Each treatment consisted of 4 to 6 minutes of clean water followed by 10 minutes or more of slurry (until the 100 kg of sand had been pumped). Each treatment required 30 to 45 minutes to carry out, which was longer than originally planned but was not the main limiting factor. During the early treatments, the life of the packers was limited to 10 to 15 treatments each. This packer life was too short to make the treatment cost effective. After several adjustments to the system, a new stronger packer was sourced and a lower packer set pressure was used. These changes resulted in extending the packer life to an average of 60 treatments per packer. Later testing demonstrated that the packers used at Dartbrook could withstand more than 200 cycles without failure if inflated in a smooth pipe. This test result strongly suggests the borehole condition is the main factor in limiting the packer life during these fracturing operations.

By the time the hydraulic fracturing in LW9 had started, the longwall was retreating. Approach of the longwall face limited the time available to fracture and drain all the holes that had been drilled so some holes were not treated and others had fractures placed every 6 m rather than every 3 m along them. Sections of the fractured holes that crossed near pre-existing fan-array holes were skipped over to avoid the possibility of the hole breaking out into the nearby existing hole. Such breakout could damage packers inflated there or lead to cross flow between the fractured hole and the older fan array hole. Table 1 contains a summary of the fracture treatments carried out in LW9 during this project.

Table 1: Summary of fractures in LW9 at Dartbrook.

Hole number	Number of fractures	Hole length fractured	Fraction of hole stimulated	Average fracture spacing	Sand per fracture
		(metres)	(percent)	(metres)	(kg)
109-13-F1	54	195	72%	3.6	78.3
109-13-F2	50	294	85%	5.9	81.2
109-13-F3	4	18	5%	4.5	100.0
109-9-F1	44	258	85%	5.9	86.1
109-9-F2	54	297	90%	5.5	84.7
109-9-F3	108	366	94%	3.4	95.6
109-9-F4	104	413	94%	4.0	83.4
109-9-F5	77	486	95%	6.3	86.8
109B-6-F1	14	110	26%	7.9	
109B-6-F3	59	373	94%	6.3	71.6
109B-6-F4	56	364	93%	6.5	75.2
Totals/ Average	**624**	**3174**			**84.3**

A total of 624 fracture treatments were carried out to stimulate 3174 metres total of 11 drain holes. Figures 6, 7, and 8 show the specific gas rate for each hole.

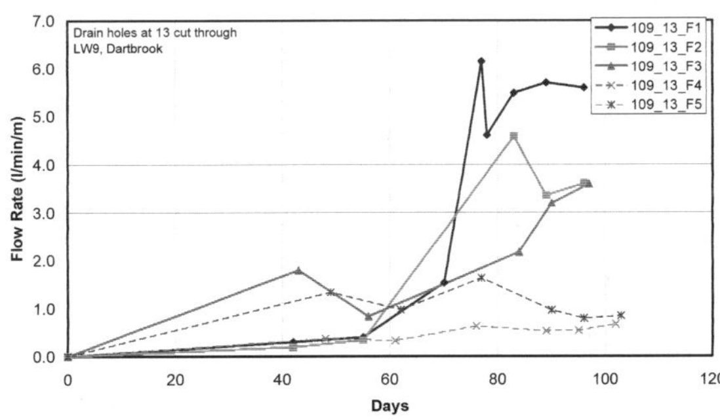

Figure 6: Specific gas rate from drain holes drilled at cut-through 13. Dashed lines indicate holes not fracture stimulated.

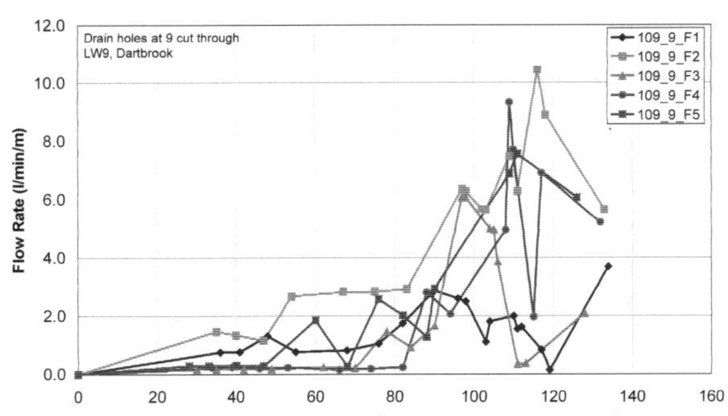

Figure 7: Specific gas rate from drain holes drilled at cut-through 9. All holes were stimulated.

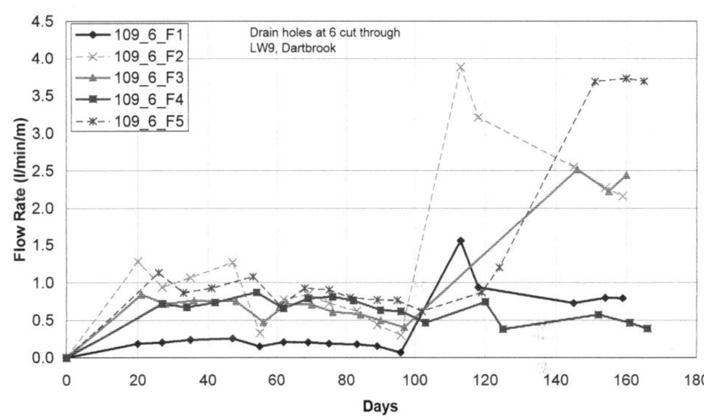

Figure 8: Specific gas rate from drain holes drilled at cut through 6. Dashed lines indicate holes not fracture stimulated

Overall the gas rate from the unstimulated holes averaged 0.74 litres per metre per minute which increased to 3.6 litres per metre per minute after stimulation. The average stimulation factor was, therefore, 4.9 for holes drilled from the three cut throughs in LW109. These averages include hole F3 at 13 cut through, which had only 5% of its length stimulated and hole F1 at 6 cut-through, which was stimulated over only 25% of its length. With more complete coverage of the holes, a stimulation ratio above 5 would be expected to result for conditions in LW9. In Figure 7, holes F1 and F3 show signs of being blocked by debris or sand which then is cleared at 110 to 120 days on the graph. Similarly, F1 and F4 at 6 cut-through (Figure 8) appear to be blocked. The gas from these holes has, evidently, found its way into holes F2 and F5, which were not stimulated but show a large increase in gas rate after the hydraulic fracturing work in the other holes at this site. Such a connection between holes can be formed either through a number of propped hydraulic fractures or directly by a hole-to-hole intersection.

Some of the sand proppant was produced back from the fractures into the drain holes. Sand production can be controlled by adding sand stabilizers to the proppant when it is pumped. It may also be worthwhile to consider cleaning the entire hole after the fractures are placed since most proppant is produced soon after the stimulation work is carried out. Sand production also becomes less of a problem later as the closure stress on the fractures increases.

DISCUSSION AND APPLICATION TO OTHER MINES

The effect of sand propped hydraulic fractures on the gas drainage rate depends on the seam parameters such as permeability and thickness and on the hydraulic fracture parameters, such as fracture conductivity, propped length and spacing between fractures. If coal seam permeability is higher, the fracture conductivity and propped length must be increased (by pumping more sand into each fracture) if the same stimulation ratio is to be obtained.

This fact partly explains why the stimulation work in LW9 provided a stimulation ratio of about 5 while at the trial site in LW8 the stimulation factor was 22 and 180. The coal in LW8 was less permeable and pre-stimulation gas rates were lower than those in LW9, but no change was made to the amount of proppant in each fracture and the average spacing between fractures was actually increased in LW9 (because of the approaching longwall face). The experience at Dartbrook has demonstrated the significant potential that sand propped fracture stimulation of drain holes has to accelerate drainage while reducing the overall amount of drilling required. The effectiveness of the fractures has been demonstrated and the only remaining issue is that of developing reliable methods to place the fractures under a range of conditions.

Alternative borehole completion methods are being investigated. The approaches that are being considered include

- Casing the hole before fracturing. Entry to the coal would then require notching the casing before fracturing. Cementing methods are a challenge from the point of view of obtaining a good bond and not pushing cement into the coal.
- The access hole can be drilled in the floor or roof rock rather than in the coal. Notching of the hole would be required to ensure fracture initiation in the desired plane and high horizontal stress conditions may cause the fracture to grow parallel to the seam rather than into it.
- Use of a system that jets the hole wall to initiate the fracture while maintaining significant pressure in the hole itself (Halliburton have developed such a method for horizontal wells in conventional formations). The variability of fracture initiation pressure along an open hole in coal may make this method difficult to use. It does have potential in holes drilled in the roof or floor rock.
- Cross measure holes can be drilled and stimulated to drain underlying seams. These holes may be drilled into the target seams or may target roof or floor rock sequences. Figure 9 shows a concept drawing for targeting an underlying seam by cross measure drilling.

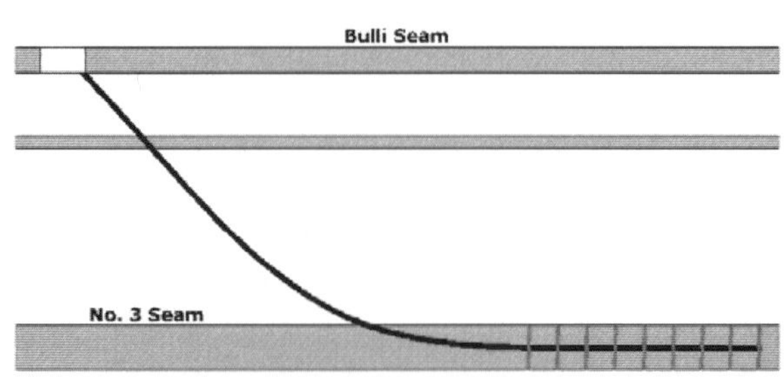

Figure 9: Cross measure drilling with sand propped fractures placed in lower seam.

CONCLUSIONS

The trial in LW8 demonstrated that sand-propped hydraulic fractures were very effective in stimulating gas drainage rates from horizontal drain holes and the best stimulation resulted for holes drilled so that the fractures extend in a direction perpendicular to the hole axis.

The sand-propped hydraulic fracture stimulation work carried out in LW9 at Dartbrook, on average, increased the rate of gas drainage by a factor of about 5. The increased rate of gas drainage allowed the gas content to be reduced in LW9 to a level that was sufficiently low so that no gas-related delays were experienced in mining the panel.

ACKNOWLEDGEMENT

The authors thank Anglo Coal and CSIRO Petroleum for supporting this work and for permission to publish this paper.

REFERENCES

Byrnes, R, 2002. Personal communication, April.

Croft, GA, 1980. Summary of research into incremental hydraulic fracturing of coal seams at the MW Haenke Colliery, Ipswich, The Research Section of the Queensland Department of Mines Report.

Gray, I, 2002. Dartbrook low permeability zones, Sigra Pty Ltd, Report to Dartbrook Coal Pty Ltd, Job number 35.

Jeffrey, RG, 1999. Stimulation of gas make from horizontal in-seam drain holes by hydraulic fracturing, CSIRO Petroleum, Confidential Report 99-061 and ACARP project C4033 report.

Jeffrey, RG and Mills, KW, 2002. Project Description, Optimising sand-propped hydraulic fracture stimulation of in-seam gas drainage holes, ACARP Project C12021.

Jeffrey, RG and Boucher, C, 2004 Sand Propped Hydraulic Fracture Stimulation of Horizontal In-Seam Gas Drainage Holes at Dartbrook Coal Mine, proceedings of the COAL 2004 Symposium, Wollongong, Feb. 4-6.

Kravits, SJ, 1993. Evaluation of hydraulically fractured in-mine horizontal wells as a completion approach for coalbed methane recovery independent of, and in conjunction with coal mining operations, REI Underground Exploration Draft Report.

Walker, RF, Ehrl, E, Arasteh, M, 1993. Simulation verifies advantages of multiple fracs in horizontal well, *Oil and Gas Journal*, Nov. 29, 1993, 66-70.

Weng, K, 1993. Fracture initiation and propagation from deviated wellbores, *in Proceedings of the 68th Annual Technical Conference and Exhibition of SPE*, Houston, paper SPE 26597.

Weijers, L, de Pater, CJ, Owens, KA, and Kogsboll, HH, 1992. Geometry of hydraulic fractures induced from horizontal wellbores, *in Proceedings of the 1992 European Petroleum Conference*, Cannes, paper SPE 25049.

2005 INTERNATIONAL COALBED METHANE SYMPOSIUM

0508

HISTORY MATCHING FOR OPTIMISATION OF GAS DRAINAGE FROM HORIZONTAL WELLS CONTAINING SAND PROPPED HYDRAULIC FRACTURES

LD Connell and RG Jeffrey
CSIRO Petroleum, Private Bag 10, Clayton South, Victoria, Australia, 3169

ABSTRACT

Stimulation of horizontal wells with sand propped fractures has been carried out at a mine in the Hunter Valley of New South Wales, Australia and shown to lead to significant increases in gas production rates. This paper presents the simulation of gas production from horizontal wells before and after fracture stimulation using the SIMED II model. As well as history matching production data and subsequent predictive use of the model for planning the fracture stimulation, the modeling work identifies the mechanism behind the substantial gas rate increases observed. While the permeability of the coal in question is very low, the gas rates observed prior to stimulation indicate the presence of substantial borehole skin. The fracture stimulation acts to bypass this skin and access the seam, leading to a dramatic increase in drainage rates. The parameter set refined through history matching is used in predictive modeling to investigate the role of fracture half-length and spacing along the borehole. It is shown that there is an optimal fracture density beyond which little gas drainage benefit is derived. The ultimate design of a drainage strategy using fracture stimulated horizontal boreholes will be determined by the relative cost of borehole drilling and the fracture treatment.

INTRODUCTION

Gas drainage using horizontal in-seam drainage wells is a common practice for coal mines around the world. While for many coal seams this is an effective strategy, problems can be encountered where coal permeabilities are low, leading to long drainage times and delays to mining. Increasing the density of the drainage wells can help alleviate this problem, however this leads to significant additional costs and for some areas the coal may still prove difficult to drain. For vertical wells fracture stimulation is a standard procedure for improving well productivity. However this technique has yet to find wide application in horizontal wells, either drilled surface to in-seam or underground in-seam. In Jeffrey et al. [1], presented at this conference, stimulation with sand propped fractures of horizontal in-seam gas drainage wells is described for a mine in the Hunter Valley of New South Wales, Australia. For this site the coal is of low permeability and proven to be difficult to drain. The fractures are shown to lead to dramatic increases in gas drainage rates.

In this paper gas drainage from sand propped fracture stimulated in-seam horizontal wells is simulated. History matching using the data presented in [1] is used to refine information on seam properties. The data obtained from the history match is then used to predict the gas rate for a range of fracture characteristics such as spacing, conductivity, and length.

SITE DESCRIPTION

The propped fracture stimulation trial was carried out at the Dartbrook mine which has been the subject of several previous studies related to stress, permeability and fracture stimulation. The directional permeability has been measured through vertical well interference tests [2]. Gas drainage from a fracture stimulated vertical well at the site has been measured and simulated [3]. Stresses were measured as part of a study on longwall design [4].

The coal seam of interest (as modeled) is 23.4m thick and divided into 14m of roof, 6m of working section and 3.4m of floor. The borehole used for this study is at 345m depth and is assumed to be in the middle of the working section. The coal contains both carbon dioxide and methane with the relative proportion of these gases varying with position. At the site of the fracture stimulation trial gas content was approximately 9 m^3/tonne with 92% CO_2 and 8% CH_4.

Coal permeability is usually a function of effective stress and therefore depth. At Dartbrook stress and permeability have been measured in a number of studies. Permeability has been estimated through surface vertical boreholes with injection-falloff and well interference tests, while stress has been measured through step-rate tests. Figure 1 presents the available measurements of permeability with respect to effective stress at Dartbrook. While there is considerable scatter in the results from the injection-falloff tests, regression through these with the regressed line forced to pass through the interference test result gives an estimate of the behaviour with respect to effective stress. Minimum total stress was estimated during the in-seam fracture stimulation for this study as 5 to 5.5 MPa [1], with a pore pressure of 1.0 to 1.2 MPa [5] this leads to an effective stress of approximately 4 MPa. The corresponding estimate of permeability using the regression presented in Figure 1 is therefore approximately 0.03 md.

Information on seam properties at Dartbrook is available from the modeling study presented in [3] which was carried out in a different area to the current work, at a shallower depth. Since coal properties are highly variable with position and depth, this information could only be used as a starting point for the seam properties at the site of the stimulation. In a manner consistent with the work presented in [3] the coal seam was also divided into the three sections described above. In terms of parameter estimation, there are a large number of model parameters involved. For the current study it is assumed that the working seam dominates the model response and it will be the parameters for this seam that the gas drainage rate is most sensitive to. In the modeling work presented here the properties of the working seam were refined through history matching against observations of gas rate, with the properties of the roof and floor sections set from the work presented in [3]. The initial value for the working seam permeability came from that estimated above. Table 1 summarises the key coal properties used in the modeling work.

The adsorption isotherms for methane and carbon dioxide had been determined in the study presented in [3]. As mentioned above this can be regarded as an initial estimate for the current work since coal properties vary significantly with position. As indicated in Table 1 the adsorption isotherm will be refined through history matching against production data in a section presented below. However the adsorption isotherms from [3] allow the desorption pressure to be determined and the initial gas desorption behaviour during gas drainage to be understood. In Figure 2 the adsorption isotherms are presented along with the estimate of the initial reservoir pressure and desorption pressure calculated using the isotherms for a gas content of 9 m^3/tonne with 92% carbon dioxide and 8% methane. This shows that the coal seam is undersaturated and that some drawdown of reservoir pressure must occur before gas desorption can take place. However, since the longwall panel was under active drainage leading up to the hydraulic fracturing effort it seems reasonable that the gas content had reached the isotherm. This suggests that the isotherms in Figure 2 may not be correct or the assumed gas content may be in error.

At the site of interest 20 sand propped fractures were placed at regular 3m intervals along a section of horizontal borehole (from 270m to 330m) within a longwall panel that had proved difficult to drain. A numerical hydraulic fracture model was used to approximately match the pressure measured during the fracture treatment [1]. This modeling predicted a propped fracture conductivity of 10 md-m extending up to 15m each side of the borehole. Even though the fracture was estimated to extend much further, up to 50m, this unpropped part of the fracture has a much lower conductivity. The observed gas rate from this borehole is presented in Figure 3.

The gas rate presented in Figure 3 shows that the fracture stimulation has a significant effect. Prior to stimulation gas rates are very low; the gas rate increased by a factor exceeding 180 with stimulation.

MODEL DESIGN

The simulator used for this study was SIMED II since this model is proven for coal gas problems and incorporates the various processes operating. This modeling study builds on the previous work conducted using SIMED at Dartbrook such as Jeffrey et al. [3] and Jeffrey et al. [2]. As described above this previous work provided valuable information on coal seam properties and this data was used as the basis for the current study.

The model grid exploited the symmetry associated with the regular 3m spacing of fractures used in the stimulation trial. The problem layout is presented in Figure 4. The model area represents one quarter of the region contributing to each fracture. The outer model boundary was selected as 40m from the borehole to minimize any interaction that this could have on model results.

Precise information on fracture geometry was not available. Jeffrey et al. [1] estimated that the propped half-length of the hydraulic fracture may be 15m, with the unpropped region extending to more than 50m from the borehole. In this study the importance of fracture size will be investigated in order to establish the likely fracture geometry.

Figure 5(a) presents an example of the model grid used. There is increasing refinement of the grid towards the borehole and the fracture to accurately represent the pressure behaviour in these regions. There is also refinement within the coal beyond the end of the fracture. Thus the grid will vary with fracture size. The grid presented in Figure 5 involves a fracture with a 20m half-length. The fracture is represented by a region of grid blocks within the model, presented in Figure 5(b). The permeability of these grid blocks is selected to be equivalent to the desired fracture conductivity; for a 10md-m conductivity fracture and a grid block width of 5cm the permeability would be 200md. Note the actual fracture is not propped to 5cm wide but this artificially large width does not affect the results produced by the reservoir model provided the conductity of the propped hydraulic fracture is represented.

The boundaries to the model are assumed to be impermeable The overlying and underlying lithologies for this site are shales with a low permeability and so the impermeable boundary assumption is reasonable for these directions. Other boundaries are determined by the problem symmetry and are thus correctly represented as impermeable.

HISTORY MATCHING

The first component of this work involved using SIMED to history matching the gas drainage data presented in Figure 3. The model parameters selected for refinement to improve the history match are identified in Table 1. This was an iterative process where parameter values were modified and model runs conducted. The final, optimal, parameter values are presented in Table 2.

Three fracture sizes were considered in the history matching and the optimal parameter values for each of these is presented in Table 2. Figure 6 presents the agreement between simulated and observed gas rates (adjusted for the borehole length represented by the model domain) for a 9.1x6m fracture. The agreement between model and observations for the other fracture sizes considered was almost identical to that seen in Figure 6. From this figure it is clear that the agreement between the two sets of results is excellent.

The large stimulation effect seen in the observations, where gas rates prior to fracture treatment are extremely low, could only be reproduced by using a large borehole skin. The fracture treatment bypasses this skin to access the coal seam directly leading to significant increases in gas rate. The shape of the gas drainage after treatment is determined by a combination of properties; in particular the permeabilities and the desorption time constant. History matching of production data can always be achieved given the large number of parameters involved, but the challenge is to produce a match while restricting the parameter ranges to measured or reasonable values. In this case, our use of a high skin and relatively small fracture conductivity might not be the only set of reasonable values that could produce a match to the gas rates measured. However other parameters, such as coal permeability and fracture size are consistent with independent souces of information.

The history matching involved refinement of the value of the CO_2 Langmuir adsorption volume. This in turn modified the initial desorption pressure to 1380 kPa for a gas content of 9.055 m^3/t, very close to the measured value. In addition the initial reservoir pressure was changed to 1500kPa. The desorption pressure for the history match is very close to the reservoir pressure and more consistent with our knowledge of the seam drainage as a result of mining than the initial values proposed in the Site Description section.

The optimal parameter values presented in Table 2 show that larger fractures required lower seam permeabilities in order to match the gas drainage data. The optimal vertical permeability was significantly lower for a fracture of 16.2m compared with a fracture of 6m; 0.04md compared with 0.008md. The X-direction permeability was 0.035md for a fracture half length of 10m and 0.02md for a fracture of 20m. The X-direction permeability will determine the connection between the fracture and coal immediately adjacent to the fracture. While the Y-direction permeability will largely determine the rate at which gas will migrate from the region of coal beyond the end of the fracture into the coal adjacent to the fracture. Since the fractures are close together, the coal adjacent to the fracture drains relatively quickly. The low vertical permeabilities will act to restrict migration from overlying or underlying areas of the seam not penetrated by the fracture.

The average horizontal permeabilities presented in Table 2 are higher than that estimated from the regression analysis based on Figure 1. However this data has a considerable spread and so the error in the line of regression could be large. In addition since the effective stress at the seam of interest is approximately 4 MPa and the measurements are between 0.9 and 1.5 MPa small errors in the slope of the regression will have large consequences in terms of the estimated value. However, the value estimated from the regression was 0.03md while that derived from the modeling for the largest fracture considered was 0.04md. The agreement between the estimate and the history match value is therefore excellent.

The ratio between average horizontal and vertical permeability varied between (approximately) 1.5 and 7.4. For the 20m x 16.2m fracture it was 5. While reliable estimates of vertical permeability are difficult to determine, but a common rule of thumb is that vertical permeability is a tenth of horizontal. The ratio for the smallest fracture is 1.5, suggesting that the vertical permeability is too high and the history match for this fracture involves significant gas migration from overlying and underlying coal seams. For the 20m x 16.2m fracture the vertical and horizontal permeabilities are much lower and closer to the value expect for this effective stress. The ratio for this case is on the high side but is within the range of possible values.

The fracture conductiviy is significantly lower than that expected; approximately 1.5md-m when 10md-m was estimated above. An efficient stimulation produces a fracture conductivity that is high enough to allow the gas to move along the fracture to the borehole with minimal pressure drop. The dimensionless fracture conductivity, F_{CD}, is a number that captures the competition between the rate that gas can be produced into the fracture and the capacity of the fracture to carry the gas to the wellbore. Dimensionless fracture conductivity is given by:

$$F_{CD} = \frac{k_f w}{k x_f}$$

where k_f is the proppant or fracture permeability, w is the fracture propped width, k is the coal seam permeability, and x_f is the half length of a constant height hydraulic fracture. Elbel (1988) found that for a constant volume of proppant, the optimum F_{CD} value was 1.26 for formations with permeabilities greater than 1 md and 3 for formations with permeabilities less than 0.1, which is the case for the coal seam modeled here. An F_{CD} of 3 is within the range for values of fracture propped length and fracture conductivity for the sand propped fractures studied here. For example, if the fracture conductivity is taken as 1 md m and the coal permeability is 0.05 md with propped fractures of 15m half length, then the value of F_{CD} is 1.33 and if the fracture conductivity is 2 this increases to 2.66 (also refer to Table 2).

The history match for the 20 x 16.2m fracture provides the parameters values that are in closest agreement to the values expected. It is also reasonably close to the fracture size estimated from the fracture modeling presented by [1].

MODEL PREDICTIONS

The history matching described above identified the parameter values that provide a good agreement between model results and field observations. In this section these parameter values are used as the basis for a series of predictive analyses investigating the role of fracture and borehole spacing on gas drainage. The parameter values derived with a 20 x 16.2m fracture were used in this analysis. Figure 7 presents the conceptual borehole and fracture layout upon which the modeling is based. It is assumed that boreholes are spaced based on the half length of the fractures and that fractures are regularly spaced along the boreholes. In this layout the model domain is equal to the fracture half-length.

Gas Content Results

Figure 8 presents the average gas content after 1 year of drainage with respect to the fracture spacing for three borehole spacings. The gas content is expressed as a fraction of the initial gas in place and calculated over the entire model domain. The trends in the results presented in Figure 8 are consistent; increased borehole spacing and fracture spacing leads to slower drainage of gas. With the 20m borehole spacing, decreasing the fracture interval below approximately 6m has little beneficial effect. With 30m borehole spacing a fracture interval of 3m has the same outcome as a fracture interval of 6m for a 20m borehole spacing. The same outcomes can be achieved using different strategies however the choice of approach will be determined by the relative costs associated with boreholes and fractures. If it is ultimately cheaper to have fewer boreholes and more fractures then the widely spaced boreholes will be used with a short interval between the fractures.

Figure 9 presents the relationship of the average gas content (calculated on the same basis as with Figure 8) with respect to the fracture density, calculated directly by converting fracture and borehole spacing into numbers of boreholes and thus fractures for a fixed area. This fracture density calculation does not take into account the varying half length of fractures. However, from this figure a general trend can be observed that is consistent across the three borehole spacings considered. There are diminishing returns beyond a fracture density of 0.008 fracs/m^2 but a strong increase in gas content remaining at fracture densities lower than this. This relationship of gas drainage behaviour with respect to fracture density has important implications for the cost effective design of fracture treatments.

In Figure 10 contour plots of gas content in a plane parallel and 1 m from the fracture within the coal seam at two times, six and twelve months, are presented. In these figures the gas content is reduced in a region around the fracture. Gas content at points higher in the seam are larger than near the fracture and the gas content is reduced at a slower rate.

Gas Drainage

Gas drainage rates with respect to time for three fracture spacings are presented in Figure 11. These calculations were for a 20m spacing between boreholes meaning a fracture half length of 10m. The gas rate has been normalized so the rate presented is on a per metre borehole length basis. Of course, the closer the fractures are the greater the gas rate, but the earlier and steeper is its decline. Up to one year of results are presented. At one year, the 3m spacing has declined to be equal to the 6m spacing and would reduce to the rate of the 9m spacing not long after. The fracture spacing would have to be determined to meet the objectives of the gas drainage; for a mining application the likely objective would be cost effective drainage of gas from the coal within a specified fairly short time frame. For a methane production scenario, it would be maximizing recovery of methane, but perhaps trying to minimize the payback time of the well. Shorter drainage times could be achieved by increasing the fracture density but this would have to take note of the constraints encountered above with respect to fracture density.

CONCLUSIONS

This paper has presented a modeling study of fracture stimulation in horizontal boreholes within coal. The work was based on measurements of gas drainage from boreholes with regularly spaced sand-propped hydraulic fractures along them. The fracture stimulation had a significant affect on gas drainage rates, with a factor of over 180 increase in the gas rate as a result of the stimulation. One consistent reason for the significant increase is postulated to be a high borehole skin that acted to restrict gas and water flow into the borehole prior to stimulation. The fractures acted to bypass this high skin, providing high conductivity pathways with large surface areas in contact with the coal.

One potential mechanism for the formation of borehole skin is associated with coal fines produced during drilling being pushed into the surrounding coal by drilling pressures and clogging the cleat system. Borehole skin could also form during gas production as a result of coal fines migration from within the seam to near the borehole. Another potential process is the precipitation of minerals within the formation near the borehole during gas and water production. Yet another mechanism relates to the stress concentration around boreholes acting to close cleats and reduce permeability. While these physical skins could potentially play a role in reducing gas drainage, the way the boreholes are brought into operation may also have an impact. Initially lowering the borehole pressure quickly could lead to gas desorption close to the hole resulting in a lowering of the water relative permeability in this region. This near borehole zone of low water permeability acts to restrict ongoing water flow from within the seam, thus delaying seam pressure drawdown and limiting the zone of gas desorption to near the borehole. While relative permeability and gas desorption are already represented in the model, if there are inaccuracies in these process a borehole skin might still result from these effects.

The predictions of gas drainage found that optimal strategies exist in relation to the design of borehole and fracture treatments. In particular, the effect of increasing fracture density with respect to gas drainage does not have a linear relationship. In fact increases in fracture density beyond a certain value give a diminishing return. Therefore investment in fracturing and drilling needs to take account of this particular optimal value. However the actual strategy used will depend on the relative costs of drilling and fracture treatments and the operational objectives. Mine drainage often has to be performed within a short time to fit in with the overall mine operational plan. For this the fracture density would be close to the optimal density. For methane production the fracture density would be related to the cost of the particular borehole/fracture strategy and the related gas production rate and its behaviour over time.

REFERENCES

1. Jeffrey, R.J., C. Boucher, and K. Mills, 2005: "Implementing Sand Propped Hydraulic Fracture Stimulation for In-seam Dranage Holes: A Demonstration of An Enhanced Drainage

Technology"; 2005 International Coalbed Methane Symposium, May 16-20, Tuscaloosa, Alabama, this issue.
2. Jeffrey, R.G., S.K. Choi, T.G. Fergusson and M. Camilleri, 1997: "Directional Permeability Well Test Results, DDH139, Dartbrook Mine"; APCRC Conf. Report No. 275, April, 17p.
3. Jeffrey, R.G., K.T.A. Meaney, and R.P. Doyle, 1999: "History matching of hydraulic fracture and production data from a vertical CO_2 and CH_4 gas drainage test well"; 1999 International Coabed Methane Symposium, May 3-7, Tuscaloosa, Alabama.
4. Enever, J.R. and R. Doyle, 1996: "The role of horizontal stress in longwall design – a case history at Dartbrook Colliery, NSW"; Proceedings of the Symposium on Geology in Longwall Mining, UNSW, Sydney, pp 143-151.
5. Gray, I. 2002, "Dartbrook low permeability zones", Report to Dartbrook Coal Pty Ltd, Job number 35, Sigra Pty Ltd.
6. Elbel, J.L. 1988. "Considerations for Optimum Fracture Geometry Design," SPE Production Engineering, August 1988, 3, No. 3, pp 323-327.

TABLES

Table 1. Parameter values for selected model parameters

Parameter	Roof	Working seam	Floor
k_x (md)	0.140		0.087
k_y (md)	0.033		0.025
k_z (md)	0.0184		0.014
CO_2 Langmuir volume (m^3)	18.41		20.3
CH_4 Langmuir volume (m^3)	12.18	14.95	13.4
CO_2 Langmuir pressure (kpa)	1120	1120.0	1120.0
CH_4 Langmuir pressure (kpa)	1950	1950.0	1950.0
CO_2 Desorption time (days)	20.1		29.0
CH_4 Desorption time (days)	58.0	58.0	58.0
Borehole skin			
Coal thickness (m)	14.0	6.0	3.4
Initial gas content (m^3/t)	9.566	9.055	10.455
Gas composition	CO_2	0.92	
	CH_4	0.08	

- parameter values refined through history matching

Table 2. Optimal values obtained through history matching for three fracture sizes.

Parameter	Fitted values		
Working seam permeability (md)	9.1m x 6m	10m x 16.2m	20m x 16.2m
X	0.035	0.035	0.02
Y	0.1	0.1	0.08
Geometric mean X-Y	0.059	0.059	0.04
Z	0.04	0.008	0.008
Ratio; vertical to horizontal	1.475	7.375	5
Fracture conductivity (md-m)	1.75	1.0	1.5
Dimensionless fracture conductivity, F_{CD}	3.2	1.7	0.8
CO_2 Desorption time (day^{-1})	10.5	10.5	10.5
CO_2 Langmuir volume (m^3/tonne)	17.0	17.0	17.0
Borehole skin	20		

FIGURES

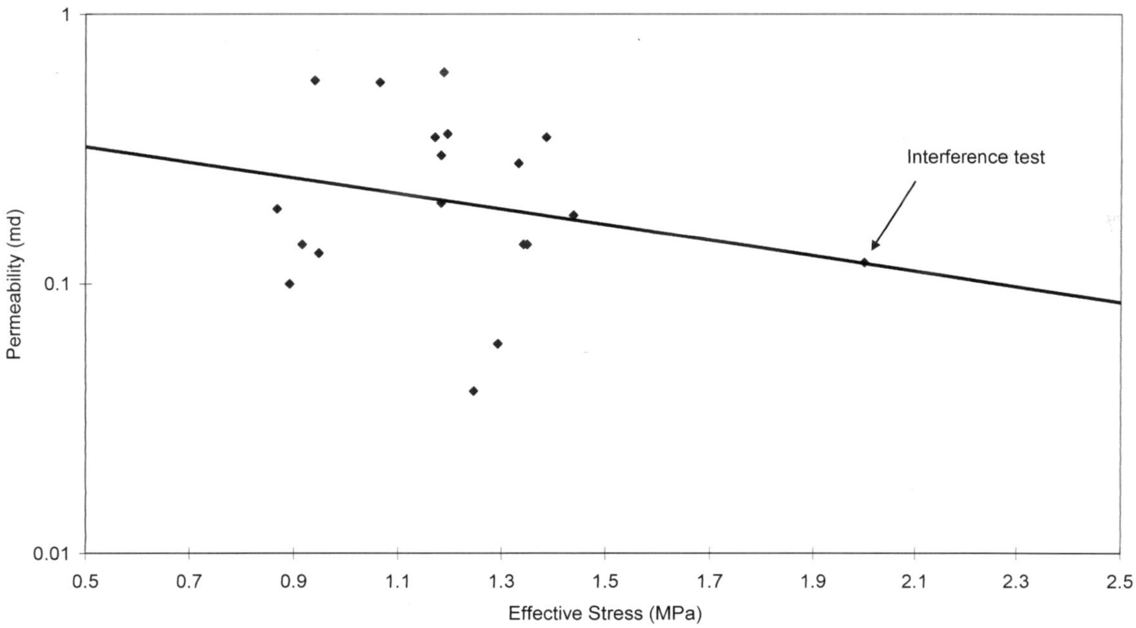

Figure 1. Permeability data from well tests in vertical boreholes at Dartbrook.

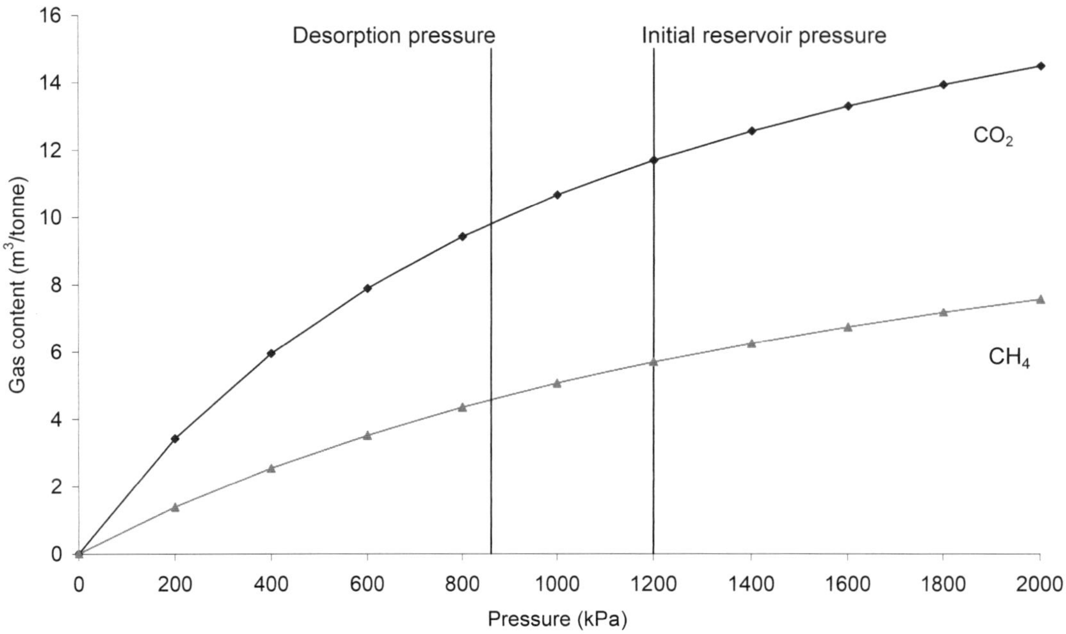

Figure 2. Adsorption isotherms for CO_2 and CH_4 with the estimate of initial reservoir pressure and desorption pressure calculated from a gas content of 9 m^3/tonne with a composition of 92% CO_2 and 8% CH_4.

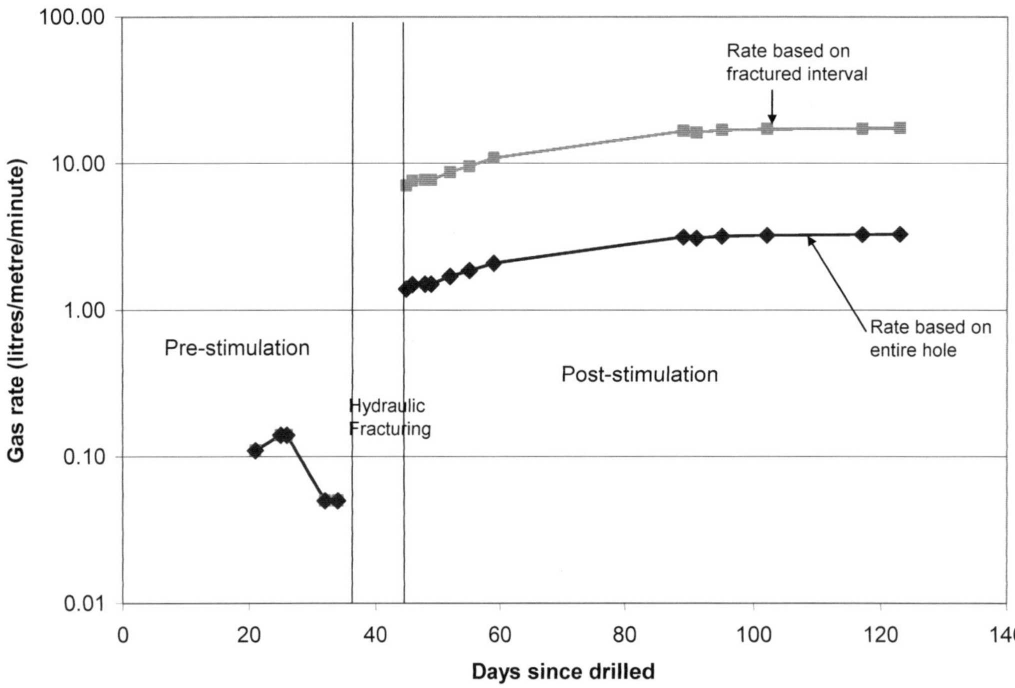

Figure 3. Observed gas rate normalized by borehole length for fracture stimulated borehole.

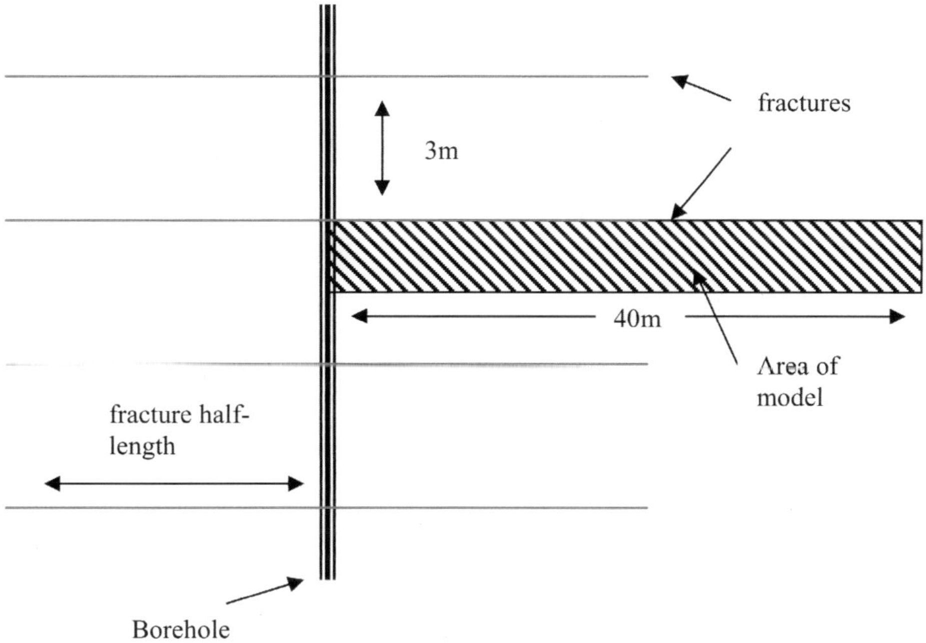

Figure 4. A plan view of the drainage borehole with regularly spaced fractures and the area represented in the SIMED model indicated.

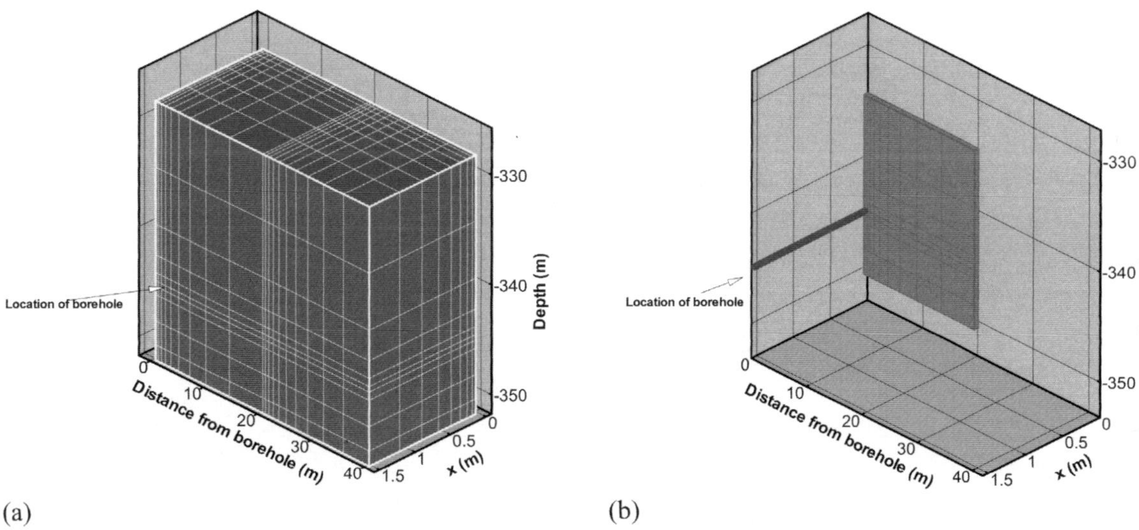

Figure 5. Example of model grid with a fracture of 20m half-length (a) model domain and discretisation (b) grid blocks representing a vertical 20x16.2m fracture with 5cm width in the y-z plane.

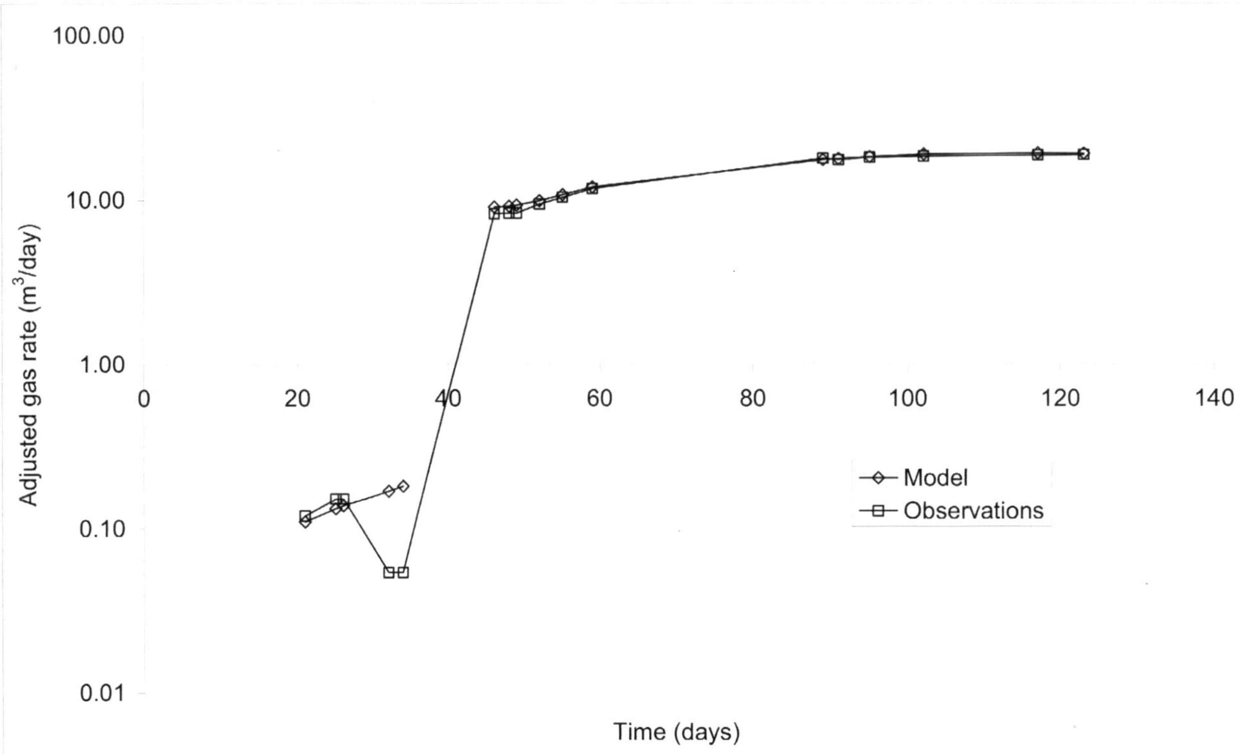

Figure 6. SIMED history match of gas rate for a 9.1m half-length x 6m high fracture.

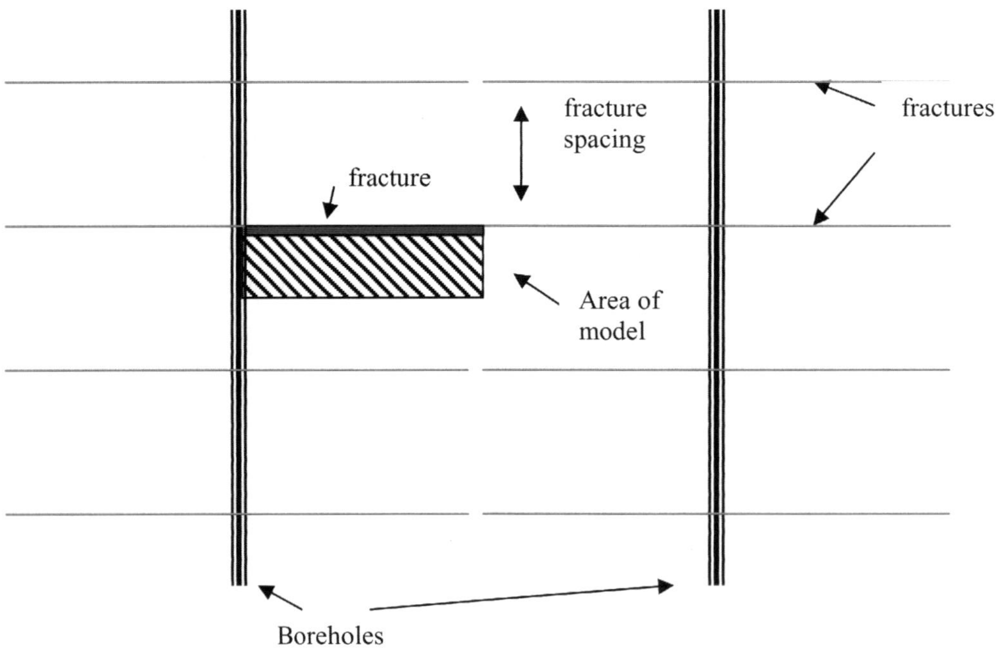

Figure 7. A plan view of the conceptual borehole and fracture layout used for the predictive SIMED modelling.

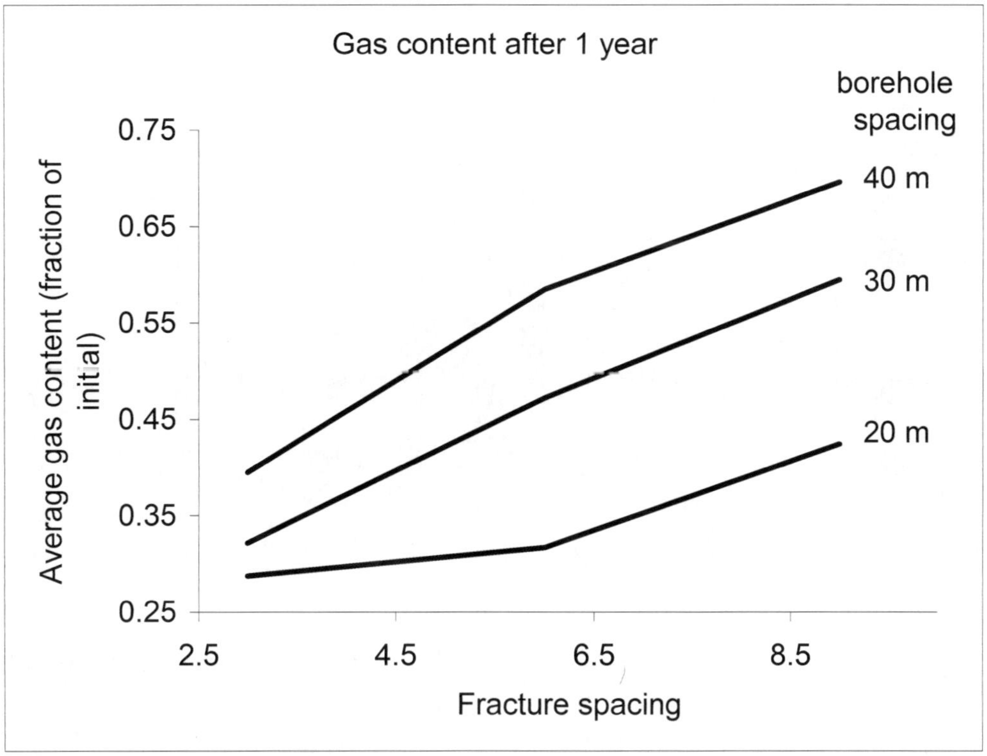

Figure 8. Average gas content calculated over model area, expressed as a fraction of the initial gas in place, after 12months of drainage.

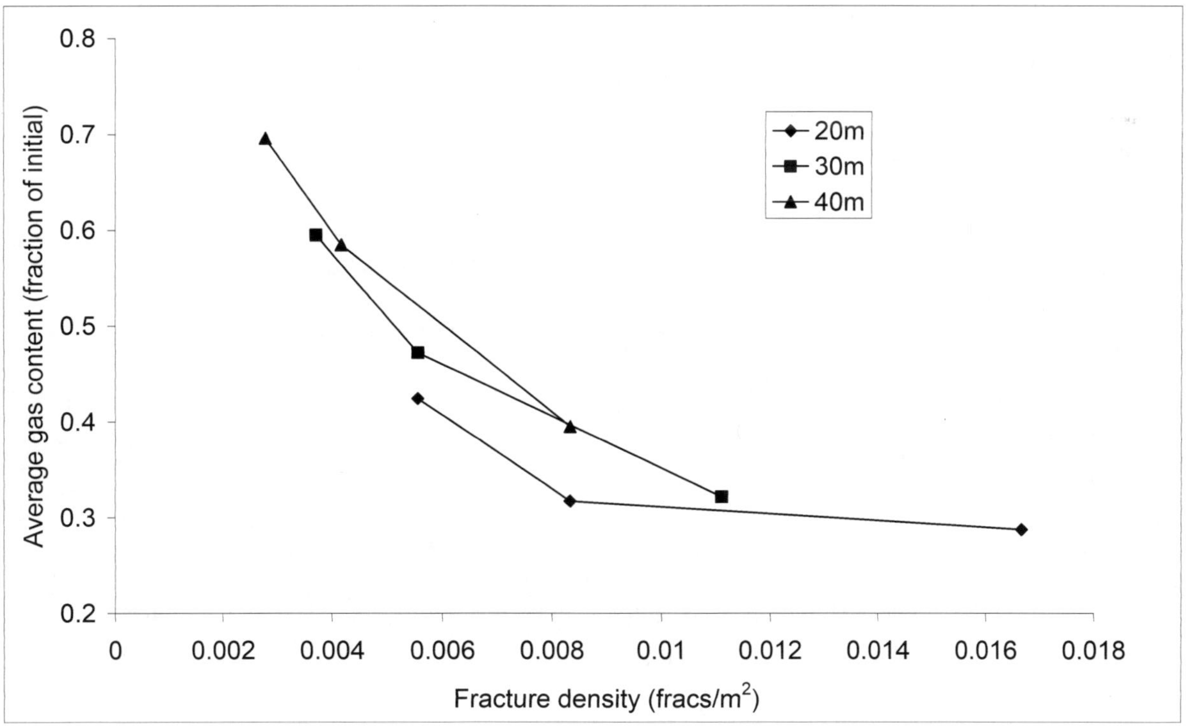

Figure 9. Average gas content calculated over model area with respect to average fracture density.

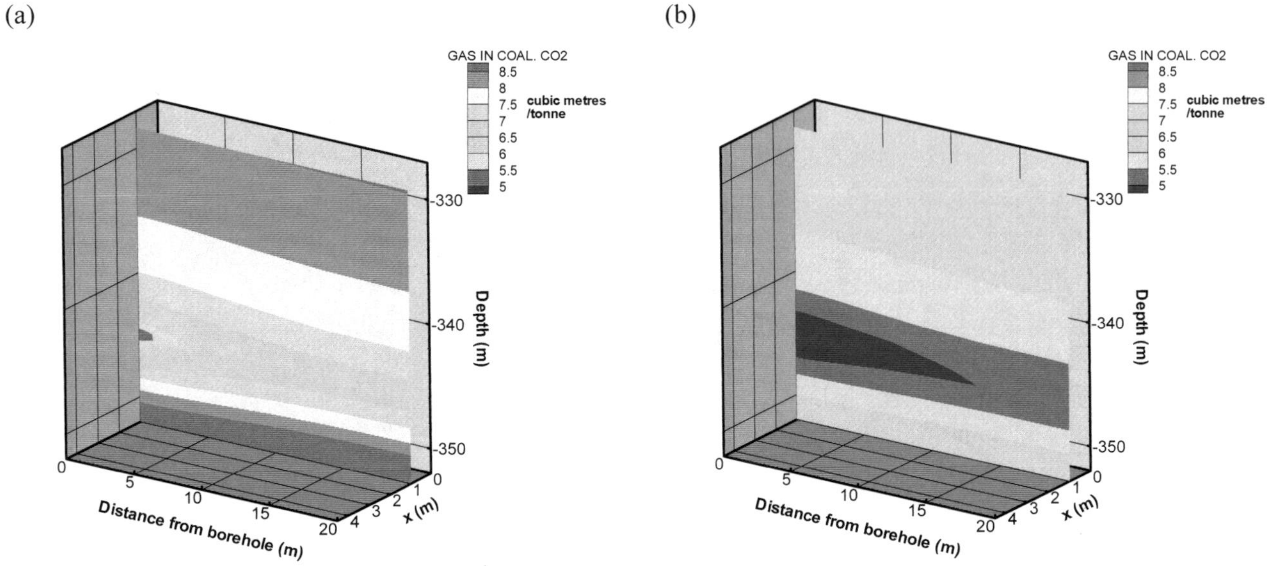

Figure 10. Contour plot of gas content in a vertical slice in the y-z plane at 1m from the fracture (a) after six months (b) after 12 months.

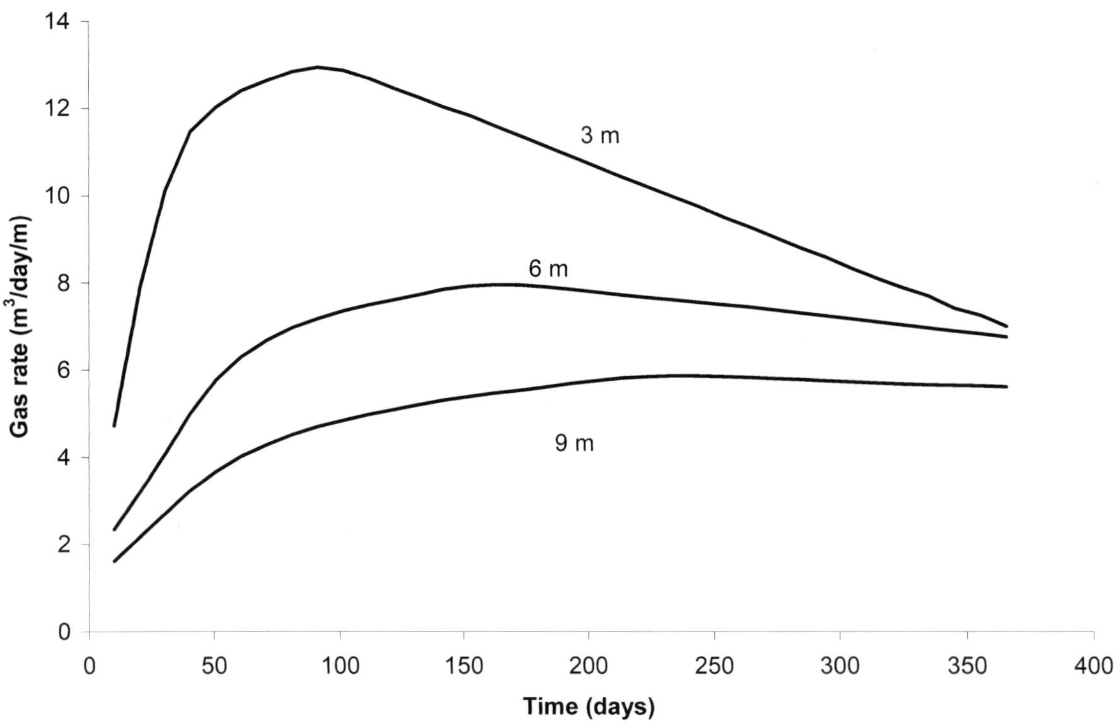

Figure 11. Predicted gas rate with respect to fracture spacing for a 20m spacing between borehole.

0516

ASSESSMENT OF THE EFFECTIVENESS OF COALBED METHANE WELL COMPLETIONS

John R. Cameron and Zissis A. Moschovidis, PCM Technical, Inc., and Ian D. Palmer, Higgs Technologies, L.L.C.

ABSTRACT

This paper presents objective methods for assessing the effectiveness of a completion program in a given CBM field. One method correlates the performance measure (e.g. maximum gas rate or cumulative production per designated time) to a comprehensive set of variables using multivariable regression coupled with statistical methods to eliminate unimportant variables. Multivariate regression analyzes the impact of all variables acting together and therefore, requires the number of CBM wells to be greater than the number of variables in order to generate optimal variable coefficients for the regression model.

For the case of a field with a limited number of CBM wells, this paper describes another method using nonparametric analysis that is useful for finding qualitative trends between the performance measure and the independent variable of interest. This methodology analyzes the impact of only one variable at a time, and thus can be used in cases where the number of wells is less than the number of variables.

Once the operator knows which variables are most important to the gas rate from the field, the operator can change the controllable variables in a way dictated by the analysis, and this should improve gas rates. Both the above methods have been applied successfully in CBM fields yielding insight as to the effectiveness of completion methods. This paper gives examples of the results of these analyses when applied in the Cherokee and Piceance basins. In fact in one field, the predictions of the analysis have been confirmed by subsequent actions to change the completion procedures.

INTRODUCTION

A survey of completion strategies in six of the major CBM basins in the U.S. has shown that there is not a consensus on the best completion to use (e.g. openhole, cased and perforated with acid breakdown, hydraulic fracture, or cavity completions) across basins or even within basins [1]. In addition, there is no consensus as to the best types or amounts of stimulation materials to use.

The production from a CBM well is affected by both non-controllable variables such as reservoir and coal properties and surrounding stratigraphy, as well as controllable variables, i.e. those that the operator can change, involved in drilling and completions operations, production operations, and workovers. Accordingly, it can be very difficult to sort out how much impact, positive or negative, a particular completions program actually has on production. It can be misleading to consider just a few selected variables, and although engineering judgment and field experience are valuable criteria to limit variables, this process may bias the results.

This paper presents objective methods for assessing the effectiveness of a completion program in a given field. One method correlates the performance measure (e.g. maximum rate or cumulative production per designated time) to a comprehensive set of both controllable and non-controllable variables using multivariable regression. Multivariate regression analyzes the impact of all variables acting together and therefore, requires the number of CBM wells to be greater than the number of variables in order to generate optimal variable coefficients for the regression model. Once these optimal coefficients are determined, the task becomes determining which of the variables are most important in the correlation. This is done using statistical tests for the significance of the coefficients (and thus the significance of the variables) coupled with a special method of variable elimination to remove variables having statistical significance less than a prescribed value (e.g. 80%).

For the case of a field with a limited number of CBM wells, another method using nonparametric analysis with the Spearman rank correlation coefficient [2] can be performed yielding a qualitative ranking for the variables giving insight about the completions program. This method analyzes the impact of only one variable at a time, and thus can be used when the number of wells is less than the number of variables.

The above two methods are described in detail in the Appendix. The following will describe the application of the former method (multivariate regression) to the Cherokee basin to evaluate completion variables important to gas production and perforation breakdown pressure, and the application of the latter method (the Spearman rank correlation coefficient) to the Piceance basin to evaluate effectiveness of selected fracture stimulation variables.

EXAMPLES OF CBM FIELD ANALYSES

Multivariate Regression with Variable Elimination: Cherokee Basin

Variable Selection and Data Accumulation. Discussions with AMVEST OSAGE, INC., a CBM E&P company operating in the Cherokee Basin, identified independent variables to be used in the analyses of two dependent variables: the maximum (peak) production normalized to feet of coal and the perforation breakdown pressure. Perforation breakdown pressure was of concern because of its wide variation, within and between fields. Tables 1 and 2 show the variables selected for each, grouped as to controllable (by the operator) or not controllable.

Results of Multivariant Linear Regression and Variable Elimination. During the analyses of the two dependent variables (maximum gas rate/ft of coal and perforation breakdown pressure), variable elimination at times showed that certain variables were not compatible with the analyses because they introduced very high multicollinearity (variables acting not as independent variables but tracking with other variables).

In the case of the correlation for **maximum gas rate/ft of coal**, data from 16 wells in Field A were used with 15 variables, but 5 of these variables (nos. 5, 10, 11, 13, and 15 in Table 1) had to be omitted from the analysis because of very high multicollinearity. Variable 15 is the *Net Pay* which at first one may consider to correlate strongly with gas rate, but in this case the dependent variable is gas rate normalized to net pay (ft of coal) which reduces the significance of the *Net Pay* variable in the correlation.

The analysis was continued with the remaining 10 variables and it was found that only one variable, *Total Frac Fluid*, had a statistical significance less than 70%. The regression was repeated using the remaining 9 variables and the model predictions are shown in Figure 1 and the model statistics in Table 3.

Even though all nine variables had statistical significances greater than 80%, variable elimination was continued until the maximum significance of the regression was attained. This occurred when the variables were reduced to four: *Sl Fld Loss, No.seams/stage, Total Sand,* and *lbs sand/ft coal,* where the regression significance increased to 95.5% from 92.4%. Note that the R^2 decreased from 0.84 to 0.56 suggesting that the eliminated 5 variables most likely impact production too.

Happily, it appears that controllable variables have the most statistical significance and suggest actions that the operator can take to improve productivity. For example, the positive correlation (i.e. the positive coefficient of the variable) of normalized production with *Sl Fld Loss* (+0.2547 Mcfd/ft coal per 1 cc increase in cement slurry leakoff) suggested using cements with higher slurry leak off. ***This seems counterintuitive, but when the operator resumed the use of its higher leakoff cements, it resulted in better production.*** Earlier, the operator had formulated the cements to reduce slurry leakoff in an attempt to reduce potential damage to the coals from invading cement filtrate. These low leakoff cements were attained using additives that resulted in a lighter cement with greater porosity. However, according to cement bond logs, these cements broke down more easily and did not give as good zonal isolation as the simpler, more basic mix, with high fluid loss. It is thought that the poor zonal isolation can reduce production by allowing water invasion from other zones and can encourage fracture breakout during fracture stimulation.

Some of the other trends predicted in this study were also validated. ***The positive correlation with higher viscosity fracturing fluids (also corresponding to cleaner fluids in this case) was seen in wells stimulated subsequent to this study***. Newer generations of cleaner foamed-fracturing fluids gave better production than gels or water fracs.

One of the strongest correlations was a counterintuitive trend in maximum normalized rate with proppant. A positive correlation of *Max rate/ft-coal* with *lbs sand/ft-coal* and a negative correlation with *Total Sand* were seen. ***This suggests inefficient fracturing when fracturing more seams per stage which was also suggested by this study in the negative trend with No.seams/stage.*** Fracturing of marginal seams requires more total proppant, but doesn't give higher gas rate/net coal thickness. But for a single seam, more proppant/coal thickness raises the gas rate, as expected from fracturing theory. In subsequent field practice, fewer marginal seams were fractured, with better well performance. Avoiding marginal seams may have avoided extra water influx too, which would have delayed dewatering.

Another trend seen from this study was that of increasing *Max rate/ft-coal* with increasing *min-ISIP grad* suggesting confined fracture stimulations are more productive in this case. This is opposite the trend reported by other observers [1, 3]. One possible explanation is that higher ISIP's indicate that stimulations have stayed in the coal seams [4] and have not broken out into lower-stress bounding layers that can reduce fracture length in coals and bring in unwanted water.

For the case of the **perforation breakdown pressure**, data from 138 wells in all the Fields (A, B, C, and D) were used with 9 variables, but 2 of these (nos. 3 and 5 in Table 2) were omitted because of very low statistical significance during variable elimination. For the remaining 7 variables, the variable elimination process showed that variables nos. 1 and 4 in Table 2 were of secondary importance with statistical significance less than 50%. The remaining 5 variables (nos. 2, 6, 7, 8, and 9 in Table 2) have statistical significance greater than 75%. The prediction of the resulting 5-variable linear regression model is shown in Figure 2 and the 7-variable regression model statistics are shown in Table 4.

When compared to analyses of breakdown pressures in the individual fields, only 3 variables shown in Table 4 had consistent trends, i.e., breakdown pressure either increasing or decreasing with a particular variable in all fields. Of these 3 variables having the most impact on breakdown pressure, *Treating Pressure* is the most statistically significant, but this cannot be controlled. ***Its optimized coefficient (+0.59 psi increase in breakdown pressure for a 1 psi increase in treating pressure) indicates that breakdown pressure increases with average fracturing treating pressure.*** This is an expected result since both perforation breakdown pressure and fracturing treating pressure increase with the in situ horizontal stress and the complexity of the frac "entry" zone.

The negative correlation of breakdown pressure with the controllable variable Perf. Interval length, indicates that this can be increased to reduce breakdown pressure, as one might expect. The negative correlation of perforation breakdown pressure with the non-controllable variable, *ISIP grad,* might be the result of cementing cleats and natural fractures. ***That is, as the ISIP grad decreases, the tendency for formation breakdown during cementing would increase promoting cement invasion making perforation breakdown more difficult because of the cemented-off fracture initiation points***.

The small R^2 of the seven-variable correlation (0.214) indicates that other, as yet uncorrelated, factors are important to the breakdown pressure. However, even though the R^2 is small, the regression is very significant as indicated by the F-Test that gives a significance of over 99.9%. Note that this high statistical significance is not surprising since the denominator of the F statistic is the mean variance of the error (between the model prediction and the actual data) and is small because of its large degree of freedom (130 vs. that of the numerator which is only 7). The mean variance is obtained by dividing the variance by the degrees of freedom. Subsequent variable elimination below 3 variables does not improve the significance of the regression.

Nonparametric Analysis Using the Spearman Rank Correlation Coefficient: Piceance Basin

Data from just a few wells were available in a study of one field in the Piceance basin and it was desired to evaluate any trends in production with fracture treatment variables including coal strength. Four independent variables were identified to correlate with the maximum gas rate/ft coal as shown in Table 5. The Spearman rank correlation coefficient was used since the data set consisted of only 7 wells and this method can work with sample sizes of less than 6 (refer to the Appendix).

Table 5 shows the ordinal ranking of the four variables with that of the dependent variable (*Max. MCFD/ ft-coal*). The calculated Spearman rank correlation coefficient shows a moderate negative correlation of *Max. MCFD /ft-coal* with *lbs proppant/ft-coal* and a weak negative correlation of *Max. MCFD /ft-coal* with total fracturing fluid suggesting that more fracturing fluid is worse, but doesn't make a lot of difference. No correlations for *Max. MCFD /ft-coal* with the coal strength or net treating pressure during fracturing were seen.

Table 5 shows that no strong correlations having greater than 95% significance ($|r_s| > 0.714$) were observed, but the following statements can be made:

- There is a weak negative correlation in *Max. MCFD /ft-coal* with total frac fluid amount indicating that conventional treatments give as good or better production than treatments with very large pad volumes.
- There is a moderately strong negative correlation in *Max. MCFD /ft-coal* with *lbs proppant /ft-coal* meaning that *Max. MCFD /ft-coal* decreased as proppant loading increased above 4000 lbs/ft-coal. This suggests a situation of diminishing returns with proppant loading, which has been seen in other basins [1].
- The negative correlation with *lbs proppant/ft coal* might be interpreted as ineffective proppant placement (e.g. fracture break out of the coal), but when frac treatments that broke out of the coal were removed from the analysis, the negative correlation still persisted. One possible explanation is that more proppant/ft of coal implies greater propped frac width, which raises the extra stress the frac places on the formation. This could lower the permeability of the formation, by stress-dependent perm, and reduce the gas rate.
- There is no correlation in *Max. MCFD /ft-coal* with coal strength. No correlation with coal strength indicates fines may not be a problem in plugging formation permeability, or the proppant pack.
- No correlation with net treating pressure indicates no correlation of production with the fracture breaking out of the coal (when considering both conventional treatments and treatments with very large pad volumes). We have observed that fracs that break out have a markedly lower frac treating pressure. Previous work has argued that confined, high-pressure fracs, have lower gas rates [1, 3].
- When considering just treatments with very large pad volumes, however, there is a negative correlation in *Max. MCFD /ft-coal* with net treating pressure, i.e., higher gas rates may be attained in wells treated with large pad volumes that have broken out of the coal. This trend has been also observed in other studies [5], and may indicate a frac that extends further in the bounding zone, and intercepts more joints from the coal that produce higher gas rates [6].

SUMMARY

We have described two methodologies for performing evaluations of CBM production to help optimize field completion strategies. The first is a ***multivariate regression analysis*** that uses statistical tests for the significance of the optimal regression model coefficients and applies methods of variable elimination to exclude unimportant variables. The second is a nonparametric method using the ***Spearman rank correlation coefficient*** that is more suited for smaller data sets where variable distributions and means are not well defined. Both methods can provide insight to the impact of completion parameters on production.

When applied to the Cherokee basin, the multivariate regression analysis coupled with variable elimination identified that normalized peak gas production could be increased by adjusting the cement properties, increasing the lbs proppant/ft coal, and by using more viscous (and cleaner) fracturing fluids. It also identified that for this particular field, production increases can be expected when fracturing fewer seams per stage and using less total sand. The analysis of the perforation breakdown pressure indicated that higher fracturing treating pressures are indicative of higher perforation breakdown pressures and possible cement invasion into fracture initiation points, and that breakdown pressures can be reduced by the use of longer perforation intervals.

When applied to the Piceance basin, the Spearman rank correlation coefficient was useful in identifying that peak production/ft coal was enhanced with decreasing lbs proppant/ft coal suggesting diminishing returns above 4000 lbs/ft-coal. There was a weak negative trend in peak production/ft coal with increasing amount of fracturing fluid indicating no real benefit by using very large pad volumes. There was no correlation in peak production/ft coal with coal strength indicating that fines may not be a problem. No correlation with net treating pressure indicates no correlation of production with the fracture breaking out of the coal (when considering both conventional treatments and treatments with very large pad volumes). However, because of the small size of the data set and because the Spearman rank correlation coefficient does not optimize all variables together, these trends should be considered as qualitative at best.

APPENDIX

Analyses Using Multivariate Regression and Variable Elimination

Multivariate regression is a common tool used to find optimal mathematical models relating an outcome to its pertinent variables. In multivariate regression, a mathematical model comprising the input variables and unknown coefficients is optimized by solving for the coefficients that minimize the error of the predicted outcome. Many times linear models are used for simplicity sake but they can yield results that reveal real trends. The form of the linear model is:

$$Y = a_0 + a_1 \times X_1 + a_2 \times X_2 + \dots\dots + a_p \times X_p \qquad (1)$$

where Y is the outcome, X_i are the independent variables, and a_i are the coefficients to be optimized.

In CBM production analysis, multivariate regression has been used with the Monte Carlo method for finding variables having the most importance to a measure of field performance [7]. The Monte Carlo method is useful when the explicit predictive formula for an outcome is known and it is desired to understand how the range or uncertainty of the input variables impacts the range of predicted results. This approach is useful for deducing what variables should be best defined when gathering data from the field.

In this study, multivariate regression is not used with the Monte Carlo method because it is desired to use actual field data and deduce trends with field and operation variables, and the explicit equations that govern the field data are not known. Independent variables used in the regression are those thought to be important to the particular measure of field performance, but it is not assumed that these comprise all the governing variables.

This methodology uses multivariate regression combined with statistical testing and a method of variable elimination to deduce the optimal set of variables that are significant in the prediction of the performance measure. It can be summarized as:

- Variable Selection (independent and dependent)
- Data Accumulation and Organization
- Multivariate Regression
- Variable Ordering (Student's t-Test)
- Variable Elimination Procedures
- Consideration of the Overall Model - goodness of fit (R^2) and statistical significance (F-Test)

Variable Selection. The first step is to compose a list of all variables that may impact the field performance measure. Some variables are not controllable by the operator, such as permeability, initial reservoir pressure, and depth, but others are controllable such as specifications for the drilling and completion design. Data for controllable variables are of particular importance to this analysis since these provide the information useful for optimizing completions. It is also important to provide data for as many non-controllable variables as possible since these can explain some of the variation of the field performance measure and make the impact of the controllable variables more readily observable. However, a complete set of the defining variables is not absolutely necessary since the statistics of the analysis can reveal likely trends.

Variable selection can be guided somewhat by examining the governing equations (e.g. that for gas deliverability) and looking at what field operations and processes impact these variables. For example, one variable that is strongly impacted by field processes (e.g. workovers and stimulations) is the permeability which is reflected in the skin factor. **Associated variables include quantities and types of acid or fracturing materials and treating pressures**. Other variables include the net coal thickness that is impacted by the success of the cement and perforation jobs. **Associated variables can include the type and properties of the cement and perforation specifications.** Care should be taken to eliminate variables that are linked or not independent.

The resulting set of independent variables should be related to a field performance measure that is uniquely quantifiable. For instance, cumulative production divided by the number of days of production or the maximum production after the well was exposed to a process in common to the other wells, or simply the current production assuming that historical processes will be included in the analysis for each well.

Data Accumulation and Organization. Perhaps the most difficult information to acquire is meaningful numbers for field permeability and reservoir pressures, production data (gas and water), and historical data for field processes. These may have to be acquired from operations or perhaps from service companies. Once acquired, these data must be entered in a database or spread sheet for the analysis along with well reservoir and completions information. This is perhaps the most time consuming step of this methodology and requires coordination with management, reservoir, and engineering personnel.

Multivariant regression. Multivariant regression is a method of optimizing the coefficients of a selected mathematical model to minimize the error of its prediction. The simplest mathematical model is one where the production is a linear function of the variables plus a constant (eq. 1). In this case, linear multivariant regression yields the optimal coefficients of each of these variables plus the optimal constant. Linear regression uses the least squares approach that adjusts the coefficients until the squares of the errors of the prediction are minimized. Linear multivariant regression is widely used and is included in most commercially available statistical software packages.

Variable ordering (Student's t-Test). Once the optimal linear model is determined, the task becomes determining which of the variables are most important in the correlation. For a linear model, this is accomplished by using the statistical test called the Student's t-Test. The 't' statistic is the ratio of the value of the variable's coefficient to the standard error of the coefficient. The standard error of the coefficient is a measure of the variation expected for this coefficient. If this ratio is large (e.g. greater than 1), the coefficient has high statistical significance and thus, the variable is considered to have high significance.

Variable Elimination Procedures. Next, the variables with lesser significance are eliminated in a statistically appropriate manner to arrive at the final set of variables with the desired significance. We use a special combination of elimination methods comprising backwards elimination, forward elimination, and stepwise elimination, combined with the all-subsets methods and are described next.

The method of backwards elimination starts with a linear model of all independent variables and discards the variable with the least significant coefficient and then re-optimizes the linear model without this variable. This process is repeated until all remaining variables have coefficients with a statistical significance greater than some defined level (e.g. > 80%). This is the opposite of the forward selection method where the procedure starts with no independent variables and successively adds the variables with the most significant coefficient until no more variables are admissible at the defined level of significance.

The method of stepwise elimination is a combination of the forward selection and backward elimination procedures. The stepwise elimination procedure begins with a forward selection step (with no variables in the model) with an entry condition of statistical significance set at some level (e.g. > 90%). After the forward step, the statistical significance of the variable coefficients are re-examined and any now insignificant variables are removed in a backward step at some removal level of statistical significance (e.g. > 80%). This process continues until no variables are either added or removed from the model.

The all-subsets method tests how well all subsets of linear models formed from the independent variables (subsets comprised from 1 to the total number of variables) predict the data. This method computes the correlation coefficient, R^2, for each optimal subset. R^2 is a measure of how closely the actual data are predicted by the linear model and is the ratio of the variation between the predicted value and the overall mean (i.e. the variation explained by the model) to the total variation between the actual value and the overall mean. R^2 can vary between 0 and 1 with 1 being a perfect correlation. The set of variables determined from the elimination procedures are compared to the set with the same number of variables from the all-subsets method to see if R^2 is near the maximum (e.g. within 10%). If it is not near the maximum, this indicates that the model is too restricted.

During the selection process it is important to watch for high instances of multicollinearity (as evidenced by the correlation matrix) when variables tend to trend with each other. If this is observed, a variable is not necessarily independent and should be eliminated from the analysis.

The remaining variables can be ranked as to their relative significance according to their "t" statistic.

Consideration of the Overall Model. In situations when the number of data points (N) is not much greater than the number of variables, R^2 can approach 1. At the limiting case, when the number of data points is reduced to equal the number of variables plus 1, R^2 can equal 1. This is true when the variables are independent, and the model can be optimized to perfectly fit the data points. However, in this case, the variables in the correlation no longer have any statistical significance. Therefore, another statistical test is desirable to evaluate the significance of the overall regression model.

The significance of the overall regression model is determined by the F-test which is the ratio of the mean square due to the variation of the regression (between the model prediction and the overall mean) to the mean square due to the variation of the residual (between the model prediction and the actual value). If this ratio is large (greater than 1), the total regression model has high statistical significance. This statistical ratio is distributed as a chi square distribution that has two degrees of freedom: 1) that equal to the number of variables (used in the numerator of F), and 2) that equal to the number of data points minus the number of independent variables minus 1 (used in the denominator of F).

Thus, using the F-test, one can deduce the effect on the significance of a regression model when changing the number of variables and/or the number of data points. For a given number of data points, as the number of variables with low statistical significance is reduced, typically the model statistical significance increases (some times to a maximum) while the goodness of fit (as measured by R^2) decreases. Note that R^2 does not necessarily have to be close to 1 for model coefficients and the total regression model to be statistically significant.

In summary, to deduce the importance of variables, the statistical significance of both the variable and of the total regression is studied along with the overall R^2 value. It is possible that the regression model can have high statistical significance with a low R^2 as well as with a high R^2. Note that a nonlinear model (e.g. a product form) can also be used for this regression optimization to test how robust the linear model is. Ideally, the nonlinear and linear regression forms will yield the same variables as being most significant.

Analysis Using the Spearman Rank Correlation Coefficient

The Spearman rank correlation coefficient, r_s, is a statistic based on ranking values of variables on an ordinal scale (1,2,3,...etc.), and therefore, is a nonparametric technique where no exact amounts or units for the variable are necessary [8]. In contrast to multivariate-based statistical techniques, this considers only associations between the dependent variable and one independent variable at a time. The r_s statistic is derived using the general form of the correlation coefficient, r, which is based on cardinal values (amounts) of x_i and y_i, except r_s is expressed in terms of their rankings denoted as Xi and Yi. The resulting formulation is shown in Table 6 for the case of no ties (i.e. where two or more observations have the same value or rank) with an example for a small data set with only 7 observations where N is the number of observations. Values of r_s may vary from –1 to 1. This example shows the relation of *peak gas rate/ft coal* with *lbs proppant/ft coal* for a hypothetical case with no ties. The resulting r_s is -0.75.

The r_s statistic can be associated with a probability by calculating the number of permutations of Xi and Yi, and tables have been published for small data sets (e.g. 4 to 30 observations) [2]. In the example in Table 6, the absolute value of r_s is 0.75 which, after referring to a one-tailed table of critical values for r_s [2], shows that the probability that *peak gas rate/ft coal* is negatively affected by *lbs proppant/ft coal* is greater than 95%. In this case, r_s is greater than the critical value of 0.714 for 7 observations and p = 0.05.

For data sets larger than 10 observations, r_s may be used in a formulation as shown in eq. 2 that is distributed as Student's t with degrees of freedom of N-2 to yield statistical significance [2].

$$t = r_s \sqrt{\frac{N-2}{1-r_s^2}} \tag{2}$$

Nonparametric statistics have certain advantages over parametric techniques in that they also work for sample sizes less than N=6, and they do not require a knowledge of the population distribution of a particular variable or the actual numeric values as long as they can be ranked by some means. Disadvantages of nonparametric statistics are that for large data sets where meaningful distributions can be obtained, nonparametric statistics are wasteful of data and they cannot test for higher-ordered interactions [8].

In the application to CBM studies, the Spearman rank correlation coefficient has been used for the identification of important CBM variables associated with a Monte Carlo generated production database [9]. In this paper, the Monte Carlo method with a predictive equation is not assumed and the Spearman rank correlation method is applied to situations where little data is available but it is desired to have a statistical assessment.

The relation for the Spearman rank correlation coefficient with ties (i.e. where two or more observations have the same value or rank) is given as:

$$r_s = \frac{\sum x^2 + \sum y^2 - \sum d^2}{2\sqrt{\sum x^2 \sum y^2}}$$

$$\sum x^2 = \sum (X - \bar{X})^2 = \frac{N^3 - N}{12} - \sum T_x$$

$$\sum y^2 = \sum (Y - \bar{Y})^2 = \frac{N^3 - N}{12} - \sum T_y$$

$$d = x - y = X - Y \text{ (since } \bar{X} = \bar{Y}\text{)}$$

$$T_x = \frac{t^3 - t}{12} \text{ and } T_y = \frac{t^3 - t}{12}$$

N = number of observations

t = number of observations tied at a given rank

X = rank of an independent variable

$\bar{X}$ = mean of ranks of an independent variable

Y = rank of the dependent variable

$\bar{Y}$ = mean of ranks of the dependent variable

NOMENCLATURE

a_i	optimized coefficient of the ith variable in the regression model
F	F statistic
N	number of data points (observations
r_s	Spearman rank correlation coefficient
R^2	correlation coefficient
t	Student's t statistic
t_x and t_y	number of ties on X_i and Y_i respectively
X_i	variable in the regression model or rank of variable X
Y_i	dependent variable in the regression model or rank of ith Y value

ACKNOWLEDGMENTS

The authors wish to thank AMVEST OSAGE, INC. for allowing us to use our analyses as applied to their wells in the Cherokee basin and Paul Bruce for his guidance during our analyses and his constructive comments on this paper. The authors acknowledge the 2005 International Coalbed Methane Symposium Executive Planning Committee for organizing this symposium, and we give our appreciation to Nova Hodo for providing guiding materials for producing this paper.

REFERENCES CITED

1. Palmer, I. and Cameron, J., 2003: "Coalbed Methane Completions: Update on Recent Practices", presented at the Coalbed Methane Symposium, Calgary, Canada, 7-9 April, 2003, 45 p.

2.Siegel, S., 1956: Nonparametric Statistics for the Behavioral Science: McGraw-Hill Book Company, Inc., p. 210 – 285.

3. Palmer, I.D., King, N.S., and Sparks, D.P., 1993: "The Character of Coal Fracture Treatments in Oak Grove Field, Black Warrior Basin", *IN SITU*, 17(3), p. 273-309.

4. Palmer, I.D., Lambert, S.W., and Spitler, J.L.1993: "Coalbed Methane Well Completions and Stimulations", Hydrocarbons from Coal, Chapter 14: (ed, B.E. Law and D.D. Rice), AAPG, Tulsa, p. 303 – 339.

5. Olsen, T.N., Brenize, G., and Frenzel, T., 2003: "Improvement Processes for Coalbed Natural Gas Completion and Stimulation," Paper 84122 presented at the SPE Annual Technical Conference and Exhibition, Denver, October 5-8, 2003.

6. Steidl, P.F., 1991: "Observations of Induced Fractures Intercepted by Mining in the Warrior Basin, Alabama", GRI Topical Report 91/0327, 38 p.

7. Clarkson, C.R. and McGovern, J.M., 2003: "A New Tool for Unconventional Reservoir Exploration and Development Applications", Paper 0336 presented at the 2003 International Coalbed Methane Symposium held in Tuscaloosa, AL, May 5-9, 33 p.

8. Siegel, S., 1956: Nonparametric Statistics for the Behavioral Science: McGraw-Hill Book Company, Inc., p. 32 – 34.

9. Roadifer, R.D., Moore, T.T., Raterman, K.T., Farnan, R.A., and Crabtree, B.J., 2003: "Coalbed Methane Parametric Study: What's Really Important to Production and When?", paper SPE 84425 presented at the SPE Annual Technical Conference and Exhibition, Denver, Oct. 5-8, p. 1-11.

TABLES

Table 1

Maximum (Peak) Gas Rate/ft of coal
(15 "independent" variables selected)

(Controllable)
1. Frac Fluid Type = type of fracturing fluid coded in order of increasing viscosity
2. No. seams/stage = number of treated seams (corrected for non-producing shales & sandstones)
3. No. Prod. Seams = number of producing coal seams at time of max. production
4. lbs sand/ft coal = total lbs sand pumped for all stages divided by total net coal pay
5. Avg. Sand Conc. = average of all average-stage sand concentrations (ppg)
6. Total Frac Fluid = total bbls of fracturing fluid pumped for all stages per well
7. Total Sand = total lbs of sand pumped into a well for all stages
8. SL Fld Loss = 30 min API cement slurry fluid loss (cc)
9. SL ECD/Min ISIP grad = cement slurry equiv. circ. density / min. Instant. shut in pressure gradient
10. Proppant Size = the size of proppant, coded as 1 = 20/40, 2 = 16/30 , and 3 = 20/40 & 12/20

(Not Controllable)
11. Lost Circ. Cement = lost circulation during cementing coded as 0 = none and 2 = reported
12. Percent stages S.O. = % of all stages screened out (0 to 100%)
13. Max ISIP Grad. = maximum instantaneous shut in pressure gradient for all stages in a well (psi/ft)
14. Min ISIP grad = minimum instantaneous shut in pressure gradient for all stages in a well (psi/ft)
15. Net Pay = sum of net coal pays for all seams treated in a well (ft)

Table 2

Breakdown Pressure
(9 "independent" variables selected)

(Controllable)
1. SL Fld Loss (cc) = 30 min API cement slurry fluid loss
2. SL Dens (ppg) = cement slurry density in pounds per gallon
3. SL ECD (psi/ft) = cement slurry equivalent circulating density
4. SL ECD/ISIP grad = ratio of cement slurry ECD to instantaneous shut in pressure gradient
5. No. Perf Int./frac = number of perforated intervals/fracturing stage
6. Perf. Interval (ft) = total length in feet of all perforated intervals treated in a particular stage

(Not controllable)
7. Treating Pressure (psi) = average treating pressure during a hydraulic fracturing stage
8. ISIP grad (psi/ft) = instantaneous shut in pressure gradient after a hydraulic fracturing stage
9. Cement Lost (bbls) = barrels of cement lost to the formation during cement displacement

Table 3

Analysis of Variables Important to Maximum Gas Rate/ft coal
(Field A: 16 wells, 9 variables)

=> 9 variables significant ; regression significant to 92.4% as per the F Test ; R^2 = 0.84

		Variable Significance	Variable Coefficient (Mcfd/ft coal/ variable units)	95% Confidence Interval
SL Fld Loss -cc	Controllable	99.44%	0.2547	0.107 to 0.403
Total Sand-lbs	Controllable	99.09%	-0.002416509	-0.00398 to -0.000856
No. seams/stage	Controllable	99.07%	-47.53	-78.4 to -16.64
No. Prod. Seams*	Controllable	97.32%	16.68	2.67 to 30.7
lbs sand/ft coal*	Controllable	97.19%	0.009655489	0.00145 to 0.0179
Frac Fluid Type	Controllable	90.82%	10.74	-2.37 to +23.8
Min ISIP grad	Not Controllable	90.85%	50.78	-11.1 to +113.
SL ECD/Min ISIP grad	Not Controllable	87.25%	25.75	-9.89 to +61.4
% stages SO	Not Controllable	81.32%	-0.3606	-0.953 to +0.232

* not consistent across other fields

Further variable elimination showed the following 4 variables gave the optimal regression of 95.5% significance as per the F test; R^2 =0.56

SL Fld Loss -cc No. seams/stage
Total Sand-lbs lbs sand/ft coal

Table 4

Analysis of Variables Important to Breakdown Pressure
(All 4 Fields: 138 wells, 7 variables)

=> 3 variables are significant and consistent; regression significant to 99.99% as per the F-Test; R^2 = 0.214

		Variable Significance	Variable Coeff. (psi/var. units)	95% Confidence Interval
significant and consistent:				
Treating Press - psi	Not Controllable	99.996%	0.59	0.32 to 0.87
ISIP grad -psi/ft	Not Controllable	81.32%	-395	-980 to +190
Perf. interval-ft	Controllable	75.82%	-26	-70 to +20
not consistent between fields:				
SL Dens - ppg	Controllable	90.00%	-384	-843 to +74.6
Cement Lost - bbls	Not Controllable	80.08%	9.49	-5.06 to +24.0
Secondary importance:				
SL Fld Loss - cc	Controllable	42.46%	-0.325	-1.47 to +0.82
SL ECD/ISIP grad	Controllable	30.09%	-216	-13212 to +889

Table 5

Correlation of treatment and formation variables with Peak Production/ft-coal										
Piceance Basin Well	lbs Proppant / ft coal	Ordinal Rank	Total Fluid (bbls)	Ordinal Rank	Est. Coal Strength UCS (psi)	Rank	Net Treating Pressure (psi)	Ordinal Rank	Max. MCFD/ft-coal	Ordinal Rank
Well A	3912	1	2854	2	1305	2	858	4	22.58	7
Well B	5736	2	1834	1	1885	6	1054	6	21.74	6
Well C	7356	6	19493	6	1250	1	120	2	13.78	5
Well D	7024	3	19638	7	1500	4	153	3	10.72	4
Well E	7304	5	16171	5	1353	3	1868	7	8.60	3
Well F	7867	7	4659	4	1305	2	91	1	8.42	2
Well G	7032	4	3388	3	1625	5	1007	5	4.40	1
r_s =	-0.643		-0.286		-0.088		0.036			
Strength of correlation	moderately strong negative correlation		weak negative correlation		no correlation		no correlation			
Implications	Suggests using less proppant/ft-coal		Does not support the use of large pads.		Fines may not be a problem		Fracture breakout from coal has no observable impact from these data			

Table 6

Spearman Rank Correlation Coefficient: Example

Formulation without ties:

$$r_s = 1 - \frac{6\sum(X_i - Y_i)^2}{N^3 - N}; \quad \text{where } N = \text{no. of observations}$$

N	**7**						
Peak gas (mcfd)/ft coal	4.4	8.42	8.6	13.78	16.05	21.74	22.57
Yi =Peak gas Ranking	1	2	3	4	5	6	7
x Variable (lbs prop./ft coal)	7032	7867	7304	7356	7024	5736	3912
Xi = Prop. Loading Ranking	4	7	5	6	3	2	1
(Xi-Yi)^2	9	25	4	4	4	16	36
tx=ties on Xi	0	0	0	0	0	0	0
ty=ties on Yi	0	0	0	0	0	0	0
rs =	**-0.750000** without ties						

FIGURES

Figure 1

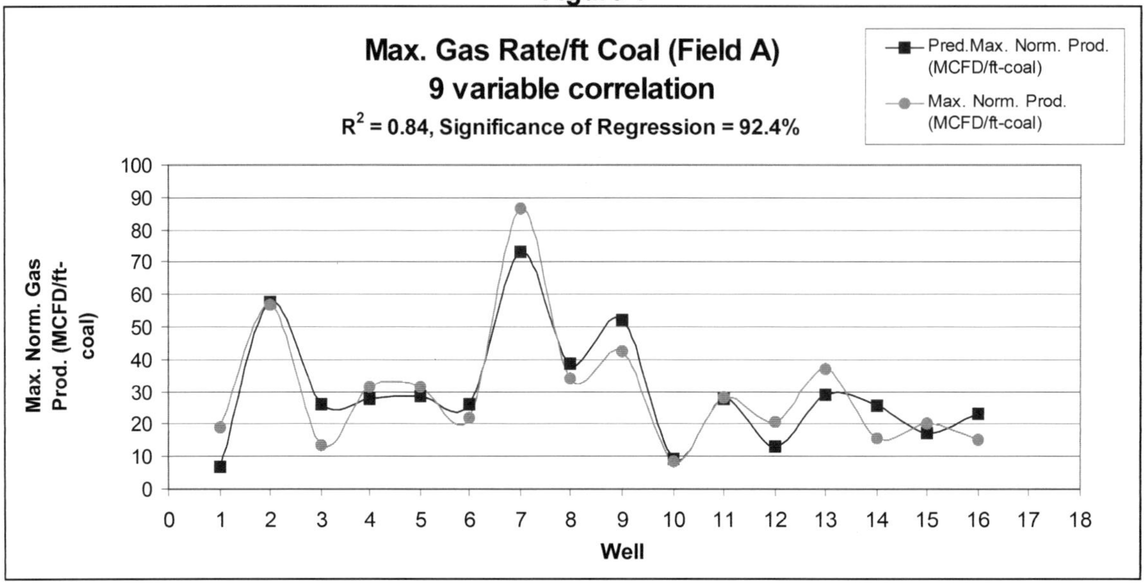

Figure 2

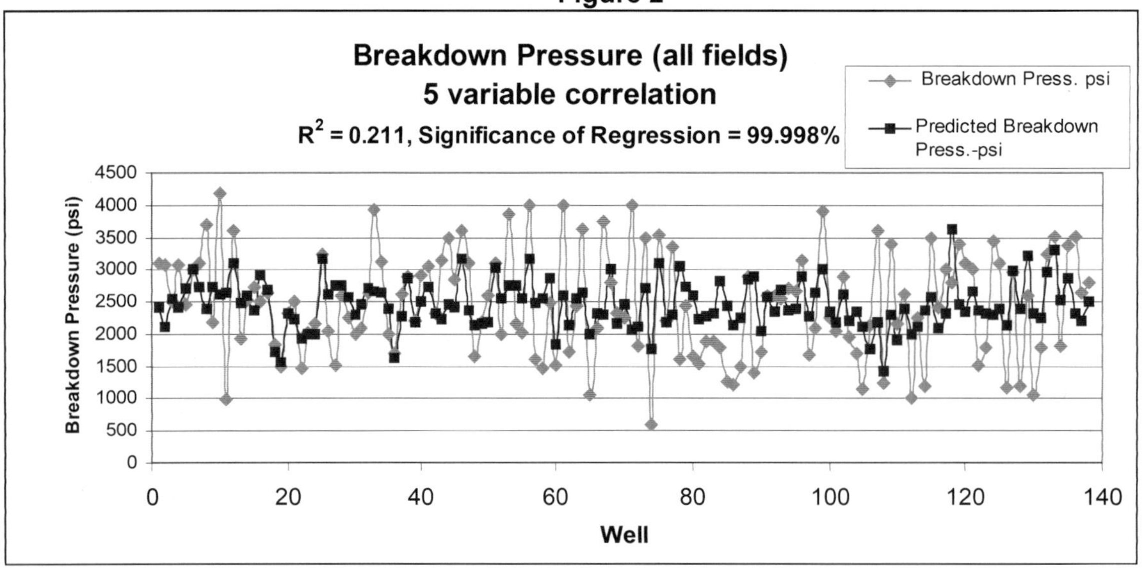

0517

METHODOLOGY AND EXAMPLES OF WELLBORE STABILITY IN COALBED METHANE WELLS

Zissis A. Moschovidis and John R. Cameron, PCM Technical, Inc., and Ian D. Palmer, Higgs Technologies, L.L.C.

ABSTRACT

Wellbore stability (WBS) and formation failure impact the economics of drilling and producing coalbed methane (CBM) wells, especially horizontal wells. The methodology for predicting WBS is based on comparing a measure of the intensity of the in situ stresses for given drilling conditions to a measure of coal strength. These calculations are most easily performed using a failure envelope, given in terms of the stress invariants, which are analogous to normal stress and shear stress. A methodology was devised to estimate the failure envelope of coals (i.e., failure at different loading conditions) from a strength parameter such as UCS (unconfined compressive strength), which can be correlated to coal rank or HGI (Hardgrove grindability index). We were not able to relate the strength of coal to logs, as could be done for sands.

The failure envelope, bottom-hole pressure, wellbore trajectory, and the in situ stresses allow the prediction of shear failure, which is associated with wellbore instability. WBS predictions for horizontal wells in coals of different ranks shows that MV and LV coals are the most susceptible to failure, but HVA coals are much less so. In general, a weak trend appears that wellbore stability deteriorates with increasing depth, but exceptions occur since weak coals seams exist in both shallow and deeper CBM plays. Furthermore, increasing well inclination and reservoir depletion tend to deteriorate CBM wellbore stability. The examples presented, are based on typical in situ stresses encountered in CBM plays in the USA. However, general application of these results in not recommended, because WBS in greatly affected by local in situ stresses and strength.

INTRODUCTION

Weak or "crumbly" coals exist in both shallow and deeper CBM plays and therefore, wellbore stability can become an issue. In particular, horizontal or highly inclined CBM wells may collapse, and it may be impossible to insert a liner. In relaxed basins, where the overburden stress is greater than the horizontal stresses, highly inclined wells tend to be less stable than vertical wells. Furthermore, CBM wells are often drilled under-balanced (pressures as low as 2.8-3.2 ppg can be reached) to avoid formation damage. This also causes WBS deterioration, since the wellbore is not sufficiently supported by the drilling fluid (e.g., when drilling with air).

An additional concern is drilling in depleted reservoirs. In weak sands, both drawdown and depletion can induce sanding, and the same principles apply in coal. When coal fails in shear, wellbore stability problems may occur. In fact it is postulated that shear failure of the coal would likely cause hole problems and the drilling mud-weight should be such as to avoid it.

Furthermore, during gas production, drawdown and depletion can cause shear failure, which could induce fines production that could contribute to operational problems. For both drilling and production stability the same methodology as described briefly in this paper is applicable. The only difference is that the strength criteria for drilling and production are different. In this paper, our discussion will be applied to drilling stability.

The purpose of the present study is to present selected wellbore stability results for horizontal (and inclined) wells drilled in coal seams of various rank, and illustrate the complexity of the drilling stability problem in coals. Typical in situ stresses encountered in five USA CBM basins were used for our results. Our in-house wellbore stability model was used to derive these results. The model requires the failure envelope, the in situ stresses, the reservoir pressure, and wellbore inclination and azimuth, as inputs. The output is the "safe mud" window of down-hole pressures during drilling to avoid shear failure in the coal. The safe mud window is defined by the upper bound corresponding to the tensile-shear failure of the formation, and the lower bound corresponding to compressive-shear failure of the formation.

The data needed for a wellbore stability study are determined from various sources:

1. Coal strength (failure envelope) from core tests, when available (ideal method) or from correlations based on coal rank (approximate method) when laboratory data are not available.
2. In situ stress magnitudes (minimum and maximum horizontal stresses, overburden and pore pressure) and directions from logs (calibrated to leak-off tests), oriented calipers or borehole images, etc.
3. Drilling conditions in vertical wells, e.g., actual mud-weight, lost returns, hole sloughing and washouts to calibrate the wellbore stability model prediction to actual field conditions.

For the present study, typical in situ stresses (overburden, equal horizontal stresses, and reservoir pressure) encountered in five CBM basins in the USA were used. The basins are denoted A, B, C, D and E in order of increasing depth. The strength of the coal was varied depending on coal rank, assuming that coals of all ranks are encountered in each basin. Specifically, for each rank considered (i.e., LV, MV, HVA, HVB), the lower bound for UCS was estimated based on correlations derived from data available in the literature based on VM-daf% (volatile matter, dry-ash-free %, which is determined from Proximate analysis). Then the failure envelope (FE) for coal was estimated using our own proprietary method (based on published coal strength data). Note that we did not mention specific basin names and depths, because wellbore stability results are sensitive to local conditions of stress and strength, and application of the results of this paper is not expected to be valid in general for these basins.

IN SITU STRESSES

The in situ stress-gradients for overburden, horizontal stress and reservoir pressure were assumed for each basin and are shown in all the figures herein (e.g., see Figs. 2, 3 and 4). Specifically we used:

- overburden gradient of 1 psi/ft (= 19.231 ppg) in all basins , which is a good average of the earth density .
- pore pressure gradient of 0.45 psi/ft (= 8.654 ppg) in all basins, which is a good average for normal pressure gradient.
- horizontal stress gradient different in each basin as shown for example in Fig. 2 (1 ppg = 0.052 psi/ft). Specifically we have used the following gradients: basin A: 0.66 to 0.69 psi/ft; basin B: 0.78 to 0.80 psi/ft; basin C: 0.63 to 0.65 psi/ft; basin D: 0.73 to 0.76 psi/ft; and basin E: 0.81 to 0.82 psi/ft.
  Note that two depths were selected in each basin, an upper depth and a lower depth (referred by indices 1, 2 respectively) representing the depth range of the basin. For each depth we used different minimum horizontal stress (Sh) to represent the range of stresses encountered in each basin.
- Furthermore we assumed that the horizontal stresses are equal in all directions, a good approximation for most CBM plays.

DETERMINATION OF COAL STRENGTH AND FAILURE ENVELOPE

UCS limits of coal based on rank

Coal strength is required to assess wellbore stability in coals, fines production in reservoir and near-wellbore, and cavity-completion stability. The UCS is often used as a strength index, since it can be used to predict the strength of coal under varying applied stress conditions (e.g., the failure envelope) required for stability analyses. Another strength index is HGI, the Hardgrove Grindability Index. The HGI is used in the mining industry to assess how easily coal can be ground for burning.

We have searched the literature in the hope of identifying data for a correlation between UCS (or HGI) and any of the usually available logs (such as density, or sonic logs); however we did not find such information. We also tried to predict UCS from the density log, estimating the ash content and relating it to strength. This attempt also failed because there is no consistent trend of strength with ash content in the published data. Ash was not an indicator of strength even within coals of a specific rank. Instead, we found a database (Gomez and Hazen [1]) that related the HGI to the Proximate analysis data, sulfur content, and heating value of coal.

We also identified a good correlation between HGI and UCS (Bell and Jones [2] and Lowry [3]). This allowed us to establish upper and lower limits of UCS based on coal rank. Table 1 shows the likely minimum UCS strength based on rank, since coals are very heterogeneous and the minimum of the UCS range is likely to be encountered in each well.

Based on this table, we considered 3 cases of UCS for our wellbore stability analyses:

- UCS = 490, applicable for LV or MV coals
- UCS = 2050, applicable for HVA coals, and
- UCS = 4800, applicable for HVB coals.

Failure Envelope of Coal

The failure envelope of rocks is normally determined by laboratory tests used to measure the strength of the rock at various conditions. Several tests are usually required, such as unconfined compression (UCS), compression at various confining pressures, and Brazilian tests.

A failure envelope capability is implemented into our wellbore stability software to predict the failure envelope for rocks and coals of different UCS. The coal failure methodology was based on coal strength data available in the public domain [4, 5] where core samples of several US mines were tested. These data were used to derive a very good correlation of the failure envelope of coals as a function of UCS. We have assumed that such general strength correlation is also applicable for this study. The following relation was implemented in our wellbore stability software:

$$y = A_0 + A_1 x + A_2 x^2 \text{, with } x = J_1 \text{ and } y = \sqrt{J_2} \qquad (1)$$

where, J_1 and J_2 are the stress invariants of the stress tensor (see Appendix) and the coefficients A_0, A_1, and A_2 are functions of UCS. Figure 1 shows examples of failure envelopes predicted by our model. Note that the failure envelope has a bigger curvature for lower UCS, while for larger values it could be approximated as linear.

WELLBORE STABILITY

Methodology

Wellbore instability occurs when the in situ stresses next to the wellbore exceed the strength of the formation. Loss of returns occurs (tensile failure) due to excessive mud-weight, and wellbore collapse (shear failure) occurs due to insufficient mud-weight. The window of safe mud-weight for drilling operations can be calculated using analyses of various complexity (e.g., elastic, elastoplastic, etc). Due to uncertainties in the estimation of in situ stresses and in the strength behavior of the formation, it is necessary to calibrate the predictions against field observations from the drilling and production history of the field. Based on our experience and from theoretical studies, we recommend the use of the simpler elastic analysis, calibrated to field observations for reliable practical wellbore stability predictions.

The objective of the present WBS analysis is to determine the safe bottom hole pressure, expressed as a safe mud-weight "window" (range) for drilling. The analysis uses software that calculates the first and second invariants (J_1 and $\sqrt{J_2}$) of the effective stresses near the wellbore for different mud-weights and compares it with a curvilinear failure envelope to determine the mud-weight range that does not induce formation failure (see the Appendix for a theoretical description). Formation failure is associated with wellbore instability. Although shear failure does not always cause catastrophic failure of the wellbore (e.g., hole collapse, stuck pipe, loss of returns) in more continuous formations, such as sands and shales, we think that shear failure is more likely to be catastrophic in coals because of the presence of the cleats, which could cause chunks of the coal to be detached.

When drilling coal wells, every effort is made not to damage the cleat system and impair production. Therefore drilling is usually done using water or air, often under under-balanced conditions. In this study, we assume the well will be drilled with water, as is often done in coals to limit the damage to sensitive cleats (fractures) which provide the gas flow into the well. Using water means that no effective filter-cake (mud-cake) is formed and that the wellbore pressure is communicated in the formation (i.e., there is no significant pressure drop at the wellbore face). Also, when wells are drilled with air or under-balanced, there is no filter-cake formed. Under these conditions the stabilizing presence of the filter-cake is absent and the wellbore stability conditions for either drilling or production are the same. Consequently our analysis is applicable to both drilling and production stability.

Usually stability calculations are performed for a formation element at a small distance from the wellbore face inside the formation, to reduce the inherent conservatism of the elastic WBS calculations. R_w / R = 0.9 is usually used, where R is radial coordinate measured from the wellbore axis and R_w is the wellbore radius. This assumes that some material on the surface, even though it may fail, remains in place and protects the material further inside the formation from failure. However for the stability of friable coals, the failed material may quickly be detached from the wellbore surface. Therefore Rw / R = 1 was used in our analysis, which means that failure of the formation at the wellbore surface implies drilling instability (most conservative assumption).

Wellbore Stability for Horizontal Wells in Coals

For the wellbore stability analysis we have used real in situ stresses encountered in the five CBM basins in USA considered, as described in a previous section. We assumed that the stability of all seams is governed by the likely weakest strength for the coal rank and estimated the failure envelope based on UCS (see Table 1).

Table 2 summarizes the results of our analysis for horizontal wells (in general the most unstable case). It shows the minimum and maximum allowable mud weight for safe drilling of horizontal wells in the CBM basins A, B, C, D and E, where the seams occur at increasing depth. For each basin the top and bottom depths in their depth range were analyzed, denoted by subscripts 1 and 2 respectively. The same results are also presented in Figs. 2, 3 and 4. The in situ stress gradients are shown in units of lbs/gallon (ppg), instead of psi/ft (1 ppg = 0.052 psi/ft). The minimum and maximum mud-weight curves define the window of safe operations. By convention in our software, when these curves simultaneously go to zero it means that there is no safe mud-weight window for stability, i.e., that the wellbore will be failed for all possible values of mud-weight. This is the case for LV or MV coals in basins C, and D (see C1, C2, D1 and D2 in Table 2). For all other cases, there is a safe mud-weight window for stable drilling operations.

The LV and MV coals could have the smallest strength, and therefore should be checked for wellbore stability. However, it is the strength in relation to the in situ stress at the wellbore wall that determines stability and general conclusions can not be reached from our results. For example horizontal wells in LV or MV coals in basins A and B (see A1, A2, B1, and B2 in Table 2) are expected to be stable according the analysis and they could be drilled under-balanced. Then the deeper basins C and D have no safe mud window. However, horizontal wells in the deepest E basin (see E1 and E2 in Table 2) should be drilled overbalanced for stability (min MW > pore pressure gradient of 8.65 ppg). In general, from Table 2 a weak trend appears that wellbore stability deteriorates with increasing depth, but numerous exceptions occur since weak coals seams do in reality exist in both shallow and deeper CBM plays, and wellbore stability depends on stress relative to strength.

Effect of Reservoir Depletion on Wellbore Stability

Figure 5 shows the variation of the safe mud-weight window for wells in the E basin (deepest basin) for LV or MV coals. The safe mud-weight window closes as the reservoir is depleted. The lower and upper mud-weight lines converge and define a "triangle" of stability. The apex of the triangle defines the critical reservoir pressure, below which any wellbore pressure (mud-weight) would cause instability. For the E1 case this occurs at a reservoir pressure gradient of 1.21 ppg (= 0.063 psi/ft) for which minMW = maxMW = 12.2 ppg (= 0.634 psi/ft).

In this particular example (E1) the stability triangle is located entirely above the reservoir pressure, which means that horizontal wells in the upper depth of the E basin (Fig. 5) should be drilled over-balanced for stability at any reservoir pressure. In general, depending on the strength and the in situ stresses, the stability triangle could be intersected by the reservoir pressure line. In this case, the stability triangle also could define the reservoir depletion beyond which under-balanced drilling may not be feasible.

Figure 6 shows the stability triangle for the lower depth of the E basin (E2). The stability triangle is very small and the critical reservoir pressure gradient is 8.31 ppg (0.432 psi/ft), slightly smaller than the current reservoir pressure (8.65 ppg = 0.45 psi/ft). A horizontal well in this case can be drilled only for undepleted reservoir pressure and with high mud-weight of 13.8 ppg (= 0.718 psi/ft) to prevent failure of the coal.

Effect of Well Inclination on Wellbore Stability

Figures 8 and 9 show the effect of well inclination of the stability of wells in the E basin for LV or MV coals. In general vertical wells are more stable than horizontal wells for the typical case when the overburden stress is larger than the horizontal in situ stress.

Figure 7 shows that the minMW is smaller than the pore pressure gradient for inclinations 0 to ~76 degrees, allowing for safe under-balanced drilling. For inclination higher than ~76 degrees under-balanced drilling is not feasible. The maxMW line is below the horizontal stress gradient (fracturing gradient) for all well inclinations, and for some inclinations it is substantially lower (e.g. an approximately 60-degree inclined well is the most susceptible to shear failure in tension). This means we cannot assume in deviated wells that mud weight can be increased to the fracturing gradient.

Figure 8 shows a very interesting example of wellbore stability in the E basin. The plot shows that there is a safe mud-weight window for inclinations 0 to ~58 and ~86 to 90 degrees. But for inclinations ~59 to ~85 degrees there is no safe mud weight window, indicating that these inclinations are the worst for stability in this deepest basin example. In this case, although a horizontal well is less stable than a well of small inclination, it is not the least stable. This type of behavior is because the shear failure of the wellbore is affected by all three principal in situ stresses, i.e., the contribution of the intermediate stress is also included in the failure criterion (see Appendix).

SUMMARY AND CONCLUSIONS

We have not discussed how wellbore stability depends on well azimuth, because we assumed the two horizontal stresses were equal. However, our model can account for unequal horizontal stresses and predict the effect of well azimuth on the mud-weight window. In general for small horizontal stress anisotropy, the well azimuth has a smaller effect as compared with well inclination. Our model predicts the mud-weight window benefits of drilling a well in the direction of minimum horizontal stress, which is the most stable azimuth direction, when the overburden stress is greater than the horizontal stresses.

The above discussion showed that wellbore stability predictions can not be scoped using heuristic arguments or analogies with field experience in other fields. The stability results can not be generalized because they strongly depend on local coal strength, in situ stresses, and wellbore inclination (and azimuth for un-equal horizontal stress).

In general, a weak trend (with several exceptions) appears to exist that deeper coals tend to be more unstable. For typical relaxed CBM plays (where the overburden stress is greater than the horizontal stresses), increasing reservoir depletion and wellbore inclination tend to deteriorate wellbore stability. It is recommended that WBS studies be performed using calibrated in situ stresses and actual coal strength for CBM wells, especially high inclination wells that intersect LV or MV coals.

APPENDIX: WELLBORE STABILITY METHODOLOGY

Wellbore stability is affected by both mechanical and chemical effects. Chemical effects are expected to be primarily significant in shales, which are chemically active and require a chemically compatible fluid to avoid hole sloughing. Sands and coals are expected to be chemically more inert than shales (depending if they contain clays) and therefore, we assume that wellbore instability in coals (and sands) is primarily a mechanical phenomenon. The present study addresses wellbore stability in coals caused by the mechanical failure of the formation, and we assume that any chemical instability effects have been eliminated by appropriate use of drilling fluid chemistry. The mechanical wellbore stability was addressed using failure theory (failure envelope) by Von Mises, which accounts for the effects of all principal stresses (see following section).

PCM Technical, Inc. has developed software for wellbore stability based on the elastic solution for the stress field induced by the wellbore (see for example the textbook by J. C. Jaeger and N. G. Cook, Fundamentals of Rock Mechanics [6]). In the following we present the essential elements of the methodology used to determine formation failure next to the wellbore. In general wellbore instability is assumed to be likely when the formation is at failure. However, formation failure does not always result in catastrophic wellbore collapse or inability to drill a well. Attempting to increase the reliability of wellbore failure predictions, complex studies using finite element analyses with elastic-plastic formation behavior were tried by numerous investigators, however these did not achieve their objective. It was found that the complexity of analysis did not increase the accuracy of the predictions of failure and the analyses were more costly and required additional material behavior parameters. Consequently, the industry widely applies elastic failure theory and calibrates the results with field observations.

To assess the mechanical failure of the formation near the wellbore at a given depth, the J_1 and $\sqrt{J_2}$ stress invariants are calculated for the in situ stress state and wellbore orientation as a function of the wellbore pressure (mud-weight) The resulting functions define a parametric curve (the effective stress trajectory) in the stress invariant space, J_1 vs. $\sqrt{J_2}$ (see schematic of Fig. 9). To determine the wellbore pressures that cause failure, the failure envelope of the formation is also plotted in the same coordinates. A stress state (J_1, $\sqrt{J_2}$) that is plotted below the failure envelope does not cause the formation to fail, and conversely a state that is plotted above the failure envelope causes formation failure.

The intersection of the stress trajectory with the failure envelope defines a “window” of wellbore pressure or mud-weight for stable wellbore (the formation does not fail) at a given depth.
Two intersection points, F and S, are generally expected:

- Point F indicates failure of the formation due to a high wellbore pressure that induces tensile hoop stresses at the wellbore, which cause either a shear or pure tensile failure (depending on the weakest formation strength, either in shear or in tension), i.e., failure due to hydraulic fracturing conditions.
- Point S indicates failure in shear due to lower wellbore pressure that induces high compressive hoop stresses, which cause shear failure (since shear strength is lower than compressive strength in rocks), i.e., failure of the formation due to insufficient mud weight during drilling or excessive drawdown during production operations.

The wellbore pressure or corresponding mud-weights within the interval defined by points S and F are safe for drilling and production operations. Depending on the operating conditions, the bottom hole pressure may never reach the points of failure F and S. For some weak formations, it is possible for the effective wellbore stress trajectory in Fig. 9 to be entirely above the failure envelope, i.e., there are no intersection points F and S, and therefore no wellbore pressure exists to avoid formation failure.

It is generally assumed that formation failure would cause wellbore problems (loss of returns or hole collapse) and should be avoided for safe drilling or production operations. However, formation failure in many cases does not cause catastrophic wellbore problems. For example the use of drilling mud aims at preventing loss of returns by forming a "thin and flexible" filter-cake. Furthermore weak ductile type formations, do not collapse when they fail but gradually deform, a problem that can be successfully addressed during the limited time they may by exposed, until installing a casing, gravel-pack, etc.

Stress Invariants

The severity of a given loading on some material point can be quantified by the invariants of the induced stress tensor (stress components). The stress invariants are independent of the coordinate system used for the calculations and allow the comparison of stress states of different stress components. The inference is made here, that stress states of the same stress invariants, although they are different, they are equally severe on the material from the point of view of failure. It is, therefore, possible to compare a given state of stress with known stress states that can be realized in the laboratory.

Failure stress states can be determined by performing simple experiments such as pure tension, compression, TWC (thick wall cylinder), and Brazilian tests. The results of these failure tests may be reported as a curve (failure envelope), which relates the normal and shear stresses that must exist at some point in the material to cause failure. The failure envelope can be determined as the osculating line to the Mohr's circles of all failure stress states. An alternative way is to report the failure envelope as a functional relation between the first invariant, J_1, and the square root of the second invariant, $\sqrt{J_2}$, of the stress tensor. Both methods are equivalent, since the Mohr's circles are also invariant with respect to coordinate transformations.

The invariants J_1 and $\sqrt{J_2}$ are related to the normal and the shear stresses on the octahedral planes, i.e., the planes that make the same angle with the principal directions of the stress tensor. The definitions used herein for the first and second invariants of the stress tensor, σ_{ij}, where i, j = 1, 2, 3 are:

$$J_1 = (\sigma_{11} + \sigma_{22} + \sigma_{33})/3$$
$$J_2 = [(\sigma_{11} - \sigma_{22})^2 + (\sigma_{22} - \sigma_{33})^2 + (\sigma_{33} - \sigma_{11})^2]/6 \quad (2)$$

REFERENCES CITED

1. Gomez Manuel and Hazen, Kathleen, "Prediction of Coal Grindability from Exploration Data", U.S. Dept. of the Interior, Bureau of Mines Report 7421.
2. Bell, G. J. and Jones, A. H. "Variation of mechanical strength with rank of gassy coals", Procs. 1989 Coalbed Methane Symp., Paper No. 8924, Tuscaloosa, AL, April 1989.
3. Lowry, H.H. Editor, Chemistry of Coal Utilization, Chapter3, John Willey &sons.
4. Jones et al "A review of the physical and mechanical properties of coal with implications for coalbed methane well completion and production", in Geology and Coalbed Methane Resources of the Northern San Juan Basin (ed. J. E. Fassett), Colorado and New Mexico Rocky Mtn. Assoc. Geologists, Denver, 1988.
5. Abass, H.H., Van Domelen, M. L., El Rabaa, W. M., "Experimental observations of hydraulic fracture propagation through coal blocks", SPE 21289, SPE Eastern Reg. Mtg., Columbus, Ohio, Oct-Nov, 1990.
6. Jaeger ,J. C. and Cook , N. G., Fundamentals of Rock Mechanics.

TABLES

Table 1: Likely min UCS of coals based on rank

Coal Rank (decreasing order)	VM-daf%	Likely min UCS (psi) Based on literature data
Anthracite	Less than 14	1780
LV	14 to 22	490
MV	22 to 31	497
HVA	31 to 39	2050
HVB	39 to 42	4800
HVC	42 to 55	5930
Lignite	Greater than 55	9400

Table 2: Safe Mud-weight Window for Horizontal Wells in CBM Basins of USA
(coded A to E in order of increasing depth; subscripts 1 and 2 denote upper and lower depth in the basin)

				LV or MV Coals (UCS=490 psi)		HVA Coals (UCS=2050 psi)		HVB Coals UCS=4800 psi	
Basin-depth	Overburden (ppg)	Res. Pressure (ppg)	Min Hor. Stress (ppg)	minMW (ppg)	maxMW (ppg)	minMW (ppg)	maxMW (ppg)	minMW (ppg)	maxMW (ppg)
A1	19.23	8.65	13.27	0.00	8.80	0.00	13.20	0.00	13.20
A2	19.23	8.65	12.69	0.00	7.80	0.00	11.80	0.00	12.60
B1	19.23	8.65	15.39	0.00	12.80	0.00	15.20	0.00	15.20
B2	19.23	8.65	15.00	0.00	12.20	0.00	14.80	0.00	14.80
C1	19.23	8.65	12.50	0 (no safe window)	0 (no safe window)	0.00	10.60	0.00	12.40
C2	19.23	8.65	12.12	0(no safe window)	0 (no safe window)	0.00	9.80	0.00	11.80
D1	19.23	8.65	14.62	0 (no safe window)	0 (no safe window)	0.00	13.60	0.00	14.60
D2	19.23	8.65	14.04	0(no safe window)	0 (no safe window)	0.00	12.80	0.00	13.80
E1	19.23	8.65	15.83	9.40	14.40	0.00	15.20	0.00	15.80
E2	19.23	8.65	15.53	13.80	14.00	0.00	14.80	0.00	15.40

FIGURES

Figure 1: Generalized Failure Envelopes for Coal Seams based on UCS

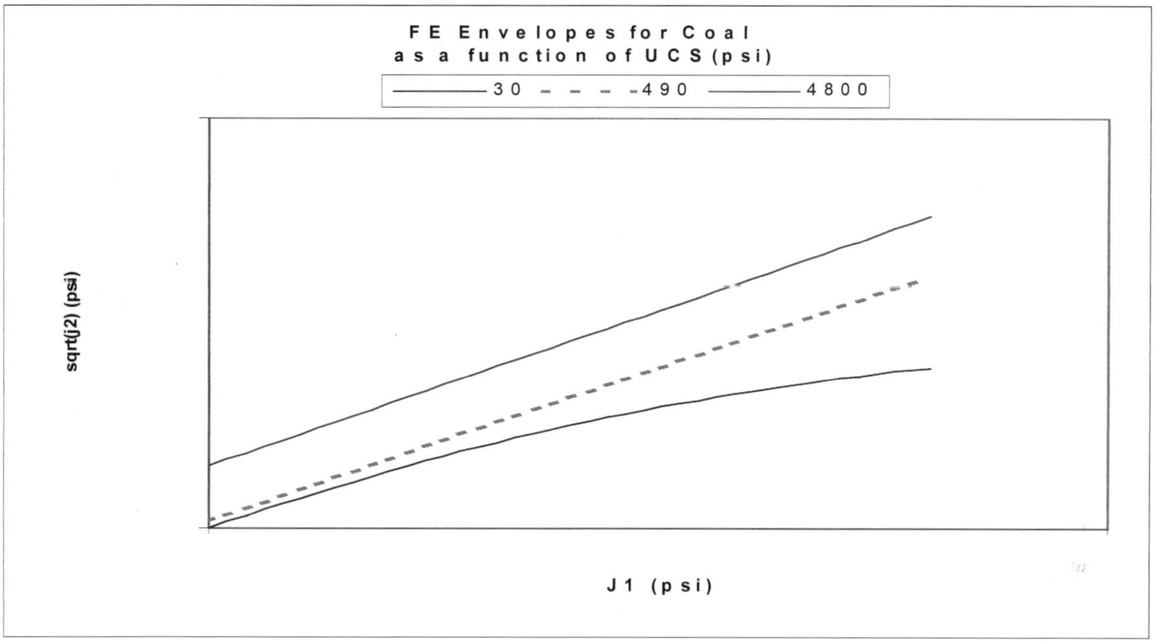

Figure 2

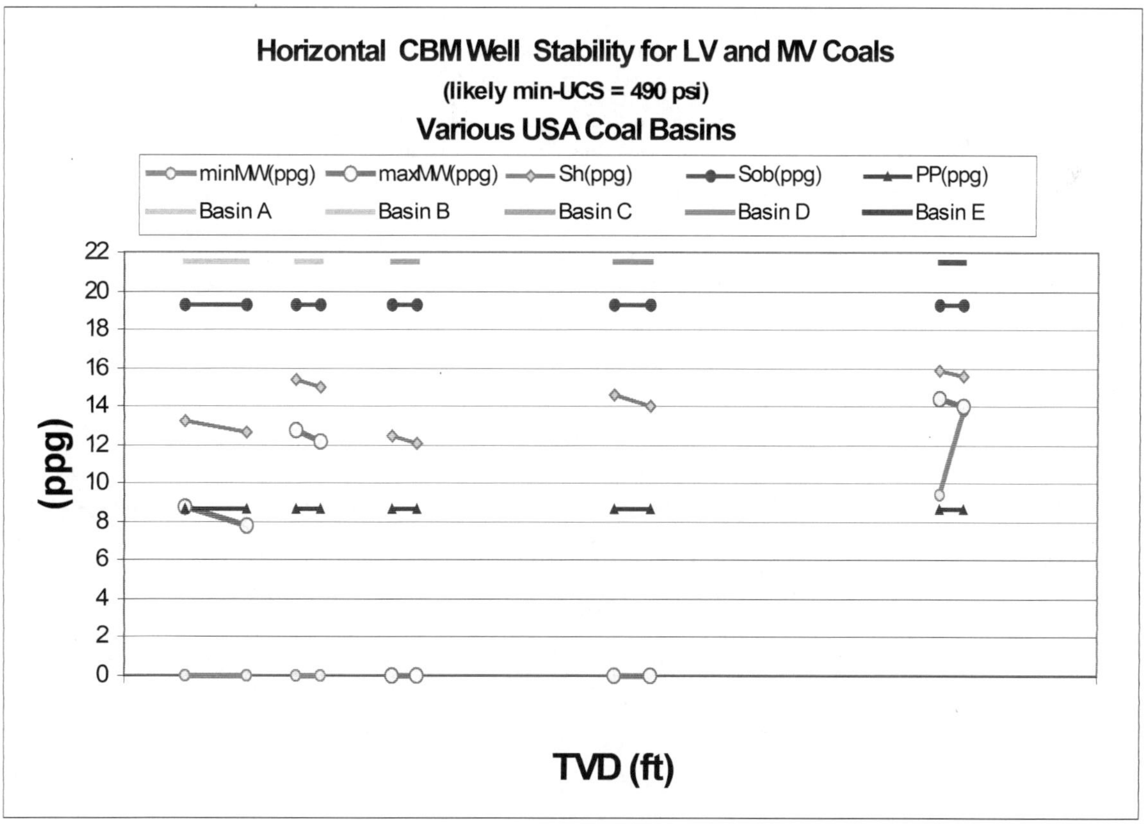

Figure 3

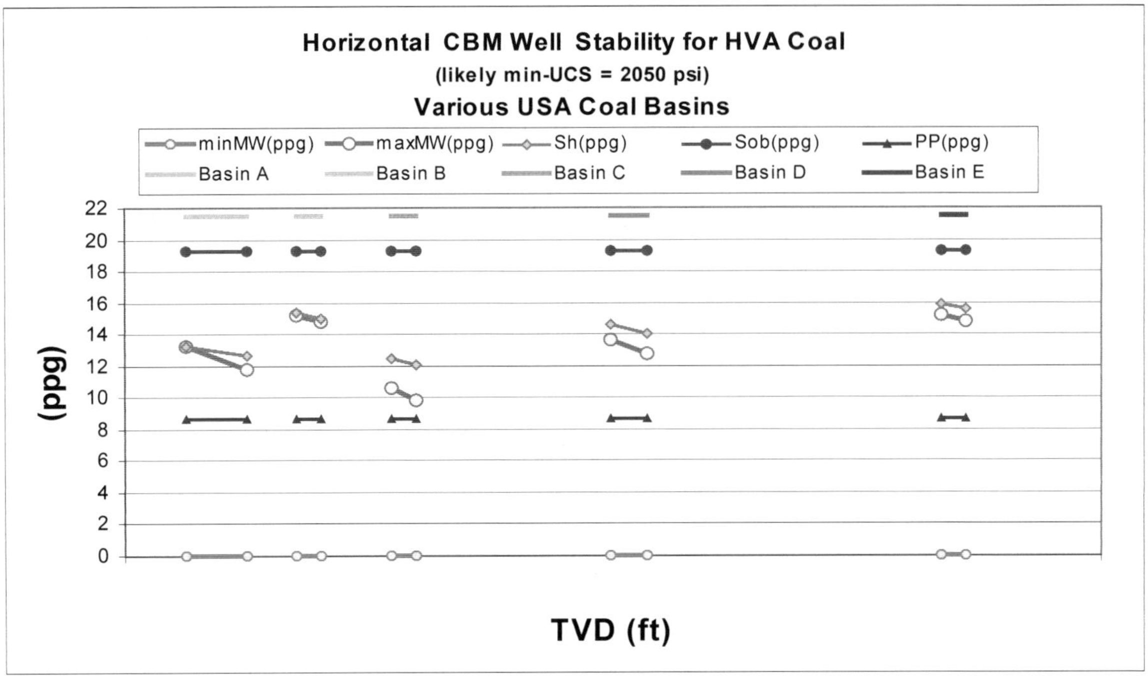

Figure 4

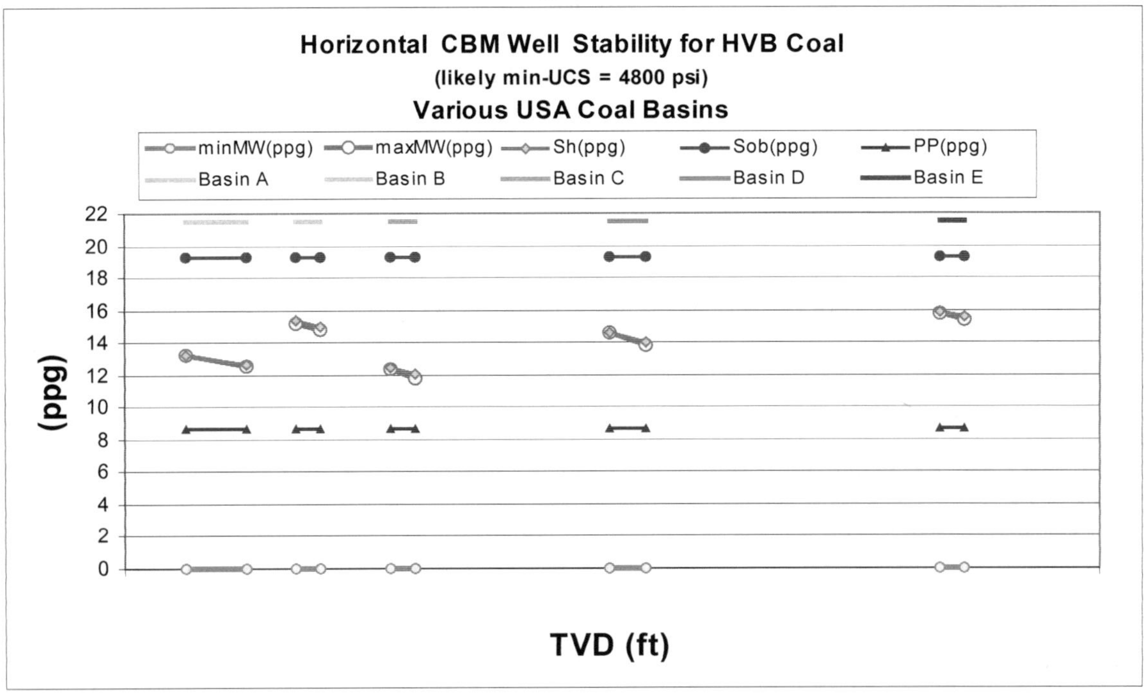

Figure 5: Effect of Reservoir Depletion on WBS
Basin E1, LV or MV Coals

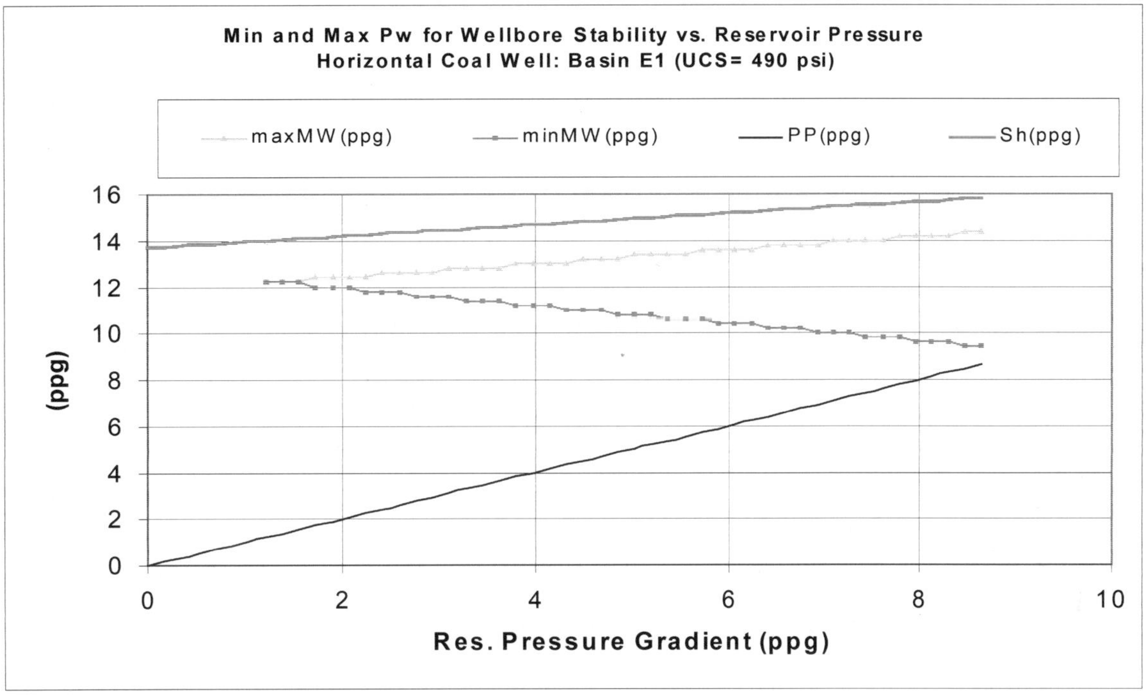

Figure 6: Effect of Reservoir Depletion on WBS
Basin E2, LV or MV Coals

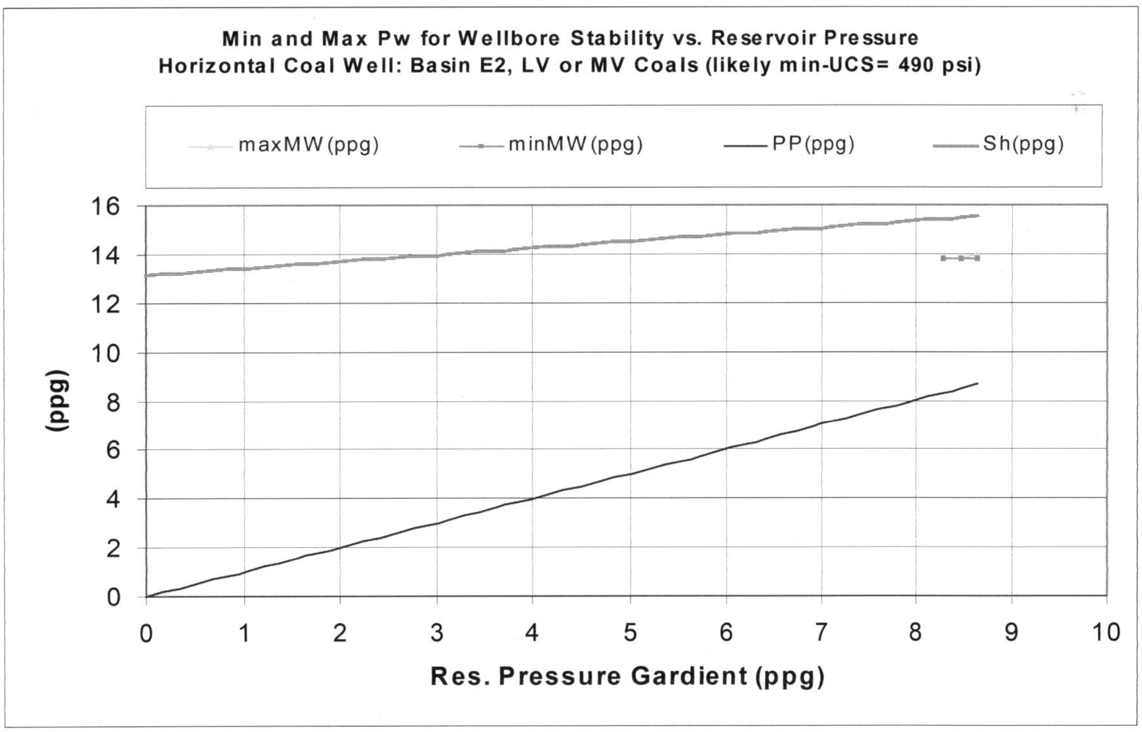

Figure 7: Effect of Well Inclination on WBS
Basin E1, LV or MV Coals

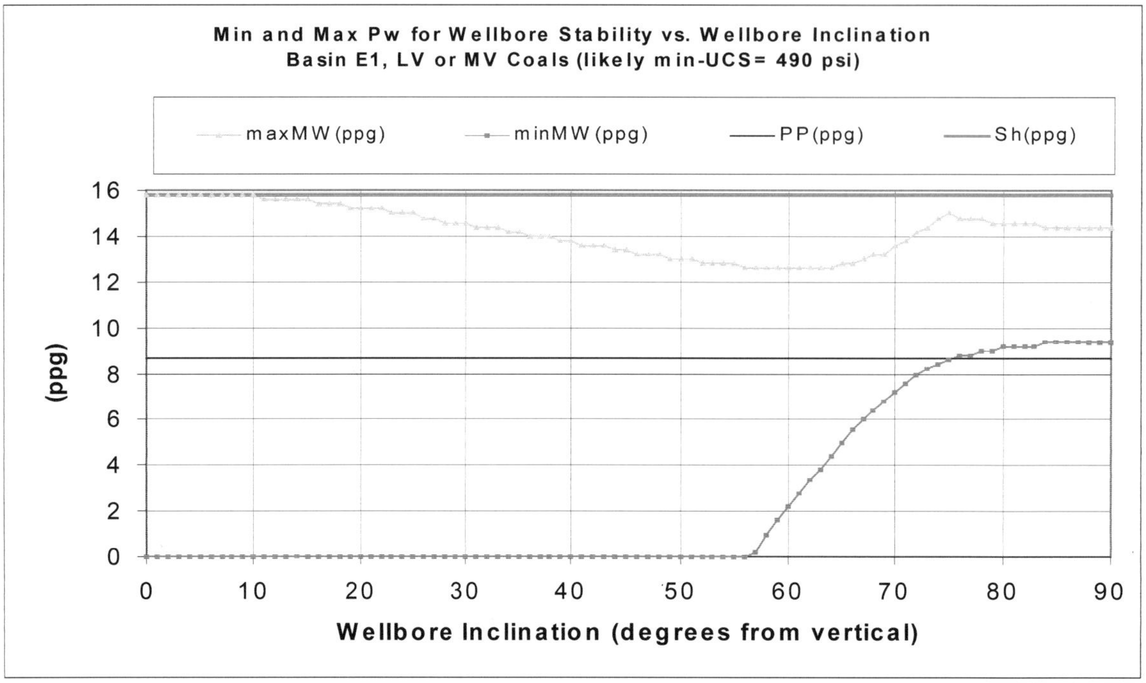

Figure 8: Effect of Well Inclination on WBS
Basin E2, LV or MV Coals

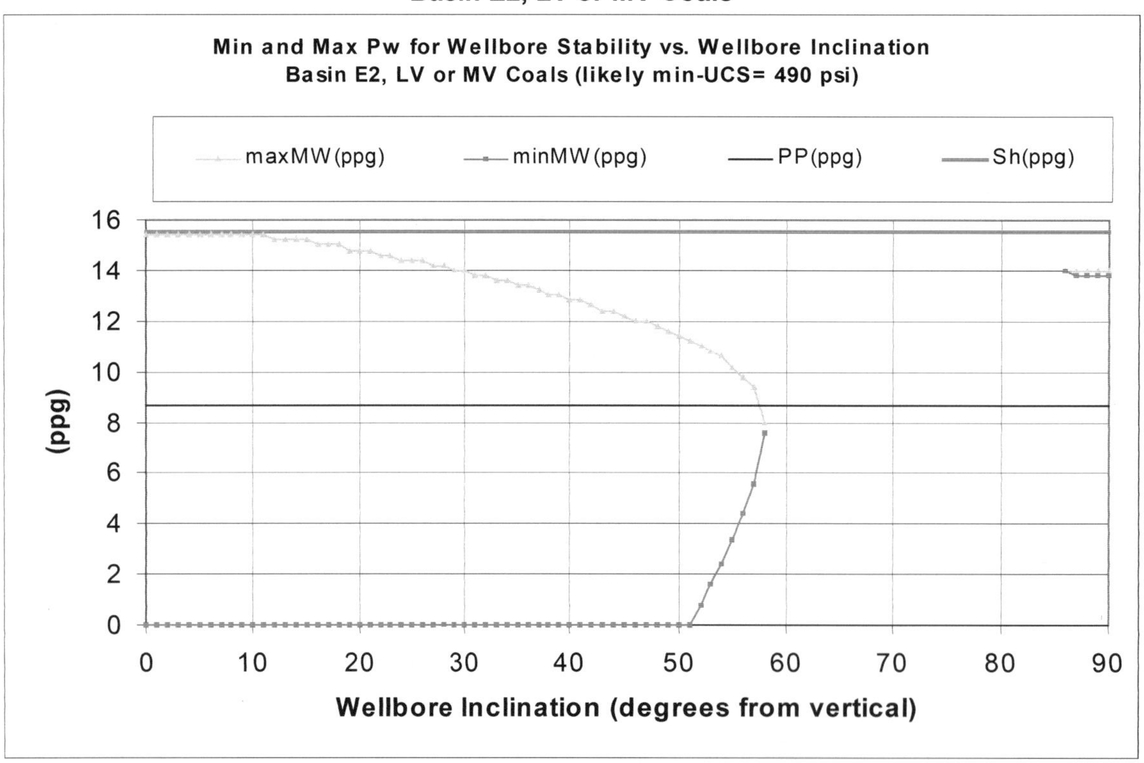

Figure 9: Wellbore Stress Trajectory in Stress Invariant Space

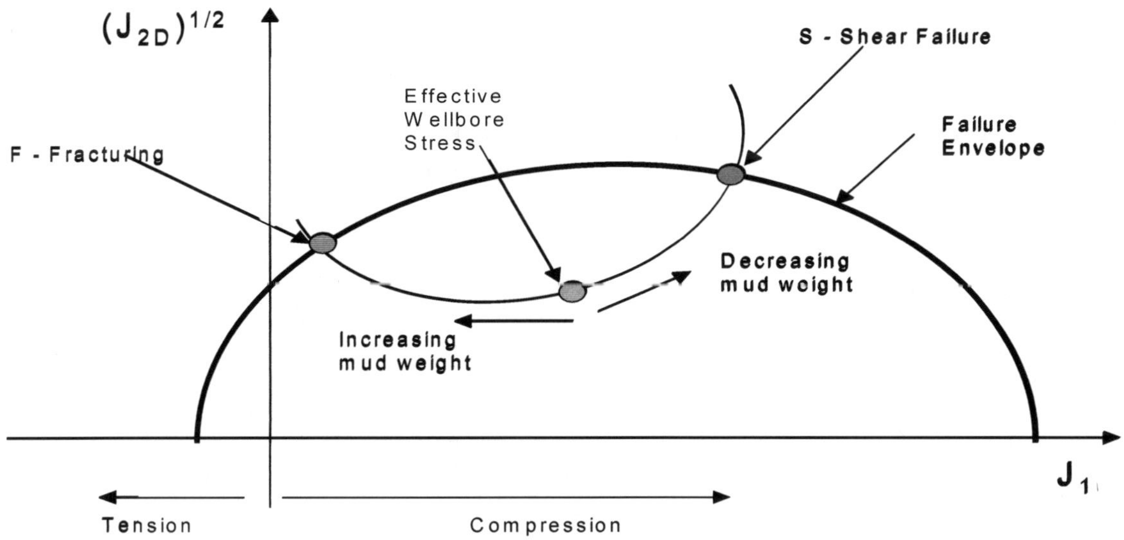

Failure Plot in Invariant Stress Space

0524

Coal Bed Methane Well Automation

R.G. Ward – Control Microsystems

ABSTRACT

Coal Bed Methane (CBM), while not invented in Wyoming, has certainly been exploited in this area during the last decade along with new advances of field automation. The purpose of implementing control on a CBM well is to de-water the well so the low pressure gas can be recovered more efficiently. The level of automation has improved since the early 1990's, evolving from simple measurement and speculative control to being completely automated with somewhat intelligent field based hardware. Other considerations in an automated CBM field are cost savings relative not, only to the control, communication and measurement devices, but also to the extent of energy conservation at the site. The remainder of this document will attempt to cover legacy techniques still being used today as well as automation strategies limited only by the users creativity. The goal of the automation project is to produce the maximum amount of gas while extracting the minimum (optimum) amount of water.

Well Automation Considerations

Coal Bed Methane wells, while simple enough to drill and complete, are not quite so simple to produce and maintain. The nature of a coal bed methane well is that it provides an extremely low down hole pressure. Producing a low pressure well is difficult in itself, add to that the high rates of water derived naturally from the coal seam and you have an opportunity for optimization. The following will provide an overview of typical considerations while automating a CBM well.

Down Hole Water Level

This is likely the single most common point of failure to a CBM automation project. The popular choice for level measurement is a submersible pressure transducer. The sensor technology is proven, the weak link is the cable from which the transducer is suspended. There are however a couple of conditions that the sensor does not have in its favor. Firstly, the fact that the sensor is being suspended in a hostile environment from several hundred foot cable that was designed for signal conduction and not mechanical support. Secondly is the reality of installation, these sensors are generally sent down hole by roustabouts that do not realize the importance of the sensor reaching bottom hole without having been sliced, gouged or severed. Further to the sensor's challenging task , is the interpretation of it's signal. A 1-5Vdc output signal intended to represent 0-500ft of level may not really be what you get. Since the water is being aerated by the natural gas, it's density changes yielding an error of thirty percent or greater. Many of the production technicians we deal with have a good handle on this factor and account for it in their logic.

Water Flow Meter

There are several water meter technologies that have been proven to work quite well. Each have their advantages and disadvantages. The most popular choice is the traditional turbine meter. This meter works on the premise that a rotor inside the body of a meter will turn a predictable number of turns per unit volume. As each blade of the rotor passes by the magnetic pickup coil, the excitation signal in the coil is amplified and an indicator or field controller interprets that pulse as a unit of measure. These pulses are gathered and used to determine flow rates and total accumulated volumes. If the signal cable is being run very far, or in an environment with a high amounts of electrical noise, a pre amplifier may be required.

A water meter technology that has not been utilized greatly in CBM automation is the magnetic flow meter. These meters are referred to in general process industry as mag-meters. This technology is extremely accurate and reliable but has traditionally not been used due to its high cost and power consumption. In the last few years a few manufacturers have released low power mag-meters but have not been successful in making them viable for small production wells. Mag-meters can only be used in processes where the liquid is conductive. The mag-meter is a non-intrusive device typically consisting of a spool with an electrode on each side. During operation a pulsed signal is sent to each electrode with the corresponding pulse reading and it's "time of travel" interpreted as a velocity measurement. Since liquids are essentially non compressible, a simple multiplication of cross sectional area and velocity yields an accurate measure of volumetric flow. This volume rate is typically provided to another field controller at a pulse per unit volume.

One of the technologies recently adapted for CBM applications are ultrasonic flow meters. These meters use time of flight to determine velocity which is then equated directly to volume. This technology has historically been quite expensive to implement. As of late, these have been manufactured in a clamp on style and advances in electronics have allowed for power and cost reductions. This too is a pulse per unit volume meter. The premise of this technology uses sound waves to determine velocity in the flow line. The lower end meters can be used only on liquids and have an accuracy nearing that of a turbine meter.

Pump Monitoring/Control

There are a number of ways the field controller can obtain process variables from the pumping system that can provide a basis for optimization. The first generation of automated CBM wells used simple pump panels to turn the pump on and off. This could be determined by monitoring the water level down hole or by measuring the current and voltage being supplied to the pump. Later techniques utilized serial communications that allowed the field controller to interrogate a device and read it's data. In the last few years variable frequency drives, with the aid of high gas prices, have made there way to the forefront of CBM automation where they are used in conjunction with a field controller to determine when and how to pump off the well. There are a few VFD manufacturers playing in this field today. One or two are touting patented applications that can provide significant production increases and energy savings to the end user.

Gas Measurement Transmitters (Secondary Device)

Coal Bed Methane gas is measured by either the traditional orifice or a V-Cone™ meter. Both meters are differential pressure producing meters and must be used in conjunction with an electronic flow measurement device (EFM). The process variables used by the EFM are the static pressure at the meter, a measurement of the differential pressure across the meter and the temperature of the process fluid at the meter. The sensors used for gas measurement have improved greatly in the last several years and quite commonly allow the user to experience accuracies of 0.05% without spending a fortune in instrumentation. Several manufacturers have created multivariable transmitters (MVT) that provide all three process signals in a single instrument. The advantage to the user is the reduced costs of the installation by taking advantage of serial or Ethernet communications while reducing the manifold/valve requirements and minimizing the points of potential fugitive emissions (leaks). Sensor technology accuracy has almost exceeded the ability for field technicians to properly calibrate them. According to API21.1, the governing document in the US, calibration equipment must be two times more accurate than the device being calibrated. This requires users be able to take to the field 0.025% test equipment.

Recent transmitter enhancements have concentrated on electronics rather than sensors. As manufacturers respond to creative producers, users are seeing innovations such as 10baseT Ethernet at the sensor level allowing IP connectivity all the way to the process measurement device.

Gas Meter (Primary Device)

As noted above, the most common methods for measuring CBM gas are orifice and V-Cone™ meters. Both meters provide a similar function, they generate a higher differential pressure as the flow rate across them increases. This differential pressure (DP), used with a static pressure (P) and temperature (T) are enough to determine compensated gas flow volumes. The orifice plate design is tried and true and is still used today because people have learned to trust it. The orifice does however have a few down sides in CBM operations. The very nature of needing to optimize a CBM well is due to water concentration in the producing coal seam. This water being extracted with the gas makes for and entirely different set of measurement practices. The traditional orifice plate offers a restriction in the line that allows flow only through the center of the line. Being in the middle of the flow line, it allows the beta edge of orifice plate to measure the high velocity regime of the flow profile which requires conditioning to insure the meter is used properly. This restriction can also act as a dam and retain water upstream of the meter. When this happens, the liquid accumulation can have an adverse effect on the differential pressure measurement by changing the effective vena contracta.

The V-Cone™ meter relies on the same process variables as the orifice meter; DP, P and T but uses a different technique to generate the DP. This design places the restriction in the center of the line so the gas flows around it, not through it. This design exhibits merit as the liquids are unable to accumulate near the beta area causing error. There are also advantages regarding range ability of the meter that can offer improvements in applications where flow rates vary offer large ranges.

Electronic Flow Measurement (EFM)

This is the device responsible for generating the volumes that will be used by accounting for royalty dispersion, field balancing and possibly revenue generation. In the US, devices are required to comply with API21.1 for complete audit trail of 35days of hourly averages, daily totals, user changes and process alarms. The physical measurements for orifice meters use AGA3 calculations, linear pulse type gas meters use AGA7 and compressibility of natural gas is determined according to AGA8. While the V-Cone™ meter is not approved for custody transfer measurement, it has been adopted by several agencies for such applications. In Wyoming, the Federal Bureau of Land Management (BLM) has a requirement that must be adhered to in several states of producing on federal lands. This requirement is Onshore Order #5 and has been amended by Notice to Lessee's (NTL) dated March 2004; this document encompasses Electronic Flow Computers (EFC).

There are some products on the market today that have integrated the functions of the multivariable transmitter and the electronic flow computer into the same device even further minimizing the cost of automation to the producer.

Programmable Logic Controller (PLC)

The PLC is where the intelligent functionality of well optimization is located. Traditional PLC's can manage the logic required at these sites, but the major consideration for determining which PLC to use is the operating environment. PLC's are traditionally considered control room equipment capable of only 32°F to 120°F (0°-49°C).

Hybrid Field Controllers

These are the products of ingenuity, the result of producers thinking outside the box, sharing their thoughts and desires with manufacturers. Several have truly accomplished the task of having a PLC and EFM in the same device. This combination allows a huge cost savings for the producer on the front of the project and even greater savings long term through reduced maintenance and field visits.

Benefits of a hybrid controller:

- Environmental ratings of the controller are similar to those of an RTU
  - Class 1 Div.2 or Class 1 Div.1
  - -40° to 158°F (-40° to 70°C) Operational
- Flexible programming options
  - IEC6-1131-3 Compliant – allows Ladder Logic, Function Block Diagram, Flow Chart, Sequential Function Charts, Structured Text and Instruction List
- Ability to remotely configure/program the controller
- Ability to remotely re-flash updates in firmware
- Ability to remotely load new gas calculations as required by governing bodies
- Independent AGA program and IEC6-1131-3 program so the control strategy can be changed without losing gas measurement data.
- Shared database so the AGA inputs and results can be used in the control strategy
- I/O Platform is typically expandable
- Open architecture protocols such as Modbus, ModbusTCP, EnronModbus and DNP3.
- Ability to display measurement and process control data locally. Display needs to take into consideration requirements of governing bodies such as BLM.
- High Speed data logging of process data allowing the reservoir engineers to see well dynamics with as frequent as one second resolution.

SCADA System Overview

The seemingly weakest link in any SCADA system is the communications. There is not a lot of rocket science here, but if communications go down, so goes the entire system. The key components of the host system are the HMI environment itself, a data historian, a polling engine and a communications network.

The HMI is the graphical interface that allows the user to easily navigate the system. Today's systems are capable of redundant servers, hot stand-by, synchronizing databases and immediate web publishing of the data for other non client users to access.

The data historian can be either an integrated function of the HMI, or a third party application capable of report generation and data management. The data is collected by a polling engine and placed on the network at some predefined location for the historian to access and possibly perform some integration functions for back office accounting systems. It is not uncommon for this to be a third party package capable of managing multiple protocols over a serial wireless communication network.

The communications system can be any combination of spread spectrum or licensed serial radios. Ethernet radios are increasingly being used in today's systems as the fear of security is being somewhat resolved by the use of pseudo-random frequency hopping techniques in the newer serial and Ethernet radios. The HMI client will request all the real time data for SCADA operations and the polling engine will manage scheduling of historical EFM data. The radio networks also have the ability to act as a slave and repeater at the same time, allowing for tolerance within difficult terrain. If the host radio cannot see one of the remote sites, there is a good chance another remote can. The host can route the message through any number of remotes to find the path to the desired device. These techniques can also be supported in several of the hybrid controllers on the market as a function called store and forward. These type of enhancements have greatly increased the viability of spread spectrum radios even further reducing the installed cost to the user.

Summary

Many components of a CBM project have been covered here, each worthy of consideration. Subjects worthy of further investigation include the capabilities of evolving hybrid controllers and their ability to communicate through serial and Ethernet communications to sensors, (including pressure and multivariable devices) and to provide the logic by which you ultimately control the well. The real intelligence is not the hardware but the intellectual property that exists within. Questions that need answering include how will you monitor each of the process variables and will you be gathering information from enough process points or are there other things to consider?

Finding the correlation between water level, pump loading, water and gas production rates are the fundamentals of optimization. The goal is to produce as little water as possible and to extract as much gas a possible while minimizing energy costs. There are hardware platforms today that offer a solution to these challenges, including the difficult tasks of interpreting well dynamics to a level where preemptive decisions can be made by the controller.

Footnotes:

1. Federal Bureau of Land Management – Onshore Order No.5 (43 CFR 3160)
   http://www.nm.blm.gov/oil_gas/onshore/ord5.html

References:

- V-Cone is a trademark of McCrometer
  http://www.mccrometer.com/products/pdwafer.htm
- AGA – American Gas Association
  http://www.aga.org
- API – American Petroleum Institute
  http://www.api.org

Technical Papers
Author Index
(by paper No.)